Semiconductor Physics

Advanced Texts in Physics

This program of advanced texts covers a broad spectrum of topics which are of current and emerging interest in physics. Each book provides a comprehensive and yet accessible introduction to a field at the forefront of modern research. As such, these texts are intended for senior undergraduate and graduate students at the MS and PhD level; however, research scientists seeking an introduction to particular areas of physics will also benefit from the titles in this collection.

Springer

Berlin
Heidelberg
New York
Hong Kong
London
Milan
Paris
Tokyo

Karlheinz Seeger

Semiconductor Physics

An Introduction

Ninth Edition
With 326 Figures

 Springer

Professor Karlheinz Seeger

Am Modenapark 13/5
1030 Vienna, Austria
and
Institut für Materialphysik der Universität
Boltzmanngasse 5
1090 Vienna, Austria

The first edition was published by Springer-Verlag Vienna.

ISSN 1439-2674

ISBN 978-3-642-06023-6

Springer-Verlag is a part of Springer Science+Business Media

springeronline.com

© Springer-Verlag Berlin Heidelberg 2010
Printed in Germany

Cover design: *design & production* GmbH, Heidelberg

Printed on acid-free paper

Preface

This book, now in its ninth edition, still has the character of a textbook with the emphasis on "Physics". The volume has increased somewhat because several improvements have been made and some new items have been included.

In Sect. 13.2 the new Quantum Cascade Laser which covers the far infrared spectral range has been added. In Sect. 14.4 the theory of the quantum Hall effect is now based on ballistic transport which in a more general respect without referring to the then still unknown quantum Hall effect was considered already by Rudolf Peierls. In the same chapter, the recent discovery of a low-temperature resistance oscillation in a very pure semiconductor under the influence of combined dc and ac electric fields in addition to a magnetic field is presented. Furthermore, quantum Hall effect observations with an unprecedented high precision are remarkable and may give a new impetus to theory. A new Sect. 15.5 presents information about coaxial carbon tubes of nanometer size diameter and how they are integrated as the current transporting element in a field effect transistor. In another new addition Sect. 15.6 with the title Molecular Electronics, the current-voltage rectifying characteristics of an organic Langmuir-Blodgett film of nanometer thickness is shown. These efforts serve to demonstrate where the ever decreasing size of electronic circuits may come to its natural limits.

The system of units preferred here is the SI system. If you want to convert e.g. the magnetic field B to its Gaussian unit you have to divide it by the velocity of light c. Vectors are typed boldface. A vector product is put in [brackets], a scalar product in (parenthesis). Notation of the form (5.7.7, 8) is used to refer to both equation (5.7.7) and equation (5.7.8) while [5.7] refers to reference 5.7.

A basic prerequisite for the existence of this book is (and always was) the excellent cooperation with Hofrat Dr.Wolfgang Kerber and his crew at the library of the Institute of Physics, University of Vienna, Austria. Dr. Hubertus von Riedesel, managing editor at Springer - Verlag, Heidelberg, and his assistant Dr. Angela Lahee deserve my heartfelt thanks for their cooperation. Last but not least my wife suffered from my occasional mental absences during this period of both writing and taming an unwilling computer, with remarkable good humour and I would like to express my appreciation for her support.

Vienna, 2004   K.S.

To Instructors who have adopted the text for classroom use, a solutions manual is available free of charge by request on departmental letterhead to Dr. A. Lahee, Springer-Verlag, Tiergartenstr. 17, D-69121 Heidelberg, Germany.

Contents

1. Elementary Properties of Semiconductors

1.1 Insulator – Semiconductor – Semimetal – Metal

A consequence of the discovery of electricity was the observation that metals are good conductors while nonmetals are poor conductors. The latter were called insulators. Metallic conductivity is typically between 10^6 and 10^4 ($\Omega\,cm)^{-1}$, while typical insulators have conductivities of less than 10^{-10} ($\Omega\,cm)^{-1}$. Some solids with conductivities between 10^4 and 10^{-10} ($\Omega\,cm)^{-1}$ are classified as *semiconductors*. However, substances such as alkali-halides whose conductivity is due to electrolytic decomposition shall be excluded. Also we restrict our discussion to chemically uniform, *homogeneous* substances and prefer those which can be obtained in monocrystalline form. Even then we have to distinguish between semiconductors and semimetals. This distinction is possible only as a result of thorough investigation of optical and electrical properties and how they are influenced by temperature, magnetic field, etc. Without giving further explanations at this stage, the statement is made that semiconductors have an *energy gap* while semimetals and metals have no such gap. However, very impure semiconductors show a more or less metallic behavior and with many substances, the art of purification is not so far advanced that a distinction can easily be made. The transition between semiconductors and insulators is even more gradual and depends on the ratio of the energy gap to the temperature of investigation. Very pure semiconductors may become insulators when the temperature approaches the absolute zero.

Typical elemental semiconductors are germanium and silicon. An inspection of the periodic table of elements reveals that these materials belong to the fourth group while typical metals such as the alkalis are in the first group and typical nonmetals such as the halogens and the noble gases which crystallize at low temperatures are in the seventh and eighth group, respectively. Other semiconducting elements in the fourth group are diamond which is a modification of carbon, while *gray* tin $(\alpha - Sn)$, which is stable only at low temperatures is a semimetal. All fourth-group semiconductors crystallize in a structure known as the diamond structure in which neighboring atoms are arranged in tetrahedral symmetry. In the third group, the lightest element boron, and in the sixth group, the heavy elements selenium and tellurium, are semiconductors. A typical semimetal is the heaviest fifth group element, bismuth, and also the lighter elements of this group, arsenic and antimony, may be classified as such although they are at present less thoroughly investigated.

Typical compound semiconductors are the III–V compounds such as gallium arsenide, GaAs, and indium antimonide, InSb, and the II–VI compounds such as zinc sulfide, ZnS (*zinc blende*). They crystallize in the zinc blende structure which can be obtained from the diamond structure by replacing the carbon atoms alternately by, e.g., zinc and sulfur atoms. These compounds have a stoichiometric composition, just as, e.g., the semiconductor silicon carbide, SiC, while germanium silicon alloys may be obtained as semiconducting mixed crystals for any arbitrary composition. Many metal oxides and sulfides are semiconductors, often with nonstoichiometric composition. Some of them are of technical importance, such as cuprous oxide, Cu_2O (formerly used for rectifiers), lead sulfide, PbS (for infrared detectors) and the ferrites (iron oxides) for their magnetic properties. Today silicon is mainly used for the fabrication of transistors which serve for amplification of electric signals. This is the most important technical application of semiconductors nowadays.

Semiconduction is specified by the following properties:

(a) In a *pure* semiconductor, conductivity rises exponentially with temperature (*thermistor* action). At lower temperatures, smaller concentration of impurities is required in order to ensure this behavior.

(b) In an impure semiconductor, the conductivity depends strongly on the impurity concentration. For example, nickel oxide NiO in a pure condition is an insulator. By doping (which means intentionally adding impurities) with 1% lithium oxide, Li_2O , the conductivity is raised by a factor of 10^{13} . In the heavily doped material, however, the conductivity changes only slightly with temperature, just as in a metal.

(c) The conductivity is changed (in general, raised) by irradiation with light or high-energy electrons or by the *injection* of carriers from a suitable metallic contact (injection will be explained in Sect. 5.1).

(d) Depending on the kind of doping, the charge transport may be either by electrons or by so-called *positive holes*. The electric behavior of positive holes is the same as that of positrons but otherwise there is no similarity. It is possible to dope a single crystal nonuniformly such that in some parts, charge transport is by (negative) electrons and at the same time in others by positive holes. Semiconductor diodes and transistors are single crystals of that kind.

Semiconducting behavior is not restricted to solids. There are liquid semiconductors. However, because of atomic diffusion, regions with different dopings will mix rapidly and a stable device with an inhomogeneous structure is not possible. Recently, attention has been paid to glassy and amorphous semiconductors which may possibly find a technical application for solar cells.

As mentioned before, semiconductors become metallic when heavily doped. Superconductivity, known for some metals and inorganic as well as organic compounds at low temperatures, has also been observed with some heavily doped semiconductors. Transition temperatures are below 150 K. Some aromatic hydrocarbons and even fullerenes, which represent besides coal and diamond a third modification of carbon, show semiconducting behavior They will be reported in Sect. 15.4. [1.1].

   Growth of single crystals of silicon in a cylindrical form of over 20 cm in diameter and a meter length with dislocation densities of less than 10^3 cm^{-3}(W.C.
Dash[1.2]) and impurity concentrations of less than one part per trillion (10^{12}) is
today's industrial standard. How are these crystals produced? In the Czochralski technique [1.3] a small seed crystal is put in contact with the molten originally polycrystalline material (melting point 1415°C) and then pulled up at a
speed of a few millimeters per minute. Slow rotation of the seed crystal ensures
that the resultant crystal is cylindrical. The crystallographic atomic structure
is, of course, the one of the seed crystal. The original material found in nature
is quartz sand (chemically contaminated silicon dioxide SiO$_2$) which is heated
with charcoal in order to remove the oxygen. Further processes of getting rid of
impurities involve the conversion of the raw silicon to trichlorosilane SiHCl$_3$, a
liquid from which by fractional distillation a boron impurity is removed which
otherwise in the final product would produce an unwanted metallic electrical
conduction. Another semiconductor of technical importance is gallium arsenide
GaAs (melting point 1238°C) which in liquid form without protection loses arsenic in the form of vapor in the Czochralski process. This decomposition of
the compound is avoided by putting a layer of molten boron oxide on top of the
gallium arsenide. The technique is known as *Liquid Encapsulation Czochralski method*, LEC. In the *Bridgman* variation of the Czochralski technique, a
temperature gradient along the length of the crucible is produced by heaters.
By gradually reducing the energy supplied to the heaters, the crystal grows
beginning at the seed crystal.

   Floating zone purification of a cylindrically shaped crystal with its axis
arranged vertically implies heating part of the cylinder by *high-frequency in-
ductive heating* to the melting temperature and gradually moving the heaters
along the crystal axis. The melt is kept in place by surface tension. The re-
crystallized section is of higher purity than the original section, i.e., the motion
of the heated section transports impurities to one end of the sample rod which
is finally deleted. This technique avoids contamination of the material by a
physical contact to a boat-shaped crucible [1.4].

   Further processing steps will finally lead to the various devices as they are
produced at present, details of which are beyond the scope of this book. Let us
just briefly mention *Molecular Beam Epitaxy*, MBE, as a method of producing a
very thin layer of one semiconductor on top of another semiconductor which has
the form of a small plate[1.5]. The source material is vaporized in a heated cell
with a very small orifice (Knudsen cell) creating a molecular beam in ultrahigh
vacuum (10^{-11} torr) thus avoiding collisions with residual gases and allowing
epitaxial growth of the layer on the target plate (*epitaxial* means an arrangement of the beam molecules as they get stuck according to the arrangement of
the atoms on the target surface). The layer quality is recorded during growth
by *Reflection High-Energy Electron Diffraction* (RHEED)[1.6] where a 10 keV
electron beam incident on the semiconductor surface and reflected at a very
large angle of (*grazing*)incidence produces a diffraction pattern on a phosphor
screen. *Quantum wells* are produced by this technique where a layer of about

1 nm thickness is sandwiched between two tin layers of another semiconductor with a larger band gap (Fig. 9.5).

1.2 The Positive Hole

As mentioned in the above section, charge transport may be due to *positive holes*. In this chapter we shall explain this idea qualitatively by considering a lattice of carbon atoms which, for simplicity, is assumed to be 2-dimensional. In Fig. 1.1, the bonds between neighboring atoms are covalent. Each C-atom contributes 4 valence electrons and receives from its 4 neighbors 1 electron each so that a *noble gas configuration* with 8 electrons in the outer shell is obtained. This is similar in a 3-dimensional lattice.

Now imagine an extra electron being transferred somehow into the otherwise perfect diamond crystal (by *perfect* we also mean that the surface with its free bonds is far enough away from the electron. Ideally let us think of an infinitely large crystal). No free bonds will be available for this electron where it could attach itself. The electron will move randomly with a velocity depending on the lattice temperature. However, if we apply an external electric field to the crystal, a drift motion will be superimposed on the random motion which, in the simplest case, will have a direction opposite to the field because of the negative charge of the electron. The extra electron which we call a *conduction electron* makes the crystal n-*type* which means that a *negative* charge is transported. In practice, the extra electron will come from an impurity atom in the crystal.

We can also take a valence electron away from an electrically neutral diamond. The crystal as a whole is now positively charged. It does not matter which one of the many C-atoms loses the electron. The main point is that this atom will now replace its lost electron by taking one from one of its neighbors. The neighbor in turn will react similarly with one of its neighbors. This process is repeated over and over again with the result that the hole produced by taking away an electron from the crystal moves in a random motion throughout the crystal just as the extra electron did in the n-type crystal. What happens if we now apply an external electric field? Wherever the hole is, a valence electron will fill it by moving in a direction opposite to the electric field with the effect that

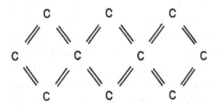

Fig. 1.1: Schematic two-dimensional representation of perfect diamond lattice. Each covalent bond (=) represents two valence electrons of opposite spin

the hole drifts in the direction of the field. This is exactly what one would expect from a positive charge. Since the crystal as a whole is charged positively, we may think of this charge as being localized at the position of the hole. In semi-conductor physics, positive holes are treated as if they were positively charged electrons. Conductivity by *positive* holes is called p-*type*. For a comparison, consider the carbon dioxide bubbles in mineral water. Instead of an electric field, there is the gravitational field and instead of an electric charge, there is the mass of the water molecules. Since the bubbles drift in a direction opposite to the field direction, they can formally be treated like negative mass particles as long as they are in the bulk of the liquid although, of course, carbon dioxide has a positive mass and is subjected to a lift only. Similarly, the assumption of positively charged particles called *holes* in semiconductors is a very simple formal description of an otherwise quite involved process. However, one should keep in mind that the hole is actually a missing valence electron. In Chap. 2 we shall derive the concept of the hole by the more rigorous quantummechanical method.

1.3 Conduction Processes, Compensation, Law of Mass Action

Before becoming involved in wave mechanics, we will continue with the classical model to investigate thermal pair generation and annihilation. Let us call ε_G the binding energy of a valence electron to an atom in the crystal (G stands for *gap* which will be explained in Chap. 2). If the energy ε_G is supplied thermally, a conduction electron may be generated which leaves a hole where the electron has been. The electron and the hole move through the crystal independent of each other. Since we consider the hole as a particle similar to the electron except for the sign of its charge, we have created an electron hole pair. Occasionally a conduction electron will recombine with a hole which actually means that it finds a free bond and *decides* to stay there. The binding energy ε_G is transformed either into electromagnetic radiation (*recombination radiation*) or atomic vibrations (*phonons*). From the particle point of view, the annihilation of the electron hole pair is usually called *recombination*. Denoting electrons by the symbol e^- and holes by e^+, a chemical reaction equation of the form

$$e^- + e^+ \rightleftharpoons \varepsilon_G \qquad (1.3.1)$$

will be an adequate description of the process. Assuming that no radiation is incident, the generation energy ε_G is taken from the lattice vibrations. There-fore, with increasing temperature, the equilibrium is shifted towards the lhs of the equation, the number of carriers and therefore the conductivity is increased which is so characteristic of semiconductors. Of course, radiation is incident on the crystal even if it is in thermal equilibrium with its environment. This *black-body radiation* compensates the recombination radiation of energy ε_G exactly (see the rhs of (1.3.1)).

It is shown in statistical mechanics that a *small system* which is in thermal contact with a *large system* can acquire an energy ε_G at a rate proportional to $\exp(-\varepsilon_G/k_B T)$, where k_B is Boltzmann's constant and T the absolute temperature (at room temperature $k_B T = 25.9\,\text{meV}$). In the present case, the *small system* is a valence electron and the *large system* the crystal. The exponential is multiplied by a power of T; however, the temperature dependence is essentially determined by the exponential, as is well known from, e.g., the law of thermionic emission from metals. For (1.3.1), the power function is T^3 if we apply the law of mass action. Denoting the concentrations of conduction electrons and holes by n and p, respectively, it is

$$n\,p = C T^3 \exp\left(\frac{-\varepsilon_G}{k_B T}\right) . \tag{1.3.2}$$

The value of the constant C depends on the semiconductor material. The form of (1.3.2) is similar to the one describing the concentrations of H^+ and OH^- in water where these concentrations are always small compared with the concentration of the neutral water molecules. In a semiconductor, the electron and hole concentrations will also be small relative to the concentrations of atoms because otherwise, the conductor would have to be classified as a metal. A rigorous derivation of (1.3.2) including a calculation of the constant of proportionality C will be given in Chap. 3 (3.1.14).

In a pure semiconductor, for every conduction electron a hole is also produced and $n = p$. We call this *intrinsic conduction* and add a subscript i to n. Equation (1.3.2) yields

$$n_i = C^{1/2} T^{3/2} \exp\left(\frac{-\varepsilon_G}{2k_B T}\right) \tag{1.3.3}$$

In Figs. 1.2a, b, n_i is plotted vs temperature for silicon ($\varepsilon_G = 1.12$ eV at 300 K) and germanium ($\varepsilon_G = 0.665$ eV at 300 K). At temperatures above 250 K (for Si) and 200 K (for Ge), ε_G varies linearly with temperature

$$\varepsilon_G(T) = \varepsilon_G(0) - \alpha T , \tag{1.3.4}$$

the coefficient α being 2.84×10^{-4} eV/K for Si and 3.90×10^{-4} eV/K for Ge. However, this does not change the exponential law (1.3.3), except for a change in the factor C, since

$$\exp\left(\frac{-\varepsilon_G + \alpha T}{2k_B T}\right) = \exp\left(\frac{\alpha}{2k_B}\right) \exp\left(\frac{-\varepsilon_G}{2k_B T}\right)$$

Therefore, in (1.3.3) we use the value of ε_G obtained by extrapolation from the range linear in T to absolute zero. At low temperatures, a T^2 term with the

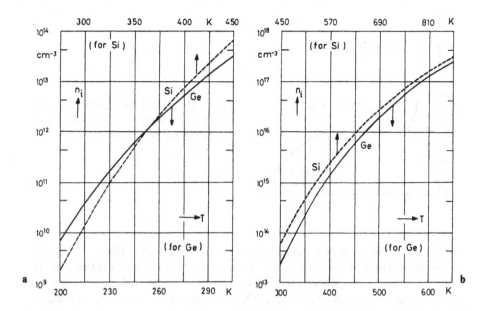

Fig. 1.2: (a) Intrinsic carrier concentration as a function of temperature for silicon and germanium: 200 to 305 K (Ge); 275 to 450 K (Si) (after[1.7]), (b) same as (a) except for temperature range: 300 to 650 K (Ge); 450 to 870 K (Si)

same sign as the αT term is found in ε_{G}. One obtains these additional terms from optical investigations.

Now we consider a doped semiconductor. Assume one atom to be replaced by an impurity atom, e.g., a 4-valent C atom by a 5-valent phosphorous atom. Only 4 of the 5 valence electrons of the phosphorous are required to bind the 4 neighboring C atoms. The fifth electron is bound very loosely. The binding energy $\Delta\varepsilon_{\mathrm{D}}$ of the fifth electron is considerably lower than the binding energy ε_{G} of a valence electron to a C atom. An impurity which releases one or several electrons in this way is called a *donor D*. If we denote a neutral donor by $D^{\times}$ and a singly-ionized donor by D^{+}, the reaction is

$$e^{-} + D^{+} \rightleftharpoons D^{\times} + \Delta\varepsilon_{\mathrm{D}} \ . \tag{1.3.5}$$

At high temperatures, all donors are thermally ionized and the concentration n of conduction electrons is equal to that of donors, N_{D}. The concentration p of holes is given by

$$n\,p = n_{\mathrm{i}}^{2} \tag{1.3.6}$$

where n_{i} is given by (1.3.3). Charge carriers which are in the minority are called *minority carriers*. At still higher temperatures, n_{i} will become larger than N_{D} and the semiconductor will then be intrinsic. The temperature range where

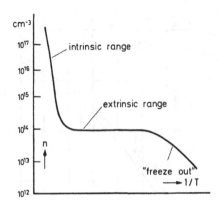

Fig. 1.3: Schematic diagram of the carrier concentration as a function of the reciprocal of the temperature

$n = N_D$ is independent of temperature is called *extrinsic*.[1] Figure 1.3 gives a schematic diagram of carrier concentration vs the inverse temperature. At the right end of the curve where the temperature is low, the balance of (1.3.5) is shifted towards the right-hand side. There is carrier *freeze-out* at the donors. The application of the law of mass action again yields an essentially exponential temperature dependence with $\Delta\varepsilon_D$ in the exponent. Because $\Delta\varepsilon_D$ is smaller than ε_G , the slope of the curve is less steep than in the intrinsic range.

If in diamond a 4-valent C atom is replaced by a 3-valent boron atom, a valence electron is lacking. Supplying an energy $\Delta\varepsilon_A$, an electron from a C atom is transferred to the B atom and a mobile hole is thus created. The B atom becomes negatively charged. Boron in diamond is an *acceptor A*. Denoting the neutral acceptor by A^x and the singly-ionized acceptor by A^- , the chemical reaction can be written in the form

$$e^+ + A^- \rightleftharpoons A^x + \Delta\varepsilon_A. \tag{1.3.7}$$

The temperature dependence of the hole concentration p is similar to that of n in n-type semiconductors.

If both donors and acceptors are distributed at random in the semiconductor we have a *compensated* semiconductor.[2] At ideal compensation there would be equal numbers of donors and acceptors. Quite often a high-resistivity material which is supposed to be very pure is compensated. A convenient method of compensation of acceptors is the lithium drift process. The positive lithium ion Li^+ has the helium configuration of the electron shell and diffuses nearly as easily through some solids as helium gas. Diffusion takes place via the interstitial mechanism. The negatively charged acceptors attract the positively charged lithium ions forming ion pairs with the result of ideal compensation. In a p-n junction during diffusion, an external electric field is applied such that ion drift

---

[1]For a *partly compensated* semiconductor N_D should be replaced by $N_D - N_A$
[2]A random distribution of a phosphorous dopant is achieved in silicon by neutron transmutation: $Si_{14}^{30}+$ thermal neutron $\rightarrow Si_{14}^{31} + \gamma\,(2.62\ \mathrm{hr}) \rightarrow P_{15}^{31} + \beta$ [1.7].

produces a p-i-n junction with a large perfectly compensated intrinsic region between the p and the n regions (e.g., a type of γ-ray counter is produced by this method [1.8, 1.9]).

The normal case is partial compensation of impurities. According to whether $N_D > N_A$ or vice versa, the semiconductor is of an n- or p-type, respectively. Compound semiconductors with nonstoichiometric composition are of an n- or p-type depending on which component of the compound is in excess. Assume that in CdS, e.g., one S^{--} ion is lacking somewhere in the crystal. There is then an excess Cd^{++} ion which will be neutralized by two localized electrons. Therefore, the lattice vacancy acts as an acceptor and makes the semiconductor p-type. In small-band-gap polar semiconductors like PbTe and HgTe, the free carriers are mainly produced by deviations from stoichiometric composition (while in large-gap ionic solids such as sodium chloride, the gap energy is larger than the energy required to generate a vacancy which compensates the ionized impurities and therefore makes the crystal insulating; this is called *auto-compensation*) [1.10].

Problems

1.1. Discuss the temperature dependence of the carrier concentration of a semiconductor (a) in the intrinsic range (b) in the extrinsic range and compare with that of a metal.

1.2. 100 g silicon are homogeneously doped with 4 μg aluminum. What is the hole concentration assuming one hole per aluminum atom? (silicon: density 2.23 g/cm^3, atomic weight 28.086; aluminum: atomic weight 26.98).

1.3. The conductivity is the product of carrier concentration, mobility, and elementary charge. Assuming a mobility of 600 cm^2/Vs, what is the conductivity of the material of Problem 1.2?

1.4. A sample of 3 cm length has been prepared from the material of Problem 1.2. A voltage of 4.5 V is applied between the ends of the sample. How much time is spent by a hole in travelling through the sample? Compare with a copper sample of equal length and for equal voltage (density 8.95 g/cm^3; atomic weight 63.57; resistivity $1.7 \times 10^{-8}\Omega$ cm; assume one electron per atom). Is the time longer than the time it takes to develop enough Joule heat to raise the temperature of the copper sample by 1°C? (apply the well known Dulong–Petit law).

1.5. Four metal pins in a linear array at equal distance D are pressed by springs against a semiconductor sample. Apply a current I through contact No. 1 and 4 and measure the voltage drop V between probes No. 2 and 3. Show that for a sample thickness $\gg$ D the resistivity is given by $\varrho = 2\pi$ D V/I. Modify the calculation for unequal distances. *Hint:* think of the current first going from No. 1 to infinity causing a voltage drop V' and then from infinity to No. 4 causing V''. Potential theory shows $V = V' + V''$.

2. Energy Band Structure

The *energy band structure* is the relationship between the energy and momentum of a carrier in a solid. For an electron in free space, the energy is proportional to the square of the momentum. The factor of proportionality is $1/(2m_0)$, where m_0 is the free electron mass. In the *simple model of band structure*, the same relationship between energy and momentum is assumed except that m_0 is replaced by an *effective mass*. This may be larger or smaller than m_0. Why this is so will be seen later in this chapter. Quite often the band structure is more complex and can only be calculated semi-empirically even with computers. A short description of some typical band structures will be given in Sect. 2.4 and used for the calculation of charge transport in Chaps. 7, 8, while in Chaps. 4, 5, the transport properties will be calculated assuming the simple model of band structure (which is quite a good approximation for most purposes).

2.1 Single and Periodically Repeated Potential Well

A charge carrier in a crystal passing an atom is first subject to an acceleration and then when leaving the atom, to a deceleration until it gets into the field of the next atom and these processes are repeated. The crystal field can be approximated by a periodic array of potential wells. The calculation is simplified if each minimum is assumed to be a square well. This is the one-dimensional *Kronig–Penney model* of the crystal [2.1]. Man-made potential structures of this kind are the semiconductor superlattices, which will be discussed in Sect. 13.3.

For an investigation of the motion of an electron through the periodic crystal lattice of atomic dimensions it is necessary to deal with non-relativistic quantum mechanics. According to Schrödinger and Dirac [1] the electron, besides being a charged particle, also has a wave-like nature. Its wave function $\Psi(r, t)$ is a solution of the Schrödinger equation

$$i\hbar\frac{\partial\Psi}{\partial t} = [-\frac{\hbar^2}{2m_0}\nabla_r^2 + V(r)]\Psi \qquad (2.1.1)$$

---

[1]Erwin Schrödinger, 1887 - 1961, Austrian physicist, and Paul Adrien Maurice Dirac, 1902 - 1984, British physicist, shared the Nobel physics award for 1933

where $\hbar$ is Planck's constant, $V(r)$ is the potential energy which may be time (t) dependent, and ∇_r^2 is the square of the partial derivative with respect to the space coordinates (x, y, z) in three dimensions

$$\nabla_r^2 = \frac{\partial^2}{\partial x^2} + \frac{\partial^2}{\partial y^2} + \frac{\partial^2}{\partial z^2} \; . \tag{2.1.2}$$

If $V(x, y, z)$ does not depend on time, (2.1.1) can be simplified

$$\varepsilon \psi(r) = (-\frac{\hbar^2}{2m_0} \nabla_r^2 + V(r))\psi(r) \tag{2.1.3}$$

where by writing

$$\Psi(r, t) = \psi(r) \exp(-i\frac{\varepsilon}{\hbar}t) \tag{2.1.4}$$

the electron energy ε and a frequency $\omega = \varepsilon/\hbar$ have been introduced and i as in (2.1.1) is the imaginary unit. The rhs of (2.1.3) is often written as $H\psi(r)$ and H is referred to as the *Hamiltonian operator*. The first term on the rhs of (2.1.3) is the kinetic energy of the electron.

As a preliminary study, we shall calculate the energy levels ε of a particle in a single potential well [2.2] (Fig. 2.1). The depth of the well is V_0 and its width b. The following notation is introduced:

$$\alpha^2 = 2m_0 \hbar^{-2}\varepsilon; \quad \beta^2 = 2m_0 \hbar^{-2}(V_0 - \varepsilon) \; , \tag{2.1.5}$$

Inside the well where $V(x) = 0$, the Schrödinger equation takes the form

$$d^2\psi/dx^2 + \alpha^2\psi = 0; \quad -b/2 \le x \le b/2 \; , \tag{2.1.6}$$

while outside the well where $V(x) = V_0$, it is

$$d^2\psi/dx^2 - \beta^2\psi = 0; \quad x < -b/2 \text{ and } x > b/2 \; , \tag{2.1.7}$$

The boundary condition is that the position probability density of the particle $|\psi|^2$ vanishes at infinity. The solutions are:

$$\psi = Ce^{i\alpha x} + De^{-i\alpha x} \text{ for } -b/2 \le x \le b/2$$

$$\psi = Ae^{-\beta x} \text{ for } x > b/2 \tag{2.1.8}$$

$$\psi = Ae^{\beta x} \quad \text{ for } x < -b/2$$

where A, C and D are integration constants which can be evaluated from the requirement of the continuity of ψ and $d\psi/dx$ at $x = \pm b/2$. The solutions for α yield discrete energy levels (in units of V_0):

$$\varepsilon/V_0 = \hbar^2\alpha^2/(2m_0 V_0) \tag{2.1.9}$$

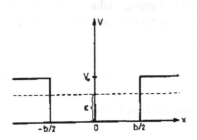

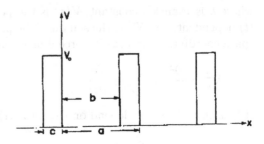

Fig. 2.1: Rectangular potential well   **Fig.2.2**: Periodic potential wells
                                                      (Kronig–Penney model [2.1])

Next we consider the Kronig–Penney model [2.1] which is shown in Fig.2.2. The
lattice constant is $a = b + c$. The potential is periodic in a:

$$V(x) = V(x + a) = V(x + 2a) = \ldots \tag{2.1.10}$$

It is reasonable to try for $\psi(x)$ a solution of the form

$$\psi(x) = u(x)\,e^{ikx} = u(x + na)\,e^{ikx} \tag{2.1.11}$$

where $n = 1, 2, \ldots$. This is called a *Bloch function*.[2]   Taking (2.1.6) for the
range $0 \le x \le b$ (inside the well), we have

$$d^2u/dx^2 + 2i\,k\,du/dx + (\alpha^2 - k^2)u = 0; \quad 0 \le x \le b, \tag{2.1.12}$$

and accepting (2.1.7) for the range $-c < x < 0$ (outside the well), we have

$$d^2u/dx^2 + 2i\,k\,du/dx - (\beta^2 + k^2)u = 0; \quad -c < x < 0. \tag{2.1.13}$$

The solutions are with constants A, B, C and D:

$$u = A\,e^{i(\alpha-k)x} + B\,e^{-i(\alpha+k)x}; \quad 0 \le x \le b, \tag{2.1.14}$$

$$u = C\,e^{(\beta-ik)x} + D\,e^{-(\beta+ik)x}; \quad -c < x < 0 \,. \tag{2.1.15}$$

The conditions of continuity for u and du/dx at $x = 0$ and of periodicity for
these quantities yield 4 equations for A, B, C and D which can be solved if the
determinant Δ vanishes. Introducing a function $L = L(\varepsilon/V_0)$ by

$$L = \frac{1 - 2\varepsilon/V_0}{2\sqrt{(\varepsilon/V_0) - (\varepsilon/V_0)^2}} \sinh\left[\sqrt{\frac{2m\,V_0}{\hbar^2}\left(1 - \frac{\varepsilon}{V_0}\right)}\,c\right] \sin\left(\sqrt{\frac{2m\,V_0}{\hbar^2}\frac{\varepsilon}{V_0}}\,b\right)$$

$$+ \cosh\left[\sqrt{\frac{2m\,V_0}{\hbar^2}\left(1 - \frac{\varepsilon}{V_0}\right)}\,c\right] \cos\left(\sqrt{\frac{2m\,V_0}{\hbar^2}\frac{\varepsilon}{V_0}}\,b\right) \tag{2.1.16}$$

---

[2] $k = 2\pi/\lambda$, where λ is the de Broglie wavelength.

the condition $\Delta = 0$ can be written in the form

$$k = k(\varepsilon) = \frac{1}{a} \arccos L(\varepsilon/V_0). \qquad (2.1.17)$$

In this way the function $k(\varepsilon)$ is calculated rather than $\varepsilon(k)$. Let us compare these results with those obtained for a single well. Assuming a value of 36 for $2m\,\hbar^{-2}\,V_0(b/2)^2$ and, in addition, the ratio $c/b = 0.1$, the function $L(\varepsilon/V_0)$ as plotted in Fig. 2.3 is obtained. In certain ranges of ε/V_0, it is larger than $+1$ or less than -1. On the other hand, the cosine can have real values only between -1 and $+1$. This means there are ranges of ε/V_0 where k is not real. These ranges are called *forbidden energy bands* or *gaps*. Between the gaps there are *allowed energy bands*. These bands arise from the discrete N-fold degenerate levels of N atoms a long way apart while they are brought together. The right ordinate scale of Fig 2.3 shows these discrete levels. In the limiting case $b = 0$ (or $c \gg b$), the bands become discrete levels, of course.

For the assumed c/b ratio, 4 gaps are obtained. The lowest gap starts at the bottom of the well ($\varepsilon = 0$) and goes up to $\varepsilon/V_0 = 0.0074$. This gap is too narrow to be shown Fig. 2.3. For smaller c/b ratios the gaps are smaller. For the case of $c/b = 1/240$, e.g., only two very narrow gaps at $\varepsilon/V_0 = 0.07$ and 0.63 are found. In the limiting case $c = 0$, $a = b$, (2.1.17) yields $\varepsilon = \hbar^2 k^2/2\,m_0$ as in free space.

For a plot of $\varepsilon(k)$, it is useful to remark that in Fig. 2.3 the function $L(\varepsilon/V_0)$ is nearly linear within a band. Denoting the band edges ε_1 (at $L = +1$) and $\varepsilon_2 > \varepsilon_1$ (at $L = -1$), (2.1.17) yields

$$\varepsilon = \frac{\varepsilon_2 + \varepsilon_1}{2} - \frac{\varepsilon_2 - \varepsilon_1}{2} \cos(ka) \quad \text{for } \varepsilon_2 > \varepsilon_1. \qquad (2.1.18)$$

The following band is passed by $L(\varepsilon/V_0)$ in the opposite direction. This changes the sign of the cosine term:

$$\varepsilon = \frac{\varepsilon_4 + \varepsilon_3}{2} + \frac{\varepsilon_4 - \varepsilon_3}{2} \cos(ka) \quad \text{for } \varepsilon_4 < \varepsilon_3, \qquad (2.1.19)$$

etc. These functions are plotted in Fig. 2.4. The dash-dotted parabola represents the function $\varepsilon = \hbar^2 k^2/2m_0$ valid for a free-electron. For an electron inside the crystal, the parabola is replaced by S-shaped parts of a sin curve which are separated from each other by discontinuities in energy at $k = n\pi/a$, where $n = 1, 2, 3, \ldots$; these are the energy *gaps* discussed before. At the lower edges of the S-shaped parts, $d^2\varepsilon/dk^2$ is positive and we shall see that it is this quantity which leads to a positive *effective mass* defined by

$$m = \frac{\hbar^2}{d^2\varepsilon/dk^2} \; . \qquad (2.1.20)$$

We will now show that this relation is meaningful.

It is shown in wave mechanics that a particle known to exist at a position somewhere between x and $x + \Delta x$ cannot be represented by a wave with a

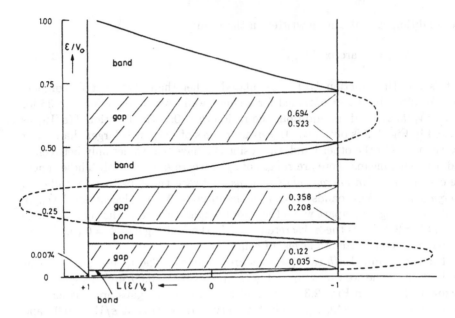

Fig. 2.3: Band structure of a one-dimensional lattice. Marks at the right ordinate scale indicate discrete levels of a single potential well

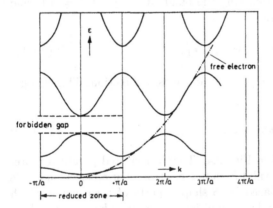

Fig. 2.4: $\varepsilon(k)$ diagram for an electron in a one-dimensional lattice

discrete k value but rather by a *wave packet* composed of waves with k values essentially between k and $k + \Delta k$ where the product $\Delta x \times \Delta k$ is of the order of 1. After introducing the momentum $\hbar k$, the relation $\Delta x \times \Delta(\hbar k) \approx \hbar$ illustrates the *Heisenberg Uncertainty Principle* which states that the position of a particle and its momentum can be observed simultaneously only with the precision given by this relation. It is the group velocity $v = d\omega/dk = \hbar^{-1} d\varepsilon/dk$ of the wave packet which can be identified with the particle velocity.

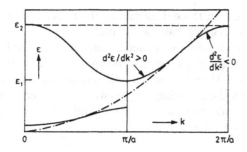

Fig. 2.5: Section of Figure 2.4 for k between 0 and $2\pi/a$

Assume that a force F is acting upon the particle in the crystal. The energy gain $d\varepsilon$ of the particle in a time dt will be $Fv\,dt$. On the other hand, $d\varepsilon = (d\varepsilon/dk)\,dk = v\,d(\hbar k)$. Hence, the force F is equal to $d(\hbar k)/dt$ which is analogous to Newton's law if we associate $\hbar k$ with a momentum as mentioned before. $\hbar k$ is called the *crystal momentum*.

The mass m of a free particle is defined by Newton's law: The derivative of v with respect to time is the acceleration of the particle by the force F which, if no other forces were acting upon the particle, would be equal to F/m. For a carrier in a solid it is possible to retain this relation by neglecting the crystal forces and introducing the *effective mass*. Hence, $Fm^{-1} = dv/dt = \hbar^{-1}(d^2\varepsilon/dk^2) \times dk/dt$ and on the other hand, $Fm^{-1} = d(\hbar k)/dt\,m^{-1}$ which yields (2.1.20). In the special case $\varepsilon = \hbar^2 k^2/2m$ valid for a free particle of mass m, (2.1.20) can be easily verified.

Now let us consider the upper edge of a band displayed in Fig. 2.5. Here the effective mass of an electron would be negative. Fortunately, it is the ratio of the charge to the mass rather than the mass itself that is important in transport theory. Consequently, one can talk either of negatively charged electrons with a negative effective mass or positively charged particles called *holes* with a positive effective mass. The convention is to adopt the latter.

For a quantitative calculation of the effective mass, (2.1.18) turns out to be too crude an approximation. However, even for the one-dimensional case considered here, the calculation is quite involved. For the special case $|V(x)| \ll \varepsilon$, it yields [2.3, 2.4]

$$m = m_0 \frac{\varepsilon_G}{2(\hbar^2/m_0)(\pi/a)^2 \pm \varepsilon_G} \tag{2.1.21}$$

at $k = \pi/a$, where in the denominator, $+\varepsilon_G$ is valid for the conduction band and $-\varepsilon_G$ is for the valence band. For three dimensions, depending on crystal structure, band calculations can be very difficult. As a general rule, the effective mass is smaller for a smaller gap. In n-InSb, e.g., at room temperature an effective mass of $0.013\,m_0$ and a gap of 0.18 eV are observed. A material with a larger gap, 1.26 eV, is n-InP where the effective mass, $0.05\,m_0$, is also larger. Typical data are compiled in Table 2.1.

These examples suggest that also in three dimensions, the relation $m \propto \varepsilon_G$ holds to a good approximation for a simple band structure.

For a more detailed calculation of the band structure such as, e.g., for InSb [2.5, 2.6], the atomic wave function as well as the electron spin have to be taken into account. With two parameters P and Δ which will not be considered in detail, Kane [2.5] obtained for InSb the following $\varepsilon(k)$ relation with energy zero assumed for the upper edge of the valence band and three kinds of holes which we will discuss in Sect. 8.1 (valid near $k = 0$)

Conduction band:

$$\varepsilon = \varepsilon_G + \frac{\hbar^2 k^2}{2m} + \frac{P^2 k^2}{3} \left(\frac{2}{\varepsilon_G} + \frac{1}{\varepsilon_G + \Delta} \right)$$

Valence band:

$$\varepsilon = -\frac{\hbar^2 k^2}{2m} \qquad \text{(heavy holes)}$$

$$\varepsilon = -\frac{\hbar^2 k^2}{2m} - \frac{2P^2 k^2}{3\varepsilon_G} \qquad \text{(light holes)} \tag{2.1.22}$$

$$\varepsilon = -\Delta - \frac{\hbar^2 k^2}{2m} - \frac{P^2 k^2}{3(\varepsilon_G + \Delta)} \qquad \text{(split − off band)} .$$

In these relations, the anisotropy of the valence band is neglected (see Sect. 2.4). Interactions which determine the heavy hole effective mass are not included in Kane's model which itself is still very simplified. The heavy hole mass m may be assumed to be infinite in this model [2.7]. The valence band consists of three subbands, two of which are degenerate at $k=0$. The third band is split-off by spin-orbit coupling of the electrons. The split-off energy at $k = 0$ is denoted by Δ. In the case of $\Delta \gg \varepsilon_G$ and $P\,k$, we find for an approximation (valid in a wider range of k):

Table 2.1. Band gaps and electron effective masses of compounds with lowest conduction band minimum at $k=0$ (see also Appendix C)

Com- pound	Optical gap (300 K) [eV]	m/m_0 (optical) (300 K)	El.gap (extrap. to 0 K) [eV]	m/m_0 (electrical)
InSb	0.180	0.0145	0.2	0.013
InAs	0.358	0.027	0.42-0.47	
InP	1.54	0.077	1.42	
GaSb	0.70	0.047	0.77-0.82	0.040-0.048
GaAs	1.43	0.067	1.52	

Conduction band:

$$\varepsilon - \varepsilon_G = \hbar^2 k^2/2m + \frac{1}{2}(\sqrt{\varepsilon_G^2 + 8P^2k^2/3} - \varepsilon_G)$$

Valence band:

$$\varepsilon = -\hbar^2 k^2/2m \quad \text{(heavy holes)}$$

$$\varepsilon = -\hbar^2 k^2/2m - 1/2(\sqrt{\varepsilon_G^2 + 8P^2k^2/3} - \varepsilon_G) \quad \text{(light holes)} \qquad (2.1.23)$$

$$\varepsilon = -\Delta - \hbar^2 k^2/2m - P^2k^2/(3\varepsilon_G + 3\Delta) \quad \text{(split − off band)} \; .$$

The conduction band and the light-hole valence band are *nonparabolic* which means that ε is not simply proportional to k^2. Introducing for simplification the quantities

$$\varepsilon' = \frac{1}{2}(\sqrt{\varepsilon_G^2 + 8P^2k^2/3} - \varepsilon_G) \qquad (2.1.24)$$

and

$$m_n = 3\hbar^2 \varepsilon_G/4P^2 \qquad (2.1.25)$$

we can solve (2.1.20) for $\hbar^2 k^2/2m_n$:

$$\hbar^2 k^2/2m_n = \varepsilon'(1 + \varepsilon'/\varepsilon_G) \; . \qquad (2.1.26)$$

If in (2.1.23) we put $m = \infty$, we get for the conduction band $\varepsilon' = \varepsilon - \varepsilon_G$ where one notices that the energy zero is shifted to the lower edge of the conduction band. For a finite value of m, (2.1.26) is essentially retained except that ε_G is replaced by an *energy parameter* depending on m. For n-InSb and n-InAs (gap 0.18 eV and 0.36 eV at 300 K, respectively; effective mass $m_n = 0.013m_0$ and $0.025m_0$), the energy parameter has the values of 0.24 eV and 0.56 eV, respectively [2.8].

Data like gap and effective masses still have to be obtained experimentally. Even in the computer age, crude approximations have to be made in the calculation of band structures. It is a normal procedure to include adjustable parameters which are determined from experimental data like the gap and the effective masses. In Sect. 11.1 we shall consider fundamental optical reflection where the band structure is of vital importance. For a clear picture of the origin of gaps, we now discuss the tight-binding approximation.

2.2 Energy Bands by Tight Binding of Electrons to Atoms

In the preceding treatment, the effect of the crystal lattice on electrons was approximated by that of a periodic potential. Another way of looking at the problem is to construct a crystal from single atoms which are infinitely far away from one another, and to see how the discrete atomic energy levels are changed during this process. When the atoms are far away from each other the electrons are *tightly bound* to each atom. At what atomic distance will an exchange of electrons between neighboring atoms begin to become important?

If the wave functions ψ of atom A and B overlap, linear combinations $\psi_A \pm \psi_B$ characterize the behavior of the compound AB. In Fig. 2.6, the energy as a function of the distance between the two atoms is plotted. The sum of the ψ-functions yields the bonding state while the difference represents the situation where the atoms repel each other (*antibonding*). A calculation is presented in monographs on quantum mechanics [2.9] and will not be attempted here. If N rather than 2 atoms come together, the atomic level is split not into 2 but into N levels. In a crystal of $N \approx 10^{23}$ atoms, the levels are so close to each other that this can be called a band. Let us again consider diamond. In a free carbon atom, the two 2s levels and two of the six 2p levels, which have a higher energy than the 2s levels, are occupied by the four valence electrons. These four filled levels per atom form the valence band if many atoms are put together to form diamond crystals, while the other four levels form the empty conduction band. At a lattice constant of 0.36 nm, there is a gap of about 5.3 eV between these two bands. At absolute zero temperature, pure diamond is an insulator since it will be shown in Chap. 3 that all electrons in the valence band are completely filling it and for a drift motion, an electron would have to increase its energy slightly (by an amount still considerably less than the gap) but there is still no unoccupied state in the valence band. Therefore, no drift is possible and diamond is thus an insulator. However, at high temperatures, a few electrons are lifted thermally into the conduction band enabling both electrons and holes to drift: diamond is then semi-conducting.

In metals there are either partly-filled bands or overlapping free and filled bands. Why is there a gap in diamond but not, e.g., in metallic tin? In fact both the crystal structure and the number of valence electrons decide upon this question: crystals of metallic Sn are tetragonal, of Pb (always metallic) face-centered cubic, while the semiconducting group-IV elements crystallize in the diamond lattice. Hund and Mrowka [2.10] and Kimball [2.11] investigated this problem very carefully with different crystal structures. The one-dimensional chain, the 2-dimensional hexagonal lattice (coordination number 3), the diamond lattice and the element-wurtzite lattice (Fig. 2.16) require only s-functions (even functions of space coordinate) and p-functions [odd functions, e.g., $xf(r)$] for the calculation of the crystal wave function while the 2-dimensional square lattice (coordination number 4) and the body-centered-cubic lattice also require d-functions [proportional to $(x^2 - y^2)g(r)$ for the square lattice]. The overlap of bands calculated for the cubic lattice yielding metallic behavior is shown in Fig. 2.7. However, the band type composed of s- and p-functions may also lead to a metal. A typical example is graphite. The bonds in the hexagonal planes are filled by 3 valence electrons per atom. The fourth valence electron of each carbon atom is located in the conduction band and produces a rather

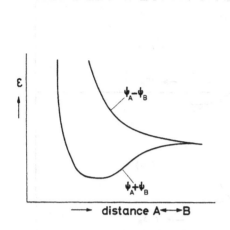

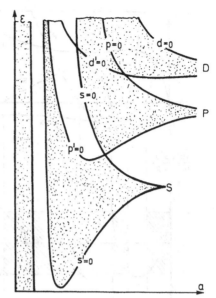

Fig. 2.6: Energy of a two-atomic molecule AB as a function of the atomic distance. There is a bonding state with a minimum in energy, and an antibonding state

Fig. 2.7: Energy bands of a cubic lattice as a function of the lattice constant (after [2.10])

metallic conductivity. Let us briefly go through the *cellular type* calculation of the diamond lattice. In two lattice points at a distance of a lattice constant, the potential has the same symmetry but a different space orientation relative to the coordinate axes. The two atoms are therefore centers of cells a and b. While the ψ-function in cell a is given by

$$\psi_a = A_a s(r) + \sqrt{3}(B_a x + C_a y + D_a z)p(r)/r \qquad (2.2.1)$$

and the ψ - function in cell b centered at $e = (a/4; a/4; a/4)$ is given by

$$\psi_b = \exp(2\pi i\ k.e) \qquad (2.2.2)$$

$$\times (A_b s(r') + \sqrt{3}[B_b(x - a/4) + C_b(y - a/4) + D_b(z - a/4)])p(r')/r'$$

Here r' is the distance from the center of cell b. The continuity condition for ψ and $d\psi/dr$ at the halfway intermediate points between the atom and its four nearest neighbors yields eight equations for the eight unknown coefficients $A_a \ldots D_b$. A solution of the homogeneous system of linear equations exists if the determinant vanishes, yielding $p^2 = 0$, $p'^2 = (dp/dr)^2 = 0$ and

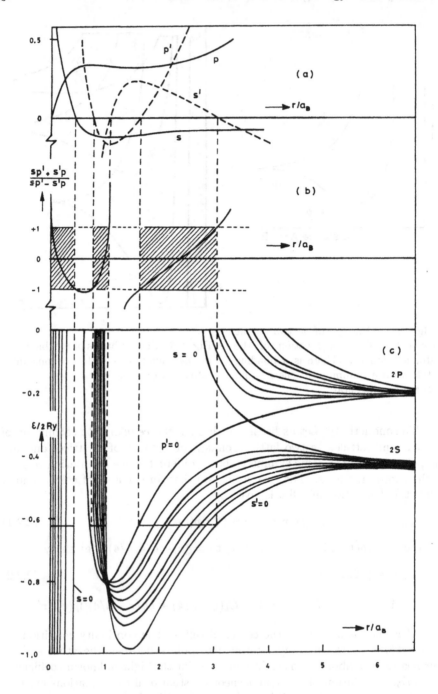

Fig. 2.8: (a.)Radial wave functions and their derivatives (multiplied by 10) for $\varepsilon/2R_y = -0.635$, where R_y is the Rhydberg energy. (**b.**) The function $(sp' + s'p)/(sp' - s'p)$ for $\varepsilon/2R_y = -0.635$. (**c.**) Calculated energy bands of diamond (a_B is the Bohr radius)

$$\frac{sp' + s'p}{sp' - s'p} = \frac{\pm 1}{2}[1 + \cos(\pi\, a\, k_x) \cos(\pi\, a\, k_y) + \cos(\pi\, a\, k_y) \cos(\pi\, a\, k_z)$$

$$+ \cos(\pi\, a\, k_z) \cos(\pi\, a\, k_x)]^{1/2} . \qquad (2.2.3)$$

The rhs of (2.2.3) can have values between +1 and -1 only. This function as well as s, p, s', and p' are plotted vs r in Fig. 2.8 by solving the Schrödinger equation numerically assuming a somewhat simplified potential. In the lower part of Fig. 2.8, the energy is plotted vs r, the distance between nearest neighbors. The numerical calculation was made for eight values of the energy. The curve representing $p' = 0$ is two-fold degenerate due to the fact that $p'^2 = 0$, as mentioned above. This is the upper edge of the valence band. This, together with the two-fold degeneracy of (2.2.3) (positive and negative values), makes room for four valence electrons per atom which is the case for diamond. The upper edge of the conduction band is also two-fold degenerate since it is formed by the solution $p^2 = 0$. If the energy comes close to the atomic d-level, the simplified calculation becomes invalid, however [2.12].

It is interesting to note that for the lattice constant of diamond at normal atmospheric pressure the upper edge of the valence band increases with r while the lower edge of the conduction band decreases with r. Decreasing the interatomic distance by external hydrostatic pressure thus increases the energy gap which yields a decrease in the number of carriers and therefore an increase in resistivity, in agreement with observations (Sect. 4.12). The cellular method of band-structure calculation goes back to Wigner and Seitz [2.13]. While it uses only one wave function for each cell, a linear combination of several of them is applied in the LCAO method (*linear combination of atomic orbitals*). The coefficients are determined by minimizing the energy which is done by a variational method resulting in the vanishing of a determinant. In contrast to this tight-binding approach, there are two methods which use plane waves: the *augmented plane wave* (APW) and the *orthogonalized plane wave* (OPW) methods. For both methods the crystal potential inside a certain radius around the atomic cores is taken as in the cellular method while it is assumed zero in the interstitial regions. This approximation is known as the *muffin-tin* potential. The APW method which goes back to Slater, has been applied mostly to metals [2.14]. In the OPW type of calculation, atomic wave functions are added to the plane waves to make them orthogonal which leads to a rapid convergence at a much smaller number of plane waves than with the APW method. It is the method applied the most to semiconductors [2.15]. In order to further simplify calculations, a *pseudopotential* is often introduced in order to remove the rapidly varying part near the atomic cores [2.16, 17]. For the diamond crystal lattice, three parameters are taken from the experimental data (optical absorption, cyclotron effective mass, etc.), while for the zincblende lattice, six parameters are required [2.18].

Another well-known method is due to Korringa [2.19] and Kohn and Rostoker [2.20], again based on the muffin-tin potential. It is a Green's function

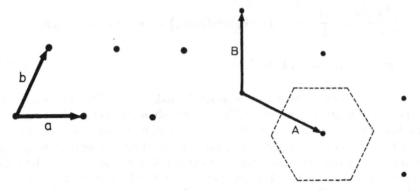

Fig. 2.9: Base vectors of a two–dimensional lattice

Fig. 2.10: Base vectors of the reciprocal lattice. The area of the hexagon is the first Brillouin zone of the real lattice shown in Fig. 2.9

method starting from the integral form of the Schrödinger equation. The approximation is to retain only a finite number of spherical harmonics in the expansion [2.21]. Finally, we mention the $k \cdot p$ method which is a perturbation type of approach around a critical point in the band structure (Sect 8.1). The *envelope function approximation* (Appendix B) is applied to heterostructure subband calculations (Sects. 9.1, 13.2).

2.3 The Brillouin Zone

As demonstrated by Fig. 2.4, the band structure given for the range $-\pi/a \le k \le \pi/a$ is repeated indefinitely. It is therefore sufficient to consider only this range which is called the *first Brillouin zone*. The neighboring regions from $-2\pi/a$ to $-\pi/a$ and from π/a to $2\pi/a$ form the *second Brillouin zone*, etc. Since for most purposes it is sufficient to consider the first Brillouin zone, this is sometimes just called the Brillouin zone. What is the Brillouin zone of a 3-dimensional lattice?

The history of Brillouin zones started with scattering processes of x-rays by a multidimensional lattice. The scattering corresponds to that of the electron waves. An electron moving in a crystal is represented by an electron wave packet continuously scattered at the crystal lattice. The scattering process obeys the Laue equation

$$\exp[\mathrm{i}(p \cdot G)] = 1,$$

where the vector

$$p = l\,a + m\,b + n\,c \qquad (l, m, n = 0, 1, 2, 3, \ldots) \qquad (2.3.1)$$

characterizes an atom of the crystal lattice which has a primitive unit cell defined by the primitive base vectors a, b and c (in two dimensions: Fig. 2.9). The vector $G = k' - k$ represents the change of wave vector due to the scattering process. The Laue equation requires $(p \cdot G) = 2\,\pi$ or an integral multiple thereof. Accordingly, vectors in k-space A, B, and C are defined such that

$$G = g\,A + h\,B + k\,C \qquad (g, h, k = 0, \pm1, \pm2, \pm3, \ldots) \qquad (2.3.2)$$

where $(a \cdot A) = (b \cdot B) = (c \cdot C) = 2\,\pi$ and other products such as $(a \cdot B) = 0$, $(b \cdot C) = 0$ etc. A, B, and C are denoted as the base vectors of the reciprocal lattice. They are perpendicular to the corresponding base vectors of the real lattice. In Fig. 2.10, the base vectors of the reciprocal lattice are shown in comparison to the real lattice of Fig. 2.9. If x_e and y_e are unit vectors in a 2-dimensional Cartesian coordinate system, the base vectors of Fig. 2.9, 10 may be represented by $a = 2\,x_e$, $b = x_e + 2y_e$, $A = \frac{\pi}{2}\,(2\,x_e\text{-}y_e)$ and $B = \pi\,y_e$. Of course, from the Laue equation the well-known Bragg equation is easily derived. The *primitive cell* contains all points which cannot be generated from each other by a translation (2.3.1). The *Wigner–Seitz cell* contains all points which are nearer to one considered lattice point than to any other one. In the

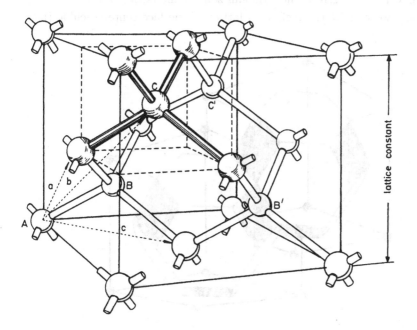

Fig. 2.11: Diamond crystal structure. The lattice constants for C, Si, and Ge are 0.356, 0.543, and 0.566 nm, respectively (after [2.22])

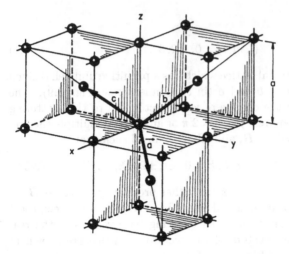

Fig. 2.12: Primitive translations in the body–centered cubic lattice

reciprocal lattice, the Wigner–Seitz cell is denoted as the *first Brillouin zone*. For a construction of the Brillouin zone, draw arrows from a lattice point to its nearest neighbors and cut these in half. The planes through these intermediate points perpendicular to the arrows form the surface of the Brillouin zone. The hexagon shown in Fig. 2.10 is the Brillouin zone of the lattice of Fig. 2.9.

It is easy to see that the reciprocal lattice of the face-centered-cubic (fcc)

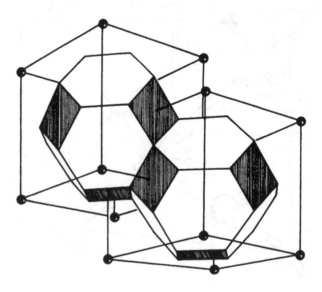

Fig. 2.13: First Brillouin zone for the face–centered cubic lattice (after [2.23])

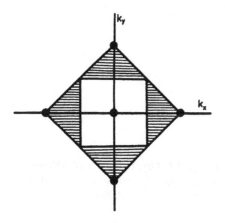

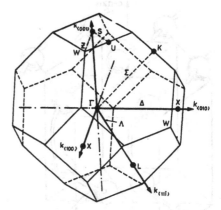

Fig. 2.14: First two Brillouin zones for the single square lattice (after [2.24])

Fig. 2.15: First Brillouin zone for the diamond and zinc-blende lattice, including important dymmetry points and lines (after [2.30])

lattice is the body-centered-cubic (bcc) lattice and vice versa. The important semiconductors of the fourth group of the periodic table crystallize in the diamond lattice shown in Fig. 2.11. Each atom is tetrahedrically bonded to four nearest neighbors, similar to the bonds in the CH_4 molecule. Let us take the atom A as the corner of a fcc lattice. In going from A along the cube diagonal halfway towards the cube center, one encounters atom B which is the corner of a second fcc lattice of which besides B only the atoms B', C, and C' are shown in Fig. 2.11. The vectors a, b, and c from A to the face centers are the base vectors. Perpendicular to these are the base vectors a', b', and c' of the reciprocal lattice, which is a bcc lattice, as shown for real space in Fig. 2.12. By calculating the scalar product of $\langle 1,1,1 \rangle$ with $\langle 1,-1,0 \rangle$, it is obvious that each cube diagonal is perpendicular to a face diagonal. The construction of the Brillouin zone is then not difficult (Fig. 2.13). The faces normal to the

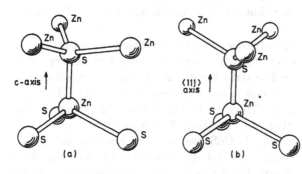

Fig. 2.16: Atomic arrangement in wurtzite (a) and zinc-blende (b) (after [2.30])

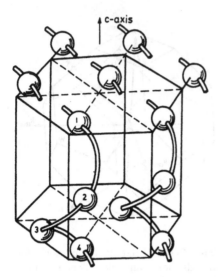

Fig. 2.17: Crystal lattice of tellurium: atomic spiral chains in trigonal arrangement (after [2.30])

cube diagonals are the hexagons while the faces normal to the lines from one cube center to the next are the hatched squares. At one of the hexagon faces, a second *truncated octahedron* is attached in order to demonstrate how k-space is filled. After suitably cutting the second truncated octahedron and attaching the parts to the first Brillouin zone, this would be the second Brillouin zone [2.25, 26]. A 2-dimensional example is shown in Fig. 2.14. The third Brillouin zone is constructed similarly, etc. All Brillouin zones of a lattice have equal volume. In Fig. 2.15, the most important symmetry points and lines in the first Brillouin zone of the fcc lattice are given. There are six equivalent X-points, eight L-points, etc. [2.27, 28].

The zinc-blende lattice so typical for III-V compounds can be generated from a diamond lattice by replacing half of the atoms by group-III atoms and half by group-V atoms. Each bond connects a group-III atom with a group-V atom. If, e.g., an InSb plate cut with (111) broad faces is etched, one side consists of In atoms and the opposite side of Sb atoms. The inversion symmetry of the diamond lattice therefore is not found in the zinc-blende lattice. (For further details see, e.g., [2.29].)

The wurtzite structure is closely related to the zinc-blende structure. In Fig. 2.16, the orientations of adjacent atomic tetrahedra in both structure types are compared [2.30]. The hexagonal wurtzite crystal has a unique crystallographic axis of symmetry, designated the c-axis, which corresponds to the $\langle 1,1,1 \rangle$ axis in the cubic zinc-blende structure. There are four atoms per primitive cell, while in the zinc-blende structure there are only two. The band structures for both types of structures are very similar. In CdS, the band gap has been determined in both modifications from the fundamental optical reflectivity. The gap of the cubic structure is only a few hundredth of an eV lower than that of the wurtzite

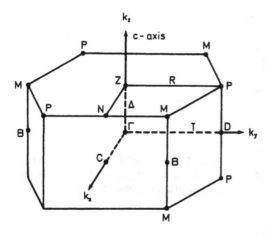

Fig. 2.18: First Brillouin zone for tellurium, including important symmetry points and lines. The trigonal symmetry about the c–axis is indicated by the distinction between the M- and P-points (after [2.30])

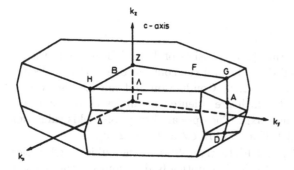

Fig. 2.19: First Brillouin zone for Bi$_2$Te$_3$, including important symmetry points and lines (after [2.30])

structure which is 2.582 eV (at 0 K). CdS ordinarily crystallizes in the wurtzite structure but can be forced to accept the zinc-blende structure by depositing it epitaxially from the vapor phase onto cubic GaAs. The Brillouin zone is the same as that of Te to be treated below, except that it has a hexagonal symmetry.

The crystal structure of Te is shown in Fig. 2.17. It consists of helical chains arranged in a trigonal structure. The chains may either be right-handed or left-handed; Fig. 2.17 shows only the right-handed version. The chains are interconnected by van der Waals and weak electronic forces. There is no center of inversion symmetry. The primitive cell contains three atoms. The Brillouin zone which is a hexagonal prism is shown in Fig. 2.18, including the group-theoretical notation of zone points [2.30]. The symmetry of the crystal about the c-axis is not sixfold but threefold (*trigonal*). Optical experiments (exciton absorption and interband magneto-absorption at 10 K, polarization parallel to the c-axis) have shown that it is a *direct semiconductor* (Sect. 11.2) with a band gap of 0.3347 ($\pm$ 0.0002) eV which increases with decreasing temperature and increasing pressure [2.31].

A semiconductor of some interest for thermoelectric applications is Bi$_2$Te$_3$.

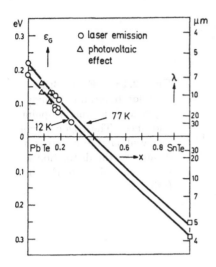

Fig. 2.20: Variation of the energy band gap of $Pb_{1-x}Sn_xTe$ alloys with composition x (after [2.34])

Its rhombohedral crystal structure [2.32] consists of Te and Bi layers in the sequence Bi-Te-Te-Bi-Te-Bi-Te-Te-Bi. The unit cell contains one Bi_2Te_3 molecule. The Brillouin zone is shown in Fig. 2.19. Conduction and valence bands seem to consist of six ellipsoids each which are oriented at angles of 76° and 56°, respectively, with respect to the c-axis. The optical absorption yields values for the band gap of 0.145 eV at 300 K and 0.175 eV at 85 K [2.33].

The lead salts PbS, PbSe, and PbTe crystallize in the cubic NaCl lattice and thus have two atoms per unit cell. The optically determined band gaps at absolute zero temperature are 0.29, 0.15, and 0.19 eV, respectively, and increase with temperature in contrast to most other semiconductors (e.g., in PbTe from 0.19 eV at 0 K to 0.32 eV at 300 K). At the application of pressure the gaps decrease.

Very interesting behavior is exhibited by PbTe-SnTe mixed crystals which have a smaller band gap than each of the components. Technological interest in narrow gap materials arises from their applications as long-wave-length infrared detectors and lasers. Similar systems are PbSe-SnSe, CdTe-HgTe, Bi-Sb, and – under uniaxial stress – gray tin (α-Sn). Figure 2.20 shows the band gap of $Pb_{1-x}Sn_xTe$ alloys as a function of composition for temperatures of 12 K and 77 K [2.34]. For example, at 77 K the gap vanishes for $x = 0.4$. At this composition the valence band edge (of PbTe) changes its role and becomes the conduction band edge (*band inversion*) (of SnTe) [2.35]. The somewhat similar situation in $Cd_xHg_{1-x}Te$ is shown in Fig. 2.21 where the band gap is plotted vs x. A second horizontal scale indicates the dependence on pressure for $x = 0.11$. The hatched area shows how the band gap widens with increasing values of x. The alloys with $x < 0.14$ are semimetals. The graph may suggest a negative band gap in the semimetal. In Fig. 2.22, where the band gap is plotted vs the average atomic number $\bar{Z}$ for five II-VI compounds with zinc-blende structure,

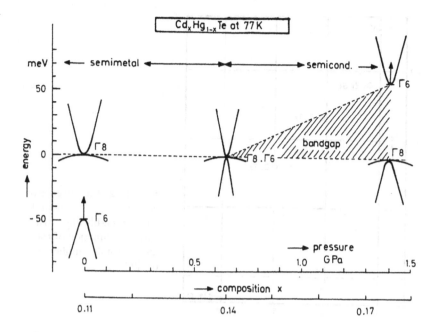

Fig. 2.21: Band structure of $Cd_xHg_{1-x}Te$ alloys near the semimetal–semiconductor transition. The vertical arrows indicate the change in band gap with increasing temperature or pressure (after [2.36, 2.38])

it is shown that with increasing values of $\bar{Z}$, one may arrive at a negative band gap. The semimetal HgTe is in fact indicated at a negative value of the band gap. Finally, band structure calculations for HgTe by the $k \cdot p$ method (Sect. 8.1) yield a gap of value ± 0.15 eV [2.36]; the negative sign seems to be more appropriate in this case. The band structure of gray tin (α-Sn) shown [2.37] in Fig. 2.23 is similar to that of HgTe. The *negative band gap* ($\Gamma_7^- - \Gamma_8^+$) is 0.3 eV. The top of the valence band is identical with the conduction band minimum at $k = 0$. However, in electrical measurements, a bandgap of 0.09 eV is observed which is the gap between the top of the valence band and the L_6^+ minima in $\langle 1,1,1 \rangle$ direction in k-space. Most conduction electrons are in these minima because of their higher density of states (see Sect. 3.1) which increases with the number of valleys (4 in $\langle 1,1,1 \rangle$ direction) and with the effective mass (which is inversely proportional to the curvature of the valley). Therefore, gray tin, although actually being a semimetal, acts in electrical measurements like a semiconductor having a gap of 0.09 eV. The fact that PbTe-SnTe mixed crystals are semiconducting on both sides of the $L_6^+ - L_6^-$ crossing point can be explained by the fact that these states have only a twofold spin degeneracy. Otherwise, as in the CdTe-HgTe crystal system, one side would be semimetallic. A review of narrow-band-gap semiconductors has been given in [2.36].

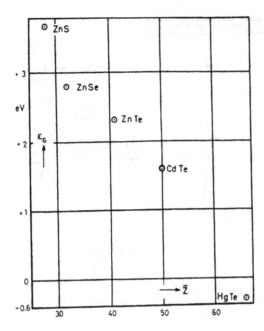

Fig. 2.22: Energy gaps of various II–VI compounds with zinc-blende structure as a function of their mean atomic number (after [2.36])

2.4 Constant Energy Surfaces

The $\varepsilon(k)$ diagrams which have been calculated for silicon [2.38] and germanium [2.6, 39] are shown in Figs. 2.24, 2.25. If in Fig. 2.24 a cut is made parallel to the k-axis not too far above the edge of the conduction band, two points are found which for silicon are on the $\langle 1,0,0 \rangle$ axis, for germanimum on the $\langle 1,1,1 \rangle$ axis. In 3-dimensional k-space, these points become surfaces of constant energy. These surfaces turn out to be ellipsoids of revolution with their long axis on $\langle 1,0,0 \rangle$ and equivalent axes in silicon and on $\langle 1,1,1 \rangle$ and equivalent axes in germanium. In silicon there are six equivalent energy minima (or *valleys*) and for a given value of energy there are six ellipsoids in k-space (Fig. 2.26). In germanium the valleys are at the intersections of $\langle 1,1,1 \rangle$ directions with the surface of the Brillouin zone. So there are eight half-ellipsoids which form four complete valleys (Fig. 2.27).

Similar cuts near the valence band edge yield pairs of constant-energy points which in 3-dimensional k-space become warped spheres around $k=0$ (Figs. 2.28a,b; the warping has been exaggerated in these figures). These warped spheres are concentric, one of them representing heavy holes and the other one light holes [in (2.1.18) the anisotropy was neglected]. For comparison it may be interesting to consider constant energy surfaces in metals [2.40–42] which are often multiply-connected at the L point. These are denoted as *open Fermi surfaces*.

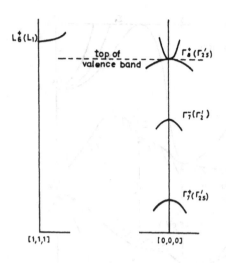

Fig. 2.23: Energy band structure of gray tin (after [2.37])

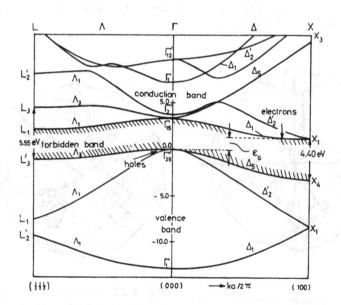

Fig. 2.24: Energy band structure of silicon (after [2.38])

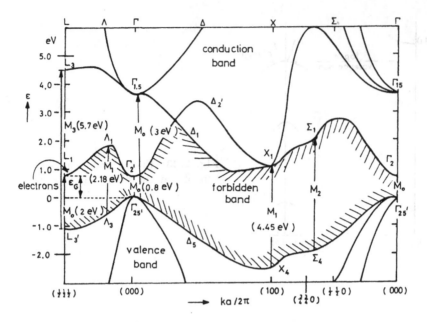

Fig. 2.25: Energy band structure of germanium. The indicated transitions will be discussed in Sect. 11.9 (after [2.39])

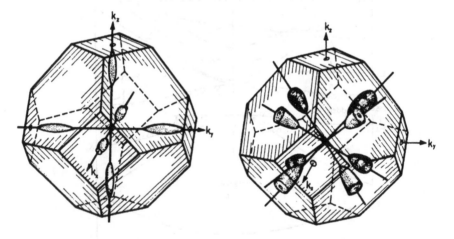

Fig. 2.26: Surfaces of constant energy in k-space for the conduction band edge of silicon. The first Brillouin zone is the same as shown in Fig.2.16 (after [2.4])

Fig. 2.27: Surfaces of constant energy in k-space for the conduction band edge of germanium: eight half–ellipsoids of revolution centered at L points on the zone boundary (after [2.4])

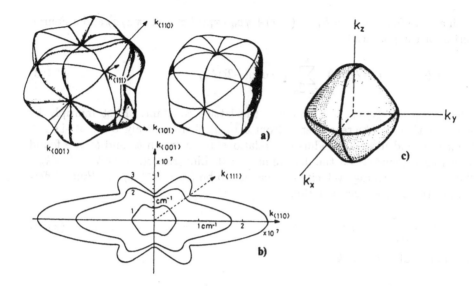

Fig. 2.28: (a) Surfaces of constant energy in k-space for the heavy-hole valence band of silicon and germanium. rhs: low energy. lhs: high energy. (b) Cross section of the constant energy surfaces in the [110] plane. (c) Surface of constant energy in k-space for the light-hole valence band edge of silicon and germanium.

Problems

2.1. Simplify the Kronig–Penney model (Fig. 2.2) by letting V_0 go to infinity and c to zero such that the product $V_0 c$ remains finite. This implies $a = b$. Assume a finite length of the atomic chain with one chain end being at $x = 0$, and a constant potential φ for $x < 0$ (vacuum level). Show that the condition of continuity for the electron wave function ψ and its derivative $d\psi/dx$ at $x = 0$ yields

$$(2p/q^2)\alpha a \cot(\alpha a) = 1 - (2p/q^2) \sqrt{q^2 - (\alpha a)^2} , \qquad (2.0.1)$$

where $p = ma V_0 c/\hbar^2$ and $q^2 = 2ma^2\varphi/\hbar^2$. Find solutions for p=7.2; q=12 of this relation in the two lowest gaps. These are *Tamm surface states*.

2.2. For a one-dimensional crystal with a periodic potential $V(x)$, calculate the effective mass m at the band edge:

$$m = m_0/(\pm 1 + 4\varepsilon_n/\varepsilon_G) , \qquad (2.0.2)$$

where the + sign refers to the conduction band and the - sign to the valence band; $\varepsilon_n = \hbar^2 k_n^2/2m_0$ where $k_n = n\pi/a$ is the edge of the n-th Brillouin zone $(n = 1, 2, 3, \ldots)$; ε_G is the energy gap and m_0 is the free electron mass. Discuss $\varepsilon(k)$ near k_n. *Hints:* solve the Schrödinger equation for an energy $\hbar^2 k_0^2/2m_0$

with a wave function $\psi = b_0 \exp(\mathrm{i}k\,x) + \gamma\, b_n \exp(\mathrm{i}\,k\,x - \mathrm{i}\,n\,2\pi\,x/a)$ and a Fourier series for the potential

$$V(x) = (-\hbar^2\gamma/2m_0) \sum_{n'=-\infty}^{\infty} c_{n'} \exp(-2\mathrm{i}\,n'x/a) \ , \tag{2.0.3}$$

where $c_{-n'} = c_{n'}^*$, $c_0 = 0$ and a is the lattice constant. First multiply by $\exp(-\mathrm{i}k\,x)$ and integrate over x from 0 to a; then multiply by $\exp(-\mathrm{i}k\,x + 2\mathrm{i}\,n\pi\,x/a)$ and integrate. Thus two relationships between b_0 and b_n are found which have solutions for the determinant $= 0$. Linearize $\varepsilon(k')$ for $k' = k - k_n$.

2.3. Show the average velocity of a carrier to be $\langle v \rangle = \hbar^{-1}d\varepsilon/dk$. *Hint:* differentiate the Schrödinger equation

$$-\frac{\hbar^2}{2m} \frac{\partial^2\psi}{\partial x^2} + V(x)\psi = \varepsilon \cdot \psi \tag{2.0.4}$$

in order to obtain $d\varepsilon/dk$.

3. Semiconductor Statistics

The periodic potential distribution of an electron in a crystal shown in Fig. 2.4 involves N discrete levels if the crystal contains N atoms, as we have seen in Fig. 2.8. A discussion of these levels can be confined to the first Brillouin zone. We saw in the last chapter that due to the crystal periodicity, the electron wave functions, which in one dimension are $\psi(x) = u(x)\exp(i\,k\,x)$, also have to be periodic (*Bloch functions*). Hence, from

$$u(x + Na) = u(x) \tag{3.0.1}$$

and

$$\exp(ikx + ik\,N\,a)\,u(x + N\,a) = \exp(ikx)\,u\,(x) \tag{3.0.2}$$

we obtain

$$\exp(i\,k\,N\,a) = 1 \tag{3.0.3}$$

or

$$k = n\,2\,\pi/N\,a; \quad n = 0, \pm 1, \pm 2, \ldots \pm N/2, \tag{3.0.4}$$

where a is the lattice constant. We notice that (3.1) is actually valid for a ring-shaped chain which means that we neglect surface states (Sect. 14.1). Since for the first Brillouin zone k has values between $-\pi/a$ and $+\pi/a$, we find that the integer n is limited to the range between $-N/2$ and $+N/2$. In Fig. 3.1, the discrete levels are given for a *crystal* consisting of $N = 8$ atoms.

How are the N electrons distributed among these levels, including impurity levels if there are any? This problem can be treated with statistical methods since a typical crystal will consist not only of $N = 8$ atoms but of $N \approx 10^{23}$ atoms, a number which is $\gg 1$, and, as a rule, the array is three-dimensional.

3.1 Fermi Statistics

The following assumptions are made:
 a) Electrons cannot be distinguished from one another.

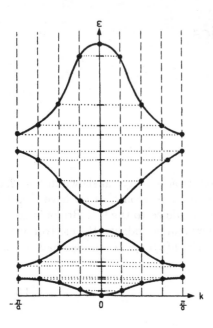

Fig. 3.1: Discrete energy levels for a linear array of eight atoms

b) Each level of a band can be occupied by not more than two electrons with opposite spin. This is due to the Pauli exclusion principle originally formulated for electrons in an atom.

c) Considering, for simplicity, only singly ionized impurities, we also postulate that each impurity level can be occupied by just one electron. We shall see later on that these considerations can be similarly applied to holes.

Bands containing n electrons may consist of N levels $\varepsilon_1, \varepsilon_2, \ldots, \varepsilon_N$ with g_1, $g_2, \ldots, g_N$ states, respectively, as shown schematically in Fig. 3.2. $n_j < g_j$ of these states are assumed to be occupied by one electron each and hence $(g_j - n_j)$ are empty. We have to calculate Boltzmann's *thermodynamic probability* W [3.1].

The most probable distribution of the electrons over all the states is obtained from

$$dW/dn_j = 0 \tag{3.1.1}$$

which is subject to the accessory conditions that both the total number of electrons n and the total energy U remain constant [3.2]. The maximum value $W_{\max}$ of W is denoted as the *thermodynamic probability*. According to Boltzmann, the entropy S is given by $S = k_B \ln W_{\max}$, where k_B is Boltzmann's constant. The *free energy* F is given by $F = U - TS$ where the temperature T is defined by

$$\frac{1}{k_B T} = \left(\frac{\partial \ln W_{\max}}{\partial U} \right)_{n = \text{const}} \tag{3.1.2}$$

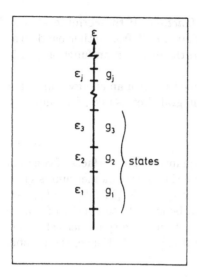

Fig. 3.2: Energy levels $\varepsilon_1, \varepsilon_2, \ldots$, in energy bands with $g_1, g_2, \ldots$, states, respectively

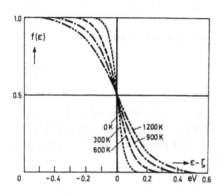

Fig. 3.3: Fermi–Dirac distribution function for various values of the temperature. For the temperature dependence of ζ which is not indicated here see (3.1.9) for extrinsic conduction and (3.1.16) for intrinsic conduction

The Fermi energy [3.1] ζ (also denoted as the *electro-chemical potential* or *Gibbs' potential*) is defined by

$$\zeta = (\partial F/\partial n)_{T=\text{const}}. \tag{3.1.3}$$

The ratio n_j/g_j is denoted as the *thermal equilibrium probability of occupancy* of a state of energy ε_j. It is found to be

$$n_j/g_j = (\exp[(\varepsilon_j - \zeta)/k_B T] + 1)^{-1} \tag{3.1.4}$$

which is called the *Fermi–Dirac distribution function* $f(\varepsilon_j)$. The function is plotted in Fig. 3.3. For a finite temperature T, the step at $\varepsilon = \zeta$ has a width of the order of $k_B T$ and the high-energy tail of the function is well approximated by an exponential

$$n_j/g_j \approx \exp[-(\varepsilon_j - \zeta)/k_B T] \propto \exp(-\varepsilon_j/k_B T). \tag{3.1.5}$$

This is the *Maxwell–Boltzmann distribution function*. If the Fermi level is located within the gap and separated by more than $4\,k_B T$ from either band edge, the semiconductor is called *nondegenerate* and the distribution function (3.1.5) may be applied to the *gas* of carriers.

The states of energy ε_j shall now be assumed to form an energy band. The sum over all states has to be replaced by an integral. Let us take, for simplicity, a parabolic band where

$$\varepsilon = \varepsilon_c + \hbar^2 k^2 / 2\,m_n \tag{3.1.6}$$

and the electron effective mass m_n is a scalar quantity. The surfaces of constant energy in k-space are concentric spheres. Since the crystal momentum is given by $\hbar k$, the volume element in phase space is $dx dy dz\, d(\hbar k_x) d(\hbar k_y)\, d(\hbar k_z)$. According to quantum statistics, phase space can be thought to consist of cells of volume $h^3 = (2\pi\hbar)^3$ with up to two electrons of opposite spins per cell. Integration over coordinate space yields the crystal volume V. Hence, the number of states in an energy range $d\varepsilon$ is given by

$$g(\varepsilon)\, d\varepsilon = 2\frac{V\hbar^3 d^3 k}{(2\pi\,\hbar)^3} = 2\frac{V}{4\pi^2}\left(\frac{2m_n}{\hbar^2}\right)^{3/2}(\varepsilon - \varepsilon_c)^{1/2} d\varepsilon \ . \tag{3.1.7}$$

The total concentration of carriers in the band is then given by

$$n = \frac{1}{V}\int\limits_{\varepsilon_c}^{\infty} f(\varepsilon)\, g(\varepsilon)\, d\varepsilon \ , \tag{3.1.8}$$

where $f(\varepsilon)$ is the Fermi-Dirac distribution given by the rhs of (3.1.4). Since this distribution decreases exponentially at large energies, the upper limit of integration may be taken as infinity without introducing appreciable error. With $g(\varepsilon)$ given by (3.1.7), we find for the integral

$$n = 2\frac{1}{4\pi^2}\left(\frac{2m_n}{\hbar^2}\right)^{3/2}\int\limits_{\varepsilon_c}^{\infty}\frac{(\varepsilon - \varepsilon_c)^{1/2} d\varepsilon}{\exp[(\varepsilon - \zeta)/k_B T] + 1} = N_c\, F_{1/2}(\zeta_n/k_B T)\,, \tag{3.1.9}$$

where we have introduced the Fermi energy relative to the band edge, $\zeta_n = \zeta - \varepsilon_c$, the *effective density of states*

$$N_c = 2\frac{1}{4\pi^2}\left(\frac{2m_n k_B T}{\hbar^2}\right)^{3/2}\left(\frac{1}{2}\right)! = 2\left(\frac{m_n k_B T}{2\pi\,\hbar^2}\right)^{3/2} \tag{3.1.10}$$

and the Fermi integral

$$F_j(\eta) = 1/j!\int\limits_{0}^{\infty}\frac{x^j\, dx}{\exp(x - \eta) + 1} \tag{3.1.11}$$

for $j = 1/2$. This integral has been tabulated in [3.3]. For $\eta > 1.25$ it is approximated by [1]

$$F_{1/2}(\eta) = (4\eta^{3/2}/3\,\pi^{1/2}) + \pi^{3/2}/6\,\eta^{1/2} \tag{3.1.12}$$

---

[1] $F_j(\eta) \approx \eta^{j+1}/(j+1)$ for $\eta \gg 1$.

with an error of $< 1.5\%$. For $\eta < -4$ which is the case of nondegeneracy (Maxwell-Boltzmann statistics), $F_j(\eta)$ is approximately

$$F_j(\eta) = \exp(\eta). \tag{3.1.13}$$

$\eta_n = \zeta_n/k_BT$ is denoted as the *reduced Fermi energy*. A corresponding result is obtained for holes in the valence band (subscript v). For $T = 300$ K and $m_n = m_p = m_0$, $N_c = N_v = 2.54 \times 10^{19}\text{cm}^{-3}$. Hence, the constant C of (1.3.2) is obtained from the product $n\,p$ for nondegeneracy:

$$C = N_cN_v/T^3 = 4\left(\frac{\sqrt{m_n m_p}\,k_B}{2\pi\,\hbar^2}\right)^3 = \left(\frac{m_n}{m_0}\frac{m_p}{m_0}\right)^{3/2}\frac{2.33 \times 10^{31}}{\text{cm}^6\text{K}^3}. \tag{3.1.14}$$

In an intrinsic semiconductor, n equals p and from

$$n = N_c\exp[(\zeta - \varepsilon_c)/k_BT]; \quad p = N_v\exp[(\varepsilon_v - \zeta)/k_BT], \tag{3.1.15}$$

the Fermi energy

$$\zeta = \frac{1}{2}(\varepsilon_c + \varepsilon_v) + \frac{3}{4}k_B\,T\ln(m_p/m_n) \tag{3.1.16}$$

tis obtained [3.4].[2]   For equal effective masses of electrons and holes, the Fermi level is independent of the temperature and is located in the middle of the gap. For $m_n \ll m_p$, the Fermi level approaches the conduction band edge with increasing temperature.

If two bands Γ and L with band edges at ε_c and $\varepsilon_c + \Delta_L$, respectively, are considered, the carrier concentration

$$n = n_\Gamma + n_L = N_{c\Gamma}F_{1/2}(\zeta_n/k_BT) + N_{cL}F_{1/2}[(\zeta_n - \Delta_L)/k_BT]; \tag{3.1.17}$$

where

$$N_{c\Gamma} = 2\left(\frac{m_\Gamma k_BT}{2\pi\,\hbar^2}\right)^{3/2} \quad\text{and}\quad N_{cL} = 2\left(\frac{m_L k_BT}{2\pi\,\hbar^2}\right)^{3/2}. \tag{3.1.18}$$

are the effective densities of states.

Figure 3.4 shows n, n_Γ, and n_L as a function of ζ_n for n-GaSb at 300 K where $m_\Gamma/m_0 = 0.05, m_L/m_0 = 0.74, \Delta_L = 101.7\text{meV}$. The total electron concentration is given by

$$n = N_{D^\times} + N_{A^-} + N_c\,F_{1/2}(\eta_n) \tag{3.1.19}$$

and the internal energy density U_n (subscript n for electrons) by

$$U_n = N_{D^\times}\varepsilon_D + N_{A^-}\varepsilon_A + \frac{3}{2}k_B\,T\,N_cF_{3/2}(\eta_n). \tag{3.1.20}$$

For the entropy density, a rather lengthy expression is obtained [Ref. 3.1, Chap. 8 h] which is simplified by neglecting all donors and acceptors:

$$S_n = -(\zeta/T)n + U_n/T + k_B\,N_c\,F_{3/2}(\eta_n) = (5\,U_n/3 - n\,\zeta)/T. \tag{3.1.21}$$

---

[2]For an inclusion of donors and acceptors as well as for ε_G in the case of degeneracy, see [3.5].

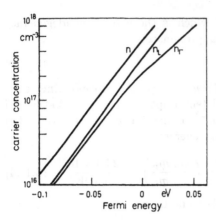

Fig. 3.4: Calculated light-(Γ) and heavy-(L) electron concentration in GaSb at 300 K, as a function of the Fermi energy (after [3.6])

From $F = U - TS$ we now find for the free energy

$$F_n = n\,\zeta - 2\,U_n/3. \tag{3.1.22}$$

Equation (3.1.2) in the form

$$1/T = (\partial S_n/\partial U_n)_{n=\text{const}} \tag{3.1.23}$$

and (3.1.3) are easily verified by taking into account that $\zeta = \zeta(n), U_n = U_n(n)$, and $\partial U_n/\partial \zeta_n = 3\,n/2$.

For the consideration of holes we introduce the concentration p_j of unoccupied states in a band

$$p_j = g_j - n_j = g_j(1 + \exp[(\zeta - \varepsilon_j)/k_B T])^{-1} \tag{3.1.24}$$

which is similar to (3.1.4) except that energies, including the Fermi energy, have the opposite sign. The free hole concentration is then given by

$$p = N_v\, F_{1/2}(\eta_p). \tag{3.1.25}$$

and the internal energy density U_p of free holes by

$$U_p = -\frac{3}{2}k_B\, T\, N_v\, F_{3/2}(\eta_p). \tag{3.1.26}$$

The entropy density is similarly

$$S_p = -(5\,U_p/3 - p\zeta)/T. \tag{3.1.27}$$

For mixed conduction, the total entropy density S is then given by [3.7]

$$S = S_n + S_p. \tag{3.1.28}$$

In deriving these equations we have assumed the density of states $g(\varepsilon)$ given by (3.1.7). In Sect. 9.2 we will consider quantum effects in a strong magnetic field which yield a different density of states.

3.2 Occupation Probabilities of Impurity Levels

For the ratio of neutral and total donor densities, we obtain from a similar calculation

$$N_{D^\times}/N_D = \left(\frac{1}{g_D} \exp[(\varepsilon_D - \zeta)/k_B T] + 1\right)^{-1} \tag{3.2.1}$$

with a spin factor g_D to be discussed in (3.2.12).

Since N_D equals $N_{D^\times} + N_{D^+}$ we have for the ratio $N_{D^\times}/N_{D^+}$:

$$N_{D^\times}/N_{D^+} = g_D \exp[(\zeta - \varepsilon_D)/k_B T]. \tag{3.2.2}$$

According to (3.1.9, 13) for nondegeneracy, the concentration of free electrons is given by

$$n = N_c \exp[(\zeta - \varepsilon_c)/k_B T] , \tag{3.2.3}$$

where N_c is the *effective density of states* in the conduction band. Introducing the donor ionization energy $\Delta\varepsilon_D$ by

$$\Delta\varepsilon_D = \varepsilon_c - \varepsilon_D , \tag{3.2.4}$$

we obtain from (3.2.2) the law of mass action

$$N_{D^\times}/(N_{D^+} n) = (g_D/N_c) \exp(\Delta\varepsilon_D/k_B T). \tag{3.2.5}$$

The energy $\Delta\varepsilon_D$ is the heat produced by the reaction

$$D^+ + e^- \rightleftharpoons D^\times + \Delta\varepsilon_D. \tag{3.2.6}$$

If in (3.2.5) $N_{D^\times}$ is replaced by $N_D - N_{D^+}$ and the equation is solved for N_{D^+}, this yields

$$N_{D^+} = \frac{N_D}{(g_D n/N_c) \exp(\Delta\varepsilon_D/k_B T) + 1} . \tag{3.2.7}$$

The charge neutrality requires the number of positive charges to be equal to the number of negative charges:

$$p + N_{D^+} = n + N_{A^-} = n + N_A - N_{A^\times} . \tag{3.2.8}$$

For an n-type semiconductor, the holes and the neutral acceptors may be neglected. Hence,

$$N_{D^+} \approx n + N_A \tag{3.2.9}$$

which with (3.2.7) yields

$$\frac{n(n + N_A)}{N_D - N_A - n} = \frac{N_c}{g_D} \exp(-\Delta\varepsilon_D/k_B T) . \tag{3.2.10}$$

From this equation and a measurement of the carrier density n as a function of temperature, it is possible to determine the *activation energy* $\Delta\varepsilon_D$ of a donor if we know how N_c and g_D depend on temperature T. According to (3.1.10), N_c is proportional to $T^{3/2}$. In Sect. 3.1, the impurity spin degeneracy g_D was introduced with a value of 2. This is true only at high temperatures. For the case of low temperatures (e.g., $< 10\,K$ for Sb in Ge), we have to consider that what we normally label as the *impurity level* in the many valley model (Chap. 7) is the donor triplet state (parallel spins), and that there is also a singlet state (antiparallel spins) at an energy level which is lower than the triplet level by $\delta\varepsilon = 0.57$ meV for Sb in Ge ($\Delta\varepsilon_D = 9.6\,\mathrm{meV}$) and $\delta\varepsilon = 4.0\,\mathrm{meV}$ for As in Ge ($\Delta\varepsilon_D = 12.7$ meV) [3.8, 9]. This energy separation is of importance, e.g., for the low-temperature ultrasonic attenuation which will be discussed at the end of Sect. 6.1. At present we notice that in (3.2.10) we have to replace $\exp(\Delta\varepsilon_D/k_B T)$ by

$$\frac{1}{4}\,\exp[(\Delta\varepsilon_D + \delta\varepsilon)/k_B T] + \frac{3}{4}\,\exp(\Delta\varepsilon_D/k_B T)$$

where $1/4$ and $3/4$ are the statistical weights of the singlet and the triplet states, respectively. Hence, (3.2.10) becomes with $g_D = 2$

$$\frac{n(n + N_A)}{N_D - N_A - n} = \frac{N_c}{2}\left[\frac{4}{\exp(\delta\varepsilon/k_B T) + 3}\right]\exp(-\Delta\varepsilon_D/k_B T). \qquad (3.2.11)$$

We can formally retain (3.2.10) if we now define g_D as

$$g_D = [\exp(\delta\varepsilon/k_B T) + 3]/2 = g_D(T). \qquad (3.2.12)$$

For such low temperatures where $\delta\varepsilon$ becomes important, we can assume $n \ll N_A$ and $N_D - N_A$ and approximate (3.2.10) by

$$\log(2\,n\,g_D\,T^{-3/2}) = \log\left[\frac{(N_D - N_A)2N_c}{N_A T^{3/2}}\right] - \frac{\Delta\varepsilon_D}{k_B T}\log(e) \quad , \qquad (3.2.13)$$

where log is to base 10 and e is the base of the natural logarithm. According to (3.1.10), $N_c/T^{3/2}$ is temperature independent. Hence, a plot of the left-hand side of (3.2.13) vs $1/T$ yields both $\Delta\varepsilon_D$ and the compensation ratio $N_A/(N_D + N_A)$.

Figure 3.5 shows experimental data of $2\,n\,g_D T^{-3/2}$ in a logarithmic scale vs T^{-1} for two samples of Sb-doped Ge in a temperature range between 4.0 and 5.1 K [3.10–12]. The straight lines have slopes corresponding to $\Delta\varepsilon_D = 9.57$ and 9.72 meV. The absolute values indicate $N_A/(N_D - N_A) = 0.036$ and 0.067, respectively. From this and the carrier density which is $n = N_D - N_A$ at high temperatures, the total impurity concentration $N_D + N_A$ could be determined. For some samples it turned out to be necessary to consider two types of donors with different energy levels, e.g., Sb and As.

For a p-type nondegenerate semiconductor, (3.2.3, 4, 10) have to be replaced by

$$p = N_v \exp[(\varepsilon_v - \zeta)/k_B T] \qquad (3.2.14)$$

$$\Delta\varepsilon_A = \varepsilon_A - \varepsilon_v \qquad (3.2.15)$$

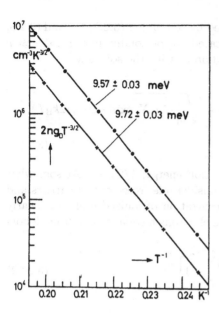

Fig. 3.5: Experimental data for $2ng_D T^{-3/2}$ as a function of the reciprocal temperature for a determination of the energy separation between the conduction band and the donor triplet level derived from the unperturbed fourfold degenerate ground state of Sb in Ge (after [3.10])

$$N_{A^-} = \frac{N_A}{(g_A p/N_v) \exp(\Delta\varepsilon_A/k_B T) + 1} \tag{3.2.16}$$

and

$$\frac{p(p + N_D)}{N_A - N_D - p} = \frac{N_v}{g_A} \exp(-\Delta\varepsilon_A/k_B T) \tag{3.2.17}$$

with quite similar meanings of the corresponding quantities . With careful experimental methods [3.13] it was possible to show in the temperature range 20–50 K that for In in Ge, the high-temperature g-factor is 4 rather than 2 which takes account of the fact that at the valence band edge, both light and heavy hole sub-bands exist and at $k = 0$ are degenerate with one another.

In obtaining (3.2.9), we have neglected the *minority carriers* which in this case are holes. These can be taken into account by calculating the product $n\,p$ from (3.2.3, 14):

$$n\,p = N_c\,N_v\,\exp(-\varepsilon_G/k_B T) \tag{3.2.18}$$

where the definition of the gap width

$$\varepsilon_G = \varepsilon_c - \varepsilon_v \tag{3.2.19}$$

has been used. The rhs of (3.2.18) is denoted by n_i^2 where

$$n_i = \sqrt{N_c\,N_v}\,\exp(-\varepsilon_G/2\,k_B T) \tag{3.2.20}$$

is the intrinsic concentration [see also (1.3.2)]. For the case of nearly but not exactly equal densities of electrons and holes, one has to solve an equation of the

4th power, in either n or p which is obtained by combining (3.2.7, 16) and (3.2.8, 18). The special case of nearly uncompensated n-type conductivity $n \gg n_i \gg p$ and N_A, yields an equation of only second order with the solution

$$n = N_{D+} \approx \frac{N_D}{(g_D n / N_c) \exp(\Delta \varepsilon_D / k_B T)} \approx \sqrt{\frac{1}{g_D} N_D N_c} \exp(-\Delta \varepsilon_D / 2 k_B T).$$

(3.2.21)

In the exponential we find *half* the activation energy, $1/2 \, \Delta \varepsilon_D$. At somewhat lower temperatures, however, there is a considerable freeze-out of carriers and hence $N_D \gg N_A \gg n \gg p$ and the concentration of ionized donors is nearly equal to that of acceptors since most of the electrons have just gone from donors to acceptors. From

$$\frac{1}{N_A} = \frac{1}{N_{D+}} = \frac{1}{N_D} \left[\frac{g_D n}{N_c} \exp \left(\frac{\Delta \varepsilon_D}{k_B T} \right) + 1 \right]$$

(3.2.22)

n is easily obtained:

$$n = \frac{N_c}{g_D} \frac{N_D - N_A}{N_A} \exp(-\frac{\Delta \varepsilon_D}{k_B T})$$

(3.2.23)

with the *full* activation energy $\Delta \varepsilon_D$ in the exponential. Similar results are obtained for p-type conductivity. Figure 3.6 shows experimental results obtained with p-type Si doped with 7.4×10^{14} acceptors cm^{-3} [3.14]. At the transition from $n \gg N_D$ (at high temperature) to $n \ll N_D$ (at low temperature), there is a kink at 10^{11}cm^{-3} which is the concentration of compensating donor impurities.

In an extrinsic semiconductor, the Fermi level is close to the impurity level which determines the type of conductivity. This is clear from (3.2.2). In order to have an appreciable amount of impurities ionized (N_{D+} of the order $N_{D\times}$), the difference $\zeta - \varepsilon_D$ must be small compared to $k_B T$.

The statistics of thermal ionization of double-donors and acceptors like, e.g., Te: Te$^+$ + e$^-$ $\rightleftharpoons$ Te^{++} + 2 e$^-$ (which is obviously a double-donor) are quite interesting. [3]   In Sect.12.2 on photoconductivity, double-impurities will be discussed.

Figure 3.7 shows energy levels of impurities observed in Ge [3.15, 16]. The numbers next to the chemical symbols are the energy intervals between the impurity levels and the nearest band edge in units of eV. Impurities known to yield singly ionized donors or acceptors have energy intervals of about 0.01 eV. These are denoted as *shallow-level impurities* or *hydrogen-type impurities*. The latter notation is derived from the fact that the Rydberg energy Ry $= m_o e^4 / [2 \, (4 \pi \kappa_o \, \hbar)^2] = 13.6$ eV, multiplied with the effective mass ratio $m/m_0 = 0.12$ and divided by the square of the dielectric constant $\kappa = 16$, yields an energy roughly equal to the energy interval of 0.01 eV as given above. In this model the

---

[3] For the occupation probability of double donors see, e.g. E. Spenke [3.1].

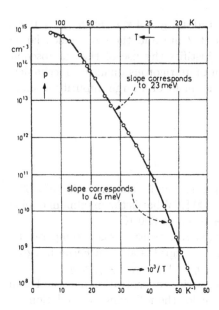

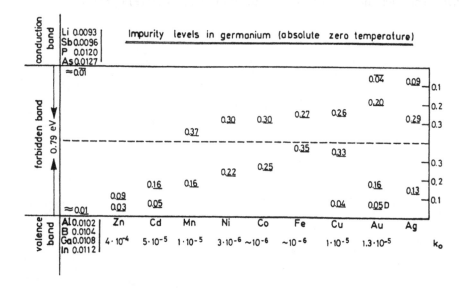

Fig. 3.6: Hole concentration as a function of the reciprocal temperature for a silicon sample with a shallow acceptor of energy 46 meV and concentration $N_A = 7.4 \times 10^{14}/\text{cm}^3$, partly compensated by $1.0 \times 10^{11}/\text{cm}^3$ donors (after Morin in Ref. [3.14])

Fig. 3.7: Impurity levels in Germanium. All deep levels indicated are acceptors except for the lowest Au level. The numbers are energies in units of eV. Distribution coefficients k_0 are given at the bottom of the figure (after [3.17])

radius of the first electron orbit is the Bohr radius $a_B = 4\pi, \kappa_o \hbar^2/m_0 e^2 = 0.053$ nm multiplied by $\kappa/(m/m_0)$ which is about 7 nm and therefore much larger than the interatomic distance. This fact justifies the use of the macroscopic dielectric constant in the atomic model. In more rigorous quantum mechanical calculations, the stress exerted by a possible misfit of a large impurity atom in the lattice has to be taken into account.

Problems

3.1. Calculate the energy gap for an intrinsic carrier concentration of $1.09 \times 10^{16} \mathrm{cm}^{-3}$ at 300 K and 5.21×10^{16} cm^{-3} at 400 K. What is the maximum doping level for intrinsic behavior at temperatures down to 200 K?

3.2. Determine the specific heat of a degenerate electron gas vs temperature. Make a plot. *Hint*: the derivative of a Fermi integral is again a Fermi integral. Determine η from $n = N_D$. Discuss the case $\eta \gg 1$ where $F_j(\eta) \approx \eta^{j+1}$.

3.3. Find the relationship between the carrier concentration and the Fermi energy in a degenerate n-type semiconductor for the following $\varepsilon(k)$ relation:

$$\varepsilon(k) = -\varepsilon_G/2 \pm (\varepsilon_G^2 + 2\hbar^2 k^2 \varepsilon_G/m_0)^{1/2}/2p. \tag{3.0.1}$$

Hint: consider the density-of-states function $g(\varepsilon) = [k^2(\varepsilon) V/\pi^2] \, dk/d\varepsilon$.

4. Charge and Energy Transport in a Nondegenerate Electron Gas

In the preceding chapters, we have seen that a mobile charge carrier in a semiconductor has an effective mass m which is different from the free electron mass m_0. The effective mass takes care of the fact that the carrier is subject to the crystal field. In discussing the velocity distribution of the *gas* of carriers, we found that the Fermi–Dirac distribution holds in general and that the Maxwell–Boltzmann distribution $f(v) \propto \exp(-m v^2/2 k_B T)$ is an approximation of the former which is valid for *nondegenerate* semiconductors. Here the carrier density is small compared with the effective density of states N_c in the conduction band and N_v in the valence band (3.1.12). For these distributions, no externally applied electric fields were assumed to be present. Instead, the calculations were based on the assumption of thermal equilibrium.

In this chapter we calculate the fluxes of charge and energy which are due to gradients of electric potential, temperature, and the concentration of carriers. These fluxes are influenced by external magnetic fields and mechanical forces such as hydrostatic pressure and uniaxial stress. Very general relationships between these fluxes and gradients were given by *Onsager* [4.1]. Rather than going into details of the thermodynamics of irreversible processes [4.2], we will take the simplified approach of the Boltzmann equation in the relaxation time approximation.

4.1 Electrical Conductivity and its Temperature Dependence

In the range of extrinsic conductivity, the current density j, which is due to an external electric field E, is given by

$$j = n e v_d , \qquad (4.1.1)$$

where v_d is the carrier drift velocity. The direction of j is from the positive to the negative electrode; for electrons with $e < 0$, the drift velocity v_d is opposite to j. The absolute value of e has the value of the elementary charge which is also assumed to be the charge of the positive hole. Neglecting the thermal motion of the carriers, the equation of motion is given by

$$d(m\,v_{\rm d})/dt + m\,v_{\rm d}/\langle\tau_{\rm m}\rangle = e\boldsymbol{E} \ , \tag{4.1.2}$$

where t is the time, $mv_{\rm d}$ the momentum of the carrier and $\langle\tau_{\rm m}\rangle$ is introduced as an average momentum relaxation time. Relaxation means *return to equilibrium*. The second term describes the *friction* which the carriers experience on drifting through the crystal. The friction depends on the vibration of the individual atoms of the crystal lattice and therefore depends on crystal temperature. This is taken care of by a dependence of $\langle\tau_{\rm m}\rangle$ on temperature.

In the steady state the first term vanishes and the drift velocity $v_{\rm d}$ is proportional to the electric field strength $\boldsymbol{E}$:

$$|v_{\rm d}| = \mu\,|\boldsymbol{E}| \ , \tag{4.1.3}$$

where the factor of proportionality μ

$$\mu = (|e|/m)\langle\tau_{\rm m}\rangle \tag{4.1.4}$$

is called the *mobility*. If this is substituted in (4.1.1) the current density $\mathbf{j}$ is given by

$$\mathbf{j} = \sigma\,\boldsymbol{E} \ , \tag{4.1.5}$$

where the conductivity σ is obtained as

$$\sigma = n\,e\,\mu = (n\,e^2/m)\langle\tau_{\rm m}\rangle \ . \tag{4.1.6}$$

The inverse conductivity $1/\sigma = \varrho$ is called *resistivity*. The unit of resistivity is $\Omega\,{\rm cm} = {\rm Vcm/A}$. Equation (4.1.6) yields the unit of mobility as ${\rm cm}^2/{\rm Vs}$, while (4.1.4) yields the unit of mass as ${\rm VA\,s}^3\,{\rm cm}^{-2}$ which is the same as $10^7\,{\rm g}$.

In numerical calculations it is quite convenient to use the electromagnetic rather than the g-unit. The ratio [1] e/m for $m = m_0$ has a value of $1.76 \times 10^{15}\,{\rm cm}^2/{\rm Vs}^2$. A typical value of $\langle\tau_{\rm m}\rangle$, which in certain simple cases is essentially the inverse vibration frequency of atoms in a crystal, has an order of magnitude value of $10^{-13}\,{\rm s}$. This yields a mobility of $176\,{\rm cm}^2/{\rm Vs}$. For electrons in n-type germanium, m is only about one tenth of m_0 and $\langle\tau_{\rm m}\rangle$ at room temperature some $10^{-13}\,{\rm s}$. This yields a mobility which is in agreement in order of magnitude with the observed mobility of $3900\,{\rm cm}^2/{\rm Vs}$.

Looking again at (4.1.2–5), it is obvious how the friction term leads to Ohm's law which states that σ is independent of the electric field strength. The notation *relaxation time* for $\langle\tau_{\rm m}\rangle$ is clarified if we assume that at $t = 0$ the electric field is switched off. The solution of (4.1.2) is easily evaluated for this case. It is:

$$m\,v_{\rm d} = (m\,v_{\rm d})_{t=0}\,\exp(-t/\langle\tau_{\rm m}\rangle) \ . \tag{4.1.7}$$

The drift momentum decreases exponentially in a time $\langle\tau_{\rm m}\rangle$ which is too short for us to be able to observe it directly. We shall see later on that in ac fields

---

[1] See the table of constants in the Appendix.

there is a phase shift between current and field which depends on the product of frequency and $\langle \tau_m \rangle$ (Sect. 4.14).

In our model where the drifting electrons occasionally collide with vibrating atoms of the crystal lattice, one would anticipate the existence of a mean free time between collisions. This *collision time* τ_c is of the order of τ_m. In Chap. 6, this model will be refined and the various collision processes, appropriately called *scattering processes*, will be discussed in detail. In general, τ_m depends on the ratio of the carrier energy ε and the average energy of vibration of an atom which is of the order of $k_B T$. In special cases, this relation takes the form

$$\tau_m = \tau_0 (\varepsilon/k_B T)^r , \tag{4.1.8}$$

where the exponent r varies between $-1/2$ (for *acoustic deformation potential scattering*) and $+3/2$ (for *ionized impurity scattering*) and τ_0 is a factor of proportionality. For acoustic scattering, the mean free path l_{ac} will be shown to be proportional to the inverse lattice temperature

$$l_{ac} \propto T^{-1} . \tag{4.1.9}$$

Thus τ_0, which is given by

$$\tau_0 = l_{ac}/\sqrt{2\,k_B T/m} , \tag{4.1.10}$$

where $\sqrt{2 k_B T/m}$ is an average velocity of the carriers in equilibrium with the lattice, is proportional to $T^{-3/2}$.

The fact that τ_m, in general, is energy dependent requires a reformulation of its definition (4.1.2). We have seen that there is a velocity distribution of carriers $f(v)$. The drift velocity v_d in an anisotropic material with an applied electric field in, e.g., the z-direction will be given by its z-component

$$v_{dz} = \int\limits_{-\infty}^{\infty} v_z f d^3 v \left/ \int\limits_{-\infty}^{\infty}\!\!\!\int f d^3 v \right. , \tag{4.1.11}$$

where $d^3 v$ stands for $dv_x\,dv_y\,dv_z$. Similarly, $\langle \tau_m \rangle$ in (4.1.2) is an *average* over the velocity distribution. Since τ_m is a function of velocity, it would be appropriate to define it by the following equation which determines the relaxation of the distribution function $f(v)$:

$$\frac{df(v)}{dt} + \frac{f(v) - f_0(v)}{\tau_m(v)} = 0 , \tag{4.1.12}$$

where f_0 is the thermal-equilibrium distribution, i.e., without field. This is the *Boltzmann equation* in the relaxation time approximation. The *drift term* df/dt can be evaluated if $dx/dt = v_x$; $dv_x/dt = (e/m)\,E_x$, etc., is taken into account:

$$\frac{df}{dt} = \frac{\partial f}{\partial t} + \frac{\partial f}{\partial x} v_x + \frac{\partial f}{\partial y} v_y + \frac{\partial f}{\partial z} v_z + \frac{\partial f}{\partial v_x}\frac{e}{m}E_x + \frac{\partial f}{\partial v_y}\frac{e}{m}E_y + \frac{\partial f}{\partial v_z}\frac{e}{m}E_z . \tag{4.1.13}$$

The first four terms on the rhs vanish for dc fields which have no temperature or concentration gradients in the crystal. For $E_x = E_y = 0$, the Boltzmann equation becomes

$$\frac{\partial f}{\partial v_z}\frac{e}{m}E_z + \frac{f - f_0}{\tau_{\mathrm{m}}} = 0 \tag{4.1.14}$$

which can be written in the form

$$f = f_0 - \frac{e}{m}\tau_{\mathrm{m}}\frac{\partial f}{\partial v_z}E_z \ . \tag{4.1.15}$$

We consider only small field intensities such that we may retain terms linear in E_z only. Therefore, in $\partial f/\partial v_z$, we may replace f by its equilibrium value f_0. In a Legendre polynomial expansion of the derivative of the distribution function, only the first two terms are retained; this is called the *diffusive approximation* of the distribution function:

$$f = f_0 - \frac{e}{m}\tau_{\mathrm{m}}\frac{\partial f_0}{\partial v_z}E_z \ . \tag{4.1.16}$$

In an iterative process in which we replace f in $\partial f/\partial v_z$ by this value, a term in E_z^2 would be obtained. This E_z^2 term is supposed to be negligible at small field intensities. The case of high field intensities will be discussed in Sect. 4.13.

The calculation of the drift velocity from (4.1.11) is straightforward . The first term in f, f_0 makes the integral in the numerator vanish since f_0 is an even function of v_z , $f_0(-v_z) = f_0(v_z)$:

$$\int_{-\infty}^{\infty} v_z f_0 \, d^3 v = 0 \ . \tag{4.1.17}$$

Similarly, since

$$-\frac{\partial f_0}{\partial v_z} = v_z f_0 \frac{m}{k_{\mathrm{B}} T} \ , \tag{4.1.18}$$

the second term in f makes the integral in the denominator vanish. Hence,

$$v_{\mathrm{d}z} = \int_{-\infty}^{\infty} v_z[-(e/m)\tau_{\mathrm{m}}(\partial f/\partial v_z)E_z]d^3 v \left/ \int_{-\infty}^{\infty} f_0 d^3 v \right. \ . \tag{4.1.19}$$

According to (4.1.3) the mobility μ is defined by the ratio $|v_{\mathrm{d}z}/E_z|$ and can be written in the form $(|e|/m)\langle\tau_{\mathrm{m}}\rangle$ where the average momentum relaxation time $\langle\tau_{\mathrm{m}}\rangle = \langle\tau_{\mathrm{m}}(v)\rangle$ is given by

$$\langle\tau_{\mathrm{m}}\rangle = \int_{-\infty}^{\infty} \tau_{\mathrm{m}}(v) \, v_z(-\partial f/\partial v_z) \, d^3 v \left/ \int\!\!\int_{-\infty}^{\infty} f_0 \, d^3 v \right. \tag{4.1.20}$$

We approximate $-\partial f/\partial v_z$ by $-\partial f_0/\partial v_z$ given for a Fermi–Dirac distribution function by

$$\partial f_0/\partial v_z = f_0(f_0 - 1)\, m\, v_z/k_B T \tag{4.1.21}$$

and obtain

$$\langle \tau_m \rangle = (m/k_B T)\int_{-\infty}^{\infty} \tau_m v_z^2 f_0(1 - f_0)\, d^3 v \left/ \int_{-\infty}^{\infty} f_0\, d^3 v \right. \tag{4.1.22}$$

We assume $\tau_m(v)$ to be an isotropic function $\tau_m(|v|)$. Since $f_0(v)$ is an isotropic function we may then replace v_z^2 in the integral by $v^2/3$. Furthermore, we assume a simple model of band structure with zero energy at the band edge, i.e., $\varepsilon = m\, v^2/2$. Moreover complex band structures will be considered in Chaps. 7 and 8. The volume element in v space is then given by $d^3 v = 4\pi v^2 dv \propto \varepsilon^{1/2} d\varepsilon$. Hence, (4.1.22) becomes

$$\langle \tau_m \rangle = -\frac{2}{3}\int_0^{\infty} \tau_m f_0' \left(\frac{\varepsilon}{k_B T}\right)^{3/2} d\left(\frac{\varepsilon}{k_B T}\right) \left/ \int_0^{\infty} f_0 \left(\frac{\varepsilon}{k_B T}\right)^{1/2} d\left(\frac{\varepsilon}{k_B T}\right) \right. \tag{4.1.23}$$

where $f_0' = \partial f_0/\partial(\varepsilon/k_B T)$. Let us now assume for $\tau_m(\varepsilon)$ the power law given by (4.1.8). For the general case of Fermi–Dirac statistics, $f_0(\varepsilon)$ is given by (3.1.4) and we obtain for $\langle \tau_m \rangle$ by partial integration

$$\langle \tau_m \rangle = \tau_0 \frac{2}{3}\left(r + \frac{3}{2}\right)\int_0^{\infty} \frac{x^{r+1/2} dx}{\exp(x - \eta_n) + 1} \left/ \int_0^{\infty} \frac{x^{1/2} dx}{\exp(x - \eta_n) + 1} \right.$$

$$= \frac{4}{3\sqrt{\pi}}\left(r + \frac{3}{2}\right)! \tau_0\, F_{r+1/2}(\eta_n)/F_{1/2}(\eta_n) , \tag{4.1.24}$$

where the Fermi integrals are given by (3.1.11) and η_n is the reduced Fermi energy. The factor $4/(3\sqrt{\pi})$ is ≈ 0.752. For a nondegenerate electron gas where the Maxwell–Boltzmann distribution (3.1.5) is valid, (4.1.24) yields

$$\langle \tau_m \rangle = \frac{4}{3\sqrt{\pi}}\left(r + \frac{3}{2}\right)! \tau_0 . \tag{4.1.25}$$

For nonpolar acoustic scattering, the exponent $r = -\frac{1}{2}$ and $\langle \tau_m \rangle$ becomes

$$\langle \tau_m \rangle = \tau_0 \frac{4}{3\sqrt{\pi}} \approx 0.752\, \tau_0 . \tag{4.1.26}$$

In a more general description it is convenient to introduce a wave vector k for the carrier by writing [see text after (2.1.16)]

$$\frac{\partial(\hbar k)}{\partial t} = e\, E , \tag{4.1.27}$$

where $\hbar \boldsymbol{k}$ is the crystal momentum vector and $\hbar$ is Planck's constant divided by 2π. The carrier velocity is given by

$$\boldsymbol{v} = \hbar^{-1} \boldsymbol{\nabla}_{\boldsymbol{k}} \varepsilon \ . \tag{4.1.28}$$

In the Boltzmann equation, the term $(\partial f / \partial v_z) e E_z / m$ is replaced by $e \hbar^{-1} (\partial f / \partial \varepsilon)(\boldsymbol{\nabla}_{\boldsymbol{k}} \varepsilon \times \boldsymbol{E})$. The effective density of states N_c in the conduction band with zero energy at the band edge defined by (3.1.9) is then given by

$$N_c = \frac{1}{4\pi^3} \int \exp(-\varepsilon / k_B T) d^3 k \tag{4.1.29}$$

and the conductivity σ for $\boldsymbol{E}$ in the z-direction by

$$\sigma = \frac{e^2}{4\pi^3 \hbar^2} \int_{-\infty}^{\infty} \tau_m \left(\frac{\partial \varepsilon}{\partial k_z} \right)^2 \left(-\frac{\partial f_0}{\partial \varepsilon} \right) d^3 k \ . \tag{4.1.30}$$

Since this calculation does not assume an effective mass, it is valid not only for the simple model of a parabolic band structure but also for a nonparabolic band structure.

There are always both electrons and holes present in a semiconductor. Since these carriers do not influence each other (except for electron–hole scattering which is not important at low carrier densities), the currents going in opposite directions are subtracted from each other to yield the total current. Due to the opposite charges that electrons and holes have, the conductivities which are always positive quantities add up to the total conductivity

$$\sigma = |e| \, (p \, \mu_p + n \, \mu_n) = |e| \, (p + n \, b) \, \mu_p \ , \tag{4.1.31}$$

where μ_p is the hole mobility, μ_n the electron mobility and b the ratio μ_n / μ_p. For intrinsic conduction, $n = p = n_i$ and therefore,

$$\sigma = |e| \, n_i \, \mu_p (1 + b) \ . \tag{4.1.32}$$

Since n_i rises exponentially with temperature according to (3.2.20), we obtain the temperature dependence of σ as shown by Fig. 4.1. The conductivity is plotted in a logarithmic scale vs the inverse temperature. An almost straight line is obtained in the intrinsic region. When the temperature is lowered, σ enters the extrinsic region where the carrier concentration is constant and σ rises, for the simplest case $\propto \tau_0 \propto T^{-3/2}$. At still lower temperatures, there is a carrier freeze-out as described by (3.2.21, 23).

4.2 Hall Effect in a Transverse Magnetic Field

Carriers which move perpendicular to the direction of a magnetic field or at an oblique angle will be deflected from the direction of motion by the Lorentz

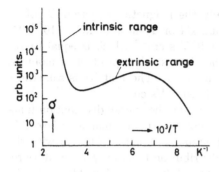

Fig. 4.1: Schematic diagram of the conductivity as a function of the inverse temperature

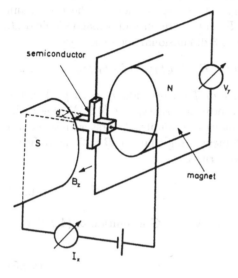

Fig. 4.2: Hall arrangement. V_y is the Hall voltage

force.[2] The deflection causes a Hall voltage V_y in the experimental arrangement shown in Fig.4.2. The voltage is measured between side arms of the filamentary sample which are opposite to each other. Its polarity is reversed by either reversing the current I_x or the magnetic induction B_z.

The Hall voltage is given by

$$V_y = R_H I_x B_z / d \ , \tag{4.2.1}$$

where d is the sample thickness in the direction of the magnetic field and R_H is called the *Hall coefficient*.

The Hall voltage is proportional to B_z if R_H in (4.2.1) is assumed to be constant. This is true only for values of B_z which are small compared with the reciprocal of the mobility of the carriers as will be shown below. The Hall field E_y is given by

$$E_y = R_H j_x B_z \ , \tag{4.2.2}$$

---

[2]For a review on Hall effect problems in semiconductors see, e.g., [4.3].

where j_x is the current density.[3]   Usually the magnetic induction B_z of an electromagnet is given in units of kilogauss kG or Tesla T (1 kG $= 10^{-1}$ T). In semiconductor physics, however, the unit V s cm^{-2} which is equal to 10^4 T is more convenient since quite often the dimensionless product of mobility, measured in units of cm^2/V s, and B_z will occur. The unit Vs is sometimes called *Weber*, "Wb" , and 1 Wb m^{-2} equals 10^{-4} Vs cm^{-2}.

For a calculation of R_H, we have to determine the energy distribution function $f(\varepsilon)$ of the carriers under the influence of electric and magnetic fields. In Sect. 4.1 we have shown that it is the Boltzmann equation (4.1.12) with the expansion (4.1.13) which determines the distribution function $f(v)$ or, if we replace v by $\hbar^{-1}\nabla_k\,\varepsilon(k)$ according to (4.1.28), $f(k)$. Now, in addition to the electric force $e\,E$ acting upon a charge carrier, we consider in the Boltzmann equation also the Lorentz force $e[v \times B]$ in the presence of a magnetic field B. For the Boltzmann equation we write in 3-dimensional vector notation

$$(v\cdot\nabla_r f)+e\,\hbar^{-1}\{(E\cdot\nabla_k f) + ([v \times B]\cdot\nabla_k f)\}+(f-f_0)/\tau_m = 0 \ .\tag{4.2.3}$$

The first term involves the dependence of f on position $r = (x,y,z)$ of the carrier which will be important for thermoelectricity, see (4.9.1), where the temperature and the Fermi level vary along the sample. At first we neglect this term and come back to it later on. Remembering that for $B = 0$ the diffusive approximation yields (4.1.16) which we write as

$$f = f_0 - \frac{df_0}{d\varepsilon}e\tau_m(v \cdot E) \ ,\tag{4.2.4}$$

we try for $B \neq 0$, introducing an unknown vector G a solution of (4.2.3) with an ansatz

$$f = f_0 + (v \cdot G) \ ,\tag{4.2.5}$$

which means that for $B = 0$ we have $G = -e\,\tau_m\,E(df_0/d\varepsilon)$. $|G|$ is a small quantity because we assume E to be small and we will see later on that also for any value of B this is true. In this approximation we may replace $(E \cdot \nabla_k f)$ by $(E \cdot \nabla_k f_0)$. For the simple model of band structure where $\varepsilon = \hbar^2 k^2/2m$, we apply

$$\nabla_k f_0 = d f_0/d\varepsilon \nabla_k \varepsilon = \frac{d f_0}{d\varepsilon}\hbar\,v\tag{4.2.6}$$

and thus obtain from (4.2.3)

$$e\frac{d f_0}{d\varepsilon}(E \cdot v) + \frac{e}{\hbar}([v \times B]\nabla_k)(v \cdot G) + \frac{1}{\tau_m}(v \cdot G) = 0 \ .\tag{4.2.7}$$

The vector G will turn out to be independent of k, and because

$$\nabla_k(v \cdot G) = (\hbar/m)\nabla_v(v \cdot G) = (\hbar/m)\,G\tag{4.2.8}$$

---

[3]A more general definition of the Hall coefficient is $R_H = (E \cdot [B \times j])/\|\ B \times j\|^2$ [Ref. 4.4, p. 25].

is valid under this condition, (4.2.7) is further simplified after multiplication with τ_m and introducing for brevity $\mu = e\,\tau_m/m$:

$$e\,\tau_m \frac{d\,f_0}{d\varepsilon}(\mathbf{v} \cdot \mathbf{E}) + \mu([\mathbf{v} \times \mathbf{B}]\mathbf{G}) + (\mathbf{v} \cdot \mathbf{G}) = 0 \ . \tag{4.2.9}$$

By applying the vectors $\mathbf{E}, \mathbf{B}$, and $[\mathbf{B} \times \mathbf{E}]$ as a triad for the representation of $\mathbf{G}$,

$$\mathbf{G} = \alpha\,\mathbf{E} + \beta\,\mathbf{B} + \gamma\,[\mathbf{B} \times \mathbf{E}] \tag{4.2.10}$$

with α, β, and γ being coefficients to be determined, we obtain applying $([\mathbf{v} \times \mathbf{B}]\mathbf{B}) = (\ \mathbf{v}[\mathbf{B} \times \mathbf{B}]) = 0$,

$$e\,\tau_m \frac{df_0}{d\varepsilon}(\mathbf{v} \cdot \mathbf{E}) + \alpha\,\mu([\mathbf{v} \times \mathbf{B}] \cdot \mathbf{E}) + \gamma\,\mu\left\{(\mathbf{B} \cdot \mathbf{E})(\mathbf{v} \cdot \mathbf{B}) - B^2(\mathbf{v} \cdot \mathbf{E})\right.$$

$$\left. + \alpha(\mathbf{v} \cdot \mathbf{E}) + \beta(\mathbf{v} \cdot \mathbf{B}) + \gamma(\mathbf{v} \cdot [\mathbf{B} \times \mathbf{E}])\right\} = 0 \ . \tag{4.2.11}$$

Separating terms in $(\mathbf{v} \cdot \mathbf{E})$, $(\mathbf{v} \cdot \mathbf{B})$, and $(\mathbf{v} \cdot [\mathbf{B} \times \mathbf{E}])$, the following relations between the coefficients are found:

$$e\,\tau_m \frac{df_0}{d\varepsilon} + \alpha - \gamma\,\mu\,B^2 = 0; \qquad \gamma\mu(\mathbf{B} \cdot \mathbf{E}) + \beta = 0; \qquad \alpha\,\mu + \gamma = 0$$

These equations are easily solved for α, β, and γ. From (4.2.10) we obtain for $\mathbf{G}$

$$\mathbf{G} = -e\,\tau_m \frac{d\,f_0}{d\varepsilon} \frac{\mathbf{E} + \mu^2(\mathbf{B} \cdot \mathbf{E})\mathbf{B} + \mu[\mathbf{E} \times \mathbf{B}]}{1 + \mu^2\,B^2} \ , \tag{4.2.12}$$

where $\mu = e\,\tau_m/m$. We notice that for $\mathbf{B} = 0$ the value mentioned in the text after (4.2.5) results, that each one of the three terms is linear in $\mathbf{E}$ and therefore $\mathbf{G}$ is small for a small field strength $\mathbf{E}$, and that this is true even for strong magnetic fields $\mathbf{B}$. And $\mathbf{G}$ is also independent of the vector $\mathbf{k}$ as was assumed for obtaining (4.2.8).

Let us now consider also the first term in (4.2.3). The derivative of f relative to r,

$$\nabla_r f \approx \nabla_r f_0 = \frac{d\,f_0}{d(\varepsilon - \zeta)/k_B T} \nabla_r \frac{\varepsilon - \zeta}{k_B T} = \frac{d\,f_0}{d\varepsilon} T\,\nabla_r \frac{\varepsilon - \zeta}{T} \tag{4.2.13}$$

does not change the results obtained so far except for an additional term to the force $e\,\mathbf{E}$ [4.5]:

$$e\,\mathbf{F} = e\,\mathbf{E} + T\,\nabla_r \frac{\varepsilon - \zeta}{T} \tag{4.2.14}$$

Since the additional term is important in the thermopower the combined field $\mathbf{F}$ is called *electrothermal* field, see (4.9.1).

In analogy to the reciprocal mass tensor

$$m^{-1} = \begin{pmatrix} m_x^{-1} & 0 & 0 \\ 0 & m_y^{-1} & 0 \\ 0 & 0 & m_z^{-1} \end{pmatrix}, \qquad (4.2.15)$$

the relaxation time τ_m may also be tensorial, i.e.,

$$\tau_m = \begin{pmatrix} \tau_x & 0 & 0 \\ 0 & \tau_y & 0 \\ 0 & 0 & \tau_z \end{pmatrix}. \qquad (4.2.16)$$

The factor μ in (4.2.11) is then also a tensor which is the product of the elementary charge, m^{-1}, and τ_m. The factor μ^2 turns out to be

$$\mu^2 = e^2 \begin{pmatrix} \tau_y\tau_z/m_ym_z & 0 & 0 \\ 0 & \tau_x\tau_z/m_xm_z & 0 \\ 0 & 0 & \tau_x\tau_y/m_xm_y \end{pmatrix}. \qquad (4.2.17)$$

For simplicity's sake, we consider m^{-1} and τ_m as scalar. In (4.2.11) the numerator consists of three terms: the first one does not contain the magnetic induction; the second term is linear in B and takes care of the Hall effect and the third terms, as well as the denominator, are quadratic in B and therefore yield the magnetoresistance.

Now we write the distribution function (4.2.5) in its general form:

$$f = f_0 - \frac{df_0}{d\varepsilon}\, e\,\hbar^{-1}\boldsymbol{\nabla}_k\varepsilon\,\tau_m \frac{\boldsymbol{F} - e[\boldsymbol{B} \times m^{-1}\tau_m\,\boldsymbol{F}] + \mu^2\,\boldsymbol{B}(\boldsymbol{F}\cdot\boldsymbol{B})}{1 + (\boldsymbol{B}\cdot\mu^2\boldsymbol{B})} . \qquad (4.2.18)$$

For the calculation of the conductivity without any magnetic fields, $\boldsymbol{B} = 0$ and $\boldsymbol{F} = \boldsymbol{E}$, we obtain

$$f = f_0 - df_0/d\varepsilon\; e\tau_m(\boldsymbol{v}\cdot\boldsymbol{E}), \qquad (4.2.19)$$

from which a conductivity tensor

$$\begin{pmatrix} \sigma & 0 & 0 \\ 0 & \sigma & 0 \\ 0 & 0 & \sigma \end{pmatrix} \qquad (4.2.20)$$

is obtained, where σ is equal to σ_0 given by

$$\sigma_0 = (n\,e^2/m)\langle\tau_m\rangle \qquad (4.2.21)$$

and the momentum relaxation time τ_m is averaged over a Maxwell–Boltzmann distribution function:

$$\langle\tau_m\rangle = \frac{4}{3\sqrt{\pi}} \int_0^\infty \tau_m\left(\frac{\varepsilon}{k_B T}\right)^{3/2} \exp\left(-\frac{\varepsilon}{k_B T}\right) d\left(\frac{\varepsilon}{k_B T}\right). \qquad (4.2.22)$$

Except for the tensor notation (4.2.20), this is of course identical to the previous result (4.1.23).

For the part of the distribution function (4.2.18) linear in $\boldsymbol{B}$, we neglect the denominator and the first and third term in the numerator and obtain for the conductivity

$$
\begin{pmatrix}
0 & \gamma B_z & -\gamma B_y \\
-\gamma B_z & 0 & \gamma B_x \\
\gamma B_y & -\gamma B_x & 0
\end{pmatrix} ,
\tag{4.2.23}
$$

where γ is equal to γ_0 given by

$$
\gamma_0 = (n e^3/m^2)\langle \tau_{\mathrm{m}}^2 \rangle .
\tag{4.2.24}
$$

For the evaluation of the Hall effect at an arbitrary magnetic field, we will introduce quantitites σ and γ defined as

$$
\sigma = (n e^2/m)\langle \tau_{\mathrm{m}}/(1 + \omega_{\mathrm{c}}^2 \tau_{\mathrm{m}}^2)\rangle
\tag{4.2.25}
$$

and

$$
\gamma = (n e^3/m^2)\langle \tau_{\mathrm{m}}^2/(1 + \omega_{\mathrm{c}}^2 \tau_{\mathrm{m}}^2)\rangle ,
\tag{4.2.26}
$$

where the *cyclotron frequency* ω_{c} has been introduced as given by

$$
\omega_{\mathrm{c}} = |e|\, B/m .
\tag{4.2.27}
$$

The quantitites σ and γ are obtained in a similar fashion to σ_0 and γ_0 except that the denominator in the distribution function is retained. The tensor (4.2.23) applied to $\boldsymbol{E}$ yields, e.g., an x-component of $\boldsymbol{j}$ given by

$$
j_x = \gamma(B_z E_y - B_y E_z) = \gamma[\boldsymbol{E} \times \boldsymbol{B}]_x = \frac{n e^2}{m}\left\langle \frac{\tau_{\mathrm{m}}[\boldsymbol{E} \times (e/m)\boldsymbol{B}\, \tau_{\mathrm{m}}]_x}{1 + \omega_{\mathrm{c}}^2 \tau_{\mathrm{m}}^2} \right\rangle
\tag{4.2.28}
$$

which is also obtained from the vector-product term in (4.2.18).

From the last term in the distribution function (4.2.18), we obtain for the conductivity

$$
\begin{pmatrix}
-\beta B_x^2 & -\beta B_x B_y & -\beta B_x B_z \\
-\beta B_x B_y & -\beta B_y^2 & -\beta B_y B_z \\
-\beta B_x B_z & -\beta B_y B_z & -\beta B_z^2
\end{pmatrix} ,
\tag{4.2.29}
$$

where β is given by

$$
\beta = -(n e^4/m^3)\langle \tau_{\mathrm{m}}^3/(1 + \omega_{\mathrm{c}}^2 \tau_{\mathrm{m}}^2)\rangle .
\tag{4.2.30}
$$

For the case of a small magnetic field intensity, $1/(1 + \omega_{\mathrm{c}}^2 \tau_{\mathrm{m}}^2)$ can be replaced by $(1 - \omega_{\mathrm{c}}^2 \tau_{\mathrm{m}}^2)$ where $\omega_{\mathrm{c}}^2 \propto B^2 = B_x^2 + B_y^2 + B_z^2$ and the $(\boldsymbol{B} \cdot \boldsymbol{E})\boldsymbol{B}$ term of (4.2.12) is taken into account. As a consequence, the tensor (4.2.29) is replaced by

$$
\begin{pmatrix}
\beta_0(B_y^2 + B_z^2) & -\beta_0 B_x B_y & -\beta_0 B_x B_z \\
-\beta_0 B_x B_y & \beta_0(B_x^2 + B_z^2) & -\beta_0 B_y B_z \\
-\beta_0 B_x B_z & -\beta_0 B_y B_z & \beta_0(B_x^2 + B_y^2)
\end{pmatrix} ,
\tag{4.2.31}
$$

where β_0 stands for

$$\beta_0 = -(n\,e^4/m^3)\langle\tau_{\mathrm{m}}^3\rangle \ . \tag{4.2.32}$$

The tensor (4.2.29) applied to F yields, e.g., an x-component of j given by

$$j_x = -\beta\,B_x(B_xF_x + B_yF_y + B_zF_z) = -\beta\,B_x(F \cdot B) \tag{4.2.33}$$

which is also obtained from the last term in (4.2.18).

For a weak magnetic field the conductivity tensor is denoted by σ_{w} and is a combination of the tensors (4.2.20, 23, 31):

$$\sigma_{\mathrm{w}} = \begin{pmatrix} \sigma_0 + \beta_0(B_y^2 + B_z^2) & \gamma_0\,B_z - \beta_0\,B_x\,B_y & -\gamma_0\,B_y - \beta_0\,B_x\,B_z \\ -\gamma_0\,B_z - \beta_0\,B_x\,B_y & \sigma_0 + \beta_0(B_x^2 + B_z^2) & \gamma_0\,B_x - \beta_0\,B_y\,B_z \\ \gamma_0\,B_y - \beta_0\,B_x\,B_z & -\gamma_0\,B_x - \beta_0\,B_y\,B_z & \sigma_0 + \beta_0(B_x^2 + B_y^2) \end{pmatrix}$$

$$\tag{4.2.34}$$

This tensor yields the Hall effect in a weak magnetic field. We choose a coordinate system such that the z-axis points in the direction of B, i.e., $B_x = B_y = 0$. From $j = \sigma_{\mathrm{w}}E$ we obtain an equation for E_z which is independent of B_z. Therefore, we may assume $E_z = 0$. For the current components j_x and j_y we find

$$j_x = (\sigma_0 + \beta_0\,B_z^2)\,E_x + \gamma_0\,B_z\,E_y \tag{4.2.35}$$

and

$$j_y = -\gamma_0\,B_z\,E_x + (\sigma_0 + \beta_0\,B_z^2)\,E_y \ . \tag{4.2.36}$$

For a steady state the component j_y vanishes if the current through the voltmeter in Fig. 4.2 measuring the Hall voltage V_y is small compared with the longitudinal current component I_x (in practice a high impedance digital voltmeter is used). The two sides of the sample perpendicular to the y-direction carry charges of opposite sign and of such magnitude that the field between these charges (which is the Hall field) counterbalances the Lorentz field. From $j_y = 0$ and (4.2.36), we obtain for E_x

$$E_x = \frac{\sigma_0 + \beta_0\,B_z^2}{\gamma_0\,B_z}E_y \ . \tag{4.2.37}$$

We eliminate E_x from (4.2.35) and obtain the Hall field E_y:

$$E_y = \frac{\gamma_0}{(\sigma_0 + \beta_0\,B_z^2)^2 + \gamma_0^2\,B_z^2}\,j_x\,B_z \ . \tag{4.2.38}$$

In comparing with (4.2.2) we find for the Hall coefficient

$$R_{\mathrm{H}} = \frac{\gamma_0}{(\sigma_0 + \beta_0\,B_z^2)^2 + \gamma_0^2\,B_z^2} \approx \gamma_0/\sigma_0^2 \ . \tag{4.2.39}$$

This approximation can be made since we have limited the calculation to the low-field case. Taking into account the definitions of γ_0 and σ_0 (4.2.24) and (4.2.21), respectively, we find for the Hall coefficient

$$R_\mathrm{H} = r_\mathrm{H}/n\,e \ , \tag{4.2.40}$$

where r_H stands for

$$r_\mathrm{H} = \langle \tau_\mathrm{m}^2 \rangle / \langle \tau_\mathrm{m} \rangle^2 \tag{4.2.41}$$

and is called the *Hall factor*. In order to get an idea of the magnitude of the Hall factor, we assume $\tau_\mathrm{m} = \tau_0 (\varepsilon/k_\mathrm{B} T)^r$ as discussed before (4.1.8) and obtain from (4.1.25) for $\langle \tau_\mathrm{m} \rangle^2$:

$$\langle \tau_\mathrm{m} \rangle^2 = \left[\frac{4}{3\sqrt{\pi}} \tau_0 (3/2 + r)! \right]^2 \ . \tag{4.2.42}$$

Similarly, by replacing τ_m in (4.2.22) by τ_m^2, the average of τ_m^2 is obtained:

$$\langle \tau_\mathrm{m}^2 \rangle = \frac{4}{3\sqrt{\pi}} \tau_0^2 (3/2 + 2r)! \ . \tag{4.2.43}$$

Hence, the Hall factor is given by

$$r_\mathrm{H} = \frac{3\sqrt{\pi}}{4} \frac{(2r + 3/2)!}{[(r + 3/2)!]^2} \ . \tag{4.2.44}$$

For acoustic deformation potential scattering where r equals $-\frac{1}{2}$, the Hall factor becomes

$$r_\mathrm{H} = \frac{3\sqrt{\pi}}{4} \frac{\sqrt{\pi}/2}{1} = \frac{3\pi}{8} \approx 1.18 \ , \tag{4.2.45}$$

while for ionized impurity scattering where $r = +\frac{3}{2}$, we find

$$r_\mathrm{H} = \frac{3\sqrt{\pi}}{4} \frac{2^{-5}\, 3^3\, 5\, 7\sqrt{\pi}}{2^2\, 3^2} = \frac{315\pi}{512} \approx 1.93 \ . \tag{4.2.46}$$

Obviously the order of magnitude of r_H is 1.

Due to the negative charge of the electron, $e < 0$, the Hall coefficient is negative for n-type conductivity while positive for p-type. Therefore, the Hall effect is an important method for the determination of the type of conductivity. The case of intrinsic or nearly intrinsic semiconductors will be discussed later on.

From R_H the carrier concentration can be determined:

$$n \text{ or } p = r_\mathrm{H}/R_\mathrm{H} e \ . \tag{4.2.47}$$

On the other hand, if R_H is known, the Hall effect is useful for an experimental determination of the magnetic induction B. In contrast to a rotating solenoid it has no moving parts. There are many more device applications of the Hall effect. One is the multiplication of electrical signals which are used for generating j_x

and B_z; the Hall voltage is determined by the product of both. A power meter has been constructed in this way.

The *Hall mobility* is defined by the product of the conductivity σ_0 and the Hall coefficient R_H:

$$\mu_H = R_H \, \sigma_0 = (r_H/n \, e) \, n \, e \, \mu = r_H \, \mu \; . \tag{4.2.48}$$

The Hall mobility is different from the drift mobility by the Hall factor. Usually it is the Hall mobility which is measured rather than the drift mobility. From the observed temperature dependence of the Hall mobility, one can get an idea of the most probable ε-dependence of τ_m from which the Hall factor r_H is calculated. In this way, the drift mobility μ_H/r_H may be determined.[4]   Since r_H is not too different from unity, the error introduced by assuming an incorrect $\tau_m \, (\varepsilon)$-dependence is not too large.

Experimentally obtained Hall mobility versus temperature curves will be presented in Fig. 5.4 for silicon and 6.25, 6.9, and 12.7 for germanium. Another quantity of interest is the Hall angle θ_H which is the angle between j and E given by

$$\tan \theta_H = \mu_H B_z = r_H \, \omega_c \langle \tau_m \rangle \; . \tag{4.2.49}$$

This relation is obtained from $[j \times E]$ and $j = \sigma_0(E + R_H[j \times B])$ for E perpendicular to B taking (4.2.48) into account. The neglect of any dependence of r_H on B, as was assumed in (4.2.40), is justified only if the Hall angle is small. In fact, whenever the carrier concentration and mobility are determined by the Hall effect, the magnetic field intensity has to be small enough so that a variation of B does not affect these quantities appreciably.

Because of the practical importance of the Hall effect we will give a numerical example. From (4.2.1, 40), the Hall voltage is given by

$$V_y = r_H \, I_x \, B_z/(n \, e \, d) \; . \tag{4.2.50}$$

Assume a semiconductor with $r_H = 1.6$ and $n = 10^{14}$ cm^{-3}. A typical sample width may be $d = 1$ mm, the sample current 1 mA and the magnetic induction 10^{-1} T $= 10^{-5}$ Vs/cm^2. The Hall voltage is 10 mV. Assume the semiconductor to be n-type germanium at room temperature with a Hall mobility of $\mu_H = 4 \times 10^3$ cm^2/V s (actually r_H is somewhat less than 1.6 in this case). The product $\mu_H B_z$ is 4×10^{-2} which is small compared to 1. Therefore, the Hall voltage will vary linearly with B_z. We shall see, however, that the criterion $\mu_H B_z \ll 1$ may not be sufficient in the case of two types of carriers such as light and heavy holes which occur in most p-type semiconductors (Sect. 8.3). If both electrons and holes are present in comparable quantities, we can add the current densities of both carrier types

$$0 = j_y = \sigma_0 \, E_y - |e| \, r_H (p \mu_p^2 - n \mu_n^2) \, E_x B_z \tag{4.2.51}$$

and since in the weak-field approximation

$$E_y = R_H \, j_x \, B_z \approx R_H \, \sigma_0 \, E_x \, B_z \; , \tag{4.2.52}$$

_____
[4] A direct method, although rarely applicable, is the Haynes–Shockley experiment (Sect. 5.2)

replacing E_y in (4.2.51) and solving for R_H yields

$$R_H = \frac{|e| \, r_H}{\sigma_0^2}(p\,\mu_p^2 - n\,\mu_n^2) = \frac{r_H}{|e|} \frac{p\,\mu_p^2 - n\,\mu_n^2}{(p\,\mu_p + n\,\mu_n)^2} = \frac{r_H}{|e|} \frac{p - n\,b^2}{(p + n\,b)^2} \,, \quad (4.2.53)$$

where the ratio of mobilities

$$b = \mu_n/\mu_p \quad (4.2.54)$$

has been introduced. R_H changes sign at $p = n\,b^2$ rather than at the intrinsic concentration $p = n$. For example, for InSb at room temperature, $b = 80$, and in intrinsic material, the Hall coefficient, is negative as in n-type InSb. This causes the *Hall overshoot* of the p-type samples over the straight line characterizing the intrinsic conductivity shown [4.6, 7] in Fig. 4.3. Similarly, one finds an overshoot in the resistivity vs temperature curves of p-type samples: it is not at $p = n_i$ but at $p = n_i\sqrt{b}$ that the maximum resistivity occurs. For $b = 80$, this is nearly an order of magnitude larger than n_i and the maximum resistivity is by a factor of $(1 + b)/(2\sqrt{b}) = 4.5$ times larger than the intrinsic value.

In the final part of this section, we consider the magnetic field dependence of the Hall coefficient. It is obvious from (4.2.39) that for not too small magnetic field intensities there is a parabolic variation of R_H with B_z.

For very strong magnetic fields, we can approximate, $\sigma, \gamma,$ and β as given by (4.2.25, 26, 30), respectively, in the following way:

$$\sigma \approx (n\,e^2/m\,\omega_c^2)\langle \tau_m^{-1} \rangle \quad (4.2.55)$$

$$\gamma \approx n\,e^3/m^2\omega_c^2 \quad (4.2.56)$$

$$\beta \approx -(n\,e^4/m^3\omega_c^2)\langle \tau_m \rangle \,. \quad (4.2.57)$$

The conductivity tensor σ_w is obtained from (4.2.20, 23, 29):

$$\sigma_w = \begin{pmatrix} \sigma - \beta B_x^2 & -\beta B_x B_y + \gamma B_z & -\beta B_x B_z - \gamma B_y \\ -\beta B_x B_y - \gamma B_z & \sigma - \beta B_y^2 & -\beta B_y B_z + \gamma B_x \\ -\beta B_x B_z + \gamma B_y & -\beta B_y B_z - \gamma B_x & \sigma - \beta B_z^2 \end{pmatrix} \quad (4.2.58)$$

For $B_x = B_y = E_z = 0$, the current components j_x and j_y are

$$j_x = \sigma\,E_x + \gamma\,B_z\,E_y \quad (4.2.59)$$

and

$$j_y = -\gamma\,B_z\,E_x + \sigma\,E_y \,. \quad (4.2.60)$$

From $j_y = 0$, we obtain

$$E_x = (\sigma/\gamma\,B_z)E_y = (m/ne^2)\langle \tau_m^{-1} \rangle j_x \,. \quad (4.2.61)$$

We eliminate E_x from (4.2.60) and obtain the Hall field

$$E_y = \frac{\gamma}{\sigma^2 + (\gamma\,B_z)^2}\,j_x\,B_z = R_H\,j_x\,B_z = (B_z/n\,e)j_x \,. \quad (4.2.62)$$

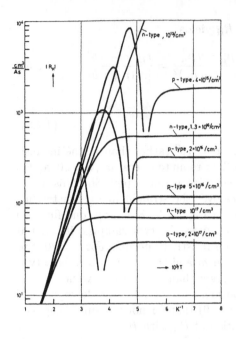

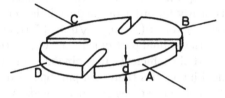

Fig. 4.3: Hall coefficient of indium antimonide as a function of the reciprocal temperature(after [4.6,7])

Fig. 4.4: Clover shaped sample for van der Pauw type Hall measurements (Sect.4.3)

We use (4.2.56–58) in order to obtain the Hall coefficient

$$R_{\mathrm{H}} = \frac{n\,e^3/m^2\omega_{\mathrm{c}}^2}{(n\,e^2/m\omega_{\mathrm{c}}^2)^2\langle\tau_{\mathrm{m}}^{-1}\rangle^2 + (n\,e^3B_z/m^2\,\omega_{\mathrm{c}}^2)^2} \approx \frac{1}{n\,e} \ . \tag{4.2.63}$$

The approximation is valid for large magnetic field strengths $(\mu_{\mathrm{H}}\,B_z)^2 \gg 1$. Assuming a value of 10 for $(\mu_{\mathrm{H}}\,B_z)^2$ and a Hall mobility of, e.g., 1000 cm^2/V s, a magnetic induction of 31.6 T would be required which can be obtained in pulsed form only, and with considerable effort. Few semiconductors have mobilities of more than 10^4 cm^2/V s where the requirements for B_z are easier to meet. In this case, however, the carrier density is obtained without having to guess the unknown Hall factor. At very large magnetic field intensities, the carrier density n may be different from its low field value, however. In this case, the energies ε_{G}, $\Delta\varepsilon_{\mathrm{A}}$ or $\Delta\varepsilon_{\mathrm{D}}$ are subject to changes induced by the magnetic field (Sect. 9.3) which can possibly be detected by optical methods. At very low temperatures in a 2-dimensional electron gas (Sect. 14.3), magnetic quantization (Sect. 9.2) yields the quantum Hall effect (Sect. 14.4). A typical experimental result for a semiconductor (InSb) with various doping levels at temperatures between 125 K and 600 K is shown in Fig.4.3.

4.3 Hall Techniques

Experimental problems in Hall measurements may arise from the difficulty in aligning the Hall arms of the sample perfectly, with the result of observing a misalignment voltage which may be of the same order of magnitude as the Hall voltage. Reversing the polarity of the magnetic field and taking the average of V_y for the two polarities eliminates the misalignment voltage. A double phase sensitive detection technique is useful, particularly for low-mobility semiconductors [4.8]: a strong electromagnet (e.g. of two kW power) is operated at 0.5 Hz from the oscillator of a phase sensitive detector via a power amplifier while the sample current I_x at e.g. 100 Hz is supplied from the oscillator of a second phase sensitive detector. The Hall voltage V_y is fed into this phase sensitive tuned to the higher frequency with an integration time constant short enough to let the 0.5 Hz modulation pass. The 0.5 Hz signal from the output of this detector is finally detected with the other phase sensitive detector tuned to this frequency. From the relative phase of the Hall signal both to the current and to the magnetic field, the type of conductivity (n- or p-type) can be determined. A misalignment voltage and a voltage induced in the Hall loop are automatically eliminated. Only the part of the Hall voltage proportional to the magnetic field is detected. An excellent signal-to-noise ratio is obtained.

The ratio of the maximum obtainable Hall signal to Johnson noise V_J (Sect. 4.15) is

$$V_y/V_J = \mu_B \, B \, (W \, V/4 \, k_B T \Delta f)^{(1/2)} \ , \tag{4.3.1}$$

where it is assumed that a current $(W \, V/R)^{1/2}$ has been applied; R is the sample resistance between the current probes, V is the sample volume, Δf is the detector band width (≈ 0.1 Hz) and W is the maximum acceptable Joule heat in the sample ($\approx 10^{-2}$ W cm^{-3}). A sample length equal to the width has been assumed. Hall mobilities down to an order of 10^{-4} cm^2/Vs can be measured at room temperature by double phase sensitive detection [4.8, 9].

For reducing the influence of the finite size of the contacts, conductivity and Hall measurements are often made with clover shaped samples according to *van der Pauw* (Fig. 4.4) [4.10], see also [4.11, 12]. If A and C are the current contacts and B and D are the Hall contacts and the magnetic field B is perpendicular to the sample of thickness d, the Hall coefficient is given by

$$R_H = [V_{BD}(B) - V_{BD}(0)]d/I_{AC} \, B \ , \tag{4.3.2}$$

where $V_{BD}(0)$ is the misalignment voltage. In a second experiment, the role of contacts A, C and B, D is exchanged and for an isotropic material the final result for R_H is the average of the two experiments.

For the resistivity measurements, A and B are the current contacts and C and D the potential probes in a first experiment yielding a resistance $R_{AB,CD} = |V_{CD}|/I_{AB}$, while in a second experiment, B and C are the current contacts and A and D the potential probes and consequently, a resistance $R_{BC,AD} = |V_{AD}|/I_{BC}$ is obtained. The resistivity ϱ is calculated from

$$\varrho = \frac{\pi}{\ln 2} d \frac{R_{AB,CD} + R_{BC,AD}}{2} f , \qquad (4.3.3)$$

where the factor f depends on the ratio $R_{AB,CD}/R_{BC,DA}$:

$R_{AB,CD}/R_{BC,DA}$	1	2	5	10	20	50	100	200	500	1000
f	1	0.96	0.82	0.70	0.59	0.47	0.40	0.35	0.30	0.26

If the resistance ratio > 2, it is common practice to renew the contacts [4.13]; another reason for a high ratio may be that the sample is inhomogeneously doped.

4.4 Magnetoresistance

The influence of a weak transverse magnetic field B_z on the current density j_x parallel to an externally applied electric field of intensity E_x is obtained from (4.2.35, 36) with $j_y = 0$:

$$j_x = \left[\sigma_0 + \beta_0 B_z^2 + \frac{(\gamma_0 B_z)^2}{\sigma_0 + \beta_0 B_z^2} \right] E_x . \qquad (4.4.1)$$

It is common practice to introduce the resistivity ϱ in this equation rather than the conductivity σ:

$$j_x = E_x/\varrho_B . \qquad (4.4.2)$$

The subscript B has been used to denote the dependence on B_z. The relative change in resistivity, $\Delta\varrho/\varrho_B = (\varrho_B - \varrho_0)/\varrho_B$, where $\varrho_0 = 1/\sigma_0$, is found from (4.4.1):

$$\Delta\varrho/\varrho_B = -B_z^2 [(\beta_0/\sigma_0) + (\gamma_0/\sigma_0)^2] = T_M (e\langle\tau_m\rangle B_z/m)^2 . \qquad (4.4.3)$$

The magnetoresistance scattering coefficient T_M is defined by

$$T_M = (\langle\tau_m^3\rangle\langle\tau_m\rangle - \langle\tau_m^2\rangle^2)/\langle\tau_m\rangle^4 . \qquad (4.4.4)$$

For the case where $\tau_m(\varepsilon)$ obeys a power law (4.1.8), T_M is given by

$$T_M = \frac{9\pi}{16} \frac{(3r + 3/2)! (r + 3/2)! - [(2r + 3/2)!]^2}{[(r + 3/2)!]^4} . \qquad (4.4.5)$$

The numerical value of T_M varies between 0.38 (for $r = -\frac{1}{2}$, acoustic deformation potential scattering) and 2.15 (for $r = +\frac{3}{2}$, ionized impurity scattering) and clearly depends more strongly on the scattering mechanism than does the Hall factor r_H, (4.2.45, 46). If we introduce the drift mobility $\mu = e\langle\tau_m\rangle/m$, the right-hand side of (4.4.3) becomes $T_M(\mu B_z)^2$. The results obtained so far are valid only if $(\mu B_z)^2 \ll 1$.[5]   Therefore, the proportionality of the magnetoresistance to B_z^2 will be correct only as long as $\Delta\varrho \ll \varrho_B$.

In a strong transverse magnetic induction B_z, the dependence of j_x on E_x is obtained from (4.2.60, 61):

$$j_x = (\sigma + \gamma^2 B_z^2/\sigma)E_x \ . \tag{4.4.6}$$

Hence, the relative change in resistivity is given by

$$\Delta\varrho/\varrho_B = 1 - (\sigma + \gamma^2 B_z^2/\sigma)/\sigma_0 \ . \tag{4.4.7}$$

In the strong field approximation, σ becomes much smaller than $\gamma^2 B_z^2/\sigma$ and can therefore be neglected. We insert for σ and γ their high-field values given by (4.2.56, 57):

$$\Delta\varrho/\varrho_B = 1 - (\langle\tau_m\rangle\langle\tau_m^{-1}\rangle)^{-1} \ . \tag{4.4.8}$$

For the case where $\tau_m(\varepsilon)$ obeys the power law mentioned earlier, we obtain for $\Delta\varrho/\varrho_B$

$$\Delta\varrho/\varrho_B = 1 - \frac{9\pi}{16(r + 3/2)!(3/2 - r)!} \ . \tag{4.4.9}$$

The numerical value is 0.116 for $r = -1/2$ and 0.706 for $r = +3/2$. For $r = 0$, where τ_m does not depend on energy and the averaging parenthesis may be omitted, the magnetoresistance effect vanishes for any value of the magnetic field. This explains why in metals the magnetoresistance effect is so small: it is only the carriers close to the Fermi surface that contribute to the conduction process. These have nearly the same energy and the same value of τ_m, although for carriers in metals, τ_m does depend on energy. However, the average of τ_m is essentially its value at the Fermi energy making it essentially a constant. At low temperatures in strong magnetic fields, quantum effects occur with the formation of *Landau levels*. In degenerate semiconductors there is an oscillatory magnetoresistance which is known as the *Shubnikov–de Haas effect* and which will be discussed in Sect. 9.2.

In intrinsic or nearly intrinsic semiconductors and in semimetals, two types of carriers, electrons and holes, contribute to magnetoresistance. It is easily verified that σ, β and γ have to be replaced by the sums of these contributions such as $\sigma_n + \sigma_p$, etc. For simplicity we omit the subscript 0 which denotes the weak field case, and obtain, instead of (4.4.3) for $\Delta\varrho/\varrho_B B_z^2$,

$$\frac{\Delta\varrho}{\varrho_B B_z^2} = -\frac{\beta_p + \beta_n}{\sigma_p + \sigma_n} - \left(\frac{\gamma_p + \gamma_n}{\sigma_p + \sigma_n}\right)^2 \ . \tag{4.4.10}$$

σ_0 (4.2.21) and β_0 (4.2.32) contain only even powers of $e, \sigma_p,$ and σ_n, β_p and β_n have the same sign, while the opposite is true for γ_n and γ_p(4.2.24). Assuming,

---

[5]μB_z is equal to the average of $\omega_c\tau_m$ over ε and in the distribution function $f(\varepsilon)$, the latter product was assumed to be small; see text after (4.2.30).

for simplicity, the same scattering mechanism for electrons and holes, (4.4.10)
yields

$$\frac{\Delta\varrho}{\varrho_B\,B_z^2} = \frac{9\pi}{16}\left\{ \frac{(3r+3/2)!}{[(r+3/2)!]^3}\frac{p\,\mu_p^3 + n\,\mu_n^3}{p\,\mu_p + n\,\mu_n} \right.$$

$$\left. - \left(\frac{(2r+3/2)!}{[(r+3/2)!]^2}\frac{p\,\mu_p^2 - n\,\mu_n^2}{p\,\mu_p + n\,\mu_n}\right)^2 \right\} \tag{4.4.11}$$

We introduce as usual the ratio $b = \mu_n/\mu_p$. For an intrinsic semiconductor $(n = p)$ and an acoustic deformation potential scattering $(r = -1/2)$ we obtain from (4.4.11)

$$\frac{\Delta\varrho}{\varrho_B\,B_z^2} = \frac{9\pi}{16}\,\mu_p^2\left\{ \frac{1+b^3}{1+b} - \frac{\pi}{4}(1-b)^2 \right\}$$

$$= \frac{9\pi}{16}\left(1-\frac{\pi}{4}\right)\mu_p^2\left[1 + \frac{\pi/2-1}{1-\pi/4}b + b^2 \right] \tag{4.4.12}$$

while for large values of b, it is reasonable to write this equation in the form

$$\frac{\Delta\varrho}{\varrho_B B_z^2} = \frac{9\pi}{16}\left(1-\frac{\pi}{4}\right)\mu_n^2\left[1 + \frac{\pi/2-1}{1-\pi/4}b^{-1} + b^{-2} \right]. \tag{4.4.13}$$

A comparison of this equation with the last one reveals that it is the type of carrier with the higher mobility which determines galvanomagnetic effects such as magnetoresistance. Considering again the case of semimetals where we may neglect the averaging procedure, both $9\pi/16$ and $\pi/4$ may be replaced by 1. We then obtain from (4.4.12)

$$\frac{\Delta\varrho}{\varrho_B\,B_z^2} = \mu_n\mu_p\,. \tag{4.4.14}$$

It is in semimetals such as bismuth and in intrinsic degenerate semiconductors that this type of behavior is found. The effect is several orders of magnitude larger than the magnetoresistance of metals with only *one* type of carrier. Before the invention of indium antimonide devices, bismuth spirals were used for magnetic field measurements. Many p-type semiconductors contain heavy and light holes, denoted by subscripts h and l. Both contribute to the conduction processes. Here all carriers have the same charge sign which is in contrast to electrons and holes in near-intrinsic semiconductors. We denote by b the mobility ratio μ_l/μ_h and by η the density ratio p_l/p_h. Assuming again a power law for $\tau_m(\varepsilon)$ with an exponent $r = -1/2$, we obtain for the magnetoresistance

$$\frac{\Delta\varrho}{\varrho_B B_z^2} = \frac{9\pi}{16}\,\mu_p^2\left[\frac{1+\eta\,b^3}{1+\eta\,b} - \frac{\pi}{4}\left(\frac{1+\eta\,b^2}{1+\eta\,b}\right)^2 \right], \tag{4.4.15}$$

where $\mu_p \equiv \mu_h$ is the mobility of the majority of holes. In, e.g., p-type germanium, $b = 8$ and $\eta = 4\%$ which yields for the factor between braces a value of

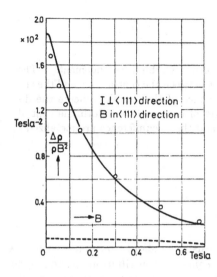

Fig. 4.5: Experimental transverse magnetoresistance of p-type germanium at 205 K (B parallel to a ⟨111⟩-direction). The dashed curve is calculated without inclusion of the high-mobility light holes (after [4.14])

10.6 instead of the value $1 - \pi/4 \approx 0.215$ for $\eta = 0$. Therefore, the one light hole out of 25 total holes raises $\Delta\varrho$ by a factor of about 50 which is equivalent to a rise of sensitivity with respect to B_z by a factor of seven. The conductivity σ and the Hall coefficient R_H contain lower powers of b and therefore are much less sensitive to the light-hole contribution (σ is raised by a factor of 1.3 and R_H by a factor of 2). Experimental results of $\Delta\varrho/\varrho B^2$ obtained with p-type germanium at 205 K with $\boldsymbol{B}$ parallel to a ⟨111⟩-direction are shown in Fig. 4.5. For a quantitative analysis, the effective mass anisotropy has to be considered [4.14]. The dashed line represents the heavy hole contribution. The observed Hall mobility (some $10^3\,\mathrm{cm}^2/\mathrm{Vs}$) would suggest that in a one-type-of-carrier model, magnetoresistance is constant at fields up to a few tenths of a Tesla. Obviously it is the light hole contribution which is large in magnetoresistance and comparatively small in the Hall mobility which makes the magnetoresistance vary strongly with $\boldsymbol{B}$ although $(\mu_H B_z)^2 \ll 1$ still holds. For more details see Sect. 8.3.

$\Delta\varrho/\varrho B^2$ is positive in the range of validity of the distribution function (4.2.18) since the Lorentz force deflects the carriers from the direction of the driving field and the drift velocity is only that component of the velocity in the direction of this field. It is mostly with *hot* carriers or in disordered conductors [4.15] that a *negative magnetoresistance* may occur.

Longitudinal magnetoresistance where $\boldsymbol{B}$ and $\boldsymbol{F}$ are parallel in (4.2.18) vanishes ($\Delta\varrho = 0$) if m^{-1}, τ_m, and μ^2 are scalar quantities. The case of non-vanishing longitudinal magnetoresistance will be discussed in Sects. 7.4, 8.3.

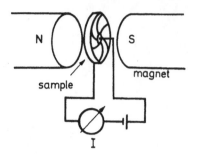

Fig. 4.6: Corbino arrangement. The curves between the central hole and the circumference of the sample indicate the direction of current flow. Both the central hole and the circumference are metalized

4.5 Corbino Resistance

Measurements of the transverse magnetoresistance should be made with long filamentary samples or else one obtains a *geometric* contribution to the magnetoresistance. This contribution is largest if the sample has the form of a circular disk with a central hole. Contacts are made at the hole and the circumference of the disk. The magnetic field is perpendicular to this *Corbino disk* [4.16, 17]. The arrangement is shown in Fig. 4.6.

The relative change in resistance of the disk upon applying a magnetic field is

$$\frac{\Delta R}{R_0} = \frac{\Delta \varrho}{\varrho_0} + \frac{\mu_H^2 B_z^2}{1 + \Delta \varrho / \varrho_0} \; , \tag{4.5.1}$$

where $\Delta \varrho / \varrho_0$ is the relative change in resistivity as measured with a long filamentary sample of the same material, and μ_H is the Hall mobility. For example, for n-type indium antimonide at room temperature in a field of 1 T, $\Delta R / R_0 = 17.7$ is obtained while $\Delta \varrho / \varrho_0$ is equal to only 0.48. There is no saturation of $\Delta R / R_0$ with increasing field intensity which makes this arrangement suitable for actual measurements of large magnetic field intensities (from the experiment a rather linear relationship between ΔR and B_z is then obtained).

The current lines in the sample are also shown in Fig. 4.6. Because of radial symmetry the electric field **E** can have only radial components. Therefore, along the x-axis the E_y-component vanishes. The current component j_y, however, does not vanish here as contrasted with conditions in magnetoresistance in filamentary samples. Hence, the resistivity cannot be defined by E_x/j_x but rather by [6]

$$\varrho_0 + \Delta \varrho = \varrho_B = (\boldsymbol{E} \cdot \boldsymbol{j})/j^2 = E_x j_x/(j_x^2 + j_y^2) \; . \tag{4.5.2}$$

From (4.2.25, 26) with the notation

$$Z = 1/(1 + \omega_c^2 \tau_m^2) \tag{4.5.3}$$

---

[6]The components of the resistivity tensor are $\varrho_{xx} = \varrho_{yy} = (\boldsymbol{E} \cdot \boldsymbol{j})/j^2$ and $\varrho_{xy} = -\varrho_{yx} = [\boldsymbol{E} \times \boldsymbol{j}]/j^2$.

and applying $B_x = B_y = 0$ to the tensors (4.2.23, 29), the current densities for $E_y = E_z = 0$ take the form

$$j_x = (n\,e^2/m)\langle Z\,\tau_m\rangle E_x; \qquad j_y = -(n\,e^3/m^3)\langle Z\,\tau_m^2\rangle B_z\,E_x \ . \qquad (4.5.4)$$

From (4.5.2) we obtain for the resistivity ϱ_B in a magnetic field

$$\varrho_B = \frac{m\langle Z\tau_m\rangle}{n\,e^2(\langle Z\tau_m\rangle^2 + \omega_c^2\langle Z\tau_m^2\rangle^2)} \ . \qquad (4.5.5)$$

The ratio of ϱ_B and the zero-field resistivity

$$\varrho_0 = 1/\sigma_0 = m/(n\,e^2\langle\tau_m\rangle) \qquad (4.5.6)$$

is given by

$$\frac{\varrho_B}{\varrho_0} = 1 + \frac{\Delta\varrho}{\varrho_0} = \frac{\langle\tau_m\rangle\,\langle Z\,\tau_m\rangle}{\langle Z\,\tau_m\rangle^2 + \omega_c^2\langle Z\,\tau_m^2\rangle^2} \ . \qquad (4.5.7)$$

On the other hand, (4.2.63) yields for R_H and thus for the product $\mu_H\,B_z$:

$$\mu_H\,B_z = \frac{R_H\,B_z}{\varrho_0} = \frac{m\,\langle Z\,\tau_m^2\rangle\omega_c}{n\,e^2(\langle Z\,\tau_m\rangle^2 + \omega_c^2\langle Z\,\tau_m^2\rangle^2)}\,\frac{n\,e^2\langle\tau_m\rangle}{m} \ . \qquad (4.5.8)$$

From the last two equations one can easily calculate the resistance ratio R_B/R_0:

$$1 + \frac{\Delta\varrho}{\varrho_0} + \frac{(\mu_H\,B_z)^2}{1 + \Delta\varrho/\varrho_0} = \frac{\langle\tau_m\rangle}{\langle Z\,\tau_m\rangle} \ . \qquad (4.5.9)$$

From (4.5.3) we obtain for this ratio

$$\langle\tau_m\rangle/\langle Z\,\tau_m\rangle = R_B/R_0 \ . \qquad (4.5.10)$$

Denoting by ΔR the difference $R_B - R_0$, we obtain (4.5.1) from (4.5.9, 10).

It may be of interest to solve (4.5.1) for μ_H and evaluate μ_H from the observed data mentioned above for $B_z = 1$ T:

$$\mu_H = \frac{1}{B_z}\sqrt{\left(1 + \frac{\Delta\varrho}{\varrho_0}\right)\left(\frac{\Delta R}{R_0} - \frac{\Delta\varrho}{\varrho_0}\right)} = 10^4\frac{\mathrm{cm}^2}{\mathrm{Vs}}\sqrt{1.48 \times 17.2} = 50500\frac{\mathrm{cm}^2}{\mathrm{Vs}} \qquad (4.5.11)$$

Experimental data at 0.1 T are $\Delta\varrho/\varrho_0 = 0.014$ and $\Delta R/R_0 = 0.25$. At such a low field intensity the product $\mu_H B_z$ for the Corbino geometry simply becomes $\sqrt{\Delta R/R_0}$, while for a filamentary sample, it would be $r_H\sqrt{\Delta\varrho/(\varrho_B T_M)}$ which contains averages over $\tau_m(\varepsilon)$. Therefore, the low-field Corbino resistance may serve in determining the low-field Hall mobility, as contrasted with magnetoresistance [4.18].

The second term on the right-hand side of (4.5.1) is sometimes called the *geometric contribution* to magnetoresistance. Another, although smaller, contribution is obtained if the magnetoresistance is measured in the conventional

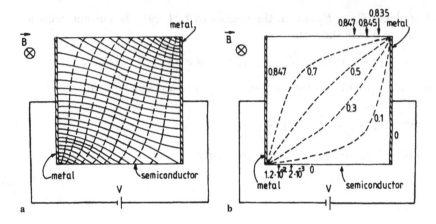

Fig. 4.7: (a) Electric field lines (–) and equipotential lines (- - -) in a square-shaped sample subject to a magnetic field B; the Hall angle at which the electric field lines begin and end at the semiconductor surface is 36° [4.19] (b) Equipotential lines in a square-shaped sample subject to a magnetic field B for an Hall angle of 89.999°; numbers specify the potential in arbitrary units. There is nearly no potential drop along most of the upper and lower sample edges. The Hall voltage across the middle of the sample is thus equal to the applied voltage

way with a short thick sample rather than a long thin sample. The electric field lines and the equipotential lines for this case are shown in Fig. 4.7. It is mainly near the end contacts of the sample that the current is not parallel to the sample axis since there the contact surface short-circuits the Hall field. While the electric field lines in a long thin sample are straight lines beginning at the positive surface charges on one Hall side of the sample and ending at the negative charges on the other side at an angle to the sample axis, which is the Hall angle θ_H (4.2.49) this is no longer true for a short sample. The results of a calculation of the equipotential lines for an Hall angle very close to 90° in a square–shaped sample are presented in Fig. 4.7b. For clarity only the equipotential lines are shown. Nearly all the voltage drop occurs in the lower left and upper right corners. Because of the strong electric field in these corners, a local electrical breakdown may occur repeatedly at short intervals which may be observed in, e.g., high-mobility n-InSb at 77 K as an emission of radiation in the microwatt range at frequencies from 1 MHz to more than 40 GHz [4.20]. In crossed electric and strong magnetic fields, a carrier performs a cycloid type of motion with frequency ω_c which is damped by the emission of radiation of this frequency because it is an oscillating dipole. This damping is, however, in semiconductors with their large dielectric constant, compared to vacuum in e.g. a pulsed magnetron, negligibly small.

The calculation for the curves shown in Fig. 4.7b was done by applying equations given by *Rendell* and *Girvin* [4.21, 22] and simplified mathematically by *Neudecker* and *Hoffmann* [4.23]. Let us briefly sketch the derivation of these equations without going into details. The boundary condition of zero current on the sample sides is replaced by the condition that the E-field make an angle θ_H with respect to the sides. If the sample geometry were that of a parallelogram with a homogeneous E-field perpendicular to the metallized ends, one could transform the rectangle of the two-dimensional Hall problem into this parallelogram by conformal mapping because the potential obeys the two-dimensional Laplace equation. *Rendell* and *Girvin* noticed that mathematically the problem equals the standard problem of finding the electrostatic potential inside a rectangle whose sides are at the potential θ_H and whose ends are at ground. In order to obtain the potential, the field is integrated first along a path across the middle of the sample thereby avoiding the corners and then going sideways. There is a singularity at a corner where the metallized surface meets the semiconducting surface. This is because the electric field lines emerge from the metal at an angle of 90° while they have to do so from the adjacent semiconductor surface at an angle which is the Hall angle < 90°. The convergence of the Fourier series presented by *Rendell* and *Girvin* must therefore be unsatisfactory close to the corners. The calculation was performed in view of the discovery of the Quantum-Hall effect (Sect. 14.3), but it is also valid for the classical Hall effect which per se is two-dimensional if we exclude the motion parallel to the magnetic field where the Lorentz force is zero. For B in the z-direction, the conductivity tensor in the plain perpendicular to B is then, combining (4.2.20, 25 and 26)

$$\sigma = \begin{pmatrix} \sigma_{xx} & \sigma_{xy} \\ -\sigma_{xy} & \sigma_{xx} \end{pmatrix} \tag{4.5.12}$$

Inverting this matrix to find the resistivity tensor, the longitudinal resistivity and the Hall resistivity, respectively are given by

$$\rho_{xx} = \frac{\sigma_{xx}}{\sigma_{xx}^2 + \sigma_{xy}^2} \qquad \rho_{xy} = \frac{-\sigma_{xy}}{\sigma_{xx}^2 + \sigma_{xy}^2} \ . \tag{4.5.13}$$

For the case of $\theta_H \to 90°$, in the middle of the sample, i.e., far enough away from the corners, the voltage drop along the sample axis and therefore σ_{xx} tends to zero. Throughout most of the sample, the current is an idle current because there is no dissipation: $(j \cdot E) = \sigma_{xx} E^2 = 0$. All of the dissipation is concentrated in the two diametrically opposed corners called *hot spots* These may be important for an understanding of the Quantum Hall Effect (Sect.14.4). The longitudinal resistance ρ_{xx} assumes the finite value $m \langle \tau_m^{-1} \rangle / n \, e^2$, and the Hall resistance $\rho_{xy} = -B/ne$ rises proportional to B as $B \to \infty$. For large values of the relaxation time τ_m, which are to be expected at very low temperatures, the longitudinal resistance becomes much smaller than the contact resistance. In a two-dimensional electron gas without any scattering events, i.e. $\tau_m \to \infty$, there is only a drift along the y axis with velocity $v_d = E/B$, but no net velocity along the x axis, hence $\sigma_{xx} = 0$ and $\sigma_{xy} = ne/B$. Because $\rho_{xx} = 0$, there is no magnetoresistance.

With l and w being sample length and width, respectively, Fig. 4.8 shows [4.17] the magnetoresistance ratio as a function of B_z for various values of the

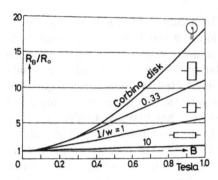

Fig. 4.8: Relative resistance as a function of magnetic field for four n-InSb samples of equal purity but different geometry (after [4.17])

ratio l/w. For a comparison, the Corbino resistance is plotted in the same figure. Only with high-mobility semiconductors is the geometric contribution to magnetoresistance important.

4.6 Transport in Inhomogeneous Samples

Quite often semiconductors are inhomogeneously doped. These inhomogeneities strongly influence the transport properties. How do they arise? The processes leading to inhomogeneous doping include clustering of dopants and are not yet well understood. When a single crystal is pulled from the melt in a Czochralski arrangement, depending on the velocity of pulling, zones of stronger and weaker doping are formed. These zones are called *striations* and are comparable in appearance to the annual rings of a tree [4.24]. Striations can be made visible by various methods including pulsed anodic etching [4.25] and selective photoetching [4.26, 27]. In a direction perpendicular to the direction of pulling, e.g., the relative magnetoresistance, $\Delta R/R_0$ may be only one tenth of its value parallel to this direction [4.28].

The explanation of this phenomenon is based on the fact that the current lines are deflected from their original direction as they enter a zone of different carrier density due to change in doping [4.29, 30]. Figure 4.9 shows a transition from a slightly n-type part of a crystal (denoted by n) to a more heavily doped n-type part (denoted by n^+). For simplicity it is assumed that the Hall mobility is independent of n and B. A coordinate system is chosen such that the x-direction is perpendicular to the n-n^+ barrier. The field component $E_x = j_x/(ne\mu)$ is proportional to $1/n$ since j_x is transmitted through the crystal by the externally applied voltage and hence is the same in both parts of the crystal. E_x is the same throughout the cross section of the crystal which can be written in the form $\partial E_x/\partial y = 0$. Since the magnetic field does not change with time, we obtain from Maxwell's equations $[\nabla_r \times E]_z = 0$ which with $\partial E_x/\partial y = 0$ yields $\partial E_y/\partial x = 0$. This means that the Hall field E_y is also the same in both parts

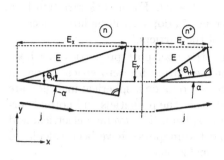

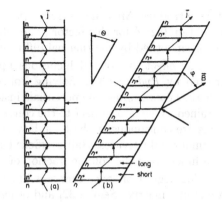

Fig. 4.9: Rotation of the current lines in a transverse magnetic field by an angle α due to a discontinuity in the doping level (e.g. n-n$^+$ junction)

Fig. 4.10: Current lines in a transverse magnetic field where the average current direction is perpendicular to the alternating n-n$^+$ junctions **(a)** or at an angle **(b)**

of the crystal. Therefore, the field direction E in Fig. 4.9 is different in the two parts. The Hall angle θ_H given by $\tan\theta_H = \mu_H B_z$ is the same, however, since μ_H is assumed to be the same. On the other hand, in a homogeneous crystal, $\tan\theta_H$ is E_y/E_x, which is no longer true here. However, if we define θ_H as the angle between E and j

$$\tan\theta_H = |[j \times E]|/(j \cdot E) \;, \tag{4.6.1}$$

this definition would cover both the case of Fig. 4.9 and that of a homogeneous crystal where $|j| = j_x$ and hence $\tan\theta_H = E_y/E_x$. The current lines j in Fig. 4.9 are deflected by the same angle α but in opposite directions on both sides of the barrier due to the magnetic field. The case of alternating n and n^+ regions is demonstrated by Fig. 4.10. In the left part of the figure, the barriers are perpendicular to the average current density while in the right part, they are at an oblique angle to the average current density. *Herring* neglected the energy dependence of the momentum relaxation time ($\tau_m = const.$) in his calculation [4.29] and still obtained a magnetoresistance effect given by

$$\Delta\varrho/\varrho_0 = \left(1 - \frac{1}{\langle n\rangle\langle n^{-1}\rangle}\right) \frac{\mu_H^2 B^2 \sin^2\varphi}{1 + \mu_H^2} B^2 \cos^2\varphi \;, \tag{4.6.2}$$

where $\langle n\rangle$ and $\langle n^{-1}\rangle$ are the values of n and $1/n$, respectively, averaged over the whole crystal, and φ is the angle between the average current and the magnetic field. Let us for simplicity assume that $\langle n\rangle = 1/2(n+n^+)$ and $\langle n^{-1}\rangle = 1/2(1/n + 1/n^+)$. For this case we obtain

$$1 - \frac{1}{\langle n\rangle\langle n^{-1}\rangle} = 1 - \frac{4n\,n^+}{(n^+ + n)^2} = \left(\frac{n^+ - n}{n^+ + n}\right)^2 = \frac{1}{4}\left(\frac{\Delta n}{\langle n\rangle}\right)^2 \tag{4.6.3}$$

if we introduce $\Delta n = n^+ - n$ and $\langle n \rangle = \frac{1}{2}(n^+ + n)$. For a 10% variation in n we obtain 0.25% for this factor. The difference in doping will be larger because Δn is smoothed by the thermal diffusion of the carriers.

The effect of isolated inclusions upon the transport properties has been treated by *Ryden* [4.31]. Solving Poisson's equation for zero space charge in spherical coordinates, the potentials inside and outside a spherical inclusion are obtained. They are subject to the condition that the potentials and normal currents are continuous at the boundary of the sphere. The average electrical and thermal conductivities, Hall coefficient and thermopower have been calculated as a function of the volume fraction of the inclusions.

The neglect of space charge is justified for metallic inclusions embedded in a metallic matrix. *Seto* [4.32] and others [4.33] calculated the effect of carriers trapped at grain boundaries in polycrystalline and powdered semiconductors on conductivity and Hall effect. Seto's calculations agree with his experimental data obtained with 1 μm thick polysilicon films doped by boron ion implantation (Sect. 5.7). The space charge yields a bending of energy bands near the surface of a grain. At a doping level of $2 \times 10^{18} \mathrm{cm}^{-3}$ there is a minimum in the Hall mobility about two orders of magnitude below that for both lower and higher concentrations. At the doping concentration of the mobility minimum there is an abrupt resistivity drop of about five orders of magnitude for a further increase in doping concentration by only a factor of 10. This behavior is explained by assuming that at low doping levels, most carriers are trapped at grain boundaries with band bending as a result and hence, low mobility of carriers between the grains. As the grain boundaries are finally filled up, the Fermi level rises, the width of the boundary space charge layer narrows and finally the space charges disappear altogether in the range of metallic conductivity resulting again in high mobilities.

In heavily doped semiconductors, e.g., n-type Si with $10^{18} - 10^{20} \mathrm{cm}^{-3}$ doping concentration, negative magnetoresistance has been observed at liquid helium temperatures [4.34]. A satisfactory explanation is still lacking.

4.7 Planar Hall Effect

In the Hall effect investigated so far, the side arms of the sample point in a direction perpendicular to that of the magnetic field. In the arrangement shown in Fig. 4.11 this is no longer the case. If the x-axis is still in the direction of current flow and the z-axis is in the direction of the Hall side arms, the magnetic induction $\boldsymbol{B}$ has components B_x and B_z in both of these directions. The *planar Hall field E_z* is given by [4.35][7]

---

[7]For the *longitudinal Hall effect* see [4.36].

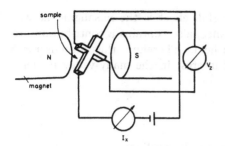

Fig. 4.11: Arrangement for planar Hall effect

$$E_z = P_H \, j_x \, B_x \, B_z \; , \tag{4.7.1}$$

where P_H is the *planar Hall coefficient* (measured in $\mathrm{cm^5/Ws^2}$). The observed voltage is

$$V_z = P_H \, I_x \, B_x \, B_z / d \; , \tag{4.7.2}$$

where d is the thickness of the crystal in the y-direction; it is a maximum for $B_x = B_z = B/\sqrt{2}$ where the angle between $\boldsymbol{B}$ and $\boldsymbol{j}$ is $45°$:

$$V_z = P_H \, I_x \, B^2 / 2d \; . \tag{4.7.3}$$

The proportionality of V_z to B^2 immediately suggests a relation to magnetoresistance. For a weak-field calculation of the coefficient P_H, we derive the current density components from the conductivity tensor given by (4.2.34) where we assume $B_y = 0$:

$$j_x = (\sigma_0 + \beta_0 \, B_z^2) E_x + \gamma_0 \, B_z \, E_y - \beta_0 \, B_x \, B_z \, E_z$$

$$j_y = -\gamma_0 \, B_z \, E_x + (\sigma_0 + \beta_0 \, B_x^2 + \beta_0 \, B_z^2) \, E_y + \gamma_0 \, B_x \, E_z \tag{4.7.4}$$

$$j_z = -\beta_0 \, B_x \, B_z \, E_x - \gamma_0 \, B_x \, E_y + (\sigma_0 + \beta_0 \, B_x^2) E_z \; .$$

We solve $j_y = 0$ and $j_z = 0$ for E_x and E_y and insert these in the equation for j_x, neglecting powers of B higher than the second. This equation is then solved for E_z:

$$E_z = \{(\sigma_0 \beta_0 + \gamma_0^2)/\sigma_0^3\} \, j_x \, B_x \, B_z \; . \tag{4.7.5}$$

We compare this equation with (4.7.1) and obtain for P_H

$$P_H = \frac{1}{\sigma_0} \left\{ \frac{\beta_0}{\sigma_0} + \left(\frac{\gamma_0}{\sigma_0} \right)^2 \right\} \; . \tag{4.7.6}$$

Introducing the resistivity ϱ_0 for $1/\sigma_0$ and the transverse magnetoresistance (4.4.3) for the term in braces, we find that the relationship between the planar Hall effect and the magnetoresistance suggested above is in fact

$$P_H = -\varrho_0 \Delta \varrho / \varrho_B B^2 \; . \tag{4.7.7}$$

An experimental problem in the planar Hall effect is the misalignment volt-age which, because of the magnetoresistance, also contains a contribution pro-portional to the square of the magnetic field. Therefore, this effect is rarely mentioned. The planar Hall effect is of interest in the many-valley model of semiconductors (Sect.7.4).

4.8 Thermal Conductivity, Lorenz Number, Comparison with Metals

Besides an electric charge, carriers also transport energy. If we set up a temper-ature gradient $\nabla_r T$ in a sample there will be a heat flow of density w which is determined by the thermal conductivity κ (not to be confused with the dielectric constant κ)

$$w = -\kappa \, \nabla_r T \tag{4.8.1}$$

if there is no electric current. In metals where practically all of the thermal energy is transported by carriers, an electric field of the right strength is set up to counteract the average carrier velocity in the direction of the temperature gradient. This is called the *thermoelectric field*. Since some carriers have a higher than average velocity in this direction due to the Fermi-Dirac distribution function, there is still some energy transport by carriers which is proportional to the density of carriers and their mobility, hence proportional to the conductivity σ. The ratio

$$L = \kappa/(\sigma T) \tag{4.8.2}$$

is denoted as the *Lorenz number*. The value calculated for a highly degenerate electron gas [4.37]

$$L = (\pi \, k_B/e)^2/3 \approx 2.45 \times 10^{-8} \text{W}\Omega/\text{K}^2 \tag{4.8.3}$$

is found experimentally to agree for most metals within 10% at room tempera-ture (W, Ω, and K are of course the units of energy, resistivity, and temperature, respectively). The temperature independence of L is known as the *Wiedemann-Franz law*. Since for metals $\sigma \propto 1/T$ to a good approximation, κ should be nearly independent of temperature.

The metal with the highest value of σ is silver; its value of κ at 273 K is 4.33 W/cm K. In silicon the experimental value of κ is 1.4 W/cm K (Fig. 4.12), a high value even though the density of carriers is many orders of magnitude less than in silver. Even in insulators there is thermal conduction. This is entirely due to energy transport by thermal lattice waves. We will not consider here the theory of this process [4.39]. It just may be worth mentioning that in an arrangement of coupled harmonic oscillators there is no interaction between lattice waves and hence no energy transfer. It is the anharmonicity of the lattice

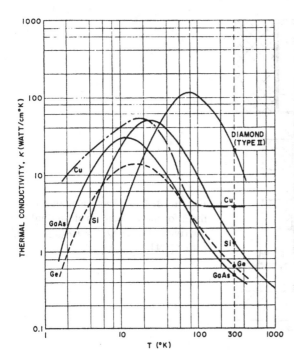

Fig. 4.12: Observed thermal conductivity as a function of temperature for various semiconductors of high purity; for a comparison data for copper are also included in the diagram compiled by S.M. Sze, Physics of Semiconductor Devices (Wiley and Sons, New York, 1969) (after [4.38])

oscillators which determines the thermal conductivity in insulating crystals at high temperatures. For example, in germanium, the fact that it contains several isotopes plays an important role in the thermal conductivity [4.40].

In nondegenerate semiconductors the lattice contribution to thermal conductivity in almost all cases is orders of magnitude larger than the electronic contribution. Therefore, we will not calculate the electronic contribution either but just comment upon the results of such calculations. Before we do this, we will discuss a typical experimental arrangement and some of the experimental results.

Methods of measuring κ were given by, e.g., *Weiss* [4.41] and *Schröder* [4.42] (for temperatures between 297 and 452 K). The arrangement by Weiss is shown in Fig. 4.13. The filamentary sample of circular cross section has a small electrically heated oven at the upper end. The power consumption of this oven is measured. The lower end of the sample is cooled by a metal plate. In order to avoid convection the arrangement is mounted in an evacuated cylinder (10^{-4} $torr$=1.3×10^{-2}Pa). Three more ovens are mounted outside the cylinder which, in order to avoid radiation losses, are kept at the same temperatures as the parts of the sample they face. Temperatures are measured by thermocouples (indicated in Fig. 4.13 by <). The accuracy of the determination of κ is 10% at normal and 20% at high temperatures.

Results obtained [4.41] with n-type and p-type InSb are shown in Fig.4.14.

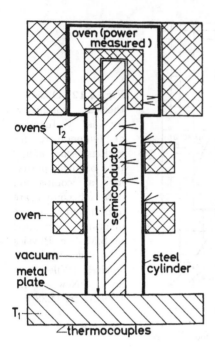

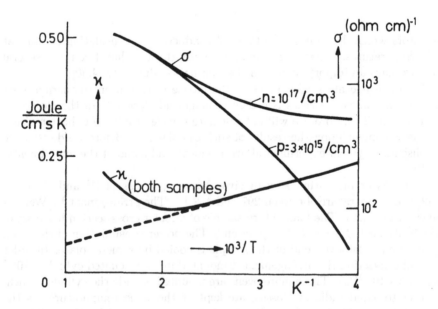

Fig. 4.13: Cross section of an arrangement for measurements of thermal conductivity. Power $Q/t = \kappa A(T_2 - T_1)/l$ (after [4.41])

Fig. 4.14: Dependence of electrical and thermal conductivities on the reciprocal of the temperature for n- and p-type InSb (after [4.41])

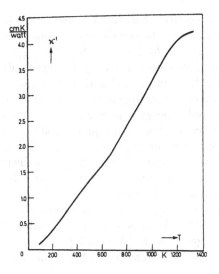

Fig. 4.15: Thermal resistivity as a function of temperature for silicon (after [4.43])

κ is independent of doping for nondegenerate semiconductors. The lattice contribution to κ (dashed curve) is roughly proportional to $1/T$. The electronic contribution is significant only above 500 K and agrees within a factor of 2 with values calculated for an intrinsic semiconductor (see below).

In Fig. 4.15, the observed [4.43] thermal resistivity $1/\kappa$ of pure silicon is plotted vs temperature T. In a first approximation, $1/\kappa$ is proportional to T which is equivalent to $\kappa \propto 1/T$ and typical for lattice conduction processes.

Bismuth telluride ($Bi_2 Te_3$) at room temperature has a thermal conductivity κ of 0.024 W/cm K and, with suitable doping, an electric conductivity σ of $10^3 (\Omega \, cm)^{-1}$. The low value of κ and the high value of σ together with the possibility of obtaining it in n- and p-type form make it an interesting material for thermoelectric cooling and energy conversion applications. Its Lorenz number, $L = 8 \times 10^{-8}$ W Ω K^{-2}, is of the order of magnitude of that of metals (4.8.3). Since the material is degenerate for the value of σ quoted above, a comparison with L of metals is justified and reveals that the electronic part of κ in $Bi_2 Te_3$ should be about 30%. Let us discuss the results of a calculation of L for nondegenerate semiconductors. If we take for κ only its electronic part and assume for $\tau_m(\varepsilon)$ a power law ε^r, we obtain

$$L = (k_B/e)^2 (r + 5/2) \qquad (4.8.4)$$

which is similar to (4.8.3) except that $\pi^2/3$ is replaced by $r+5/2$. For acoustic deformation potential scattering ($r = -\frac{1}{2}$) the latter factor becomes 2 while $\pi^2/3$ is about 3.29. Obviously degeneracy does not change the order of magnitude of L. In intrinsic or nearly intrinsic nondegenerate semiconductors L becomes [4.44][8]

$$L = (k_B/e)^2 \{r + 5/2 + [2r + 5 + \varepsilon_G/k_B T]^2 \sigma_n \sigma_p/\sigma^2\} \, , \qquad (4.8.5)$$

---

[8]This does not include energy transport by excitons.

where the same scattering mechanism for electrons and holes is assumed and σ_n and σ_p are the electron and hole contributions to the electric conductivity $\sigma = \sigma_n + \sigma_p$. For intrinsic conduction, $\sigma_n\sigma_p/\sigma^2 = b/(1+b)^2$, where $b = \mu_n/\mu_p$ is the mobility ratio. The term containing the band gap energy ε_G represents the generation of electron hole pairs at the hot sample end and recombination at the cold sample end. Each pair carries an energy of magnitude ε_G in addition to its energy of motion. The latter is the only energy converted to heat upon recombination in a zero-gap intrinsic semiconductor with equal mobilities of electrons and holes [9] (b=1) and is taken care of by a term $(r + 5/2)^2$ in addition to the *one-band* term $r + 5/2$ in this case. For large-gap intrinsic semiconductors (which will occur at high temperatures only), the *ambipolar* part of L and therefore also κ will be quite large, although it may still be much smaller than the lattice contribution. This is the case, e.g., in silicon at 800 K where κ calculated from (4.8.2, 5) is 10^{-3} W/cm K while the observed value of κ is more than two orders of magnitude larger (4×10^{-1} W/cm K).

Finally, let us study the *speed* of heat flow. The equation of continuity for w in one dimension is given by

$$\frac{\partial w_x}{\partial x} = -\frac{\partial}{\partial t}(c_v \varrho T) , \tag{4.8.6}$$

where c_v is the specific heat per mass unit, ϱ is the mass density, and the product $c_v \varrho$ the specific heat per unit volume. This, together with (4.8.1), yields the *diffusion equation*

$$(\kappa/\varrho c_v)\, \partial^2 T/\partial x^2 = \partial T/\partial t , \tag{4.8.7}$$

where we may introduce a *diffusion constant of temperature* (thermal diffusivity) [4.45]

$$D_T = \kappa/\varrho c_v \tag{4.8.8}$$

which, e.g., in silicon at room temperature has a value of

$$D_T = 1.4 \text{W/cm K}/(2.33\,\text{g cm}^{-3}\,0.76\,\text{Ws/g K}) = 0.79\,\text{cm}^2/\text{s} . \tag{4.8.9}$$

(The product ϱc_v is also the ratio of the specific heat per mole and volume per mole; for solids according to Dulong and Petit, the former has a high-temperature value of 25 Ws/mole K which may be useful for estimates of D_T.)

A periodic solution of (4.8.7) is given by

$$T \propto \exp(-x/l)\, \exp[\mathrm{i}\omega(t - x/v)] , \tag{4.8.10}$$

where ω is the angular frequency, $v = \sqrt{2\,D_T\,\omega}$ is the velocity and $l = \sqrt{2\,D_T/\omega}$ is the range of the temperature wave. At a higher frequency the wave is faster but does not get as far. At a given minimum range l the maximum frequency is $\omega/2\pi = D_T/\pi l^2$ which, e.g., in a silicon sample of $l = 1$ mm length requires a frequency of $\omega/2\pi = 25$ Hz and a transmission time of $1/\omega \approx 6 \times 10^{-3}$ s. Contrary

---

[9]In this case there will be no thermoelectric field since electrons and holes travelling at the same speed neutralize each other.

to these *slow* temperature waves, acoustic lattice waves have a propagation velocity in solids of about 10^5 cm/s which is nearly independent of frequency. It is mainly because of the *slowness* of temperature waves that thermal effects do not attract much attention in modern electronics while acoustoelectric effects have become much more interesting.

An electric device based on heat dissipation is the thermistor which is useful for surge suppression and measurements of temperature and radiation. The rise of temperature T above ambient temperature T_0 due to the generation of Joule heat VI is taken proportional to this quantity:

$$C(T - T_0) = VI ,\tag{4.8.11}$$

with a dissipation constant C depending both on the material and the geometry of the sample. For an intrinsic conductor the resistance is

$$R_0 \exp[B(1/T - 1/T_0)] = V/I ,\tag{4.8.12}$$

where R_0 is the resistance at ambient temperature and B is a constant. Eliminating T from the two equations yields an S-shaped $I = I(V)$ characteristic with an abrupt transition to a high-current state when the voltage reaches a maximum value

$$V_\mathrm{m} = \sqrt{C\,R_0(T_\mathrm{m} - T_0)\exp(-T_\mathrm{m}/T_0)} ,\tag{4.8.13}$$

where T_m is the temperature T at $V = V_\mathrm{m}$:

$$T_\mathrm{m} = \frac{1}{2}B(1 - \sqrt{1 - 4T_0/B}) .\tag{4.8.14}$$

The time constant of the thermistor is the product of its mass and its specific heat divided by the dissipation constant C and varies from about one second to one minute in practical devices.

4.9 Thermoelectric (Seebeck) Effect

The thermoelectric field mentioned is Sect. 4.8 will be calculated now. In solving the Boltzmann equation for the case of a temperature gradient we have introduced an electrothermal field $\boldsymbol{F}$ given by (4.2.14). Since $\boldsymbol{F}$ depends on the carrier energy ε we cannot take it outside the integral in calculating $\boldsymbol{j}$, but obtain instead

$$\boldsymbol{j} \propto \langle \tau_\mathrm{m} e \boldsymbol{F} \rangle = e\,\boldsymbol{E}\,\langle \tau_\mathrm{m} \rangle + T\langle \tau_\mathrm{m} \boldsymbol{\nabla}_r (\varepsilon - \zeta)/T \rangle .\tag{4.9.1}$$

Since the zero of energy is arbitrary, we introduce an electron kinetic energy

$$\varepsilon_n = \varepsilon - \varepsilon_\mathrm{c}\tag{4.9.2}$$

which is zero at the conduction band edge ε_c, and a hole kinetic energy

$$\varepsilon_p = -(\varepsilon - \varepsilon_v) \tag{4.9.3}$$

which is zero at the valence band edge. It is likewise convenient to introduce Fermi energies ζ_n and ζ_p:

$$\zeta_n = \zeta - \varepsilon_c \tag{4.9.4}$$

and

$$\zeta_p = -(\zeta - \varepsilon_v) \ . \tag{4.9.5}$$

In nondegenerate semiconductors the Fermi level ζ is located in the gap and both ζ_n and ζ_p are therefore negative ($\zeta_n + \zeta_p = -\varepsilon_G$). If we subtract (4.9.2) from (4.9.4) we obtain

$$\varepsilon_n - \zeta_n = \varepsilon - \zeta \ . \tag{4.9.6}$$

Similarly, (4.9.3, 5) yield

$$-(\varepsilon_p - \zeta_p) = \varepsilon - \zeta \ . \tag{4.9.7}$$

In this section we omit, for simplicity, the subscripts n and p except when we consider simultaneous conduction by electrons and holes; in the case of holes, we also omit the negative sign in front of $(\varepsilon_p - \zeta_p)$.

For a calculation of the thermoelectric field $\boldsymbol{E}$ we consider the case where there is no current flow through the conductor (*open circuit*). Equation (4.9.1) yields for $j=0$

$$e\,\boldsymbol{E} = -\frac{T}{\langle \tau_m \rangle} \left\langle \tau_m \nabla_r \frac{\varepsilon - \zeta}{T} \right\rangle = \frac{1}{T} \left(\frac{\langle \tau_m \varepsilon \rangle}{\langle \tau_m \rangle} - \zeta \right) \nabla_r T + \nabla_r \zeta \ . \tag{4.9.8}$$

We introduce an *entropy transport parameter* S/e by [4.46]

$$S/e = \frac{1}{eT} \left(\frac{\langle \tau_m \varepsilon \rangle}{\langle \tau_m \rangle} - \zeta \right) \tag{4.9.9}$$

which should not be confused with the entropy mentioned in Sect. 3.1. Now we have from (4.9.8)

$$e\,\boldsymbol{E} = S\,\nabla_r T + \nabla_r \zeta \ . \tag{4.9.10}$$

In principle, a measurement of $\boldsymbol{E}$ involves the arrangement given in Fig..4.16.

The filamentary sample has equal metal contacts at both ends, B and C, which are held at different temperatures, T_B and T_C. A voltmeter is connected to the end contacts. The voltage V indicated by the voltmeter is given by

$$V = \int_A^D (\boldsymbol{E}\,d\boldsymbol{r}) = \frac{1}{e} \int_A^D S(\nabla_r T\,d\boldsymbol{r}) + \frac{1}{e} \int_A^D (\nabla_r \zeta\,d\boldsymbol{r}) \ . \tag{4.9.11}$$

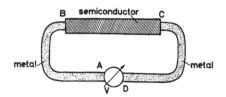

Fig. 4.16: Principal arrangement for measurement of the Seebeck effect

A and D are the voltmeter contacts assumed to be at the same temperature and made from the same metal. Then, the last integral which is ζ/e between limits A and D, vanishes. Equation (4.9.11) is simplified to

$$V = \frac{1}{e} \int_A^D S(\nabla_r T \, dr) \ . \tag{4.9.12}$$

From the path $A \to D$ we split the path $B \to C$ which is the semiconductor part while the rest, which is $A \to B$ plus $C \to D$, represents the metal part. Since the path through the voltmeter does not contribute to the integral and the parts $C \to D$ and $A \to B$ can thus be combined to give $C \to B$ which makes the integral the negative of that over $B \to C$, we obtain

$$V = \frac{1}{e} \int_{T_B}^{T_C} (S \, dT)_{\text{Semiconductor}} - \frac{1}{e} \int_{T_B}^{T_C} (S \, dT)_{\text{Metal}} \ . \tag{4.9.13}$$

This is the thermoelectric force of the thermocouple consisting of the semiconductor and the metal. The effect is known as *Seebeck effect*. We denote by $d\Theta/dT$ the *thermoelectric power* of a material:

$$\frac{d\Theta}{dT} = \frac{S}{e} = \frac{k_B}{e} \left(\frac{\langle \tau_m \varepsilon / k_B T \rangle}{\langle \tau_m \rangle} - \frac{\zeta}{k_B T} \right) , \tag{4.9.14}$$

where $k_B/|e|$ is about 86 μV/K.

The *absolute thermoelectric force* Θ as a function of temperature T is given by

$$\Theta = \frac{1}{e} \int_0^T S \, dT \ . \tag{4.9.15}$$

From (4.9.13) we obtain the observed voltage V in terms of $\Theta(T)$:

$$V = [\Theta(T_C) - \Theta(T_B)]_{\text{Semiconductor}} - [\Theta(T_C) - \Theta(T_B)]_{\text{Metal}} \ . \tag{4.9.16}$$

The Thomson effect, discussed in Sect. 4.10, offers a possibility for an experimental determination of $\Theta(T)$.

In the case where the momentum relaxation time $\tau_m(\varepsilon)$ obeys a power law ε^r, we obtain for the thermoelectric power of a nondegenerate n-type semiconductor

$$d\Theta/dT = -(k_B/|e|)(r + 5/2 - \zeta_n/k_B T) \tag{4.9.17}$$

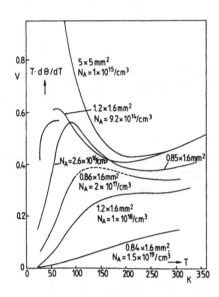

Fig. 4.17: Product of temperature and thermopower for p-type silicon samples of various cross-sections and acceptor concentrations (after[4.49])

while for the degenerate electron gas in a metal [4.47]

$$d\Theta/dT = -(k_B/|e|)(r + 3/2)(\pi^2/3)(k_B T/\zeta_n) . \qquad (4.9.18)$$

is obtained. At room temperature and below, $T \ll \zeta_n/k_B$, the thermoelectric power of a metal (typically a few $\mu V/K$) [4.48] is much smaller than that of a nondegenerate semiconductor and in (4.9.16), the term concerning the metal can therefore be neglected in most practical cases. The product of the thermoelectric power and the temperature for p-type silicon samples of various doping levels and dimensions is shown in Fig.4.17. Values for $d\Theta/dT$ range from 10 mV/K for one sample at 80 K to 0.5 mV/K for another sample at 300 K [4.49].

Since $d\Theta/dT$ in (4.9.14) depends on the first power of e, it is different in sign for electrons and holes. Figure 4.18 shows the product of T and $d\Theta/dT$ as a function of T for two n-type and two p-type germanium samples differing in doping [4.50]. The dashed lines represent $Td\Theta/dT = 2k_B T/e$ for both signs of e which is obtained from (4.9.17) for $r = -1/2$ (acoustic deformation potential scattering) and $\zeta = 0$. Hence, the difference between an experimental curve and the dashed line with the same sign of e gives the negative Fermi energy in units of e. The vertical arrows in Fig. 4.18 demonstrate these quantities for an arbitrarily chosen temperature of 275 K where the arrows have approximately equal lengths of nearly 0.3 V. In spite of the fact that the n and p-type samples are differently doped and the Fermi energies ζ_n and ζ_p are therefore not comparable in magnitude, the sum of ζ_n and ζ_p is about equal to the band gap energy ε_G which in germanium is about 0.7 eV. This indicates that to a first approximation, the term $\langle \tau_m \varepsilon/k_B T \rangle/\langle \tau_m \rangle$ in (4.9.14) may be neglected and the Fermi energy determines the thermoelectric power.

A typical arrangement for a thermoelectric determination of the conduction

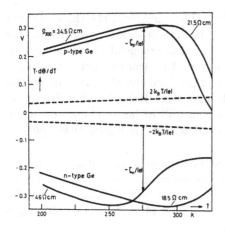

Fig. 4.18: Thermoelectric power for n- and p-type germanium (experimental data and calculated curves). The vertical arrows are Fermi energies in units of the elementary charge (after ([4.50])

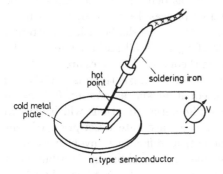

Fig. 4.19: Hot-probe arrangement for a thermoelectric determination of the conductivity type of a semiconductor (for n-type the polarity of the voltage is as shown; for p-type the polarity is reversed)

type of a semiconductor is shown in Fig. 4.19. The sample is put on a cold metal plate while the hot point of a soldering iron is pressed against the top surface of the sample. A voltmeter is connected to the plate and the soldering iron. For an n-type sample the polarity of the voltmeter is as indicated. For a p-type sample the polarity has to be reversed for normal deflection of the mV-meter. An arrangement where a hot plate and a cold point are used has been suggested [4.51] for high-resistivity material such as silicon carbide.

For intrinsic or near-intrinsic semiconductors, both electrons and holes have to be considered. Since $j = j_p + j_n = 0$, we obtain instead of (4.9.8) for this case:

$$|e|\,\boldsymbol{E}\,(\sigma_p + \sigma_n) = -\frac{T}{\langle \tau_p \rangle}\left\langle \tau_p \boldsymbol{\nabla}_r \frac{\varepsilon_p - \zeta_p}{T} \right\rangle \sigma_p - \frac{T}{\langle \tau_n \rangle}\left\langle \tau_n \boldsymbol{\nabla}_r \frac{\varepsilon_n - \zeta_n}{T} \right\rangle \sigma_n$$

(4.9.19)

which yields

$$\frac{d\Theta}{dT} = \frac{d\Theta_n}{dT}\frac{\sigma_n}{\sigma} + \frac{d\Theta_p}{dT}\frac{\sigma_p}{\sigma} \;,$$

(4.9.20)

where $\sigma = \sigma_n + \sigma_p$. For $\sigma_n = \sigma_p$ and $d\Theta_n/dT = -d\Theta_p/dT$, the thermoelectric power vanishes. Although this is an unrealistic case, it shows, nevertheless, that the thermolectric power in the *intrinsic* temperature range of a semiconductor is smaller than that in the *extrinsic* range. In Fig. 4.18 near the high-temperature ends of the curves, germanium becomes intrinsic and the absolute value of the thermoelectric power is decreased. For the curves marked p-Ge, there is even an indication that a reversal in sign will occur at still higher temperatures which would be due to the fact that the electron mobility is higher than the hole mobility and hence $\sigma_n > \sigma_p$ in (4.9.20).

The thermopower depends also on the dimensions of the sample, in particular at low temperatures. Reducing the cross-section of a certain sample by a factor of 36 reduces the thermopower by 32% at 55K. A qualitative explanation was given by *Herring* [4.52] and *Frederikse* [4.53, 54] by considering a non-equilibrium *phonon* distribution.

So far we have considered the electron *gas* as being in equilibrium with the oscillations of the atomic lattice. The interaction was described by the assumption of a momentum relaxation time τ_{m}. In Chap. 6 we shall consider scattering processes which determine the magnitude of τ_{m} in detail and assume that the lattice waves are in equilibrium. In quantum theory, lattice waves are quantized, the quanta being called *phonons*. Besides the electron gas there is also the gas of *phonons*, and the electron-Boltzmann equation and a *phonon-Boltzmann equation*[10] are actually coupled integro-differential equations which have to be solved simultaneously in order to obtain an accurate description of conduction processes. Because of the mathematical difficulties involved we have not attempted to do this, but rather assumed the equilibrium distribution of phonons as given by the well-known Planck equation. This has been the procedure followed so far. However, in the present case of a temperature gradient in a semiconductor yielding the thermoelectric effect, phonons travel preferentially from the hot to the cold sample end. Due to the electron–phonon interaction which will be discussed in detail in Sect. 7.7 dealing with the acousto-electric effect, carriers are dragged by the phonons (in the acousto-electric effect, carriers are transported by *coherent* sound waves, while the *phonon drag* effect is caused by incoherent waves of thermal conduction [4.52, 53, 54]). In this way more carriers are accumulated at the cold end than in the normal process described by (4.9.14).

In a very simplified approach to the problem, we consider a filamentary sample of a length equal to two phonon mean free paths $2L$ with temperature T and energy density $U(T)$ in the middle of the filament, and temperature $T - L\, dT/dx$ and energy density

$$U(T - L\, dT/dx) \approx U(T) - L(dT/dx)\, dU/dT \qquad (4.9.21)$$

at the hot end ($dT/dx < 0$), while for the cold end the minus signs should be replaced by plus signs. $dU/dT = c_v \varrho$ is the specific heat per unit volume where

---

[10]There is no particle conservation for phonons, in contrast to electrons.

ϱ is the mass density. The power absorbed per unit volume is given by

$$\Delta U = U(T - L\, dT/dx) - U(T + L\, dT/dx) \approx -(dT/dx)\, 2\, Lc_v\, \varrho \; . \quad (4.9.22)$$

For simplicity the electron-phonon interaction will be assumed to be strong enough to ensure a complete energy transfer to the carriers. With no current flow there will be an electric field E_x, and the force $e\,E_x$ acting on the $n\,2\,L$ carriers will counterbalance that due to the phonons:

$$n\,2\,L\,e\,E_x = (dT/dx)\,2\,L\,c_v\,\varrho \; . \quad (4.9.23)$$

This yields a thermoelectric power of magnitude

$$\frac{d\Theta}{dT} = \frac{E_x}{dT/dx} = \frac{c_v\,\varrho}{n\,e} \; . \quad (4.9.24)$$

The specific heat per mole of solids at high temperatures according to Dulong and Petit is 25 Ws/mole K and equals $c_v\,\varrho$ times the volume per mole. The latter is 12 cm^3/mole in silicon which yields about 2 W s/K cm^3 for $c_v\,\varrho$. If the material contains, e.g., $n = 10^{18}$ carriers cm^{-3}, the product $n\,e = 0.16$ Ws/V cm^3 and the ratio $c_v\,\varrho/n\,e$ has an order of magnitude of 10 V/K. Since the specific heat at temperatures much lower than the Debye temperature is proportional to T^3, the same law should hold for $d\Theta/dT$.

In fact, in Fig. 4.17 there is an increase of $d\Theta/dT$ with T for the heavily doped samples. The increase, however, is only a modest factor of two from 40 K to 300 K for the lowest curve which can be explained by assuming that only a small fraction of the directional phonon power is transferred to the carriers. In fact, the effect is increased by decreasing the carrier concentration n as may be expected from (4.9.24).

4.10 Thomson and Peltier Effects

In contrast to the previous section, we now permit an electric current density j to exist in addition to a temperature gradient $\nabla_r T$ in the semiconductor sample. For this case, in (4.9.10) j/σ must be added:

$$E = j/\sigma + \frac{1}{e}(S\nabla_r T + \nabla_r \zeta) \quad (4.10.1)$$

which for $\nabla_r T = \nabla_r \zeta = 0$ yields Ohm's law, of course. We are interested in the heat generated in the sample which for thermal equilibrium is naturally the well-known Joule heat. Let us first calculate the heat flow density w. Carriers not only transport charge e but energy ε as well. In order to obtain w one might think of replacing e in j by ε. From (4.10.1) and (4.9.9) we obtain j in the form

$$j = \frac{n}{m}[\langle \tau_m e\rangle(e\,E - \nabla_r \zeta) - \langle \tau_m(\varepsilon - \zeta)e\rangle\, T^{-1}\nabla_r T] \; . \quad (4.10.2)$$

The heat flow density $\boldsymbol{w}$, however, is given by

$$\boldsymbol{w} = \frac{n}{m}[\langle\tau_\mathrm{m}(\varepsilon - \zeta)\rangle\,(e\,\boldsymbol{E} - \boldsymbol{\nabla}_{\boldsymbol{r}}\,\zeta) - \langle\tau_\mathrm{m}(\varepsilon - \zeta)^2\rangle\,T^{-1}\boldsymbol{\nabla}_{\boldsymbol{r}}\,T]\ . \tag{4.10.3}$$

A comparison of these two equations reveals that to obtain $\boldsymbol{w}$ from $\boldsymbol{j}$, e has to be replaced by $(\varepsilon - \zeta)$ except for e in the combination $e\,\boldsymbol{E}$ which is the driving force operating on the charged particle in an electric field $\boldsymbol{E}$. Why should one replace e by $(\varepsilon - \zeta)$ rather than by just ε ? It is known from thermodynamics that an increase in heat δQ is given by a change in internal energy dU minus a change in free energy (Helmholtz function) dF. The Fermi energy ζ was introduced by (3.1.3) as the change in free energy dF with carrier concentration at constant temperature while ε is the change in U with n. Therefore, in the heat flow density $(\varepsilon - \zeta)$ is effective.

The signs in (4.10.2, 3) are typical for electrons with charge $e < 0$. In this case, the Fermi energy ζ is given by (4.9.4) and the difference $(\varepsilon - \zeta)$ by (4.9.6). For holes ($e > 0$) only the second term in $\boldsymbol{j}$ and the first term in $\boldsymbol{w}$ reverse sign because the energy $(\varepsilon - \zeta)$ becomes negative according to (4.9.7). Assuming for simplicity

$$\langle\tau_\mathrm{m}\varepsilon\rangle_n/\langle\tau_\mathrm{m}\rangle_n = \langle\tau_\mathrm{m}\varepsilon\rangle_p/\langle\tau_\mathrm{m}\rangle_p = k_\mathrm{B}T(r + 5/2)$$

$$\langle\tau_\mathrm{m}\varepsilon^2\rangle_n/\langle\tau_\mathrm{m}\rangle_n = \langle\tau_\mathrm{m}\varepsilon^2\rangle_p/\langle\tau_\mathrm{m}\rangle_p = (k_\mathrm{B}T)^2(r + 7/2)(r + 5/2) \tag{4.10.4}$$

with the same scattering mechanism for both types of carriers, and introducing

$$\sigma_n = n\,e^2\langle\tau_\mathrm{m}\rangle_n/m_n \qquad \sigma_p = p\,e^2\langle\tau_\mathrm{m}\rangle_p/m_p\ , \tag{4.10.5}$$

the current and heat flow densities in the case of mixed conduction are given by

$$\boldsymbol{j} = \frac{\sigma_n + \sigma_p}{|e|}(|e|\,\boldsymbol{E} - \boldsymbol{\nabla}_{\boldsymbol{r}}\,\zeta)$$

$$+ \left[\frac{\sigma_n - \sigma_p}{|e|}k_\mathrm{B}T\left(r + \frac{5}{2}\right) - \frac{\sigma_n\zeta_n - \sigma_p\zeta_p}{|e|}\right]\frac{1}{T}\boldsymbol{\nabla}_{\boldsymbol{r}}T \tag{4.10.6}$$

and

$$\boldsymbol{w} = -\left[\frac{\sigma_n - \sigma_p}{e^2}k_\mathrm{B}T\left(r + \frac{5}{2}\right) - \frac{\sigma_n\zeta_n - \sigma_p\zeta_p}{e^2}\right](|e|\,\boldsymbol{E} - \boldsymbol{\nabla}_{\boldsymbol{r}}\,\zeta)$$

$$-\left[\frac{\sigma_n + \sigma_p}{e^2}(k_\mathrm{B}T)^2\left(r + \frac{7}{2}\right)\left(r + \frac{5}{2}\right)\right.$$

$$\left.-2\frac{\sigma_n\zeta_n + \sigma_p\zeta_p}{e^2}k_\mathrm{B}T\left(r + \frac{5}{2}\right) + \frac{\sigma_n\zeta_n^2 + \sigma_p\zeta_p^2}{e^2}\right]\frac{1}{T}\boldsymbol{\nabla}_{\boldsymbol{r}}T\ . \tag{4.10.7}$$

From these equations (4.8.5) is easily obtained by assuming $\boldsymbol{j} = 0$ and eliminating $\boldsymbol{E}$. Only the contributions to $\boldsymbol{j}$ and $\boldsymbol{w}$ by carriers have been considered above. An additional term $+\kappa'\boldsymbol{\nabla}_{\boldsymbol{r}}T$ on the rhs of (4.10.6) takes care of the phonon drag contribution while additional terms $-\kappa_L\nabla_r T$ and $-\,T\,\kappa' E$ on the rhs of (4.10.7) take care of heat conduction by phonons and the *electron drag* effect, respectively.

It may be interesting to investigate the diffusion of carriers in a temperature gradient. We shall see from (4.10.2) that the terms depending on the Fermi level ζ account for this phenomenon [4.55, 4.56]. In $\nabla_r\zeta$ we replace ζ by $\zeta_n + \varepsilon_c$ according to (4.9.4) while in $\langle\tau_m(\varepsilon-\zeta)\rangle$, we replace $(\varepsilon-\zeta)$ by $(\varepsilon_n-\zeta_n)$ according to (4.9.6). From (3.1.42) we obtain for $n \gg N_{D^x},\ N_{A^-}$

$$n = N_c\, F_{1/2}(\eta_n)\ , \tag{4.10.8}$$

where $n_c \propto T^{3/2}$ and $\eta_n = \zeta_n/k_B T$. Since $\nabla_r\zeta_n = (\partial\zeta_n/\partial T)\nabla_r T$, we calculate $\partial n/\partial T$ from (4.10.8) and solve for $\partial\zeta_n/\partial T$:

$$\frac{\partial n}{\partial T} = \frac{3}{2}\frac{N_c}{T}F_{1/2}(\eta_n) + N_c\left(\frac{1}{k_B T}\frac{\partial\zeta_n}{\partial T} - \frac{\zeta_n}{k_B T^2}\right)F_{-1/2}(\eta_n)\ , \tag{4.10.9}$$

$$\frac{\partial\zeta_n}{\partial T} = \frac{\zeta_n}{T} + \frac{k_B T}{N_c F_{-1/2}}\frac{\partial n}{\partial T} - \frac{3}{2}k_B\frac{F_{1/2}}{F_{-1/2}}\ , \tag{4.10.10}$$

where $\partial F_{1/2}(\eta_n)/\partial\eta_n = F_{-1/2}(\eta_n)$ has been applied. Since $(\partial n/\partial T)\nabla_r T = \nabla_r n$, we obtain for the current density from (4.10.2)

$$\boldsymbol{j} = n|e|\mu_n\boldsymbol{E} - \frac{\langle\tau_m e\rangle}{m}k_B T\frac{F_{1/2}}{F_{-1/2}}\nabla_r n$$

$$+\frac{n}{m}\left(\langle\tau_m e\rangle\frac{3}{2}k_B\frac{F_{1/2}}{F_{-1/2}} - \frac{\langle\tau_m\varepsilon_n e\rangle}{T}\right)\nabla_r T \tag{4.10.11}$$

The second term on the right-hand side is the diffusion current density usually written as $-e\,D_n\nabla_r n$, where D_n is the diffusion coefficient for electrons. A comparison yields for D_n

$$D_n = \mu_n\frac{k_B T}{|e|}\frac{F_{1/2}(\eta_n)}{F_{-1/2}(\eta_n)} \approx \mu_n\frac{k_B T}{|e|}\ , \tag{4.10.12}$$

where we have introduced the mobility $\mu_n = (|e|/m)\langle\tau_m\rangle$. The approximation is valid for a nondegenerate electron gas.

This is known as the *Einstein relation*. The ratio $F_{1/2}(\eta_n)/F_{-1/2}(\eta_n)$ is about 3 for $\eta_n = 4$ and about 6.9 for $\eta_n = 10.$[11]

Now back to our original problem: for a calculation of heat transport in an n-type semiconductor, we replace $e\,\boldsymbol{E} - \nabla_r\zeta$ in (4.10.3) from (4.10.2) and remember that for $\boldsymbol{j} = 0$, the heat current density $\boldsymbol{w}$ is given by $-\kappa\nabla_r T$. With S/e given by (4.9.9), we obtain

$$\boldsymbol{w} = \boldsymbol{j}\,T(S/e) - \kappa\,\nabla_r T = \Pi\boldsymbol{j} - \kappa\,\nabla_r T \tag{4.10.13}$$

---

[11] $F_j(\eta_n) \approx \eta_n^{j+1}/(j+1)!$ for $\eta_n \gg 1$ yields $F_{1/2}/F_{-1/2} \approx 2\zeta_n/3k_B T$ and $D_n = 2\mu_n\zeta_n/3|e|$.

where the *Peltier coefficient* Π has been introduced.

$$\Pi = T S/e = T \, d\Theta/dT \; . \tag{4.10.14}$$

The relationship between the Peltier coefficient and the thermoelectric power is one of the *Onsager relations* and is called the *second Kelvin relation*. The heat Q generated per unit volume and time is given by

$$Q = (\boldsymbol{j} \cdot \boldsymbol{E}) - \boldsymbol{\nabla}_r(w - \zeta \, \boldsymbol{j}/e) \; , \tag{4.10.15}$$

where the second term is due to energy transport [see text after (4.10.3)]. Eliminating $\boldsymbol{E}$ and w from (4.10.1, 13), respectively, we find

$$Q = j^2/\sigma - \mu_{\mathrm{Th}}(\boldsymbol{j} \cdot \boldsymbol{\nabla}_r T) + \boldsymbol{\nabla}_r(\kappa \, \boldsymbol{\nabla}_r T) \; , \tag{4.10.16}$$

where a *Thomson coefficient* μ_{Th} has been introduced by

$$\mu_{\mathrm{Th}} = T \, d(S/e)/dT = T d^2 \Theta/dT^2 \; . \tag{4.10.17}$$

The second term in (4.10.16) is the Thomson heat while the first and third terms are Joule heat and heat transport by thermal conduction, respectively. The Thomson heat reverses sign on the reversal of either $\boldsymbol{j}$ or $\boldsymbol{\nabla}_r T$.

A measurement of the Thomson heat at various temperatures allows the determination of the absolute thermoelectric power, which is important for metallic conductors. From (4.10.17), the *first Kelvin relation*

$$\frac{d\Theta}{dT} = \int_0^T \frac{\mu_{\mathrm{Th}}}{T} dT \tag{4.10.18}$$

is obtained. If μ_{Th} as a function of T is known experimentally, the integral may be calculated. This relation may also be obtained by thermodynamic arguments [4.46, 56].

At a constant temperature throughout the semiconductor, the Thomson heat at first seems to vanish since $\boldsymbol{\nabla}_r T = 0$. However, a reversible heat is still generated at places where the Fermi level changes with position such as at n n^+ junctions or a junction of two different materials (*heterojunction*) if a current of intensity I flows through the junction. However, if we bring Q_{Th} into the form

$$Q_{\mathrm{Th}} = -T\frac{d(S/e)}{dT}(\boldsymbol{j} \cdot \boldsymbol{\nabla}_r T) = -T\boldsymbol{j} \cdot \boldsymbol{\nabla}_r(S/e) \tag{4.10.19}$$

and denote the junction cross section by A, the heat generated per unit time is given by

$$-A \int_1^2 T\boldsymbol{j} \cdot \boldsymbol{\nabla}_r(S/e)dr = I T \, (S_1 - S_2)/e = \Pi_{(1 \to 2)}I \tag{4.10.20}$$

and called the *Peltier heat*, where the Peltier coefficient of the junction is

$$\Pi_{(1 \to 2)} = \Pi_1 - \Pi_2 \tag{4.10.21}$$

and Π of each conductor is defined by (4.10.14). The Peltier heat is very large at a p-n junction with both sides degenerate. The difference in Fermi levels on both sides is about ε_G, and if this is, e.g., 1 eV, a current of 1A will generate a Peltier heat of 1 W.

In connection with the Peltier effect, a *vaccum model of a semiconductor* has been suggested by *Herring* [4.52] which consists of two disk shaped parallel-plate metal electrodes at temperatures T and $T+dT$, respectively, with T low enough that thermionic emission causes no appreciable space charge. The work function φ is assumed to be temperature-independent. We apply the Richardson equation and obtain for the current equilibrium

$$A\,T^2 \exp\left(-|e|\varphi/k_B T\right) = A(T + dT)^2 \exp\left[-|e|(\varphi + d\Theta)/k_B(T + dT)\right]$$

$$(4.10.22)$$

where the hotter electrode has gained a potential higher than the cooler one by an amount $d\Theta$ due to the emission of (negatively charged) electrons. Solving for $d\Theta/dT$ yields a thermoelectric power of magnitude

$$\left|\frac{d\Theta}{dT}\right| = 2\frac{k_B}{|e|} + \frac{\varphi}{T} \ . \tag{4.10.23}$$

This agrees with (4.9.17) for $\zeta_n = -|e|\varphi$ and $r = -1/2$, the latter being typical for an energy-independent mean free path. If with both electrodes at the same temperature T a current is passed through the diode, the electrode which looses electrons suffers a heat loss. Both the incoming and the outgoing electrons have a half-Maxwellian distribution and therefore each electron transports $2k_B T$ on the average. The heat loss per electron is $|e|\,\varphi + 2k_B T$ on the average. This yields a Peltier coefficient

$$\Pi = \frac{|e|\,\varphi + 2k_B T}{|e|} = T\frac{d\Theta}{dT} \tag{4.10.24}$$

in agreement with (4.10.14).

At low temperatures the Peltier effect may be subject to *electron drag* [see text after (4.10.7)]: drifting carriers *drag* phonons which increases the heat flow.

Peltier cooling devices are occasionally used for laboratory purposes. For large-scale energy conversion, thermoelectric devices with their low efficiency have not been able to compete with conventional devices even though they do not contain moving parts and therefore have a nearly unlimited lifetime. Much research has been devoted to the search for a suitable semiconductor material. A semiconductor is characterized by a *thermoelectric figure of merit*

$$Z = \frac{\sigma}{\kappa}\left(\frac{d\Theta}{dT}\right)^2 \ . \tag{4.10.25}$$

This combination of material constants results from the following considerations.

Assume a voltage V_0 being applied to a bar-shaped sample of length l and cross section A. The current

$$I = V_0 \sigma A/l \qquad (4.10.26)$$

through the sample causes a Peltier heat at the metal contacts of magnitude

$$\Pi I = T\frac{d\Theta}{dT}I = T\frac{d\Theta}{dT}V_0\sigma A/l \qquad (4.10.27)$$

neglecting the small contribution by the metal. The Peltier heat causes a temperature difference ΔT which can be used in a refrigerator. There is a heat loss by thermal conduction

$$Q = \kappa\,\Delta T\,A/l \qquad (4.10.28)$$

which should be kept small by choosing a material with a low thermal conductivity κ. The ratio of Peltier heat and Q should be large:

$$\Pi\,I/Q = T(d\Theta/dT)V_0\sigma/(\kappa\Delta T)\gg 1 \ . \qquad (4.10.29)$$

A second effect of ΔT is the thermoelectric voltage which for no electric current, $I = 0$, would be simply

$$V = (d\Theta/dT)\Delta T \ . \qquad (4.10.30)$$

We eliminate ΔT from (4.10.29, 30) and by introducing Z from (4.10.25), we obtain

$$V_0 Z \gg V/T \ . \qquad (4.10.31)$$

In this relation all the material constants form the *figure of merit* Z which should be as large as possible. In a more accurate calculation, Joule heat and Thomson heat should be taken into account. Without going into more details, we will just give some data on a well-known thermoelectric semiconductor, bismuth telluride (Bi_2Te_3).

By appropriate doping, Bi_2Te_3 can be obtained in n and p-type form with the same absolute value of the thermoelectric power (but, of course, differing in sign): $|d\Theta/dT| = 2 \times 10^{-4}$ V/K, while the electrical conductivity σ equals $10^3\ \Omega^{-1}cm^{-1}$ and the thermal conductivity κ equals 1.5×10^{-2} W/cm K. With these data one obtains from (4.10.25), $Z = 3 \times 10^{-3}$/K which happens to be T^{-1} with T being room temperature where the data have been taken [4.57–59]. Hence, (4.10.31) requires $V_0 \gg V$. Often thermocouples consisting of n- and p-type Bi_2Te_3 are connected to form a *battery*.

4.11 Thermomagnetic Effects

The thermomagnetic effect which is easy to observe is the *Nernst effect*. All others are more difficult to investigate because the energy transport by carriers

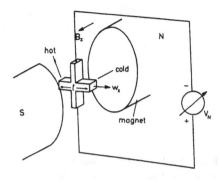

Fig. 4.20: Nernst arrangement; w_x is the heat flow density. For the polarity of the Nernst voltage as shown, the Nernst coefficient is positive

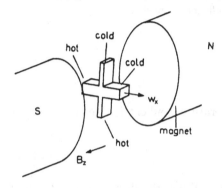

Fig. 4.21: Righi-Leduc arrangement; w_x is the heat flow density. For the sign of the temperature gradient shown, the Righi-Leduc coefficient is positive

in semiconductors is usually many orders of magnitude smaller than the heat conduction by the crystal lattice and the voltages developed are correspondingly smaller.

A sample of the same shape as used for the Hall measurements may be used for Nernst measurements (Fig. 4.2). A heat flow density w_x is transmitted (instead of an electric current j_x) by holding the sample ends at different temperatures and thus introducing a thermal gradient $\partial T/\partial x$. Just as in Hall measurements a transverse voltage V_y is developed between the side arms of the sample. The intensity of the *Nernst field* is given by

$$E_y = Q_N \frac{\partial T}{\partial x} B_z \ . \tag{4.11.1}$$

The *Nernst coefficient* Q_N is measured in units of cm^2/s K. It is taken as positive when the directions of $\boldsymbol{E}$, $\boldsymbol{B}$, and the temperature gradient are as indicated in Fig. 4.20. The *isothermal* Nernst effect is subject to the condition $\partial T/\partial y = 0$. However, the *adiabatic* Nernst effect is what one normally observes. The difference between the two effects is quite small, and therefore the isothermal effect is calculated from the observed adiabatic effect by adding to Q_N a small correction term [Ref. 4.4, p. 84] $S_{RL} \, d\Theta/dT$, where S_{RL} will be given below by (4.11.10).

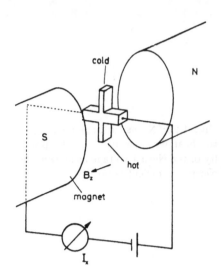

Fig. 4.22: Ettingshausen arrangement; for the sign of the temperature gradient shown, the Ettingshausen coefficient is positive

The occurrence of a transverse temperature gradient is due to the Righi-Leduc effect:

$$\frac{\partial T}{\partial y} = S_{RL} \frac{\partial T}{\partial x} B_z \ , \tag{4.11.2}$$

where S_{RL} is measured in units of cm^2/Vs. When the directions of $\mathbf{B}$ and of the temperature gradients are as indicated in Fig.. 4.21, $S_{RL} > 0$.

Hall measurements are occasionally subject to an error caused by the occurrence of the *Ettingshausen effect*. The heat transported by carriers which are deflected from the $\mathbf{E}$-direction by a magnetic field generates a transverse temperature gradient

$$\frac{\partial T}{\partial y} = P_E \, j_x \, B_z \ , \tag{4.11.3}$$

where the *Ettingshausen coefficient* P_E is measured in units of cm^3 K/Ws. It is positive for the directions indicated in Fig. 4.22. The Ettingshausen effect is called *isothermal* if $\partial T/\partial x = 0$. The temperature difference between the side arms of the sample yields a thermoelectric voltage between the metal-semiconductor contacts which adds to the Hall voltage. Since heat transport is a slow process in semiconductors (Sect. 4.8), the Ettingshausen effect in Hall measurements can be eliminated by applying a low-frequency ac current instead of a dc current and measuring the ac Hall voltage. The Ettingshausen effect may be appreciable (up to 10% of V_{Hall}) in low-resistivity (e.g., 10^{-3}Ωcm) semiconductors with low thermal conductivity (e.g., 5×10^{-2} W/cm K). For certain laboratory applications, Ettingshausen cooling shown by Fig. 4.23 may be of interest [4.60]. With a Bi/Sb alloy in a magnetic induction of 1.5 T at a temperature of 156 K at the sample base, the temperature at the top of the sample is 102 K which is less by 54 K.

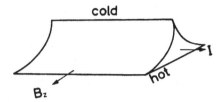

cold

Fig. 4.23: Sample shape for Ettingshausen cooling

The temperature differences in the Righi-Leduc and the Ettingshausen effects are given by

$$\Delta T_y = S_{RL}\, \Delta T_x\, B_z\, b/l \tag{4.11.4}$$

and

$$\Delta T_y = P_E\, I\, B_z/d \ , \tag{4.11.5}$$

respectively, while the Nernst voltage is given by

$$V_N = Q_N\, \Delta T_x\, B_z\, b/l \ , \tag{4.11.6}$$

where l is the length of the filamentary sample, d its thickness (in the **B**-direction) and b its width (in the transverse direction).

A relation between Ettingshausen and Nernst effects

$$P_E = Q_N T/\kappa \tag{4.11.7}$$

called the *Bridgman relation* is obtained from the thermodynamic arguments (the thermal conductivity κ includes the lattice contribution). It is convenient to measure the Nernst rather than the Ettingshausen effect and calculate P_E from (4.11.7). Typical data, e.g., for InSb with 8×10^{17} electrons cm^{-3} at 600 K are $Q_N = 0.3$ cm^2/s K and $\kappa = 0.08$ W/cm K [4.61]. The Bridgman relation yields for this material $P_E = 2.25 \times 10^3$ cm^3 K/Ws. Assuming $j_x =$ 1mA mm^{-2} $= 10^{-1}$A cm^{-2} and $B_z = 0.1$ T $= 10^{-5}$ Vs cm^{-2}, a temperature gradient $\partial T/\partial y = 2.25 \times 10^{-3}$ K/cm is calculated. For a sample width of 1 mm and assuming a thermoelectric power of 0.5 mV/K, a voltage of about 10^{-7} V is to be expected. This is many orders of magnitude smaller than the Hall voltage and in this case would be difficult to distinguish from the latter in an actual experiment.

Besides the thermomagnetic effects mentioned so far, there is, of course, also an influence of the magnetic field on the thermoelectric effects. These effects have to be taken into account in measurements of, e.g., the Nernst effect: due to a misalignment which can hardly be avoided, the side arms of the sample will have somewhat different temperatures if there is a temperature gradient along the sample filament. A thermoelectric voltage will occur in addition to the Nernst voltage. However, in contrast to the Nernst voltage it will keep its polarity if the magnetic field is reversed and can thus be eliminated.

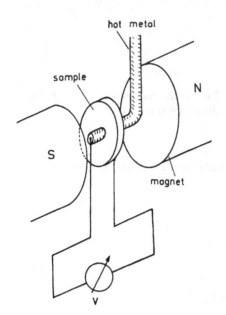

hot metal

sample

N

S

magnet

V

Fig. 4.24: Corbino thermopower arrangement. The voltage is measured between the hot metallic rod and the circumference of the semiconducting disk. N and S are the magnet poles

Of the magneto-thermoelectric effects, the easiest to calculate is the Corbino thermopower [4.62]. The experimental set-up is shown in Fig. 4.24. A semiconducting Corbino disk is mounted on a heated metal rod. A thermoelectric voltage is measured between the rod and the metal contact along the circumference of the disk. For small magnetic field intensitites, the change in thermoelectric power $\Delta(d\Theta/dT)$ is proportional to B^2.

The saturation value of the transverse thermoelectric power with increasing magnetic field intensity may be used for determining the non-parabolicity of the energy bands [4.63]. For the Nernst coefficient we find from (4.10.6) together with (4.11.1)

$$Q_N = \frac{k_B}{|e|}\left[\left(\frac{\sigma_n}{\sigma}\mu_{Hn} + \frac{\sigma_p}{\sigma}\mu_{Hp}\right)r + \frac{\sigma_n\sigma_p}{\sigma^2}(\mu_{Hn} + \mu_{Hp})(2r + 5 + \frac{\varepsilon_G}{k_BT})\right]$$

(4.11.8)

For an extrinsic semiconductor where either σ_n or σ_p vanishes, the ambipolar term containing the factor $\sigma_n\sigma_p$ also vanishes. The sign of Q_N does *not* depend on the carrier type but on the sign of r, in contrast to the Hall effect:

$$Q_N = \frac{k_B}{|e|}\mu_H r \approx \mu_H r \times 86\,\mu V/K \ .$$

(4.11.9)

In the intrinsic region of temperature, the ambipolar term dominates Q_N if electron and hole mobilities are about equal. Usually at temperatures somewhat below this temperature range, lattice scattering dominates with $r < 0$ in the

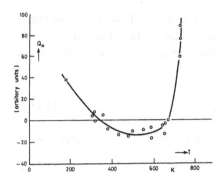

Fig. 4.25: Nernst coefficient for p-type GaAs with 1.0×10^{17} holes cm^{-3} as a function of temperature (after [4.64])

extrinsic range. Hence, a change of sign of Q_N can be expected at the transition from extrinsic to intrinsic behavior. At very low temperatures, ionized impurity scattering dominates in semiconductors with not too large a dielectric constant. There is another sign reversal at the transition from ionized impurity scattering ($r = +3/2$) to lattice scattering ($r < 0$). This behavior is shown in Fig. 4.25 where experimental results of Q_N obtained on highly doped p-type GaAs are plotted versus temperature. In the temperature range from 300 to 700 K, lattice scattering is dominant while below this range, there is ionized impurity scattering and above this range the material becomes intrinsic.

The Nernst voltage is usually of the order of μV since r in (4.11.9) is of the order of unity and $\mu_H B \ll 1$. For $dT/dx = 10$ K/cm and a sample width of 1 mm, the above order of magnitude for V_y is obtained.

The different behavior of the Hall and Nernst effects discussed above can be understood qualitatively by considering the drift motion of the carriers. In an electric field, electrons and holes drift in opposite directions while in a temperature gradient, they both drift in the same direction. Both types of carriers are deflected by a magnetic field to the same side of the sample in the first case (Hall effect) and to opposite sides in the second case (Nernst effect). In an extrinsic semiconductor due to the opposite charges of the carriers, the direction of the Hall field depends on the charge of the carrier while the direction of the Nernst field does not. The sensitivity of the Nernst effect on the scattering mechanism is just as easy to understand. Carriers drifting from the hot to the cold end of the sample have to return after cooling off in order to maintain charge neutrality. Hence, there are two opposite flows of the same type of carrier, one of the *hot* carriers and one of the *cool* carriers. By a magnetic field, hot carriers are deflected from their drift direction to one side of the sample while cool carriers are to the other side. Scattering decreases the influence of the magnetic field. For $\tau_m \propto \varepsilon'$ and $r > 0$, the hot carriers are less often scattered than cool carriers and the direction of the Nernst field is determined by the Lorentz force acting upon them. For $r < 0$, the same is true for the cool carriers instead of the hot carriers. Therefore, the direction of the Nernst field depends on the sign of r.

In an intrinsic semiconductor with equal mobilities of both types of carriers, there is no *return flow* of carriers since at the cold sample end, electrons and holes recombine and release an energy of magnitude ε_G. It is this energy rather than the scattering processes which determines the Nernst effect in this case.

The Righi-Leduc effect is calculated assuming $j_x = j_y = w_y = 0$ where for a weak magnetic field

$$S_{RL} = -s_{RL}\,\mu_H \left(\frac{k_B}{e}\right)^2 \frac{T\sigma}{\kappa} \ , \tag{4.11.10}$$

where s_{RL} is the *Righi-Leduc factor* which for the case $\tau_m \propto \varepsilon'$ is given by

$$s_{RL} = r(r+2) + 5/2 \tag{4.11.11}$$

with values of 1.75, 2.5, and 7.75 for $r = -1/2, 0$ and $+3/2$, respectively; κ is the total thermal conductivity. If the lattice contribution to κ were negligible, the factor $(k_B/e)^2 T\sigma/\kappa$ would be the inverse Lorentz constant, given by $1/(r+\frac{5}{2})$, and S_{RL} would be of the same order of magnitude as the Hall mobility μ_H. However, in most semiconductors the carrier contribution to κ is many orders of magnitude smaller than κ itself and S_{RL} is smaller than μ_H by this ratio. Assuming a ratio of, e.g., 10^{-5}, a value of 2×10^{-2} for the product $\mu_H B_z$ and a temperature gradient $\partial T/\partial x = 10$ K/cm, we find $\partial T/\partial y = 2 \times 10^{-6}$ K/cm. Considering a sample width of 1 mm and a thermoelectric power of 0.5 mV/K, a voltage of 10^{-10} V would have to be measured in order to observe the Righi-Leduc effect. Since this voltage is additive to the Nernst voltage which is many orders of magnitude larger, it could hardly be detected. Only in semiconductors with a large figure of merit Z, given by (4.10.25), is the Righi-Leduc effect important.

The minus sign in (4.11.10) characterizes an n-type semiconductor while a plus sign would be typical for a p-type semiconductor since μ_H contains the first power of the electronic charge, while all other factors on the rhs of (4.11.10) either contain e^2 or are independent of e.

Now we consider the Corbino thermopower mentioned above. In Sect. 4.5 we noticed that due to the radial symmetry of the Corbino arrangement along the x-axis, $E_y = 0$. The same is, of course, true for $\partial T/\partial y$ and $\partial \zeta/\partial y$. From $j_x = 0$ where j_x is given by

$$j_x = \sigma\left(E_x + \frac{1}{e}\frac{\partial \zeta}{\partial x}\right) + \sigma'\frac{1}{T}\frac{\partial T}{\partial x} \ , \tag{4.11.12}$$

we obtain with a little algebra for n-type semiconductors

$$\Delta\frac{d\Theta}{dT} = -2r\frac{k_B}{|e|}\mu_{Mn}^2 B^2 \approx -(86\,\mu\,\text{V/K})\frac{2r(3r+3/2)!}{[(r+3/2)!]^3}\frac{9\pi}{16}(\mu B)^2 \ , \tag{4.11.13}$$

where μ_{Mn}^2 is given by

$$\mu_{Mn}^2 = \mu^2(r_H^2 + T_M) = \mu_H^2 + T_M\,\mu^2 \tag{4.11.14}$$

and T_M is given by (4.4.4). For a p-type semiconductor the sign in (4.11.13) is $+$. The r-dependent fraction in (4.11.13) has values of $+1$ for $r = -\frac{1}{2}$ and -30 for $r = +\frac{3}{2}$: its sign and magnitude depend strongly on the type of scattering mechanism.

For the numerical example given in the discussion of the Nernst effect ($\mu = 10^4$ cm^2/Vs; $B = 0.1T$; $r = -\frac{1}{2}$), the value of $\Delta(d\Theta/dT)$ is $1.5\mu V/K$ which is of the same order of magnitude as the Nernst effect (of course, the geometry of the sample is different). As mentioned before, the sign of $\Delta(d\Theta/dT)$ does not change when B is reversed.

An interesting result is obtained for the Corbino thermopower in the limit of a strong magnetic field:

$$\Delta\frac{d\Theta}{dT} = -2r\frac{k_B}{|e|} = -2r \times 86\mu V/K \ . \tag{4.11.15}$$

For a p-type semiconductor the sign is $+$. The saturation value of the Corbino thermopower allows a direct determination of the exponent r for an assumed energy dependence of the momentum relaxation time given by $\tau_m \propto \varepsilon'$. If this assumption is not valid, we obtain instead of (4.11.15)

$$\Delta\frac{d\Theta}{dT} = -\frac{k_B}{|e|} \left(\frac{\langle \tau_m \varepsilon/k_B T \rangle}{\langle \tau_m \rangle} - \frac{\langle \tau_m^{-1} \varepsilon/k_B T \rangle}{\langle \tau_m^{-1} \rangle} \right) \ , \tag{4.11.16}$$

where the integrals in the averages over the distribution function have to be evaluated numerically. Equations (4.11.15, 16) are valid only if the condition $\mu_{Mn}^2 B^2 \gg 1$ is fulfilled which for most semiconductors requires magnetic fields which can be obtained in pulsed form only, if at all. This poses a severe limit to the applicability of these equations. At low temperatures where mobilities may be high enough to meet the requirement, magnetic quantum effects may invaliditate the present classical treatment (Sect. 9.2).

4.12 Piezoresistance

The change of the electrical resistivity upon the application of an external uni-axial stress or hydrostatic pressure is called *piezoresistance*. Figure 4.26 shows the observed resistance of an n-type silicon sample as a function of hydrostatic pressure X transmitted by an electrically insulating liquid [4.66]. Up to $X = 20$ GPa (1 kbar = 0.1 GPa) there is a slight linear decrease of the resistance with X in the semilog plot which is followed by a drop of more than six orders of magnitude. This drop is due to a phase transition of the silicon lattice and will not be discussed here. We focus our attention on the slight initial drop where $\Delta \log R \propto \Delta\varrho/\varrho \propto X$. Later on we will introduce a tensor π_{ik} of which two coefficients π_{11} and π_{12} will suffice to describe the relationship between $\Delta\varrho/\varrho$ and X in the hydrostatic-pressure experiment:

$$\Delta\varrho/\varrho = -\Delta\sigma/\sigma = (\pi_{11} + 2\pi_{12}) X \ . \tag{4.12.1}$$

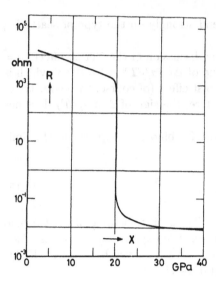

Fig. 4.26: Resistance of an n-type silicon sample as a function of hydrostatic pressure (after [4.65]). 1 GPa = 10 kbar

These coefficients and one other, π_{44}, are needed to describe the results of uniaxial-stress experiments with possibly different directions of stress X and current j in a cubic semiconductor. The Table 4.1 gives the combinations of the coefficients for some crystallographic directions of X and j where X is counted as positive for a tension and negative for a compression.

All three components of the tensor can be determined from three measurements in different directions.

The complete piezoresistance tensor π_{ijkl} is given by

$$\frac{\Delta\varrho_{ij}}{\varrho_0} = -\frac{\Delta\sigma_{ij}}{\sigma_0} = \sum_{k,l=1}^{3} \pi_{ijkl} X_{kl} \; , \tag{4.12.2}$$

where X_{kl} is the stress tensor and σ_{ij} is the conductivity tensor. It is well known from the theory of stress that X_{kl} is symmetrical and therefore has only six independent components which can be combined formally to give a vector in six dimensions [4.66]:

$$X_1 = X_{11}; \quad X_2 = X_{22}; \quad X_3 = X_{33};$$
$$X_4 = X_{23}; \quad X_5 = X_{31}; \quad X_6 = X_{12}$$

The same is true for the ϱ_{ij} tensor (assuming no magnetic field). The piezoresistance tensor π_{ij} is then in six *dimensions*:

$$\Delta\varrho_i/\varrho_0 = \sum_{j=1}^{6} \pi_{ij} X_j; \quad i = 1, 2, \ldots, 6 \; . \tag{4.12.3}$$

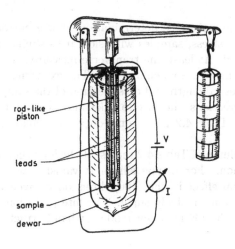

rod-like piston

leads

sample

dewar

Fig. 4.27: Arrangement for low-temperature measurements of the piezoresistance

The *tensor of elastic constants* c_{ij} in six dimensions is given by

$$X_i = \sum_{j=1}^{6} c_{ij}\, e_j \ . \tag{4.12.4}$$

Table 4.1. Components of the piezoresistance tensor

	X	j	$\Delta\varrho/(\varrho X)$
longit.	$\langle 100 \rangle$	$\langle 100 \rangle$	π_{11}
	$\langle 110 \rangle$	$\langle 110 \rangle$	$(\pi_{11} + \pi_{12} + \pi_{44})/2$
	$\langle 111 \rangle$	$\langle 111 \rangle$	$(\pi_{11} + 2\pi_{12} + 2\pi_{44})/3$
transv.	$\langle 110 \rangle$	$\langle 010 \rangle$	π_{12}
	$\langle 110 \rangle$	$\langle 1\bar{1}\,0 \rangle$	$(\pi_{11} + \pi_{12} - \pi_{44})/2$

Here the vector e_j contains the six components of the symmetrical *deformation tensor* which will be defined in Sect. 7.1. In (4.12.3) we substitute for X_j its value given by (4.12.4):

$$\Delta\varrho_i/\varrho_0 = \sum_{j,k=1}^{6} \pi_{ij}\, c_{jk}\, e_k = \sum_{k=1}^{6} m_{ik}\, e_k \ , \tag{4.12.5}$$

where we have introduced a tensor m_{ik} of *elastoresistance*. For more details of these tensors see [4.67-70]. An experimental set-up for measurements of the piezoresistance is shown in Fig. 4.27. Assume the weight of a mass of e.g. three kg acting on an n-type silicon sample of cross section 2mm^2 via a lever with an arm ratio of 5:1. This results in a stress of 75 MPa and a relative resistance change $\Delta\varrho/\varrho_0$ of 7.5% at room temperature (since $\pi_{11} = -102.2 \times 10^{-11}\mathrm{Pa}^{-1}$ assuming $\Delta\varrho/\varrho_0 \propto X$ up to this large value of $\Delta\varrho/\varrho_0$). In germanium the linear relationship is valid up to a stress of about 10 MPa.

Potential probes are used for the resistance determination. For a current density perpendicular to the applied stress, samples with side arms similar to those for Hall measurements are used. At least one transverse measurement is required for a determination of all three components of π_{ik}. The experimental data have to be corrected for changes in length and cross section of the sample under stress. In making these corrections one has to take into account the anisotropy of the resistivity [4.71]. Table 4.2 gives some data on germanium and silicon.

The data in the second to last column of Table 4.2 represent the longitudinal piezoresistance in the $\langle 111 \rangle$ direction. For n-type silicon it vanishes within experimental error. The longitudinal effect in the $\langle 100 \rangle$ direction is given by π_{11} which for n-type germanium is comparatively small. The interpretation of these results requires knowledge of the many-valley model of band structure which will be discussed in Chap. 7.

Table 4.2 Piezoresistance $\Delta\varrho/(\varrho X)$ in units of 10^{-2} GPa at room temperature

	ϱ_0 [Ω cm]	π_{11}	π_{12}	π_{44}	$\dfrac{\pi_{11} + 2\pi_{12} + 2\pi_{44}}{3}$	$\dfrac{\pi_{11} + \pi_{12} + \pi_{44}}{2}$
n-Ge	16.6	-5.2	-5.5	-138.7	-96.9	-74.7
p-Ge	15.0	-10.6	5.0	98.6	65.5	41.5
n-Si	11.7	-102.2	53.7	-13.6	-0.7	-31.1
p-Si	7.8	6.6	-1.1	138.1	93.5	71.8

The temperature dependence of the piezoresistance of n- and p-type silicon is plotted in Fig. 4.28 [4.72]. In n-type silicon, the resistance change is proportional to $1/T$ over a large range of the abscissa. This can be explained by a *repopulation of valleys* in the many-valley model (Sect. 7.5) which yields a change in the *conductivity effective mass*. In addition *inter-valley scattering* which determines τ_{m} in some semiconductors is subject to change upon application of stress to the sample (see also Sect. 7.5).

According to (1.3.3), the carrier concentration in intrinsic semiconductors is proportional to $\exp(-\varepsilon_{\mathrm{G}}/2k_{\mathrm{B}}T)$ where the band-gap energy ε_{G} depends on the atomic distance and is changed by stress. This yields

$$\Delta\varrho/\varrho \propto (X/2\,k_{\mathrm{B}}T)d\varepsilon_{\mathrm{G}}/dX \ . \tag{4.12.6}$$

In this way a value for $d\varepsilon_{\mathrm{G}}/dX$ of 5 meV/10^8 Pa was determined from experimental data on intrinsic germanium at room temperature [4.73]. Equation (2.1.17) shows that the effective mass depends on the gap energy. Hence, stress also affects the mobility via the effective mass. This is observed with *direct* III-V compounds like, e.g., InSb, InAs, GaAs, and InP where the conduction band minimum and the valence band maximum are at the same k-value ($k = 0$). Figure 4.29 shows the piezoresistance of n-GaAs and n-InP. For example for n-GaAs the initial slope of the curve is 9.6%/GPa. Since the band gap energy is $\varepsilon_{\mathrm{G}} = 1.53$ eV and the compressibility $\kappa = \Delta V/(V\,X) = 1.38 \times 10^{-2}$/GPa, a value of 11 eV is found for the product $\varepsilon_{\mathrm{G}}\Delta\varrho/(\varrho X\kappa)$. For n-InSb the corresponding

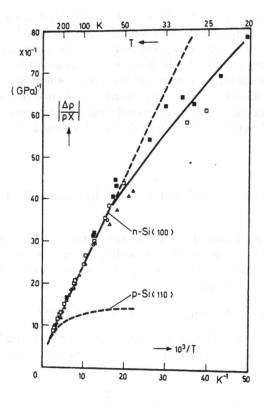

Fig. 4.28: Longitudinal piezoresistance for n- and p-type silicon as a function of the reciprocal temperature (after [4.72])

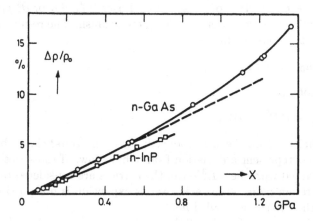

Fig. 4.29: Piezoresistance for n-type GaAs and n-type InP as a function of hydrostatic pressure (after [4.73])

values are 60%/GPa, ε_G =0.27 eV, $\kappa = 2.3 \times 10^{-2}$/GPa, $\varepsilon_G \Delta\varrho/(\varrho X\kappa) = 7$ eV. The value of this product is 9 eV for InAs and 8 eV for InP. The variation between 11 and 7 eV is small compared to that of the gap energy [4.73].

The mechanisms involved in the influence of stress on the resistance partly counterbalance each other, and the explanation of an observed stress dependence is quite often a difficult problem. We therefore will not consider here piezo-galvanomagnetic effects. The comparatively simple interpretation of a stress-induced shift of the optical absorption edge will be given in Sect. 11.2.

4.13 Hot Electrons and Energy Relaxation Time

In the approximate solution of the Boltzmann equation (4.1.16) we have considered only small field intensities $\boldsymbol{E}$. For qualitative considerations, this solution can be used for evaluating the term $\partial f/\partial v_z$ on the rhs of (4.1.15). If this iteration is continued, we obtain for f

$$f = f_0 - (\tau_\mathrm{m}\, e\, E_z/m)\partial f_0/\partial v_z + (\tau_\mathrm{m}\, e\, E_z/m)^2\, \partial^2 f_0/\partial v_z^2 + \cdots. \quad (4.13.1)$$

For not too large values of E_z where the drift velocity is much smaller than the thermal velocity ($\approx \sqrt{k_B T/m}$), the series expansion converges. The drift velocity

$$v_{\mathrm{dz}} = \int\limits_{-\infty}^{\infty} v_z\left(-\frac{e}{m}\tau_\mathrm{m}\frac{\partial f}{\partial v_z}E_z\right) d^3v \bigg/ \int\limits_{-\infty}^{\infty} f_0 d^3v \quad (4.13.2)$$

contains only odd powers of E_z if τ_m is independent of the direction of v; integrals where an even function of v_z is averaged over $\partial f_0/\partial v_z$ or $\partial^3 f_0/\partial v_z^3$, etc., or an odd function averaged over $\partial^2 f_0/\partial v_z^2$, etc., vanish. The result for the drift velocity $v_{\mathrm{dz}} = \mu\, E_z$ has the form

$$\mu\, E_z = \mu_0(E_z + \beta\, E_z^3 + \cdots) \quad (4.13.3)$$

and for the mobility

$$\mu = \mu_0(1 + \beta\, E_z^2 + \cdots)\,, \quad (4.13.4)$$

where a coefficient β has been introduced; μ_0 is the zero-field mobility. The terms $\beta E_z^3 + \cdots$ represent a deviation from Ohm's law. If the series expansion may be terminated with the $\beta\, E_z^3$-term, the carriers are considered to be *warm*. If one has to retain more terms or if a series expansion is not possible such as for $\mu \propto E_z^{-1}$ the carriers are called *hot*.

It is only if a current flows that the carrier temperature rises. In a built-in field, which is due to a doping inhomogeneity (Sect. 5.3), the carriers remain in thermal equilibrium with the lattice if there is no current as a result of an applied voltage.

For a quantitative calculation of, e.g., the coefficient β, the scattering theory which will be treated in Chap. 6 has to be applied. Often for hot carriers, and sometimes also for warm carriers, a Maxwell–Boltzmann distribution with an *electron temperature* T_e which is larger than the lattice temperature is assumed in order to simplify the calculations:

$$f \propto \exp(-\varepsilon/k_B T_e) \ . \tag{4.13.5}$$

This distribution function has been justified by considering the energy gain per unit time of carriers from the field; it is the scalar product of the force $e\,E$ and the drift velocity $\mu\,E$ which is $\mu\,e\,E^2$. In equilibrium this is equal to the energy loss by collisions:

$$\mu\,e\,E^2 = -\langle \partial\varepsilon/\partial t \rangle_{\mathrm{coll}} \ . \tag{4.13.6}$$

If the energy gain of a particular carrier in the field direction is rapidly distributed in all other directions due to carrier-carrier interaction, the distribution given by (4.13.5) is a good approximation. Its main advantage is that integrals which contain this distribution function can be solved analytically. Since most of the experimentally observable quantities change only quantitatively if a more realistic distribution is used in the calculation, we will restrict analytical warm and hot-carrier calculations to the Maxwell–Boltzmann distribution (4.13.5).

It is useful to introduce an energy relaxation time τ_ε. For the equilibrium case it is defined by

$$\mu\,e\,E^2 = \frac{3}{2}\,k_B (T_e - T)/\tau_\varepsilon = [\langle \varepsilon(T_e) \rangle - \langle \varepsilon(T) \rangle]/\tau_\varepsilon \ , \tag{4.13.7}$$

where $\langle \varepsilon(T_e) \rangle = 3\,k_B\,T_e/2$ is valid for a nondegenerate carrier gas.[12]   A schematic representation of this equation is given in Fig. 4.30 where the energy flow is demonstrated by arrows and the hatched areas indicate the energies contained in the carrier gas and in the crystal lattice.

In general, the momentum relaxation time τ_m depends on the carrier energy ε; the mobility $\mu = (e/m)\langle \tau_m \rangle$ is then a function of the electron temperature T_e; this function we will denote as $g(T_e)$. The ohmic mobility μ_0 is given by $g(T)$. A series expansion of the ratio μ/μ_0 yields

$$\mu/\mu_0 = 1 + (T_e - T)g'/g + \cdots \ , \tag{4.13.8}$$

where g' stands for $\partial g/\partial T_e$ at $T_e = T$ and g stands for $g(T)$. For the case of warm carriers we terminate the expansion after the linear term and replace $T_e - T$ in (4.13.7) by $(g/g')(\mu - \mu_0)/\mu_0$:

$$\mu\,e\,E^2 = \frac{3}{2}\,k_B (g/g')\,\beta\,E^2/\tau_\varepsilon \ , \tag{4.13.9}$$

where $(\mu - \mu_0)/\mu_0$ for warm carriers has been replaced by $\beta\,E^2$ according to (4.13.3). In this approximation we may replace μ by μ_0 for calculating the energy relaxation time.

---

[12]For degeneracy see (6.5.29, 32)

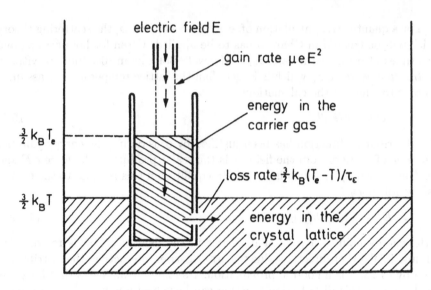

Fig. 4.30: Schematic representation of the carrier energy balance indicating the energy gain from an applied electric field E and the energy loss to the crystal lattice. At equilibrium for $E \neq 0$ the mean carrier energy is $[(3/2)k_B T_e]$. Notice the small specific heat of the carriers and the large specific heat of the crystal lattice

$$\tau_\varepsilon = \frac{3}{2}\frac{k_B T}{e}\beta \Big/ \left\{ \mu_0 \left(\frac{d\ln g}{d\ln T_e}\right)_{T_e=T}\right\} = \frac{T}{7740\,\mathrm{K/V}}\frac{\beta}{\mu_0}\Big/\left(\frac{d\ln g}{d\ln T_e}\right)_{T_e=T}$$

$$(4.13.10)$$

For the case of $\tau_m \propto \varepsilon^r$, the factor $(d\ln g/d\ln T_e)^{-1}$ becomes $1/r$ which is of order unity. Assuming at $T = 77\,\mathrm{K}$ (liquid nitrogen temperature) a mobility μ_0 of $10^4\,\mathrm{cm^2/Vs}$ and a value of $10^{-4}\,\mathrm{cm^2/V^2}$ for $|\beta|$, we find τ_ε to be of the order of magnitude 10^{-10} s. This is the typical value for the time constant of a relaxation of the deviations from Ohm's law; it can be measured at a frequency of the order of magnitude $1/2\pi\tau_\varepsilon \approx 1$ GHz which is in the microwave range of frequencies.

For a decrease of τ_m with ε, i.e., $r < 0$, the factor $d\ln g/d\ln T_e$ is negative. Since the time constant τ_ε has to be a positive quantity, the coefficient $\beta < 0$. Therefore, in this case the mobility μ decreases with increasing electric field intensity E. Similarly, for an increase of τ_m with ε, i.e., $r > 0$, a positive sign of β can be deduced. Hence, for the limiting case of τ_m independent of ε, one would expect β to vanish even though the expansion (4.13.8) is not feasible. However, the present treatment is of a qualitative nature only and the scattering theory (Chap. 6) has to be applied for a quantitative evaluation of β.

For a degenerate semiconductor we have to replace (4.13.7) by

$$\mu\,e\,E^2 = [\langle\varepsilon(T_e)\rangle - \langle\varepsilon(T)\rangle]/\tau_\varepsilon \ ,$$

$$(4.13.11)$$

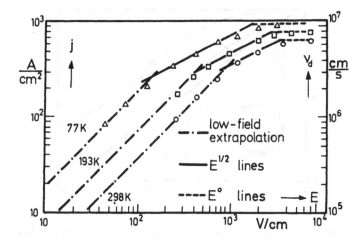

Fig. 4.31: Current density in n-type germanium as a function of electric field intensity at various lattice temperatures (after [4.75])

where $\langle \varepsilon(T_e) \rangle$ is given by

$$\langle \varepsilon(T_e) \rangle = \frac{3}{2} k_B T_e F_{3/2}(\zeta_n/k_B T_e) N_c(T_e)/n \ . \tag{4.13.12}$$

In this relation the function $F_{3/2}(x)$ is the Fermi-Dirac integral for $j = 3/2$ (3.1.11), and $n_c(T)$ is given by (3.1.10). For a quantitative calculation of the energy relaxation time τ_ε, the scattering theory to be discussed in Chap. 6 has to be applied.

Experimental data on deviations from Ohm's law and on energy relaxation times in extrinsic semiconductors are quite often obtained in field regions where the carrier density does not change with field intensity. The current voltage characteristic then reflects the variation of the drift velocity with field intensity. First measurements on n-type Ge at 77, 193, and 298 K were made by *Ryder* and *Shockley* [4.74, 75] and are shown in Fig. 4.31 in a log-log plot where Ohm's law is represented by straight lines rising at an angle of 45° . Depending on the lattice temperature, deviations from these lines are significant at field strengths between 10^2 and 10^3 V/cm. Current saturation occurs above about 2 kV/cm at drift velocities of about 10^7 cm/s and is nearly independent of the lattice temperature. A convenient way of showing the deviations from Ohm's law is a plot of the conductivity ratio σ/σ_0 where σ_0 is the zero-field conductivity. The full curves in Fig. 4.32 are valid for a field applied in a $\langle 100 \rangle$ direction while the dashed curves are for a $\langle 111 \rangle$ direction. The inset in Fig. 4.32 shows the shape of the n-type Ge sample [4.76]. The positive contact is large in order to prevent minority carrier injection (Sect. 5.1). The data are corrected for the small voltage drop across the large-area part of the sample. The homogeneity of the field in the filamentary part of the sample has been questioned [4.77] even though the material was homogeneously doped.

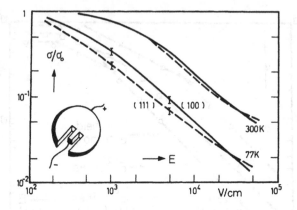

Fig. 4.32: Conductivity of n-type germanium as a function of electric field intensity for two crystallographic directions of the applied field and for two temperatures. The inset shows the sample shape (after [4.76])

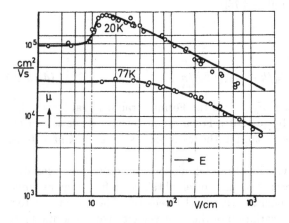

Fig. 4.33: Mobility of n-type germanium as a function of electric field intensity at 20 and 77 K (after [4.75])

At low temperatures (20 K in n-type Ge), positive deviations from the zero-field mobility may occur as indicated in Fig. 4.33 [4.75]. These are considered to be due to ionized impurity scattering where $r > 0$ (Sect. 6.6).

It may be interesting to look for deviations from Ohm's law in degenerate semiconductors or even metals. As will be seen in Sect. 6.6, particularly Fig. 6.8, deviations from Ohm's law decrease as the concentration of donors in n-type Ge increases due to ionized impurity scattering. In Fig. 4.34, the current voltage characteristics for various metals at room temperature have been plotted up to fields of several kV/cm where the drift velocity is of the order of the sound velocity [4.78]. The data have been taken by applying sub-nanosecond pulse techniques. Joule heating was still negligible. Ohm's law is perfectly valid in metals, i.e., the mobility μ is independent of the applied field E. In fact, a calculation of $\langle \tau_m \rangle$ from (4.1.24) for $r = -1/2$, $\eta_n = \zeta_n/k_B T_e$ and

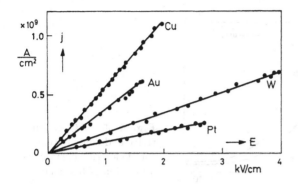

Fig. 4.34: Current voltage characteristics of metals at room temperature (after [4.78])

$\tau_0 = l_{ac}/\sqrt{2k_B\, T_e/m}$ yields for $\eta_n \gg 1$, where $F_j(\eta_n) \approx \eta^{j+1}/(j+1)$, an independence of $\langle\tau_m\rangle$ from T_e. The resistivity is proportional to $1/l_{ac}$ which is proportional to the lattice temperature T as is well known for metals at not too low temperatures. The situation may, however, be different for *impure metals* where scattering processes different from acoustic-phonon scattering, such as impurity band conduction, may prevail.

A technique of measuring the mobility and particularly that at high field intensities in high-resistance material is the time-of-flight experiment: A pulsed beam of ≈ 50 keV electrons is focussed on one side of a trap-free conductor and the time of arrival of the carriers at the opposite side of the conductor is observed [4.79]. For poor-quality conductors, the electron beam is microwave-frequency modulated and the phase and amplitude of the microwave output current is observed [4.80].

4.14 High-Frequency Conductivity

The Boltzmann equation with df/dt given by (4.1.13) contains a term $\partial f/\partial t$ which in the case of a sinusoidal time variation of the electric field intensity E depends on time t with the same frequency. Since $\partial f/\partial t$ is linear in E and $\partial f_0/\partial t = 0$, we have for $E \propto \exp(i\omega t)$,

$$\partial f/\partial t = i\omega(f - f_0) \ . \tag{4.14.1}$$

Therefore,

$$\partial f/\partial t + (\partial f/\partial v_x)\, e\, E_x/m + \cdots = -(f - f_0)/\tau_m \tag{4.14.2}$$

may be written in the form

$$[(\partial f/\partial v_x)\, e\, E_x/m + \cdots]/(1 + i\omega\, \tau_m) = -(f - f_0)/\tau_m \ . \tag{4.14.3}$$

Hence, the dc formulas developed so far may also be applied to the present case if we replace E by $E/(1 + i\omega\, \tau_m) = [E/(1 + \omega^2\tau_m^2)](1 - i\omega\, \tau_m)$. In discussions

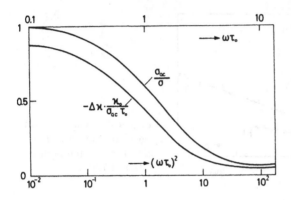

Fig. 4.35: AC conductivity and contribution to the dielectric constant as a function of frequency

of Maxwell's equations the imaginary part of this expression is found to be a contribution to the dielectric constant which is essential for the optical properties of semiconductors. The ac mobility μ_{ac} is then given by

$$\mu_{ac} = (e/m)\langle \tau_m/(1 + \omega^2 \tau_m^2)\rangle \ . \tag{4.14.4}$$

For the special case of acoustic deformation potential scattering where $\tau_m = \tau_0(\varepsilon/k_B T)^{-1/2}$, the averaging procedure can be performed analytically. For this purpose we introduce $q = \omega \tau_0$; $y = q^2 + \varepsilon/k_B T$ and obtain

$$\frac{\mu_{ac}}{\mu} = \int\limits_{q^2}^{\infty} \frac{(y - q^2)^2 \exp(-y + q^2)}{y} \, dy$$

$$= \exp(q^2) \int\limits_{q^2}^{\infty} y \exp(-y) \, dy - 2q^2 \int\limits_{q^2}^{\infty} \exp(-y) \, dy + q^4 \int\limits_{q^2}^{\infty} y^{-1} \exp(-y) \, dy$$

$$= 1 - (\omega \tau_0)^2 - (\omega \tau_0)^4 \exp(\omega^2 \tau_0^2) \, \mathrm{Ei}(-\omega^2 \tau_0^2) \ , \tag{4.14.5}$$

where $\mathrm{Ei}(x)$ is the exponential integral function [6.27].

For $\omega \tau_0 = 1$, the above is equal to 0.596 while without averaging, we get $(1 + \omega^2 \tau_0^2)^{-1} = 0.500$. For $\omega \tau_0 = 4$ the result (4.14.5) is by about a factor of 2 larger than $(1 + \omega^2 \tau_0^2)^{-1}$. The ratio $\sigma_{ac}/\sigma \propto \mu_{ac}/\mu$ as a function of $\omega \tau_0$ is shown in Fig.4.35; also shown is the contribution to the dielectric constant $-\Delta\kappa\kappa_0$ in units of $\sigma_{ac} \tau_0$ which has been calculated numerically. It seems worth mentioning that the integrals involved in the averaging procedures (4.14.4) and a similar equation for $\Delta\kappa$ are the same as in Corbino resistance (4.5.5) except that the cyclotron frequency $\omega_c = e B/m$ is replaced by ω.

The momentum relaxation time is usually of the order of magnitude 10^{-12} to 10^{-13} s. The product $\omega \tau_m$ equals unity at terahertz frequencies. The maximum sample dimension is usually about one wavelength or less, and it is therefore necessary to make measurements by waveguide- optical techniques. In high-frequency and microwave measurements both the real and imaginary parts of the conductivity are usually measured simultaneously. The imaginary part, which determines the dielectric constant, can be observed with high accuracy by varying the frequency of the incident radiation and observing interference fringes in the radiation transmitted through and reflected from the semiconductor. For silicon, the relative dielectric constant at 26.5 - 40 GHz was found to be 11.7, and for gallium arsenide, 12.8. For the experimental arrangement and the evaluation of the observations see Ref. [4.81]. It should be noted that the frequency-dependent conductivity and the frequency-dependent dielectric constant are interrelated by the Kramers–Kronig relations [11.6]. These relations are usually important only in the range of the fundamental optical absorption (Chap. 11) but they may be useful even in the microwave range if there is a strongly nonlinear current voltage characteristic, as for charge density wave conductors [4.82]. (The same is certainly true for semiconductor superlattices, see Sects. 9.1 and 13.3.)

Let us consider the $\varepsilon(k)$ diagram Fig. 2.4 for an electron in a one-dimensional lattice and investigate the hot-electron high-frequency conductivity for this case [4.83]. In addition to the ac field $E_1 \cos(\omega t)$ we apply a dc field E_0 to the sample.

$$E(t) = E_0 + E_1 \cos(\omega t) \tag{4.14.6}$$

As discussed after (2.1.16), the force F on the particle of charge e is

$$F = d(\hbar k)/dt = eE_0 + eE_1 \cos(\omega t) \tag{4.14.7}$$

Integration over time t, multiplication with the lattice constant a, and introducing for brevity $\omega_B = eE_0 a/\hbar$ and $\mu = eE_1 a/\hbar\omega$, yield

$$k a = \omega_B(t - t_0) + \mu[\sin(\omega t) - \sin(\omega t_0)] \tag{4.14.8}$$

where t_0 is the time when $E(t)$ is applied to the sample. The wavy $\varepsilon(k)$ structure of an energy band can be written

$$\varepsilon = const. - (\Delta/2)\cos(ka) \tag{4.14.9}$$

and, according to text after (2.1.16), the velocity v which is the group velocity of an electron wave packet, is $d\varepsilon/d(\hbar k)$, and according to (4.14.9) this is

$$v = \frac{1}{\hbar}\frac{d\varepsilon}{dk} = \frac{\Delta}{2}\frac{a}{\hbar}\sin(ka) = v_0 \sin(ka) \tag{4.14.10}$$

by introducing $v_0 = \Delta a/2\hbar$ for brevity where the argument of the sin function is given by (4.14.8). The electron wave packet performs a phase-modulated *Bloch oscillation*. For electron transport a damping of the oscillation is required which is taken care of by a relaxation time τ in the following averaging procedure [4.84]:

$$v_d^{HF} = \int_0^{2\pi/\omega} \frac{dt}{2\pi/\omega} \int_{-\infty}^t \frac{dt_0}{\tau} \exp\left(-\frac{t - t_0}{\tau}\right)$$

$$\times v_0 \sin\left[\omega_B(t - t_0) + \mu\left(\sin(\omega t) - \sin(\omega t_0)\right)\right] \tag{4.14.11}$$

Introducing $\Omega_B = \omega_B \tau$; $\Omega = \omega \tau$; $x = t/\tau$; $y = t_0/\tau$ and writing $\sin(z_1 + z_2) = $ Im $(\exp[i\,z_1]\exp[i\,z_2])$ where i is the imaginary unit and z_1 and z_2 are arbitrary we obtain

$$\frac{v_d^{HF}}{v_0} = \frac{\Omega}{2\pi} \int_0^{2\pi/\Omega} dx \int_{-\infty}^t dy \, \exp(y) \exp(-x)$$

$$\times \mathrm{Im}[\exp(i\Omega_B x)\exp(i\mu\sin(\Omega\,x))\exp(-i\Omega_B y)\exp(-i\mu\sin(\Omega y))] \quad (4.14.12)$$

By performing the integration over y and simplifying the integral over x by introducing Bessel functions $J_n(\mu)$:

$$\exp(i\,\mu\sin(\Omega\,x)) = \sum_{n=-\infty}^{\infty} J_n(\mu) \, \exp(i\,n\Omega\,x) \quad (4.14.13)$$

where $\mu = eE_1 a/\hbar\omega$ and taking into account that

$$\frac{\Omega}{2\pi} \int_0^{2\pi/\Omega} dx \, \exp[i(n - m)\Omega\,x] \quad (4.14.14)$$

is equal to 1 for $n = m$ and otherwise zero, the final result is

$$\frac{v_d^{HF}}{v_0} = \sum_{n=-\infty}^{\infty} J_n^2\left(\frac{eE_1 a}{\hbar\omega}\right) \frac{\omega_B\tau + n\omega\tau}{1 + (\omega_B\tau + n\omega\,\tau)^2} \quad (4.14.15)$$

Taking for a first approximation only the largest one of the Bessel functions for small values of the argument into account which is J_0 it is accompanied with a factor proportional to $E_0/[1 + (e\,E_0 a\,\tau/\hbar)^2]$ while the argument of the Bessel function is independent of E_0. With increasing dc electric field, the drift velocity rises to a maximum and then falls off with $1/E_0$. The calculation neglects elastic scattering processes. It may be applied to low-dimensional conductors. Notice that it does not involve the energy relaxation time which in the classical Boltzmann theory determines the behavior of hot electrons. The energy relaxation time is very often determined by an *Harmonic Mixing* experiment. Let us therefore see how the theory developed above can be applied to this case. The electric field is assumed to consist of two ac components including a phase shift ϕ and no dc component

$$E(t) = E_1 \cos(\omega t + \phi) + E_2 \cos(n\omega t); \quad n = 2, 3, \ldots \quad (4.14.16)$$

Let us assume a nonlinear current - field characteristic with arbitrary coefficients $a, b, c, \ldots$.

$$j = a\,E + b\,E^2 + c\,E^3 + d\,E^4 + e\,E^5 + \ldots \quad (4.14.17)$$

Of course, the ohmic term $a\,E$ will not contribute to mixing. Assuming for simplification $n = 2$ and in (4.14.16) $E_2 = E_1$ a dc current component is obtained:

$$j_{dc} = (\frac{3}{4}\,c\,E_1^3 + \frac{25}{8}\,e\,E_1^5 + \ldots)\cos(2\phi) \quad (4.14.18)$$

The ϕ dependence serves to identify it as harmonic mixing while rectification is independent of ϕ. Since odd values of n in (4.14.16) do not produce a dc current we consider now $n = 4$:

$$j_{dc} = (\frac{1}{16} eE_1^5 + \ldots) \cos(4\phi) \qquad (4.14.19)$$

By observing j_{dc} for various values of ϕ it is of course possible to distinguish the contributions from rectification. The largest one as a rule comes from $n = 2$. For this case a similar calculation as done above for the combination of a dc field and an ac field yields

$$\frac{v_{dc}}{v_0} = \sum_{m,n=-\infty}^{\infty} \left[J_m(\mu) J_{m+2}(\mu) J_n(\nu) J_{n+1}(\nu) \frac{(m+2+2n)\omega\tau \cos(2\phi)}{1 + [(m+2+2n)\omega\tau]^2} \right]$$

$$(4.14.20)$$

where $\nu = eE_2a/2\hbar\omega$ and $\mu = eE_1a/\hbar\omega$. If both ν and μ are $\ll 1$, (4.14.20) is simplified to

$$\frac{v_{dc}}{v_0} = \frac{\mu^2 \nu}{4} \left(\frac{1}{1 + 4\omega^2\tau^2} - \frac{1}{1 + \omega^2\tau^2} \right) \omega\tau \cos(2\phi) \qquad (4.14.21)$$

The largest value of the average drift velocity v_{dc} is obtained for $\nu = \mu$. In this case $E_2 = 2E_1$ is required, and the dc current increases with the third power of E_1 for small field strengths and goes through a maximum for larger values, in agreement with observations in e.g. n-type germanium at microwave frequencies. These effects applied to quantum structures (Sect. 14.5) are known as *photon assisted tunneling* (PAT). J.R. Tucker treated them under the aspect of *tunnel junction mixing*[4.85]. Data reproduced in Fig. 6.19 have been obtained experimentally by the mixing method and evaluated in terms of the energy relaxation time τ_ε [4.83].

4.15 Noise

The output of a device such as a resistor or a p-n structure with zero signal input is called *noise* [4.86, 87]. Even in the absence of current flow, the random thermal motion of carriers in a resistor produces noise across the terminals of the resistor which is known as *Johnson noise*; its power P is given by the Nyquist formula ([4.88] and [Ref. 4.56, Chap. 85])

$$P = \langle \Delta V^2 \rangle \mathrm{Re}\{1/Z\} = 4k_BT\Delta f , \qquad (4.15.1)$$

where ΔV is the noise voltage, Z is the impedance and Δf is the band width of the detector used for the noise measurement. For a band width of 100 kHz, a resistor of impedance $10^4 \Omega$ at room temperature produces a rms noise voltage of 4 μ V.

Besides fluctuations in the velocity of carriers, v_i, there are also fluctuations in the number of carriers n due to the detailed balance between generation and recombination of electrons and holes. If for simplicity we consider only one dimension, the current I is given by

$$I = e \sum_{i=1}^{n} v_i = \bar{n} e \bar{v} + e \sum_{i=1}^{\bar{n}} (v_i - \bar{v}) + e \bar{v} (n - \bar{n}) , \qquad (4.15.2)$$

where a bar indicates the average value of the quantity. We denote $v_i - \bar{v}$ as Δv_i and n $- \bar{n}$ as Δn and obtain for the mean square of the fluctuation of the current, $\Delta I = I - \bar{n} e \bar{v}$,

$$\Delta \bar{I}^2 = \left(e \sum_{i=1}^{\bar{n}} \Delta v_i \right)^2 + e^2 \bar{v}^2 \bar{\Delta n^2} , \qquad (4.15.3)$$

where the first term on the rhs represents the Johnson noise and the second represents the *shot noise*; the latter depends on the average current $\bar{n} e \bar{v}$. There are many more noise sources such as the crystal surface [4.89] (surface recombination, surface states, see Sect. 5.8, Chap. 14) and the metallic contacts.

For a p-n junction or a transistor, a *noise figure* F is defined as the ratio of the total noise power and P given by (4.15.1). Usually 10 $\log_{10} F$ is quoted rather than F itself which is indicated by the addition of the symbol dB (decibel) to the number. For a low-noise transistor a noise figure of 2-3 dB is common.

If the noise power is independent of frequency (within a certain range of frequencies), the noise is called *white*. The noise power of an etched Ge filament decreases with frequency f roughly as $1/f$ (called $1/f$ *noise*) in the range of 10 to 10^4 Hz [4.90, 91]. This can be explained by generation and recombination processes at localized energy levels in the band gap which are distributed over a considerable energy range [4.92].

Experimental investigations about the noise of hot carriers have been made in the range of low-temperature impact ionization (Sect. 10.1) by *Lautz* and *Pilkuhn* [4.93] and at liquid-nitrogen temperature by *Erlbach* and *Gunn* [4.94], and *Bryant* [4.95]. While the breakdown current increases by two orders of magnitude, $(\Delta I)^2/\Delta f$ increases by a factor 10^8 up to a sharp maximum [4.93]. The idea underlying the Erlbach-Gunn experiment is to have more direct access to the electron temperature T_e by taking the noise power P as $4 k_B T_e \Delta f$. A theoretical investigation by *Price* [4.96] shows, however, that this has to be done with some caution. Calculations for the case of acoustic-phonon scattering yield a ratio of noise temperature to electron temperature which increases with field strength from unity up to about 1.14 in the high-field limit. Erlbach and Gunn measured noise temperatures at 420 MHz in n-type Ge at 77 K up to 3600 K at 1.4 kV/cm; it depends strongly on orientation. The calculations of Price, however, may not be applicable because of the strong influence of optical phonon scattering. Besides, Price has indicated a contribution to noise from inter-valley scattering in a many-valley semiconductor [4.96].

Problems

4.1. Plot the Maxwell – Boltzmann distribution of free carriers $f(v) \propto \exp(-m v^2/2 k_B T)$ for 77 K and 300 K for velocities v up to 3×10^7 cm/s. What is the mean free width half maximum of the distribution? At what velocity v has (a) this distribution, (b) the Fermi–Dirac distribution $f(v) = \{\exp[(\varepsilon - \zeta)/k_B T] + 1\}^{-1}$ for $\zeta=0.1$ eV and T=300 K dropped to 1% of its value at $v=0$? ($\varepsilon = m v^2/2$). Assume $m = m_0$.

4.2. Intrinsic Ge at room temperature has a resistivity of 47 Ωcm. The mobility ratio $\mu_n/\mu_p = 2$. What is the maximum room temperature resistivity at what type of doping?

4.3. Multiply the Boltzmann equation by the momentum $m v_z$ of a carrier in the z-direction and integrate over velocity space. Show that the equation of motion emerges: $d(m v_d)/dt + m\langle v_z/\tau_m \rangle = e E_z$, where v_d is also in the z-direction, and calculate $\langle v_z/\tau_m \rangle$ for a shifted Fermi–Dirac distribution function $f_0(v) - m v_z v_d\, \partial f_0/\partial \varepsilon$ for $\tau_m \propto \varepsilon^r$ in terms of a ratio of Fermi integrals, $F_{1/2-r}(\eta_n)/F_{1/2}(\eta_n)$.

4.4. A semiconductor sample of conductivity $10^{-3}\,\Omega^{-1}\,\mathrm{cm}^{-1}$ may have zero Hall voltage at small magnetic field strengths. Assume the same Hall factor for electrons (mobility 1300 cm^2/V s) and holes (mobility 300 cm^2/Vs) and determine the carrier concentrations.

4.5. A permanent magnet is rotated around a semiconductor sample and an induction solenoid. The solenoid ends are connected to the sample such that the induced current through the sample is perpendicular to the magnetic field parallel to its axis of rotation. The Hall side arms of the sample are parallel to the solenoid axis. Show that the Hall voltage is partly dc. How does the dc Hall voltage depend on the magnetic field strength? What side effect of Hall measurements is eliminated?

4.6. The resistivity of a semiconductor in a magnetic field B is

$$\varrho = \begin{pmatrix} \varrho_{11} & \varrho_{12} & 0 \\ -\varrho_{12} & \varrho_{11} & 0 \\ 0 & 0 & \varrho_{33} \end{pmatrix} \tag{4.0.1}$$

where $\varrho_{11} = \varrho_{33}$ for $B = 0$. Calculate E, $R_H = (E \cdot [j \times B])/[j \times B]^2$, and $\Delta\varrho/\varrho = (E \cdot j)/(E \cdot j)_{B=0} - 1$ for B in the z-direction, and j in (a) the x-direction and (b) in the xy-plane at an angle of 45° relative to the axes.

4.7. Calculate the electronic contribution to the Lorenz number for n-type semiconductors. Explain in simple terms why the gap energy enters for mixed conduction. Hint: Use (4.10.6, 7).

4.8. Show that for an intrinsic semiconductor, the absolute thermopower is given by

$$\frac{d\Theta}{dT} = \left(\frac{d\Theta}{dT}\right)_n \frac{\sigma_n}{\sigma_n + \sigma_p} + \left(\frac{d\Theta}{dT}\right)_p \frac{\sigma_p}{\sigma_n + \sigma_p}. \tag{4.0.2}$$

4.9. Calculate the Nernst coefficient Q_N for mixed conduction in a weak magnetic field. Hint: Take into account (4.10.6) for j_x (neglecting the product of

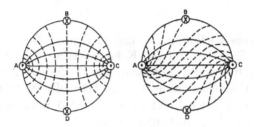

Fig. 4.36: Current streamlines (continuous lines) and equipotentials (dashed lines) in a disc; current electrodes are identified by dots, voltage probes by crosses. On the left $B = 0$, on the right B is normal to the plane of the sample (see Problem 4.13 (after H.H. Wieder, Thin Solid Films **31** 123 (1976))

small quantities E_y and B_z) and

$$j_y = (\sigma_n + \sigma_p)E_y - \frac{\sigma_n \mu_{Hn} - \sigma_p \mu_{Hp}}{|e|} B_z \left(|e|E_x - \frac{\partial \zeta}{\partial x} \right) - B_z \frac{\gamma'}{T} \frac{\partial T}{\partial x} \quad (4.0.3)$$

where

$$\gamma' = \frac{1}{|e|} [(\sigma_n \mu_{Hn} + \sigma_p \mu_{Hp}) k_B T(2r + 5/2) - (\sigma_n \mu_{Hn} \zeta_n + \sigma_p \mu_{Hp} \zeta_p)] (4.0.4)$$

and solve $0 = j_x = j_y$.

4.10. Prove that the Corbino thermopower for $\tau_m \propto \varepsilon^r$ and a strong magnetic field is given by $A d\Theta/dT = 2r \, k_B/e$.

Hint:Combining (4.9.1) and (4.5.4) yields

$$\frac{d\Theta}{dT} = \frac{k_B}{e} \left(\frac{\langle Z\tau_m \varepsilon / k_B T \rangle}{\langle Z\tau_m \rangle} - \frac{\zeta}{k_B T} \right) . \quad (4.0.5)$$

4.11. At 77 K you measure the sample resistance R as a function of the applied voltage V and find 100.00 Ω for 0.1 V; 100.25 Ω for 7 V; 100.52 Ω for 10 V; 101.34 Ω for 16 V. Determine the non-ohmicity coefficient β and estimate the energy relaxation time from (4.13.10). The sample length is 0.8 cm, the mobility 3.78×10^4 cm^2/Vs, and you assume acoustic deformation potential scattering.

4.12. Show the Nyquist relation $\langle \Delta I^2 \rangle \, \text{Re} \, \{Z\} = 4 \, k_B \, T \, \Delta f$ to be a consequence of the Johnson noise equation

$$\langle \Delta I^2 \rangle = e^2 \left(\sum_i \Delta v_i \right)^2 . \quad (4.0.6)$$

Hints: statistical mechanics proves $\frac{1}{2} m \langle \Delta v^2 \rangle = \frac{1}{2} k_B T$. The Fourier analysis yields for a finite frequency range $\Delta \omega = 2 \pi \, \Delta f$,

$$\langle \Delta v^2 \rangle = \int_0^\infty v_\omega^2 \, d\omega = v_\omega^2 \, \Delta \omega \quad (4.0.7)$$

and for $\omega^2 \tau_m^2 \ll 1$

$$v_\omega^2 = \langle \Delta v^2 \rangle \frac{2}{\pi} \frac{\tau_m}{1 + \omega^2 \tau_m^2} \approx \langle \Delta v^2 \rangle \frac{2}{\pi} \tau_m \ . \tag{4.0.8}$$

Because of the statistical independence of the Δv_i, we have [Ref. 4.56 b, p. 277]

$$\sum_{i \neq j} \Delta v_i \Delta v_j = 0; \quad \left(\sum \Delta v_i \right)^2 = n \cdot \langle \Delta v^2 \rangle \ . \tag{4.0.9}$$

4.13. Fig. 4.36 (p.116) shows the distribution of streamlines (continuous lines) and equipotentials (dashed lines) in a van der Pauw type of Hall effect measurement with a circular semiconductor sample. Compare with the situations in Figs. 4.7 (a) and (b). For a good mathematician: try to apply the Neudecker and Hoffmann method of calculation to this geometry.

5. Carrier Diffusion Processes

The discussion of (4.10.8–11) has shown that a temperature gradient in a conductor yields a concentration gradient ∇_n with the effect of a diffusion current $j = -e\,D_n\nabla_r n$, where D_n is proportional to the electron mobility due to the Einstein relation (4.10.12). In this chapter we will investigate the diffusion of *injected* carriers in local variations in the type of doping, which is typical for p-n junctions and bipolar transistors.

5.1 Injection and Recombination

In Sect.1.2 we considered the transfer of an electron into a semiconductor from outside. Normally, however, electron-hole pairs rather than carriers of one type are *injected* into a semiconductor. One injection technique which is easy to understand is pair generation by absorption of light quanta $\hbar\omega \geq \varepsilon_G$ due to the *internal photoelectric effect* where ε_G is the gap energy. Let n_0 and p_0 be the dark-concentrations of electrons and holes, respectively. At a given light intensity these concentrations will be increased by $\Delta n = n - n_0$ and $\Delta p = p - p_0$, respectively. In an n-type semiconductor $n_0 \gg p_0$. At a not too low light intensity, $n_0 \gg \Delta n = \Delta p \gg p_0$. The equals sign is valid for no recombination with impurities or trapped carriers. The holes, which are called *minority carriers* in this case, have strongly increased in relative number while the electrons (called *majority carriers*) have only weakly increased:

$$\Delta p/p_0 \gg \Delta n/n_0 \ . \tag{5.1.1}$$

Obviously the opposite is true for a p-type semiconductor. Because of the strong relative increase of the minority carrier concentration, the process is called *minority carrier injection*.

Of course ionizing radiations other than light, such as a beam of high-energy (several keV) electrons, will act similarly. For practical purposes it is convenient to inject carriers by pressing a metal pin on the semiconductor and applying a voltage of an appropriate magnitude and polarity between the pin and another metal contact at the semiconductor sample. The production of a p-n junction at the semiconductor surface by diffusion of impurities at an elevated temperature yields a device with greater stability than the previous method of injection, as we shall see in Sect. 5.3.

The electron-hole pairs produced in this way gradually disappear by *recombination* such that a conduction electron becomes a bound valence electron. The binding energy ε_G is released in the form of either a photon or a quantum of lattice vibrations called a *phonon*. In a special kind of semiconductor p-n junction called a *semiconductor laser*, even coherent recombination radiation is generated (Sect. 13.2). In most cases, however, the phonon generation is predominant, especially in those semiconductors where there is an energy level for an impurity or a lattice defect near the middle of the gap. In this case, the carrier may first be trapped at this level and then recombine in a second process. The surface of the crystal contains a large number, up to $10^{15}/cm^2$, of such states, which is the order of magnitude of the concentration of surface atoms (Sect. 14.1).

If the carrier densities n and p are not too large, the recombination rate R' is proportional to the product np:

$$R' = C\,np = C(n_0 + \Delta n)(p_0 + \Delta p) , \tag{5.1.2}$$

where C is a factor of proportionality. In equilibrium with no injection ($\Delta n = \Delta p = 0$) we have only thermal recombination and the rate of thermal generation $C\,n_0\,p_0$ which, substracted from R', gives us the recombination rate due to injection R:

$$R = C[(n_0 + \Delta n)(p_0 + \Delta p) - n_0 p_0] \approx C(n_0\Delta p + p_0\Delta n) . \tag{5.1.3}$$

The product $\Delta n \times \Delta p$ of two small quantities has been neglected in this approximation. This can be justified if Δn and Δp are much smaller than the majority carrier density. For the case $\Delta n \approx \Delta p$ in an n-type semiconductor, $p_0\Delta n \ll n_0\Delta p$ and (5.1.1) can be approximated by

$$R \approx C\,n_0\,\Delta p , \tag{5.1.4}$$

while in a p-type semiconductor $n_0\Delta p \ll p_0\Delta n$ (5.1.3) can be approximated by

$$R \approx C\,p_0\,\Delta n . \tag{5.1.5}$$

This shows us that the recombination rate is proportional to the density of excesss minority carriers. For example, in an n-type semiconductor, we denote the lifetime of an excess hole by τ_p which is defined by

$$R = -\frac{d\Delta p}{dt} = \frac{\Delta p}{\tau_p} , \tag{5.1.6}$$

where t is the time. The solution

$$\Delta p(t) = \Delta p(0)\,\exp(-t/\tau_p) \tag{5.1.7}$$

is valid for the case of switching off the injection at $t = 0$. A comparison of (5.1.4, 6) yields for the constant C

$$C = 1/n_0\tau_p . \tag{5.1.8}$$

Similarly, for a p-type semiconductor we find

$$C = \frac{1}{p_0 \tau_n} .$$
(5.1.9)

After this short introduction to injection [5.1] and recombination processes we turn our attention to the diffusion of excess carriers which, in general, are generated somewhere at the crystal surface.

5.2 Diffusion and the Einstein Relation

If we increase the carrier concentration by injection somewhere at the crystal surface there will be a diffusion of the excess carriers throughout the crystal which is assumed to be homogeneous. The diffusion current j is proportional to the concentration gradient

$$j = -e\, D_n \nabla_r n ,$$
(5.2.1)

where D_n is the diffusion coefficient and the negative sign takes care of the fact that the carriers move *away* from regions of large concentrations. The dimension of D_n is cm^2/s. In (5.2.1), $e < 0$ for electrons, while for holes, $e > 0$ and D_n and n have to be replaced by D_p and p, respectively.

In the presence of an electric field a drift current adds

$$j = n\,|e|\,\mu_n \mathbf{E} - e\, D_n \nabla_r n$$
(5.2.2)

We introduce the potential Φ of the field $\mathbf{E} = -\nabla_r \Phi$. Assuming thermal equilibrium with no current flow in the crystal we obtain

$$0 = n\,|e|\,\mu_n \nabla\Phi + e\, D_n \nabla n$$
(5.2.3)

which in the form

$$0 = \frac{e\,\mu_n \nabla\Phi}{|e|\, D_n} + \nabla n / n ,$$
(5.2.4)

can be integrated to yield

$$\text{const} = \frac{e\,\mu_n}{|e|\, D_n}\Phi + \ln(n) .$$
(5.2.5)

The zero-point of the potential Φ is arbitrary and can be chosen to give for the lhs $const = \ln(n_i)$, where n_i is the intrinsic carrier density. Solving (5.2.5) for n yields

$$n = n_i \exp\left(-\frac{e\,\mu_n}{|e|D_n}\Phi\right) .$$
(5.2.6)

If we introduce the Einstein relation for a nondegenerate electron gas (4.10.12)

$$D_n = \mu_n \, k_B \, T/|e|$$

(5.2.7)

we obtain

$$n = n_i \, \exp(-e \, \Phi/k_B T) \; ,$$

(5.2.8)

where $e < 0$ is valid for electrons and $e > 0$ for holes. The nonhomogeneous carrier distribution due to a potential barrier $e \, \Phi$ (e.g., in a p-n junction) causes diffusion to take place. Equation (5.2.8) is equivalent to Perrin's barometric equation if $e \, \Phi$ is replaced by the potential energy of a molecule in the gravitational field. The exponential in (5.2.8) is frequently called *Boltzmann's factor*. The treatment is valid only if the particles are independent of each other which is neither true, e.g., in a liquid nor in a degenerate electron gas (Sect. 4.10).

At a temperature of 300 K, the value of $k_B T/|e|$ is 25.9 mV; for a mobility of 10^3 cm^2/Vs, D_n is 25.9 cm^2/s. The Einstein relation allows the calculation of the diffusion coefficient from the value of the mobility. If the temperature dependence of the mobility is given by $\mu \propto T^{-1}$, the diffusion process is independent of the temperature.

We now assume that equal numbers of electrons and holes are injected into a semiconductor sample: $\Delta n = \Delta p$. The variation of the carrier densities with time and position is described by the continuity equation. The recombination rate is given by (5.1.6). The continuity equation reads

$$\frac{\partial \Delta n}{\partial t} + \frac{1}{e}(\boldsymbol{\nabla_r} \cdot \boldsymbol{j_n}) = G - \frac{\Delta n}{\tau_n} \; ,$$

(5.2.9)

where $\boldsymbol{j_n}$ is given by (5.2.2) with $n \, |e| \, \mu_n = \sigma_n$.

The electric field $\boldsymbol{E}$ can be an internal field set up due to different diffusion coefficients of electrons and holes. Eliminating $\boldsymbol{E}$ from (5.2.9) and a corresponding equation for holes, one obtains (the subscript r will be omitted)

$$(\boldsymbol{j_n} - |e| \, D_n \boldsymbol{\nabla} n)/\sigma_n = (\boldsymbol{j_p} + |e| \, D_p \boldsymbol{\nabla} p)/\sigma_p \; .$$

(5.2.10)

Since $\sigma_n \propto n \, \mu_n \propto n \, D_n$ and $\sigma_p \propto p \, D_p$, we find from this equation

$$\boldsymbol{j_n}/n \, D_n - \boldsymbol{j_p}/p \, D_p = |e| \, (\boldsymbol{\nabla} n/n + \boldsymbol{\nabla} p/p) \; .$$

(5.2.11)

Let us calculate the *ambipolar* diffusion by assuming that the total current density vanishes

$$\boldsymbol{j} = \boldsymbol{j_n} + \boldsymbol{j_p} = 0$$

(5.2.12)

and that at the same time, $\boldsymbol{\nabla} \, n = \boldsymbol{\nabla} \, p$. We introduce an *ambipolar diffusion coefficient* D by the definition

$$\boldsymbol{j_n} = |e| \, D \boldsymbol{\nabla} n$$

(5.2.13)

for which a value of

$$D = \frac{(n + p) D_n D_p}{n \, D_n + p \, D_p}$$

(5.2.14)

is readily obtained. In an n-type semiconductor $p \ll n$ and therefore $D \approx D_p$. Ambipolar diffusion is therefore determined by *minority* carriers. We introduce a *diffusion length* L_n by the ansatz

$$\Delta n(x) = n(x) - n_0 = \Delta n(0) \exp(-x/L_n) \tag{5.2.15}$$

and obtain

$$L_n = \sqrt{D_n \tau_n} \tag{5.2.16}$$

from (5.2.2, 9) by assuming $\partial \Delta n/\partial t = (\nabla_r \cdot \mathbf{E}) = G = 0$:

$$\Delta n/\tau_n = D_n(\nabla)^2 n = D_n \Delta n/L_n^2 \ . \tag{5.2.17}$$

Again taking a value of 25.9 cm^2/s for D_n and a typical value of 100 μs for the lifetime τ_n, a diffusion length of 0.5 mm is calculated. When the shortest distance from the location of the generation to a surface is less than the diffusion length calculated with a bulk lifetime of τ_n, surface recombination will be important (5.8.1).

In our discussion on lifetime and diffusion length, let us for completeness also include the *dielectric relaxation time* τ_d and the *Debye length* L_D. Consider a disturbance of the equilibrium distribution of *majority* carriers as, e.g., in a plasma wave. The continuity equation for no generation and recombination and $\mathbf{j}_n = \sigma_n \mathbf{E}$ is given by

$$\frac{\partial \Delta n}{\partial t} + \nabla_r (\sigma \mathbf{E})/e = 0 \ . \tag{5.2.18}$$

The conductivity σ is assumed to be the same everywhere. Poisson's equation yields a second relationship between $\mathbf{E}$ and Δn

$$(\nabla_r \cdot \mathbf{E}) = e\Delta n/\kappa \kappa_0 \ , \tag{5.2.19}$$

where κ is the relative dielectric constant and κ_0 the permittivity of free space. Equation (5.2.18) can be solved by integrating over time t

$$\Delta n \propto \exp(-t/\tau_d) \ , \tag{5.2.20}$$

where the *dielectric relaxation time* τ_d is given by

$$\tau_d = \kappa \kappa_0/\sigma = 8.8 \times 10^{-14} \ \text{s} \cdot \kappa \varrho/\Omega \ \text{cm} \tag{5.2.21}$$

and ϱ is the resistivity of the semiconductor. N-type germanium ($\kappa = 16$) with an electron density of 10^{14}/cm^3, which at room temperature is equivalent to a resistivity of 15 Ω cm, yields $\tau_d = 2.2 \times 10^{-11}$ s. If in (5.2.16) we introduce for τ_n the dielectric relaxation time τ_d and denote L_n by L_D, the *Debye length*, we obtain

$$L_D = \sqrt{D_n \tau_d} = \sqrt{k_B T/e^2} \cdot \sqrt{\kappa \kappa_0/n} \tag{5.2.22}$$

which at room temperature equals

$$L_D(300\text{K}) = \sqrt{(\kappa/n) \cdot 1.42 \times 10^4/\text{cm}} \ . \tag{5.2.23}$$

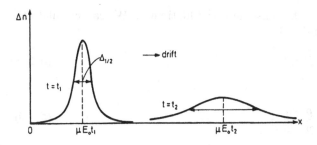

Fig. 5.1: Diffusion profiles of carriers injected at x=0 into a long filamentary sample at which an electric field E_0 is applied, after elapsed time intervals t_1 and t_2 (after [5.2])

For the case of germanium mentioned above, the Debye length L is 4.8×10^{-5} cm with, on an average, 1730 germanium atoms and only 2 conduction electrons along one Debye length. In more heavily doped semiconductors, inhomogeneities of doping will not seriously disturb the homogeneity of the majority carrier distribution along one Debye length [1].

Let us now return to *minority* carriers and consider simultaneous diffusion and drift in an externally applied electric field of intensity E_0. We introduce a new space coordinate

$$x' = x - \mu E_0 t , \qquad (5.2.24)$$

where μ is an *ambipolar drift mobility* defined by

$$\mu = |e|(n-p)\mu_n\mu_p/\sigma = (n-p)\mu_n\mu_p/(n\,\mu_n + p\,\mu_p) \qquad (5.2.25)$$

[note the minus sign in the numerator, in contrast to the otherwise similar expression for the ambipolar diffusion coefficient, (5.2.14)]. For example, for $p \ll n$, the ambipolar drift mobility is essentially equal to μ_p which is then the minority mobility. The solution of (5.2.9) for the present case is given by

$$\Delta n = \Delta n_0 \exp(-t/\tau_n)/2\sqrt{\pi Dt}\exp(-x'^2/4Dt) , \qquad (5.2.26)$$

where Δn_0 is the number of injected pairs per sample cross section. Diffusion profiles are given in Fig. 5.1 at two different times, t_1 and t_2. During the course of the drift motion, the distribution of the excess carriers over the space coordinate becomes broader due to diffusion. The half-width $\Delta_{1/2}$ of the pulse at a time t is given by

$$\Delta_{1/2} = 4\sqrt{Dt \ln 2} = 3.33\sqrt{Dt} . \qquad (5.2.27)$$

For example, for $D = 31$ cm^2/s, valid for electrons in p-type silicon and $t = 60~\mu$ s, the half-width is 1.7 mm. For an electron drift of 1 cm a field intensity E_0 of 14 V/cm is then required. Assume there is a collector at a distance d from the point of injection which allows the determination of the drift time t and,

---

[1]$L_D = \sqrt{2\,\zeta_n/3\,e^2} \cdot \sqrt{\kappa\,\kappa_0/n}$ for strong degeneracy; $1/L_D = k_{TF}$ is called the *Thomas-Fermi wave vector*; $\zeta_n = (\hbar^2/2m)(3\pi^2\,n)^{2/3}$.

from the total area under the pulse, also the lifetime τ. We can calculate the ambipolar mobility μ from [5.3]

$$\mu = (\sqrt{1 + a^2} - a)d/E_0 t ,\qquad(5.2.28)$$

where a stands for

$$a = \left(1 + \frac{2t}{\tau}\right)\frac{n + p}{n - p}\frac{k_B T}{e E_0 d} .\qquad(5.2.29)$$

The a-dependent factor in (5.2.28) is a consequence of the influence of carrier diffusion on the drift time.

An experiment of the kind described above was first reported by *Haynes* and *Shockley* [2] [5.4]. A schematic diagram of the arrangement is given in Fig. 5.2. The ends of the filamentary sample are connected to a pulse generator. Holes are injected into the n-type germanium sample by an emitter contact mounted at one side of the sample. The applied pulsed field causes the carriers to drift towards the second side contact which is the *collector*. In series with the collector there is a resistor R of about 10^4 Ω, and the collector current is measured by observing the voltage drop across this resistor with an oscilloscope. The collector contact is rectifying and biased in the reverse direction (Sect. 5.3). The small reverse current strongly depends on the hole concentration in the vicinity of the contact area and is increased when the drifting carriers arrive at the contact.

The accuracy of the measurement is limited by the ratio of the half-width $\Delta_{1/2}$ and the drift distance $d = \mu E_0 t$. Since the field intensity E_0 is essentially the applied voltage V_0 relative to d, d becomes

$$d = \sqrt{\mu V_0 t}\qquad(5.2.30)$$

and the ratio $\Delta_{1/2}/d$ is given by

$$\Delta_{1/2}/d = 3.33\sqrt{D/\mu V_0}\qquad(5.2.31)$$

which is proportional to $V_0^{-1/2}$. For greatest accuracy the applied voltage should be as large as possible. Because of Joule heat generation the voltage has to be pulsed, as is the case in Fig. 5.2. At too large a field strength E_0, the mobility depends on E_0 (*hot electrons*, Sect. 4.13).

By comparing the integrated pulses from different collectors mounted at various distances from the emitter, one can determine the lifetime τ using (5.2.26): the ratio of the *areas* under the curves, giving the time variation of pulses obtained at average drift times t_1 and t_2, is equal to $\exp[(t_2 - t_1)/\tau]$. The ratio of the pulse *amplitudes* $\sqrt{t_2/t_1}\exp[(t_2 - t_1)/\tau]$ is less suitable for a lifetime determination since t_2/t_1 is not as well known experimentally as $t_2 - t_1$.

The electron mobility in Ge observed with the Haynes–Shockley arrangement at various temperatures is shown in Fig. 5.3. Below room temperature the

---

[2] A movie showing *Haynes* and *Shockley* performing the experiment, was commercially available under the title "Minority Carriers in Semiconductors" as reported in W.R. Riley, Am. J. Physics **36**, 475 (1968).

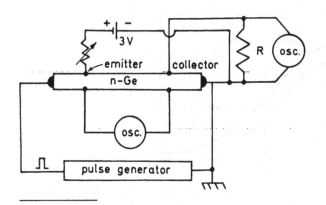

Fig. 5.2: Haynes-Shockley arrangement

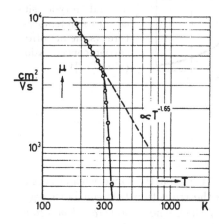

Fig. 5.3: Observed drift mobility of electrons in p-Ge as a function of temperature (after [5.3])

mobility is proportional to $T^{-1.65}$. Above room temperature the semiconductor becomes intrinsic and the ambipolar mobility, proportional to n–p, strongly decreases.

For electrons in silicon $\mu \propto T^{-2.5}$, for holes in n-type germanium and silicon $\mu \propto T^{-2.33}$ and $T^{-2.7}$, respectively. In Fig. 5.4 the room temperature drift mobilities of minority carriers in n- and p-type silicon as well as the drift mobilities of majority carriers (calculated from the Hall mobilities) are plotted vs the carrier densities. The carrier densities are about equal to the densities of ionized impurities. The decrease of mobility with increasing impurity concentration is due to ionized impurity scattering (6.3.22). The differences in mobilities of one type of carrier in either n- or p-type material may be partly due to carrier-carrier interaction.

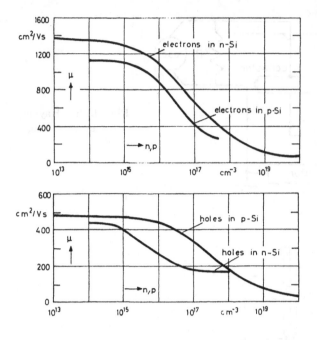

Fig. 5.4: Mobility of majority and minority carriers in n- and p-type silicon at room temperature, vs the majority-carrier concentration (after [5.2])

5.3 The p-n Junction

Consider a semiconducting single crystal which is partly doped with donors and partly with acceptors. In actual practice the transition between both regions will be gradual as indicated in Fig. 5.5a by the dashed line (*graded junction*). In order to simplify the calculation we will assume an *abrupt* junction where an n-type region containing a constant net donor concentration is adjacent to a region with a constant net acceptor concentration as indicated by the full line. The idealized picture of the abrupt junction will give us the essential features of the physical behavior of the junction.

In a treatment of the problem by *Schottky* [5.5], Poisson's equation

$$\frac{d^2\Phi(x)}{dx^2} = \left\{ \begin{array}{l} |e|N_{\mathrm{A}}/\kappa\,\kappa_0 \text{ in the p-region of the transition} \\ -|e|N_{\mathrm{D}}/\kappa\,\kappa_0 \text{ in the n-region of the transition} \end{array} \right\}$$

$$(5.3.1)$$

is solved in one dimension neglecting the space charge due to the carriers which obviously is much less than that due to ionized impurities for the following reason. Since everywhere in the transition region for the case of no current flow the product $np = n_{\mathrm{i}}^2$ is a constant , the sum $n + p = n + n_{\mathrm{i}}^2/n$ has a minimum value $2\,n_{\mathrm{i}}$ at $n = n_{\mathrm{i}}$ which has to be orders of magnitude less than both N_{D} and N_{A} otherwise no junction would be possible by definition. With the minimum being so small, there is a deficiency of carriers throughout the

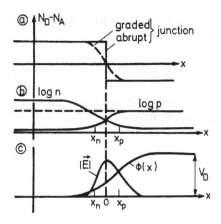

Fig. 5.5: Abrupt p-n junction in thermal equilibrium. (**a**) Spatial distribution of donors and acceptors. (**b**) Spatial distribution of electrons and holes. (**c**) Potential distribution and electric field intensity; for a discussion of the energy band edges see Fig. 5.9

junction. Therefore the carriers do not contribute significantly to the space charge in the junction.

The potential Φ as a function of the space coordinate x is obtained by integrating (5.3.1).

$$\Phi(x) = \begin{cases} (|e|/2\kappa\,\kappa_0)N_A(x - x_p)^2 + \Phi_p & \text{in the p - region} \\ -(|e|/2\kappa\,\kappa_0)N_D(x - x_n)^2 + \Phi_n & \text{in the n - region} \end{cases}$$

$$(5.3.2)$$

The constants of integration Φ_p and Φ_n are related because of the condition of continuity at $x = 0$:

$$(|e|/2\kappa\,\kappa_0)\,N_A\,x_p^2 + \Phi_p = -(|e|/2\kappa\,\kappa_0)\,N_D\,x_n^2 + \Phi_n \ . \tag{5.3.3}$$

The internal potential difference $\Phi_n - \Phi_p = V_D$ is called the *diffusion voltage* or sometimes the *built-in potential*:

$$V_D = (|e|/2\kappa\,\kappa_0)\,(N_A\,x_p^2 + N_D\,x_n^2) \ . \tag{5.3.4}$$

A second boundary condition is that the field which is the first derivative of $\Phi(x)$ must also be continuous at $x = 0$. This condition yields a relationship between the second set of constants of integration x_n and x_p:

$$-N_A\,x_p = N_D\,x_n \ . \tag{5.3.5}$$

From (5.3.2) it is obvious that the transition region extends into the p-region essentially for a distance x_p and into the n-region for a distance x_n. Since $x = 0$ between the two regions, one of the constants, x_n or x_p, is negative. The total width of the junction is thus given by

$$d = |x_n - x_p| \ . \tag{5.3.6}$$

From (5.3.4, 5) it is easy to calculate d. We add $N_A x_n$ on both sides of (5.3.5) and multiply by $N_D (x_n - x_p)/(N_D + N_A)$:

$$\frac{N_D N_A}{N_D + N_A}(x_n - x_p)^2 = N_D x_n(x_n - x_p) = N_D x_n^2 + N_A x_p^2 , \tag{5.3.7}$$

where (5.3.5) has again been used. Except for a factor $|e|/2\kappa \kappa_0$, the rhs of (5.3.7) is V_D according to (5.3.4). Hence, we obtain for $d = |x_n - x_p|$:

$$d = \sqrt{\frac{2\kappa \kappa_0}{|e|}(N_D^{-1} + N_A^{-1}) V_D} . \tag{5.3.8}$$

[If in addition to the internal diffusion voltage we apply a bias V_B, d becomes

$$d = \sqrt{(2\kappa \kappa_0/|e|) \times (N_D^{-1} + N_A^{-1})(V_D + V_B)} \tag{5.3.9}$$

and is proportional to $\sqrt{V_D + V_B}$].

How large is the diffusion voltage V_D? The potential energy $-|e|\Phi(x)$ for an electron as given by (5.3.2) is large in the p-type region and small in the n-type region, the difference being $-|e|V_D$. For a nondegenerate electron gas we may assume (5.2.8) with $\Phi = \Phi_n$ and $n = n_n$, characteristic of the n-type region, and a corresponding equation with $\Phi = \Phi_p$ and $n = n_p$, to be valid. The ratio of n_p and n_n for no external voltage is thus given by

$$n_p/n_n = \exp[-|e|(\Phi_p - \Phi_n)/k_B T] = \exp(-|e|V_D/k_B T) . \tag{5.3.10}$$

A similar equation holds for holes with $e = |e|$:

$$p_p/p_n = \exp[-|e|(\Phi_p - \Phi_n)/k_B T] = \exp(|e|V_D/k_B T) . \tag{5.3.11}$$

In equilibrium (no current flow), we have from (1.3.6) the product

$$n_p p_p = n_n p_n = n_i^2 . \tag{5.3.12}$$

The intrinsic concentration n_i at a given temperature is a material constant. Therefore the ratio n_p/n_n can be expressed as $n_i^2/n_n p_p$. In the case of complete ionization of impurities and no compensation, the electron density n_n in the n-type region is equal to the net donor concentration N_D and likewise $p_p = N_A$. Solving (5.3.10) for V_D yields

$$V_D = \frac{k_B T}{|e|}\ln\frac{n_n}{n_p} = \frac{k_B T}{|e|}\ln\frac{N_D N_A}{n_i^2} = \frac{k_B T}{|e|}\left(\ln\frac{N_D}{n_i} + \ln\frac{N_A}{n_i}\right) . \tag{5.3.13}$$

It is convenient to introduce the decimal logarithm. For T equal to room temperature (300 K), we obtain

$$V_D = 59.6\,\mathrm{mV}[\log(N_D/n_i) + \log(N_A/n_i)] . \tag{5.3.14}$$

For example, for a germanium sample where $n_i = 2.4 \times 10^{13}/\mathrm{cm}^3$, we take e.g., $N_D = 2.4 \times 10^{16}/\mathrm{cm}^3$ and $N_A = 2.4 \times 10^{14}/\mathrm{cm}^3$ (the p-n junction as a rule is *not* symmetrical); a diffusion voltage V_D of 0.24 V is calculated from (5.3.14).

A measurement of V_D as a function of temperature yields $n_i(T)$ from which, according to (1.3.3), the gap energy ε_G can be obtained. For this purpose we write (5.3.13) in the form

$$|e|V_D = k_B T \ln(N_D N_A) - k_B T \ln n_i^2 = k_B T \ln \frac{N_D N_A}{C} - 3 k_B T \ln T + \varepsilon_G$$

(5.3.15)

In a small temperature range the arguments of the logarithms may be taken as constants. V_D is then a linear function of T. The extrapolation of this line to $T = 0$ yields ε_G.

How can V_D be measured? The p-n junction may be considered to be a capacitor since, as shown above, there are fewer carriers of either sign in the transition region than outside this region. To a first approximation we consider the p-n junction to be a parallel plate condenser. If we denote the junction cross section by A, the capacity C is given by $C = \kappa \kappa_0 A/d$. The variation of the free carrier contribution with the dielectric constant κ in the junction may be neglected. According to (5.3.9), a plot of $1/C^2$ vs the bias V_B yields a straight line which can be extrapolated to cut the abscissa at $-V_D$.

For a numerical example we consider silicon at room temperature with $N_D \gg N_A$ and a resistivity ϱ_p on the p-type side of the junction. If we express C in units of nanofarad (nF), we obtain

$$\left(\frac{A \cdot 33\,\text{nF/cm}^2}{C} \right)^2 = \varrho_p \cdot (V_D + V_B)/(\Omega \cdot \text{cm} \cdot \text{V}) \ .$$

(5.3.16)

Quite often this relationship between C and V_B cannot be verified by an experiment. The reason is that the junction is graded rather than abrupt. A more realistic calculation can be based on the assumption that the transition is *linear* with a constant gradient of the doping concentrations given by

$$d(N_D - N_A)/dx = \text{const} \ .$$

(5.3.17)

In this case, Poisson's equation becomes

$$d^2 \Phi(x)/dx^2 \propto -(|e|/\kappa \kappa_0)x \ .$$

(5.3.18)

The integration yields for the capacity of the junction

$$C \propto (V_D + V_B)^{-1/3}$$

(5.3.19)

which means a linear relationship between C^{-3} and V_B. In a typical diode the exponent of C is somewhere between -2 and -3. For a determination of the value of this exponent, it is therefore useful first to plot the experimental data of C vs V_B on a log-log scale. For a measurement of the capacity, a resonant circuit in an ac bridge is used with the ac voltage being much smaller than the dc voltage applied via a large series resistor [5.6] (see Fig. 5.6). A p-n junction used as an electrically variable capacitor is called a *varactor diode*; this device has been used for the automatic fine adjustment of a resonant circuit.

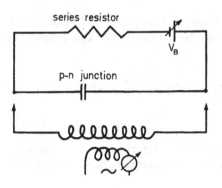

Fig. 5.6: Arrangement for measurements of the p-n junction capacity as a function of the bias voltage V_B; the reverse biased p-n junction is shown as a capacitor. The meter in the primary coil indicates resonance when the frequency is varied

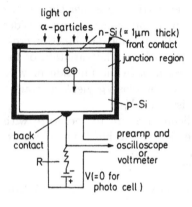

Fig. 5.7: Reverse biased p-n junction used for the detection of high-energy particles or light. The junction region may be made intrinsic by lithium drift compensation (Sect. 1.3)

Another device application is the high-energy particle counter shown in Fig. 5.7. For example, 8.78 MeV α-particles of a thorium C' source have a range of 60 μm in silicon. The junction width d in silicon with $N_D \gg N_A$ at room temperature according to (5.3.16) is given by

$$d = 0.33 \ \mu\text{m} \sqrt{\varrho_p / \Omega \ \text{cm}} \sqrt{(V_D + V_B)/V} \ . \tag{5.3.20}$$

Assuming typical values of $\varrho_p = 1000 \, \Omega$ cm and $V_B = 40$ V, we obtain $d = 65 \, \mu$m which for this purpose is chosen larger than the range of the α–particles. Each α–particle generates electron-hole pairs by impact ionization with an energy consumption of 3.5 eV per pair which is more than three times the gap energy $\varepsilon_G = 1.1$ eV of silicon (for a discussion of impact ionization in semiconductors see Chap. 10). If all the electron hole pairs are separated by the field, the p-n junction acting as a capacitor obtains an additional charge of $(8.78 \times 10^6/3.5) \times 1.6 \times 10^{-19}$ As $= 4 \times 10^{-13}$ As. If we assume a junction area $A = 1$ cm^2 and a bias voltage $V_B = 40V \gg V_D$, we obtain from (5.3.16) a capacitance of 165 pF. Hence, the pulse height across the junction is 4×10^{-13} As/165 pF $= 2.4$ mV per α-particle. At a larger bias voltage the capacity is smaller and the pulse height larger. An upper limit for the bias voltage is given by the onset of

electrical breakdown which may eventually result in permanent damage of the
p-n junction. However, already at a much lower bias there is a *dark current*
through the junction which causes noise. The dark current and the noise level
increase with bias. Hence, there is an optimum bias depending on the particular
type of detector and the bandwidth of the amplifier used in connection with the
detector.

The life time of a particle detector is limited by radiation damage. Instead
of material particles light quanta may also be detected, possibly without even
using a bias (*photocell*, Sect.5.9).

So far, the calculation has not taken into account any current through the
junction (apart from, e.g., the few carriers generated by radiation). Now we will
investigate the rectifying property of a p-n junction by considering the *diffusion*
and *recombination* of carriers in the junction. We first consider the case where
the generation rate G can be neglected and any recombination is small (diffusion
length large compared to Debye length). Another case to be investigated later
is that of a strong generation of carriers which dominates the reverse current of
the junction diode.

For equilibrium ($\partial \Delta n/\partial t = 0$) and no generation ($G = 0$), the continuity
equation (5.2.9) for holes ($e > 0$) is

$$\frac{1}{e}(\nabla_r \boldsymbol{j}_p) = -\frac{p - p_n}{\tau_p} \ . \tag{5.3.21}$$

To a first approximation the drift current $\sigma \boldsymbol{E}$ may be neglected in (5.2.10) if
the applied voltage is less than the diffusion voltage:

$$\boldsymbol{j}_p = -e\, D_p \nabla_r p \ . \tag{5.3.22}$$

For this case, the solution of (5.3.21) is given by

$$p(x) - p_n = [p(x_n) - p_n]\exp[-(x - x_n)/L_p] \ , \tag{5.3.23}$$

where the diffusion length $L_p = \sqrt{D_p \tau_p}$ has been introduced and the boundary
condition $p(\infty) = p_n$ has been taken into account.

We now apply a voltage V to the junction with a polarity opposite to the
reverse bias V_B of (5.3.9): $V = -V_B$ (*forward bias*). While for $V = 0$ the hole
concentration at $x = x_n, p(x_n)$, is equal to p_n, it is now raised by a *Boltzmann
factor*

$$p(x_n)/p_n = \exp(|e|\, V/k_B T) \ . \tag{5.3.24}$$

This is valid for not too large a voltage V such that $p(x_n) \ll p_p$. We substitute
for $p(x_n)$ in (5.3.23) the value given by (5.3.24):

$$p(x) - p_n = p_n[\exp(|e|V/k_B T) - 1]\exp[-(x - x_n)/L_p] \ . \tag{5.3.25}$$

Equation (5.3.22) yields for $\boldsymbol{j}_p$ in the x-direction (we omit the subscript x)

$$j_p = (|e|p_n D_p/L_p)[\exp(|e|V/k_B T) - 1]\exp[-(x - x_n)/L_p] \ . \tag{5.3.26}$$

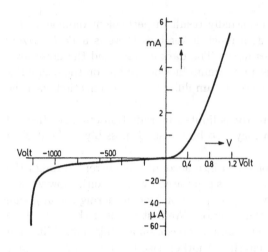

Fig. 5.8: Current voltage characteristics of a germanium p-n junction (after[5.9])

The diffusion length of holes L_p is assumed to be much larger than the transition region $(x_p - x_n)$. Therefore the last exponential may be omitted:

$$j_p = (|e|p_n D_p/L_p)[\exp(|e|V/k_B T) - 1] \ . \tag{5.3.27}$$

A similar expression is obtained for the electron current density j_n. The total current density $j = j_p + j_n$ is given by [5.7, 5.8]. [3]

$$j = j_s[\exp(|e|V/k_B T) - 1] \ , \tag{5.3.28}$$

where j_s stands for

$$j_s = |e|(p_n D_p/L_p + n_p D_n/L_n) = |e|n_i^2(N_A^{-1}\sqrt{D_n/\tau_n} + N_D^{-1}\sqrt{D_p/\tau_p}) \tag{5.3.29}$$

The dimension of $\sqrt{D/\tau}$ is that of a velocity. For example, in germanium where $D_n = 100\,\mathrm{cm^2/s}$, $D_p = 49\,\mathrm{cm^2/s}$, $\tau_n = \tau_p = 100\,\mu s$, we obtain $\sqrt{D_n/\tau_n} = 10^3\,\mathrm{cm/s}$ and $\sqrt{D_p/\tau_p} = 700\,\mathrm{cm/s}$; $n_i = 2.4 \times 10^{13}/\mathrm{cm^3}$. Assuming as before $N_D = 2.4 \times 10^{16}/\mathrm{cm^3}$ and $N_A = 2.4 \times 10^{14}/\mathrm{cm^3}$, we obtain $j_s = 3.8 \times 10^{-4}\,\mathrm{A/cm^2}$.

Figure 5.8 shows the current voltage characteristics of a germanium p-n junction. In the *forward* direction $(V > 0)$, the current increases exponentially in agreement with (5.3.28). At higher current densities the series resistance, which is due to the part of the sample outside the p-n junction, causes the characteristics to deviate from an exponential behavior towards a more linear behavior. In the reverse direction , (5.3.28) suggests a saturation of the current

---

[3]The application of this type of law to actual $j - V$ characteristics yields a factor of $1/n$ with n between 1 and 2, in the argument of the exponential where n is the *"diode ideality factor"*. For higher doping levels and lower temperatures, the value of n is closer to 2.

density at a value of j_s. Figure 5.8 shows the reverse characteristics to be very flat up to voltages of about 10^3 V (note the difference in scale between forward and reverse direction) and a strong increase beyond this voltage which is typical of an electrical breakdown as we discussed earlier.

The calculated j_s-value is in agreement with observations made on germanium diodes but not with those on silicon diodes. In silicon n_i^2 is about 6 orders of magnitude smaller than in germanium and j_s should be smaller by the same factor. This is not the case, however, and besides, the reverse current does not saturate. This is due to the thermal generation of electron-hole pairs in the junction region at centers with energy levels near the middle of the band gap. Let us denote by f_t the occupation probability of such a center by an electron; $1 - f_t$ is then the probability of occupation by a hole. We assume the rate of electron generation G to be proportional to f_t:

$$G = a\,f_t \;, \tag{5.3.30}$$

where a is a factor of proportionality. In equilibrium, G must be equal to the recombination rate of electrons $n(1 - f_t)/\tau_n$, which is proportional to the electron density, to the probability that a center is occupied by a hole, and inversely proportional to the lifetime of excess electrons:

$$n(1 - f_t)/\tau_n = a\,f_t \;. \tag{5.3.31}$$

For an intrinsic semiconductor, $n = n_i$ and the Fermi level is located near the middle of the gap. To a good approximation we may assume the Fermi level to be at the energy level of the recombination center which yields $f_t = 1/2$ and from (5.3.31), $a = n_i/\tau_n$. Hence, the generation rate is given by

$$G = f_t n_i / \tau_n \;. \tag{5.3.32}$$

Under nonequilibrium conditions, the increase of electron density per unit time is thus

$$\frac{\partial n}{\partial t} = \frac{n_i}{\tau_n} f_t - \frac{n(1 - f_t)}{\tau_n} \;. \tag{5.3.33}$$

A similar equation holds for holes:

$$\frac{\partial p}{\partial t} = \frac{n_i}{\tau_p}(1 - f_t) - \frac{p}{\tau_p} f_t \;. \tag{5.3.34}$$

The condition for pair generation

$$\frac{\partial n}{\partial t} = \frac{\partial p}{\partial t} \tag{5.3.35}$$

yields for f_t:

$$f_t = \frac{n\tau_p + n_i\tau_n}{(n + n_i)\tau_p + (p + n_i)\tau_n} \tag{5.3.36}$$

and for the rate of pair generation

$$\frac{\partial n}{\partial t} = \frac{\partial p}{\partial t} = \frac{n_i^2 - np}{(n + n_i)\,\tau_p + (p + n_i)\tau_n} \tag{5.3.37}$$

At equilibrium, $\partial n/\partial t = 0$, we find $n_i^2 = np$ which in (1.3.6) has been obtained by the law of mass action.

Since the electrons and holes are usually separated by the electrical field before they can recombine, n and p will be small in the junction region ($\ll n_i$) and thus can be neglected in (5.3.37):

$$\frac{\partial n}{\partial t} + \frac{\partial p}{\partial t} = \frac{n_i}{\tau} \ , \tag{5.3.38}$$

where

$$\tau = \frac{1}{2}(\tau_n + \tau_p) \tag{5.3.39}$$

is an average lifetime. The reverse current density j_s given by

$$j_s = |e| \left(\frac{\partial n}{\partial t} + \frac{\partial p}{\partial t} \right) d = \frac{|e| n_i d}{\tau} \ . \tag{5.3.40}$$

The thermal generation in the junction is proportional to its width and hence also the reverse current. For example, for silicon at room temperature, $n_i = 1.3 \times 10^{10}/\text{cm}^3$. Assuming as above $d = 60\,\mu\text{m}$ and $\tau = 100\,\mu\text{s}$, we calculate from (5.3.40) $j_s = 1.3 \times 10^{-7}\,\text{A/cm}^2$ which is 3 orders of magnitude larger than j_s from (5.3.29).

Since d in (5.3.40) increases with the applied bias voltage, j_s also increases and there is no current saturation as observed in silicon [5.7].

The dependence of the junction width d and hence of the capacitance C on the doping concentrations N_D and N_A according to (5.3.9) has been applied for a determination of trap levels. The technique is known as deep level transient spectroscopy (DLTS) [5.10]. For simplicity, let us take an $n^+ p$ diode where $N_D \gg N_A$ and therefore $C \propto N_A^{1/2}$ which also includes traps since these also contribute to the space charge. A small change in the capacitance ΔC is obtained by filling or emptying the traps by $dN = 2N_A dC/C$. By a short pulse in the forward direction at the time $t = 0$, carriers are injected and flood the traps. Return to equilibrium is observed by recording the voltage in a time interval from t_1 to t_2 where t_2 is, e.g., 10 t_1 and t_1 is, e.g., 100 μs. This is performed applying a double box car integrator. The procedure is repeated at short intervals while the temperature of the sample is raised or lowered at a constant rate. The occupation probability $f(t_1, \varepsilon)$ of a trap at the time t is

$$\exp \left[-\lambda_0 \int\limits_0^\infty \exp(-\frac{\varepsilon}{k_B T}) dt \right]$$

at a temperature $T(t)$ where ε is the trap energy and λ_0 the product of the carrier capture cross section, the mean thermal velocity of the carriers and the effective density of states in the band of the carriers [5.7]. For a continuous distribution of traps $N = N(\varepsilon)$, the signal is proportional to

$$\int N(\varepsilon)[f(t_1, \varepsilon) - f(t_2, \varepsilon)]d\varepsilon$$

which may be approximated by $N(\varepsilon_0)[f(t_1, \varepsilon_0) - f(t_2, \varepsilon_0)]$ where ε_0 is the energy for a maximum of the term in brackets. In making thermal scans to determine trap energies, it is convenient to vary t_1 and t_2 but keep t_1/t_2 fixed. As in *glow curves* of thermoluminescence, signal peaks are observed at certain temperatures which are only shifted somewhat by a variation of t_1 and t_2 at a constant ratio t_1/t_2. Besides bulk trap levels, interface defect states at oxidized silicon surfaces have also been determined in this way [5.11]. For a determination of trap levels in an electric field which may be, e.g., a space charge field, the *Poole-Frenkel effect* has to be taken into account [5.12]. It consists of a lowering of the Coulombic potential barrier by the electric field similar to the well-known field emission of electrons from a metal surface, and it can occur only in a trap which is neutral when filled, as it depends on the existence of the Coulombic potential barrier [5.13]. In contrast to the field emission where the image charge is mobile, the positive charge is fixed at the trap when an electron moves away. For a distance r between the two charges, the potential energy is given by $\varepsilon_I - e\,E\,r - e^2/(4\pi\,\kappa\,\kappa_0\,r)$ and has its maximum value of $\varepsilon_I - (e^3\,E/\pi\,\kappa\,\kappa_0)^{1/2}$ for $r = (e/4\pi\,\kappa\,\kappa_0\,E)^{1/2}$. Hence, the ionization energy ε_I is lowered by the amount of $(e^3 E/\pi\,\kappa\,\kappa_0)^{1/2}$. This one-dimensional model has been extended to the three-dimensional case by *Hartke* [5.14] and corrected by *Keller* and *Wünstel* [5.15] for the influence of neighboring charged centers, recapture during the emission process, and tunneling through the barrier. DLTS experimental data obtained with silicon containing oxygen and thermal donors generated by an anneal at 723 K thus yield levels which agree with data obtained by Hall-effect measurements.

5.4 Quasi-Fermi Levels

The current voltage characteristics of a p-n junction have been calculated without considering the energy distribution of the carriers. The current has been related to thermal diffusion of the carriers with diffusion constants which are determined by the carrier mobilities due to the Einstein relation. The voltage applied to the junction changes the potential difference between both sides of the junction. Depending on the polarity of the applied voltage, this change in potential difference causes more or less carriers to diffuse to the opposite side than in thermal equilibrium.

The potential distribution is shown in Fig. 5.9a. In going from the p- to the n-side of the junction, both the conduction band edge and the valence band edge are lowered by $|e|V_D$ with no applied voltage, by $|e|(V_D - V)$ with a forward

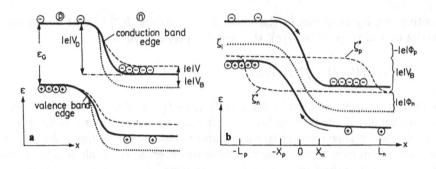

Fig. 5.9: (a) Energy band edges in a p-n junction at thermal equilibrium (full curves), with forward bias applied (dashed) and with reverse bias applied (dotted). (b) Quasi-Fermi levels in a reverse biased p-n junction

bias voltage V and by $|e|(V_D + V_B)$ with a reverse bias voltage V_B. In Fig. 5.9b we consider the case of a reverse bias voltage V_B in more detail. In the regions $x > L_n$ and $x < -L_p$, the equilibrium carrier densities are hardly affected by the junction. There the densities of both carrier types can be described by the same Fermi level, denoted by ζ_p^* on the p-side and by ζ_n^* on the n-side [these should not be confused with ζ_p and ζ_n given by (4.9.5) and (4.9.4), respectively], since any voltage applied to the junction will cause an additional potential drop in the region of the junction width $-x_p \le x \le x_n$, while the field outside this region, though necessary for a current flow, is negligibly small to a first approximation. As we approach the junction from either side, the *majority* carrier densities are not seriously perturbed from their equilibrium values by the current flow, but the *minority* carriers are. Hence, there is no longer a unique Fermi level for both carrier types. Therefore, we denote ζ_p^* and ζ_n^* as *quasi-Fermi levels* (sometimes called *imrefs*). In Fig. 5.9 b, ζ_p^*, ζ_n^*, and the intrinsic Fermi level ζ_i, which is given by (3.1.16), are shown. ζ_i is close to the middle of the gap, depending on the effective mass ratio, but independent of doping and serves as a reference level throughout the junction. The constants Φ_n and Φ_p defined by (5.3.2) are also indicated. Far away from the junction they are related to the quasi-Fermi levels by

$$|e|\Phi_n = \zeta_n^* - (\zeta_i)_{n-side} \tag{5.4.1}$$

$$|e|\Phi_p = \zeta_p^* - (\zeta_i)_{p-side} \tag{5.4.2}$$

Due to the potential barrier across the junction, electrons on the p-side slide down the potential hill, thus being converted from minority carriers to majority carriers. The same holds for the hole transfer from the n-side to the p-side (see arrows in Fig. 5.9b). In the case of germanium, it is exactly this charge transfer which causes the reverse current (in silicon, thermal generation of electron hole pairs in the junction is predominant). At an average distance of a diffusion length away from the junction, the excess majority carriers recombine with minority carriers. Consequently, at distances greater than the diffusion length, the thermal equilibrium distribution is present.

The decrease in minority carrier densities in the regions contiguous to the junction is described by a decrease of ζ_n^* on the p-side and an increase of ζ_p^* on the n-side, by an amount $|e|V_B$ as shown in Fig. 5.9b. For no applied bias V_B, the Fermi level $\zeta = \zeta_n^* = \zeta_p^*$ is the same everywhere in the crystal. For example, the replacement of ζ_n by ζ_n^* in (3.1.9) takes care of the excess carriers in n; otherwise this equation is valid only in thermodynamic equilibrium, i.e., for zero applied voltage.

It is obvious from Fig. 5.9b that the intrinsic Fermi level ζ_i drops across the junction by an amount

$$|e|\Phi_n + |e|V_B - |e|\Phi_p = |e|(V_B + V_D) , \qquad (5.4.3)$$

where $V_D = \Phi_n - \Phi_p$ is the diffusion voltage. The drop of ζ_i is equal to the drop of the conduction band edge. Hence, the electron density in the p-type region n_p is lower than in the n-type region n_n by a factor of

$$\exp[-|e|(V_D + V_B)/k_B T] .$$

In comparison with (5.3.10) which holds for $V_B = 0$, we find the sum of V_D and V_B to be the quantity which determines the carrier density ratio and therefore also the width of the junction given by (5.3.9). A treatment of $n - n^+$ and $p - p^+$ junctions has been given by *Gunn* [5.16] and will not be discussed here.

5.5 The Bipolar Transistor

A power amplifying p-n structure is called a *transistor*. A typical example is the p-n-p transistor shown schematically in Fig. 5.10. The middle part of the single crystal is n-type; it is called the *base*. The adjacent p-type parts are the emitter and the collector which have been introduced in the discussion of the Haynes–Shockley experiment (Sect. 5.2).

For the emitter base junction we obtain from (5.3.21, 22)

$$D_p d^2 p/dx^2 = (p - p_b)/\tau_b , \qquad (5.5.1)$$

where we have introduced the symbols p_b and τ_b for p_n and τ_p, respectively, in the base. By

$$L_b = \sqrt{D_p \tau_b} \qquad (5.5.2)$$

we denote the base diffusion length. The solution of (5.5.1) for the region of the base $x_e \leq x \leq x_c$, where $w = x_c - x_e$ is the base width, is given by

$$p(x) = p_b + \frac{p_e - p_b}{\sinh(w/L_b)} \sinh \frac{x_c - x}{L_b} + \frac{p_c - p_b}{\sinh(w/L_b)} \sinh \frac{x - x_e}{L_b} , \qquad (5.5.3)$$

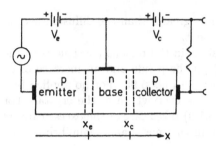

Fig. 5.10:   p-n-p transistor connected in common-base configuration (schematic diagram)

where we denote by p_e the hole density $p(x_e)$ in the emitter and by p_c the hole density $p(x_c)$ in the collector. The boundary conditions $p(x_c) = p_c$ $p(x_e) = p_e$ are fulfilled. Due to the voltages V_e and V_c applied between the emitter and base, and the collector and base, respectively, the following relationships exist between p_b, p_e, and p_c:

$$p_e = p_b \exp(|e| V_e/k_B T) \tag{5.5.4}$$

and

$$p_c = p_b \exp(|e| V_c/k_B T) \ . \tag{5.5.5}$$

We substitute for p_e and p_c in (5.5.3) their values given by (5.5.4, 5) and calculate j_p at $x = x_e$:

$$j_p(x_e) = -|e| D_p \left(\frac{dp}{dx}\right)_{x_e} = \frac{|e| D_p p_b}{L_b}$$

$$\times \frac{\cosh(w/L_b) \left[\exp(|e| V_e/k_B T) - 1\right] - \left[\exp(|e| V_c/k_B T) - 1\right]}{\sinh(w/L_b)} \tag{5.5.6}$$

This is the hole contribution to the emitter current density. For a calculation of the electron contribution we consider the ratio of electron densities in the p-type emitter n_e and in the n-type base n_{e0} as a function of the emitter voltage V_e:

$$n_e(x_e)/n_{e0} = \exp(-|e| V_e/k_B T) \ . \tag{5.5.7}$$

The x-dependence of the electron density is similar to (5.3.25) seen to be

$$n(x) = n_{e0}[\exp(-|e| V_e/k_B T) - 1] \exp[(x - x_e)/L_e] + n_{e0} \tag{5.5.8}$$

We find for the electron contribution to the emitter current density an equation similar to (5.3.27) which represents a diode characteristic:

$$j_n(x_e) = (|e| n_{e0} D_n/L_e) [\exp(|e| V_e/k_B T) - 1] \ , \tag{5.5.9}$$

where we have introduced an electron diffusion length $L_e = \sqrt{D_n \tau_e}$ in the emitter. The total emitter current density is given by

$$j_e(x_e) = j_p(x_e) + j_n(x_e)$$

$$= \frac{|e|\, D_p\, p_b}{L_b \sinh(w/L_b)} \left[\cosh \frac{w}{L_b} \left(\exp \frac{|e|\, V_e}{k_B T} - 1 \right) - \left(\exp \frac{|e|\, V_c}{k_B T} - 1 \right) \right]$$

$$+ \frac{|e|\, D_n\, n_{e0}}{L_e} \left(\exp \frac{|e|\, V_e}{k_B T} - 1 \right) \; . \tag{5.5.10}$$

Similarly, the total collector current density j_c can be calculated:

$$j_c = \frac{|e|\, D_p\, p_b}{L_b \sinh(w/L_b)} \left[- \cosh \frac{w}{L_b} \left(\exp \frac{|e|\, V_c}{k_B T} - 1 \right) + \exp \frac{|e|\, V_e}{k_B T} - 1 \right]$$

$$- \frac{|e|\, D_c\, n_{c0}}{L_c} \left(\exp \frac{|e|\, V_c}{k_B T} - 1 \right) \; . \tag{5.5.11}$$

The *current amplification factor* α is given by

$$\alpha = \left(\frac{\partial j_c}{\partial j_e} \right)_{V_c = \text{const}} = \left(\frac{\partial j_c / \partial V_e}{\partial j_e / \partial V_e} \right)_{V_c = \text{const}} \tag{5.5.12}$$

In this definition it is assumed that the emitter input is a small signal and that the output is obtained at the collector. We obtain $\partial j_c / \partial V_e$ from (5.5.11)

$$\left(\frac{\partial j_c}{\partial V_e} \right)_{V_c = \text{const}} = \frac{|e|\, D_p\, p_b}{L_b \sinh(w/L_b)} \frac{|e|}{k_B T} \exp \left(\frac{|e|\, V_e}{k_B T} \right) \tag{5.5.13}$$

and $\partial j_e / \partial V_e$ from (5.5.10)

$$\left(\frac{\partial j_e}{\partial V_e} \right)_{V_c = \text{const}} = \left[\frac{|e|\, D_p\, p_b \cosh(w/L_b)}{L_b \sinh(w/L_b)} + \frac{|e|\, D_n\, n_{e0}}{L_e} \right] \frac{|e|}{k_B T} \exp \frac{|e|\, V_e}{k_B T} \tag{5.5.14}$$

α is obtained from (5.5.12)

$$\alpha = \left[\cosh(w/L_b) + \frac{D_n\, L_b\, n_{e0}}{D_p\, L_e\, p_b} \sinh(w/L_b) \right]^{-1} \tag{5.5.15}$$

Since all quantities are positive, α is < 1 which means there is *no* current amplification in the circuit shown by Fig. 5.10 (however, in the common-emitter configuration, there is a current gain equal to $\alpha/(1 - \alpha)$ which is $\gg 1$)

In order to have α as close to 1 as possible, the base width should be much less than the diffusion length and the base heavily doped: $w \ll L_b; p_b \gg n_e$. Assuming $L_b \approx L_e$ and $D_n \approx D_p$, we obtain for α :

$$\alpha = 1 / \cosh(w/L_b) \approx 1 - w^2/2 L_b^2 \; . \tag{5.5.16}$$

It is possible to have $\alpha = 0.98$ in germanium with a lifetime τ_b of 50 μs, a diffusion length of 0.5 mm and a base width of 0.1 mm. In actual practice, an even smaller base width is used.

In the differentiations in (5.5.13, 14) the base width w was assumed constant. The treatment of the p-n junction in Sect. 5.4 has shown, however, that the junction width d depends on the applied voltage. Since the emitter voltage V_e is quite small, the error made by neglecting this dependence is not very large. The collector voltage V_c, however, is relatively large and therefore should be taken into account according to (5.3.9):

$$\frac{\partial w}{\partial V_c} = \frac{\partial d}{\partial V_c} = \frac{\partial}{\partial V_c}(\sqrt{V_D + V_c}\ d_0/\sqrt{V_D}) \approx \frac{d_0}{2\sqrt{V_D\,V_c}}\ , \qquad (5.5.17)$$

where d_0 is the base collector junction width for $V_c = 0$. This enables us to calculate the inverse collector impedance $1/r_c$:

$$\frac{1}{r_c} = \left(\frac{\partial I_c}{\partial V_c}\right)_{V_e = \text{const}} \approx \frac{I_c}{w}\frac{\partial w}{\partial V_c} = \frac{I_c d_0}{2\,w\,\sqrt{V_D\,V_c}}\ . \qquad (5.5.18)$$

Typical values are $d_0/w = 10^{-2}$, $V_D = 0.25\,\text{V}$, $V_c = 4\,\text{V}$ and $I_c = 2\,\text{mA}$. From (5.5.18), a collector impedance of value $10^5\,\Omega$ is calculated.

The inverse emitter impedance

$$\frac{1}{r_e} = \left(\frac{\partial I_e}{\partial V_e}\right)_{V_c = \text{const}} \qquad (5.5.19)$$

is obtained from (5.5.14) and (5.5.10), neglecting to a first approximation the -1 terms and the $\exp(|e|\,V_c/k_B\,T)$ in (5.5.10):

$$\frac{1}{r_e} \approx I_e\frac{|e|}{k_B\,T} = \frac{I_e}{25.9\,\text{mV}}\ , \qquad (5.5.20)$$

where the equals sign is valid for $T = 300K$. Assuming the emitter current to be the same as the collector current which is 2 mA, the order of magnitude obtained for the value of r_e is $10\,\Omega$ which is 4 orders of magnitude smaller than the collector impedance. Hence, the power amplification factor in this case is given by

$$I_c^2\,r_c/I_e^2\,r_e = r_c/r_e \approx 10^4\ . \qquad (5.5.21)$$

This is also the factor of voltage amplification since $\alpha \approx 1$.

The operating power of the transistor is essentially the product $I_c V_c$ which in our case is of the order of a few mW. In vaccum tubes used before the invention of the transistor in 1947 by *Bardeen* and *Brattain*[5.17](for the historical development that led to the invention of the transistor as well as to integrated circuits, see [5.18]), the filament current power was a hundred times larger. A fast computer with about 10^7 transistors, which nowadays is quite common, would have been practically impossible with vacuum tubes because of the enormous heat production at the short distances required for fast signal transfer.

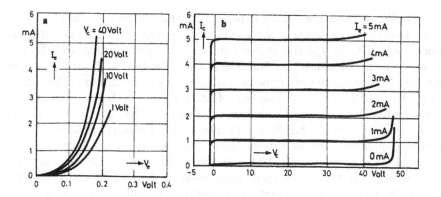

Fig. 5.11: Emitter and Collector characteristics of a p-n-p transistor

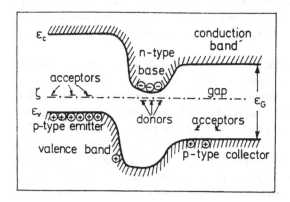

Fig. 5.12: Energy band edges in a p-n-p transistor at thermal equilibrium

The characteristics of a typical transistor are shown in Fig. 5.11 [5.19]. The emitter characteristics I_e vs V_e are, of course, the forward characteristics while the collector characteristics I_c vs V_c are the reverse characteristics of a p-n junction diode. In the latter, the emitter current I_e is a parameter.[4] Up to quite large values of V_c, I_c is constant and equal to I_e, in agreement with $\alpha \approx 1$. If the base is extremely thin and only lightly doped, it is possible that at a large collector voltage, a galvanic connection between the collector and the emitter develops shorting the base. There is of course no amplification of signals in this case.

So far we have investigated the electrical behavior of the p-n-p transistor by very formal mathematical methods. We will now consider qualitatively the behavior of individual electrons and holes in order to gain a more *physical* understanding of transistor action.

The potential distribution in a p-n-p transistor with no voltages applied is

---

[4]The shift of the V–I characteristics of a p-n junction by injection will also be treated in Sect. 5.9 where injection is due to absorption of light.

shown in Fig. 5.12. In the heavily doped p-type emitter the Fermi level is locked at the acceptor levels, while in the heavily doped n-type base, it is locked at the donor levels. The p-type collector is lightly doped. Since it may then be considered to be near-intrinsic, the Fermi level is close to the middle of the gap [see (3.1.16)]. The potential well at the base is a barrier to the diffusion of holes from the emitter to the collector. The flow of holes across this barrier is controlled by the emitter-base voltage V_e as shown in Fig. 5.11 (left diagram), while the collector-base voltage V_c has little influence on I_c. The small difference of the current amplification factor α from unity is due to the loss of holes, on their way from the emitter through the base towards the collector, by recombination with electrons. In this rather qualitative consideration, we have neglected the current contribution by electrons which may not be quite as apparent as the hole contribution. The reverse current of the base collector junction is due to electrons which are the minority carriers. Because the depletion region of this junction has a high impedance which manifests itself by flat collector characteristics, a small increase in collector current causes a large increase in collector voltage if the battery voltage is supplied via a series resistor having not too small a resistance.

The n-p-n transistor is similiar to the p-n-p transistor except that the roles of electrons and holes are exchanged; it will not be discussed here.

Let us again consider the Haynes–Shockley experiment discussed in Sect. 5.2. The part of the germanium sample between the emitter and the collector contacts may be considered as a base. In contrast to the transistor discussed above, the base is much longer than the diffusion length of minority carriers. This requires that the minority carriers drift rather than diffuse through the base; this is achieved by the application of a drift field. It is the action of emitter and collector which is the same as in a transistor. Similar to the Haynes-Shockley arrangement is the *filament transistor* where the metal-semiconductor contact close to the collector of the Haynes-Shockley type sample is used as a collector.

A feature of the Haynes-Shockley experiment is that the emitter and collector contacts are metal pins rather than p-n junctions. In the following chapter we shall explain how a metal-semiconductor contact under proper conditions behaves like a p-n junction. The first type of transistor invented made use of such contacts.

5.6 The Metal-Semiconductor Contact

A metal-semiconductor contact can be used as a rectifier for an ac voltage like a p-n junction.[5]   Rectification is improved by *forming* which is a local heating of an extremely small contact area by a large current pulse of about one

---

[5]For a review see [5.20].

second duration. During the pulse, metal atoms diffuse from the metal pin into the semiconductor thus forming a p-n junction. This method is not applicable in the case of large-area contacts such as those used in solar cells [5.21] and *surface barrier counters* for high-energy particles in nuclear and high-energy physics [1.6, 7]. For this purpose, by electron beam heating gold is evaporated on silicon to form a thin layer at one side of the crystal surface. By adsorption of oxygen molecules during a few days of exposure to room air, an electrical dipole layer is formed which rejects majority carriers from the surface towards the bulk of the semiconductor thus forming a carrier depletion region at the surface called a *Schottky barrier*. With Pd, Pt, and Ti at 400–600 °C, silicide formation also yields stable Schottky rectifiers.

For the theoretical treatment we consider two cases of practical interest: in the first we entirely neglect the diffusion contribution to the current while in the second case, the field contribution is neglected.

Let us consider the first case applicable to silicon and germanium. By n_b we denote the carrier density in the semiconductor bulk. If V is the voltage applied to the contact and V_D is a diffusion voltage characterizing the band bending near the surface due to carrier rejection, the current density contribution due to holes flowing from the semiconductor into the metal, or electrons flowing in the opposite direction, is given by

$$j_1 = e\, n_b \sqrt{k_B T/2\pi\, m}\, \exp[-|e|(V_D - V)/k_B T] \tag{5.6.1}$$

The exponential represents the familiar *Boltzmann factor*. We introduce the surface concentration of carriers n_s for no applied voltage ($V = 0$) given by

$$n_s = n_b \exp(-|e|V_D/k_B T) \ . \tag{5.6.2}$$

and obtain for j_1:

$$j_1 = e\, n_s \sqrt{k_B T/2\pi\, m}\, \exp(|e|V/k_B T) \ . \tag{5.6.3}$$

This, together with the contribution j_2 due to carriers going in the opposite direction, yields the total current j:

$$j = j_1 + j_2 \ . \tag{5.6.4}$$

The magnitude of j_2 is easily obtained from the fact that $j = 0$ for no applied voltage $V = 0$, and we obtain for the total current density

$$j = e\, n_s \sqrt{k_B T/2\pi\, m}[\exp(|e|V/k_B T) - 1] \ . \tag{5.6.5}$$

There is a saturation of the current density for a large reverse bias voltage.

For the second case (no field contribution), one obtains just as in the case of the p-n junction

$$j = j_s[\exp(|e|V/k_B T) - 1] \ , \tag{5.6.6}$$

where the current density for a large reverse voltage equals

$$j_s = |e|\mu_n n_s E_s \tag{5.6.7}$$

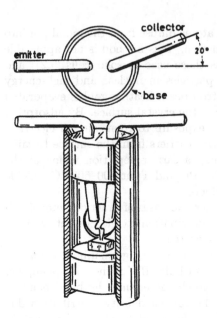

Fig. 5.13: Cut-away view of a point-contact transistor, a historical reminiscence (after [4.85])

and E_s is the electric field strength at the surface given by

$$E_s = \sqrt{(2k_B T/eL_D^2)(V_D - V)} \; , \tag{5.6.8}$$

where L_D is the Debye length given by (5.2.24). Since this factor depends on V there is no current saturation .

Advantages of Schottky diodes are low temperature fabrication, fast switching (10^{-9} s), low forward and large reverse impedances and the fact that they can easily be integrated in *integrated circuits*. The barrier height is lower than the built-in potential of the p-n junction. Most Schottky diodes are used as clamping diodes in transistor-transistor logic (TTL) integrated circuits.

5.7 Various Types of Transistors Including MOSFET

The first commercial transistor was a point-contact transistor. A later version is shown in Fig. 5.13 in a cut-away view ([5.22],[4.87]). On top of a brass plug a germanium wafer 0.5 mm thick is soldered. Against the wafer two phosphor bronze springs (0.01 mm diameter) with sharp points are pressed at a distance of 0.02 mm which is much less than the diffusion length. The springs are welded to nickel mounting wires. After an electrical forming treatment one spring works as an emitter and the other as a collector, while the plug is the base. Typical voltages and currents applied are: emitter 0.7 V, 0.6 mA; collector 40 V, 2 mA. A large diffusion length and a good rectifying property are absolutely necessary for the qualification of a semiconductor as a transistor material. Prior to the invention of the germanium transistor, semiconductors used for rectification purposes were cuprous oxide and selenium. Lifetime and mobility in these semiconductors are too small, however, to yield transistor action. In a lead sulfide transistor, a current amplification has not been observed [5.23]; the power amplification factor was only 4 which is small compared to that in a germanium or silicon transistor. Even in these materials, purification, which nowadays is done by zone-refining [5.24], is essential for obtaining the large diffusion length necessary for transistor action. The point contact transistor is of historical interest only. Then came the alloy transistor [5.25], usually a p-n-p type made of germanium.

Two small pellets of indium metal are placed on either side of a semiconductor wafer of about 0.1 mm thickness. While the assembly is heated, (to 820 K in the case of germanium) an alloying process takes place which results in the formation of the emitter and collector contacts. The base contact is a dot contact at one edge of the wafer. By making the collector somewhat larger than the emitter, good collection with a minimum of surface recombination around the emitter is guaranteed. The alloy process is not sufficiently controllable to yield the extremely small base width, required for a high-frequency transistor, which is of the order of 10^{-3} mm.

After 1960 the *planar transistor* [5.26] became the most frequently fabricated type of transistor. By thermal oxidation (for 1 hour at 1500 K or 10 hours at 1200 K in steam) a thin ($\approx 1\,\mu m$) layer of silicon dioxide is grown on silicon which is free of holes and highly insulating (*quartz*) [5.27]. With the help of the *photoresist process* [5.28, 29], small holes can be cut into this layer. For this process the surface of the silicon wafer should be flat (*planar*), otherwise imperfections result. The quartz layer is covered with a thin coat of *Kodak Metal-Etch-Resist* which is baked and exposed to UV light through a mask [5.29]. The exposed parts of the etch-resist solidify by polymerization due to the exposure while the unexposed parts can be removed by an organic solvent. The resist is baked again. The parts of the quartz layer not protected by the polymerized etch-resist are then etched away by hydrofluoric acid. The remaining etch-resist is afterwards dissolved. The leftover parts of the quartz layer are then used as a mask in a subsequent process of diffusion of impurities like, e.g., boron into the n-type silicon substrate. On top of this structure , another quartz layer is generated and perforated with holes smaller than the first ones but at the same locations. Through this second set of holes, phosphorus is diffused into the silicon substrate to a depth that is less than the previous boron diffusion depth. Phosphorous overcompensates the existing boron doping and yields a heavily doped n-type layer. A cross section of the n-p-n structure so obtained is shown in Fig. 5.14. At the position indicated by an arrow, another hole is then etched through the second quartz layer. Fine metal wires

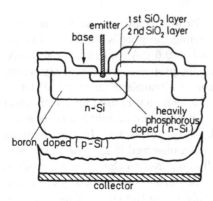

Fig. 5.14: Silicon planar transistor (cross section)

are pressed through the holes and alloyed with the silicon thus serving as connections between the emitter (n-Si, phosphorous doped) and the emitter power supply, and between the base (p-Si, boron doped) and the base power supply, respectively. A third alloy contact at the back of the n-type silicon wafer serves as the collector . Since the base is only a few μm thick the carrier diffusion time is quite small and high-frequency operation is possible. Another advantage of this process is that the p-n junctions at the surface are protected against atmospheric influences by the grown oxide layers with the result being a smaller leakage current and better stability.

With the help of the masking and diffusion techniques, complete circuits with resistors, capacitors, transistors and rectifiers are fabricated in a single silicon chip with very small space requirements. These *integrated circuits* (IC) became important for use in computers. Up to a million transistors per mm^2 are being made on a single chip . Precise photographic masks are required to define the microscopic areas to be exposed to etchants [5.30]. For producing features with minimum dimensions (less than 100 nm) [5.31], pattern delineation uses a focused beam of 10 keV electrons. Modulation and deflection of a beam of charged particles can be accomplished with high speed and great precision and diffraction phenomena are avoided. A promising novel technique is ion beam lithography [5.32]. In the *bipolar technology* discussed above, from five to seven masking and etching steps are needed.

In a MOS technology [5.33, 34] only four such steps are required. A MOSFET is a *m*etal *o*xide *s*emiconductor *f*ield *e*ffect *t*ransistor [5.34, 35]. A cross section is given in Fig. 5.15. A voltage is applied between the electrodes labelled *source* and *drain*. The circuit thus contains two p-n junctions, one of which is biased in the reverse direction while the other is biased in the forward direction. The reverse biased junction has a high impedance. A positive potential is applied to the electrode called *gate* which is insulated from the semiconductor by a quartz layer. The positive potential at the gate induces a negative charge at the boundary between the semiconductor and the insulating quartz layer. Due to the induced negative space charge the semiconductor, although doped p-type, converts to n-type at the boundary thus forming an n-type channel between the

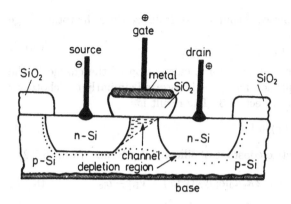

Fig. 5.15: Metal oxide semiconductor field effect transistor (MOSFET), cross section

two n-type source and drain regions. A strong current flows through the channel which can be controlled by the gate potential. The small leakage currents to the gate electrode are nearly independent of temperature. Since the breakdown field intensity in quartz is about $5 \times 10^6 \, V/cm$, gate voltages up to $10^3 \, V$ may be used.

For a calculation of the current voltage characteristics of a MOSFET, it is convenient to introduce a *base* as a fourth probe although the actual MOSFET does not require it for operation. At the interface between silicon and the oxide SiO_2 close to the gate electrode, there is a localized interface charge q_{ic} due to dangling silicon bonds (Sect. 14.1). If an appropriate voltage V_{GB} is applied between the gate and the base, a charge $-q_{ic}$ is induced at the gate and a space charge in the silicon is avoided. V_{GB} must equal $-q_{ic}/C_{ox}$, where C_{ox} is the capacitance of the oxide per cm^2 ($35nF/cm^2$ for a thickness of the oxide layer of 100 nm). However, if V_{GB} is reduced the gate charge is also reduced and in the silicon, a space charge of negatively charged acceptors not compensated by positive holes builds up, i.e., there is then a depletion layer. As a consequence of the potential resulting from the space charge, the energy bands of the silicon are bent down which may be so strong that the conduction band approaches the Fermi level and the electron concentration $n = n_i^2/p$ becomes larger than the hole concentration: there is an inversion layer of electrons connecting the n-type source with the n-type drain as indicated in Fig. 5.15. On the application of a small voltage V_{DS} between drain and source, there is a current I_D which is proportional to the field dV/dx and the charge density in the inversion layer q_{inv}:

$$I_D = \mu_n \, q_{inv} \, w \, dV/dx \; , \tag{5.7.1}$$

where μ_n is the electron mobility and w the channel width. For a calculation of q_{inv} we consider the total charge q_{tot}. It consists of the localized interface charge q_{ic}, the mobile charge in the inversion layer q_{inv} and the localized charge in the depletion region q_{dep}:

$$q_{tot} = q_{ic} + q_{inv} + q_{dep} \; . \tag{5.7.2}$$

As the potential $V(x)$ varies between source and drain, so does the total charge

$$q_{tot} = C_{ox}[V_{GS} - V(x)] \; , \tag{5.7.3}$$

where V_{GS} is the gate source voltage. For a calculation of the charge in the depletion region, Poisson's equation has to be integrated similarly to Schottky's treatment of the p-n junction in Sect. 5.3. The result is

$$q_{dep} = \sqrt{2\kappa_s\kappa_0|e|\,N_A}[V(x) + V_B] \; , \tag{5.7.4}$$

where κ_s is the relative dielectric constant of silicon . [6]   Hence, from (5.7.2),

$$q_{inv} = C_{ox}[V_{GS} - V(x)] - \sqrt{2\,\kappa_s\kappa_0|e|\,N_A[V(x) + V_B]} - q_{ic} \tag{5.7.5}$$

and from (5.7.1), the drain current I_D is calculated. Integration of I_D over the channel length L yields

$$I_D\,L = \mu_n w \left\{ C_{ox} \left[(V_{GS} - V_t')\,V_{DS} - \frac{1}{2}\,V_{DS}^2 \right] - \frac{2}{3}\sqrt{2\,\kappa_s\,\kappa_0|e|\,N_A} \right.$$

$$\left. \times \left[(V_{DS} + V_B)^{3/2} - V_B^{3/2} \right] \right\} \; , \tag{5.7.6}$$

where $V_t' = q_{ic}/C_{ox}$ has been introduced for simplicity. A linear approximation of the $3/2$ power term relative to V_{DS} results in

$$I_D = \beta\,V_{DS}(V_{GS} - V_t - \frac{1}{2}V_{DS}) \quad \text{for} \quad 0 \le V_{DS} \le V_{GS} - V_t \; , \tag{5.7.7}$$

where $\beta = \mu_n w C_{ox}/L$ and $V_t = V_t' + \sqrt{2\kappa_s\kappa_0|e|\,N_A\,V_B}$ have been introduced. So far we have tacitly applied the *gradual channel approximation* valid up to the maximum value for I_D at $V_{DS} = V_{GS} - V_t$. This maximum value is

$$I_D = \frac{\beta}{2}(V_{GS} - V_t)^2 \; . \tag{5.7.8}$$

It can be shown that this is approximately valid also for $0 \le V_{GS} - V_t \le V_{DS}$. At these high values of V_{DS}, the inversion layer is shorter than the source drain distance L and electrons are injected from the inversion layer into the depletion region on their way towards the drain. In the depletion region the field strength is large resulting in the carriers becoming *hot* (Sect. 4.13) and their drift velocity approaching a saturation value of about 10^7 cm/s. The beginning pinch-off of the inversion channel is shown in Fig. 5.15. At a still higher voltage the inversion layer disappears completely. Typical characteristics are shown in Fig. 5.16. As (5.7.7) indicates for a small but fixed voltage V_{DS}, the drain current $I_D \approx 0$ for $V_{GS} = V_t$. Hence, V_t is a threshold voltage. A short wide channel yields a large drain current and a large upper frequency limit for ac operation [given by $\mu_n(V_{GS} - V_t)/2\pi L^2$]. By means of "VMOS" techniques, channel lengths down to $L = 1\,\mu\text{m}$ at widths of $w = 25\,\mu\text{m}$ have been produced. Switching times

---

[6]V_B can be approximated by the sum of the source base voltage and twice the energy difference between the bulk Fermi level and the midgap position, in units of $|e|$.

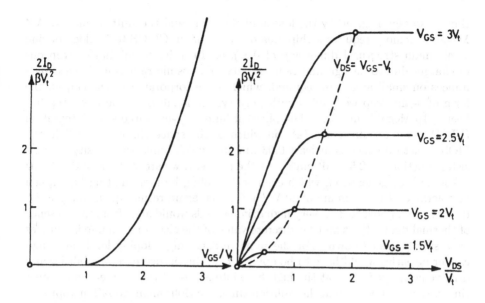

Fig. 5.16: Current voltage characteristics for an n-channel depletion type MOSFET

are then as low as 2 ns. The gate resistance is about 10^{13} to $10^{14}\,\Omega$ due to the excellent insulating properties of the quartz layer. It takes great efforts to avoid getting mobile alkali ions into the quartz which ruin the transistor effect [5.36]. Top layers of Si_3N_4 or Al_2O_3 guard the SiO_2 film against contamination. Since even high purity silicon contains measurable amounts of sodium, the crystal itself may be a source of alkali contamination of the oxide. A *getter*[7] treatment with glassy phosphorous oxide and the manufacture of extremely smooth surfaces paved the way to present-day silicon technology. Silicon single crystals are now pulled from the melt at diameters up to 30 cm and several meters in length and sliced into wafers of a few tenths of a mm in thickness. Each wafer is subdivided into chips of 30 mm^2 area. As mentioned above, integrated circuits are produced on these chips for computers and microprocessors which include memories. Charge-coupled devices (CCD) are gates arranged in linear arrays without any sources and drains [5.34]. Positive gate potentials produce depletion layers in the p-type silicon. Electrons deliberately introduced collect under the most positive gate. Gate potentials are changed in steps with 120° phase shifts between adjacent gates with the effect that the electrons are moved along the array. A charge packet might contain 10^7 electrons over a time period of 10^{-7} s. Collecting the charge in a reverse biased p-n junction results in a current of $(10^7 \times 1.6 \times 10^{-19}/10^{-7})$A $= 16\mu$A. In a binary memory, the injected charge represent s a "1" while the absence of a charge represents a "0". The

---

[7]Gettering is a removal or preventing of metal precipitates from surface layers, as described by A. Goetzberger, W. Shockley, J. Appl. Phys. **31**, 1821 (1960).

time of storage is given by the length of the array and the shift frequency. A 4 Megabit memory requires a chip area of 91.3 mm^2 in CMOS technology having a minimum structural dimension of 0.8 μm. In order to minimize erroneous discharges due to background radioactivity and cosmic radiation, large capacitances on small areas are required, which are incorporated in the chip in the form of 4-μm deep wells filled with polycrystalline silicon. The next step has been a 16 Megabit memory. The physical limits to very large scale integration (VLSI) are determined by: (a) the width of the space charge channel in the MOS structure which is at least 0.03 μm and which limits the density of transistor functions to 2.5×10^7 cm^{-2}. At the moment, we are still above this limit; (b) joule heat generation, which even with cooling by a circulating liquid, can be tolerated at a maximum of 20 W/cm^2 which again results in the integration density given above [5.37]. Very high driving fields would arise from the presence of thermal fluctuations and the discrete value of the electronic charge if voltages were simply scaled down. For the same reasons, high doping levels and high carrier densities would have to be applied resulting in decreased mobilities and drift velocity saturation at the high field intensities. If the source-drain transit time becomes shorter than the collision time, the Boltzmann equation approach to transport fails and tunnel phenomena may become important [5.38]. Pattern replication on the chip may result in superlattice effects (Sect. 9.1). Of course the speed of signal transmission which determines the practical usefulness of computers increasing in size is limited finally by the velocity of light.

Both on silicon and on gallium arsenide, MESFET structures have been realized for operation in the 10-GHz range of frequencies. In a MESFET structure the gate is a Schottky barrier diode, i.e. a metal semiconductor rectifying contact which in contrast to the MOSFET does not contain an oxide layer. Gallium arsenide offers an advantage over silicon because of the higher electron mobility: In short-channel ($\sim 1\,\mu m$) GaAs MESFET's the drain currents are about six times those for silicon n-channel FET's at equivalent gate biases resulting in much higher switching speeds [5.39, 40]. Another advantage is the availability of semi-insulating GaAs for a substrate. Using electron beam lithography, gate lengths of 0.5 μm and propagation delays in circuits below 20 ps would be possible. These speeds are about the same as those achieved with superconducting Josephson junction technology but without liquid helium cooling. If cooling with liquid nitrogen ($\sim 77\,K$) would be anticipated, even higher performance might be obtained by going to InAs or InSb with electron mobilities up to 10^6cm$^2/Vs$ at this temperature. Besides excellent switching speeds very low operating voltages and dynamic switching energies would be obtained.

The ultimate transistor will be based on single electron tunneling SET (see (14.5.16) and text thereafter).

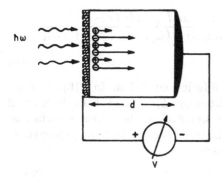

Fig. 5.17: Dember arrangement

5.8 Dember Effect and PEM Effect

This and the next section are devoted to a further discussion of the diffusion of carrier pairs generated by the absorption of light. Details of the absorption process will not concern us here and will be discussed in Chap. 11.

Figure 5.17 shows a semiconductor sample with metal contacts on opposite sides which are connected by a voltmeter. One of the contacts is semitransparent to light of a frequency high enough to generate electron hole pairs close to the surface. Let us assume a penetration depth α^{-1} of the incident light in the semiconductor having a generation rate G for a given light intensity (α: absorption coefficient). The product $\alpha^{-1}G$ is denoted by G_0 and measured in units of $\mathrm{cm^{-2}\,s^{-1}}$. The electron hole pairs generated at one side of the sample diffuse towards the opposite side. Since, in general, the diffusion coefficients of both carrier types are different, a voltage V is indicated by the voltmeter. The effect is called *photodiffusion* or the *Dember effect* [5.41–43]. If one of the contacts is somewhat non-ohmic or even rectifying, a photovoltaic effect is measured instead of the Dember effect; this will be discussed in the following section.

The recombination of carriers at the surface will be taken into account. A *surface recombination velocity* s is defined by considering a current density perpendicular to the surface

$$j_n = |e|\, s\, \Delta n \tag{5.8.1}$$

and a current density of holes $j_p = -j_n$, where $\Delta n = \Delta p$ is the excess of carriers over the equilibrium values; s is the velocity of pair disappearance in the surface *drain* .

The solution of the diffusion equation (5.5.1) which we write in the form

$$D\, d^2p/dx^2 = \Delta p/\tau \tag{5.8.2}$$

is given by

$$\Delta p(x) = \frac{G_0 L}{D} \frac{\cosh[(x-d)/L] - L/L_{sd} \sinh[(x-d)/L]}{(1 + L^2/L_{sd}L_{s0}) \sinh(d/L) + (L/L_{s0} + L/L_{sd}) \cosh(d/L)}$$

(5.8.3)

where $L = \sqrt{D\tau}$, D is the ambipolar diffusion coefficient (5.2.16), τ is the lifetime of holes and electrons assumed to be equal, $L_{s0} = D/s$ at the illuminated surface $(x = 0)$, L_{sd} is the corresponding quantity at the opposite surface at $x = d$, and d is the sample dimension in the direction of light propagation, assumed to be $d \gg \alpha^{-1}$. The current density

$$j_{nx} + j_{px} = \sigma E - |e| (D_p - D_n)dp/dx$$

(5.8.4)

vanishes and the Dember voltage is given by

$$V = \frac{|e| (D_n - D_p)}{\sigma}[\Delta p(0) - \Delta p(d)] .$$

(5.8.5)

We obtain $\Delta p(0) - \Delta p(d)$ from (5.8.3):

$$V = k_B T/|e|$$

(5.8.6)

$$\times \frac{G_0 L(\mu_n - \mu_p) [\cosh(d/L) + (L/L_{sd}) \sinh(d/L) - 1]}{D(n\mu_n + p\mu_p) [(L/L_{s0} + L/L_{sd}) \cosh(d/L) + (1 + L^2/L_{sd}L_{s0}) \sinh(d/L)]}$$

For the simplifying case of $d \gg L$ and $L_{s0} = L_{sd}$

$$V = G_0 \frac{k_B T}{|e|} \frac{\mu_n - \mu_p}{n\mu_n + p\mu_p} \frac{1}{\sqrt{D/\tau} + s} .$$

(5.8.7)

Obviously the method is applicable for a determination of the surface recombination velocity s if the mobilities and the lifetime of minority carriers are known. For a photon flux of $G_0 = 10^{18}/cm^2$ s and an electron density of $2.6 \times 10^{13}/cm^3$ (and practically no holes), a hole mobility $\mu_p = 1/2\,\mu_n$, a D/τ ratio of 10^6 cm^2/s^2 at room temperature and neglecting surface recombination, a Dember voltage of 0.5 V would be obtained. A surface recombination velocity s of 10^6 cm/s, which is typical of a sandblasted surface, would reduce the effect to 500 μV. For an etched surface where s is of the order of magnitude of $10^2\,cm/s$, surface recombination would not appreciably influence V.

The application of a magnetic field perpendicular to the direction of light propagation yields the Hall equivalent of photodiffusion called the *photoelectromagnetic effect* (PEM effect). The arrangement is shown in Fig. 5.18. The polarity of the voltage V_y is reversed if B_z is reversed. The effect has been thoroughly treated theoretically by *van Roosbroeck* [5.44]. A simplified treatment neglecting surface recombination and assuming sample dimensions much larger than a diffusion length will be presented here. Only the case of a weak magnetic field in the z-direction will be considered. The current density component in the y-(*Hall*) direction is given by

$$j_y = \sigma E_y + (j_{nx} \mu_{Hn} - j_{px} \mu_{Hp}) B_z ,$$

(5.8.8)

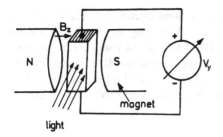

Fig. 5.18: PEM arrangement (schematic)

where the subscript H indicates the Hall mobilities. Since no current is assumed to exist in the x-direction ($j_{px} = -j_{nx} = -|e|\, D\, dp/dx$), we obtain for the PEM short-circuit current density ($E_y = 0$)

$$j_y = |e|\, D(\mu_{Hn} + \mu_{Hp})\, B_z\, \Delta p/L \ , \tag{5.8.9}$$

where in the approximation mentioned above, dp/dx has been replaced by $\Delta p/L$. Occasionally, the PEM effect is combined with a measurement of the photo-conductivity to yield the minority carrier lifetime τ. (Actually, the electrical injection method as applied in the Haynes-Shockley experiment yields more accurate results for τ, and the PEM effect is more often used for a determination of the surface recombination velocity.) We apply an electrical field E_y but no magnetic field. The change in current density upon illumination of the sample Δj_y is given by

$$\Delta j_y = |e|\, (\mu_n + \mu_p)\, E_y\, \Delta p \ , \tag{5.8.10}$$

and the ratio of photo and PEM current densities is given by

$$(\Delta j_y)_{\text{Photo}}/(j_y)_{\text{PEM}} = \frac{E_y}{B_z\, r_H}\frac{L}{D} = \frac{E_y}{B_z\, r_H}\sqrt{\frac{|e|\tau}{k_B T \mu}} \ , \tag{5.8.11}$$

where the Hall factor r_H has been assumed to be the same for electrons and holes and the same illumination is applied to the measurement of both effects.

The open-circuit PEM field E_y is obtained by replacing j_y in (5.8.9) by $-\sigma E_y$ and averaging over the sample cross section:

$$E_y = -\frac{|e|D(\mu_{Hn} + \mu_{Hp})\, B_z}{\sigma L}\frac{1}{d}\int_0^d \Delta p\, dx \ . \tag{5.8.12}$$

The integral in the approximation mentioned above is given by $\Delta p \cdot L$. The PEM voltage is $V_y = E_y l$ where l is the length of the crystal. Hence,

$$V_y = -\frac{|e|D(\mu_{Hn} + \mu_{Hp})\, B_z}{\sigma}\frac{l}{d}\Delta p = -\frac{k_B T}{|e|}\mu_{Hp}\, B_z\left[\frac{b(b+1)\,(\eta+1)}{(\eta b + 1)^2}\frac{\Delta p}{p}\right]$$

$$\tag{5.8.13}$$

where $\eta = n/p$ and $b = \mu_n/\mu_p = \mu_{Hn}/\mu_{Hp}$ have been introduced. For η and $b \gg$ 1, the factor in braces becomes $\Delta p/n = \Delta n/n$. Assuming for this factor a value of 10% and for $\mu_{Hp}B_z$ a value of 10%, the PEM voltage at room temperature is seen to be quite small, namely, $260\,\mu V$. It is quite sensitive to inhomogeneities in doping since internal $p - n$, n^+-n or p^+-p barriers yield a photovoltaic effect much larger than the PEM effect. The former will be described in the following section.

5.9 Photovoltaic Effect

When light, X rays, β rays or other radiation, whose quantum energy exceeds a threshold of the order of the gap energy, ε_G, ionizes the region in or near a potential barrier, a photovoltaic potential is generated. The most important device application of the photovoltaic effect is the solar cell for conversion of solar radiation into electrical energy. At the surface of the atmosphere of the earth or at the moon, the solar power density is $0.135\,W/cm^2$. Assuming a conversion efficiency of 10%, a 5 kW power consuming lunar vehicle motor requires 30 m^2 of effective solar cell surface for operation.

For equilibrium $(\partial \Delta n/\partial t = 0)$, the continuity equation (5.2.9) becomes for holes $(e > 0)$

$$\frac{1}{e}(\nabla_r \cdot j_p) = G - \frac{p - p_n}{\tau_p} \ . \tag{5.9.1}$$

which, except for the term G, is equal to (5.3.21). With (5.3.22) we obtain for holes at the n-side of the junction

$$-D_p(\nabla_r)^2 p = G - \frac{p - p_n}{\tau_p} \ . \tag{5.9.2}$$

The boundary conditions are $p = p_n$ and $G = 0$ at $x \gg x_n$ and p at $x = x_n$ as given by (5.3.24). The p-side of the junction is assumed to be close to the illuminated surface as shown in Fig. 5.7. An equation similar to (5.9.2) holds for electrons at the p-side. For the boundary condition at the illuminated surface, we assume, for simplicity, surface recombination to be very large so that we may assume $G = 0$ at the p-side. Otherwise, the conditions are the same as those given in Sect. 5.3. We also assume that the incident radiation is converted into electron hole pairs only in the transition region and the diffusion lengths $L_p = \sqrt{D_p \tau_p}$; $L_n = \sqrt{D_n \tau_n}$, are much larger than the junction width $d = |x_p - x_n|$. The solution (5.9.2) is then given by

$$p(x) - p_n - G\tau_p = [p(x_n) - p_n - G\tau_p]\exp[-(x - x_n)/L_p] \tag{5.9.3}$$

which yields for the hole component of the current density

$$j_p = \frac{|e|\,D_p\,p_n}{L_p}[\exp\frac{|e|V}{k_B T} - 1] - G\,|e|\,L_p \ , \tag{5.9.4}$$

where the relation $D_p \tau_p = L_p^2$ has been taken into account. The electron component is similar except for the term $G|e| L_p$. Hence, the total current density is given by

$$j = j_s[\exp(|e|V/k_BT) - 1] - j_L , \tag{5.9.5}$$

where

$$j_L = G|e| L_p = \frac{\Delta p}{\tau_p}|e|\sqrt{D_p \tau_p} = |e| \Delta_p \sqrt{D_p/\tau_p} \tag{5.9.6}$$

is the current component due to irradiation of the junction. A more complete expression for j_L, which includes the finite penetration depth of light and a finite surface recombination velocity, is found in [Ref. 5.44, p. 163].

The short-circuit current density $j = -j_L$. The open-circuit voltage V_{oc} is given by

$$V_{oc} = \frac{k_BT}{|e|} \ln(1 + j_L/j_s) . \tag{5.9.7}$$

The output power per front area is $j V$. It is a maximum when $\partial(j V)/\partial V = 0$; this yields for the maximum power (subscript mp) compared with the open-circuit value (op)

$$V_{mp} = V_{oc} - \frac{k_BT}{|e|} \ln(1 + |e|V_{oc}/k_BT) \tag{5.9.8}$$

assuming $V_{oc} \gg k_BT/|e|$, and a maximum power per front area relative to the product of the short-circuit current and the open-circuit voltage *fill factor* FF:

$$FF = \frac{j_{mp}V_{mp}}{j_L V_{oc}} = \frac{V_{mp}}{V_{oc}} \exp(|e|V_{mp}/2k_BT)\frac{\sinh\{|e|(V_{oc} - V_{mp})/2 k_BT\}}{\sinh(|e|V_{oc}/2 k_BT)} \tag{5.9.9}$$

In Fig. 5.19, typical current-voltage characteristics of a silicon p-n junction under illumination is shown. The shaded area represents the maximum power rectangle. A typical value for V_{oc} at room temperature is 0.57 V. The fill factor would be 82% according to (5.9.9). If, however, the voltage drop, particularly at the front area of the device outside the actual p-n junction, is taken into account, a more realistic value of 75% is obtained.

The history of the silicon photovoltaic solar cell began in 1954 [5.46]. A large-scale application, however, had to wait until the advent of the space age three years later. The low weight, its simplicity and its reliable performance in utilizing a source of energy which is available practically everywhere in space were the basic factors that made it a success. Telstar in 1963 had already 3600 cells on board with 14 W of power output, while Skylab ten years later used 500 000 cells with 20 kW power output, just to name two out of more than thousands of spacecrafts powered by solar cells nowadays.

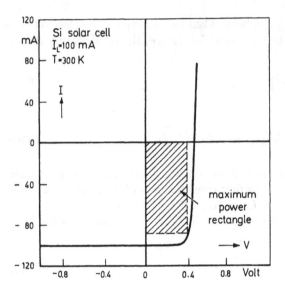

Fig. 5.19: Current-voltage characteristics of a solar cell under illumination (after [5.45])

When an oil crisis hit the world [8] in 1973, research efforts were directed towards the development of a solar cell for terrestrial use which was low in production cost and had a high yield and long lifetime in order to be able to compete with conventional energy production techniques [5.47].

An interesting problem for terrestrial use is the amount of energy required for producing a cell, relative to its energy output during its lifetime, which depends, of course, on the availability of sunshine. Concentrators comparable to magnifying lenses have been developed for both direct and diffuse incident light. Most efforts have been devoted to silicon p-n junctions and Schottky barriers even though silicon is an indirect semiconductor and therefore cannot provide the highest possible yield. It is produced from quartz which is available in large enough quantities if necessary to meet present-day energy production requirements all over the world. It is nontoxic and its corrosion product is again nontoxic quartz. Its technology is well known from transistor production. Its band gap energy is close enough to the photon energy for maximum sunlight intensity: yields of up to 24% in cells made from single crystal silicon and 10% in much less expensive polycrystalline cells have been obtained. A major cost reduction has been achieved by applying material with crystallites which are not much larger than the diffusion length of minority carriers, and by limiting the expensive highly purified silicon to a thin layer, with a thickness almost equal to the penetration depth of electron hole pair producing photons, on top of a cheap substrate. Since solar cells are subject to wide variations in temperature, particularly if used in conjunction with concentrators, the thermal expansion of the substrate must match that of the top layer. For this reason, cheap metallurgical grade silicon has been applied for a substrate [5.48].

---

[8]The heyday of the solar cell will have to wait until the decline of terrestrial oil fields.

Depositing a 100 μm thick layer of pure silicon on a substrate is easily done by chemical vapor deposition (CVD) from a gaseous SiHCl$_3$/H$_2$ mixture at about 1300°C. The p-n junction is produced by an admixture of diborane B$_2$H$_6$ for the p-side or phosphine PH$_3$ for the n-side. A different technique is based on *phosphorus – ion* implantation of p-type silicon and subsequent laser annealing of the lattice defects produced by the implantation process [5.49, 50]. Schottky barrier diodes instead of p-n junctions have been successfully applied as well, particularly in amorphous silicon which has been produced by a glow discharge in SiHCl$_3$ and treated with hydrogen for a saturation of dangling bonds [5.51]. Problems of low-resistance contacts, encapsulation and antireflex coating, as well as the storage of electrical power, cannot be discussed here.

Solar cells competitive with those of silicon have been developed from layers of cadmium sulphide CdS produced by evaporation on a substrate and subsequent conversion of the surface of CdS to Cu$_2$S by either dipping in a copper salt solution or by a solid state process at elevated temperatures [5.52]. The energy gap of Cu$_2$S is similar to that of silicon. The Cu$_2$S – CdS transition acts as a heterojunction. However, for a large scale production, both the toxicity and the availability of cadmium may possibly turn out to be adverse factors. Also, the interface between two semiconductors of different lattice constants will contain a mismatch dislocation field with compensating lattice defects surrounding each line of dangling bonds [5.48]. This causes a decrease in the open circuit voltage. Replacing CdS by Cd$_{0.75}$Zn$_{0.25}$S or Cu$_2$S by CuInSe$_2$ may be advantageous [5.53].

Various other processes such as, e.g., producing thin ribbons of single crystal pure silicon by Czochralski crystal growing, or ingots by casting and subsequent slicing into wafers by inner-diameter sawing with diamond abrasives, are being pushed forward. 20% efficiency at 100°C makes GaAs the front runner in the race for concentrated-light conversion, although gallium is very expensive and arsenic is highly toxic. Even Cu$_2$O, formerly used for rectifiers, is being considered as a possible solar cell material in view of the abundance and relative cheapness of copper compared with other semiconductor materials [5.54–56].

Problems

5.1. Solve the Boltzmann equation for the case of a carrier concentration $n(z)$ and a distribution function $f(v, z) = n(z) \cdot \Phi(v)$. The carrier mean free path λ is assumed to be $\ll n/(dn/dz)$ independent of v. Show that the diffusion coefficient $D = \lambda\langle v\rangle/3$ where

$$\langle v\rangle = \int\limits_0^\infty v^3 f_0(v)\,dv \left/ \int\limits_0^\infty v^2 f_0\,dv \right. , \tag{5.0.1}$$

and $\langle v_z\rangle = \frac{1}{3}\lambda\langle v\rangle n^{-1}(-dn/dz)$. Derive the Einstein relation from these equations.

5.2. Assume a current I_x in a homogeneously doped filamentary intrinsic semiconductor sample in the x direction. There is a magnetic field B_z in the z-direction, which by means of the Lorentz force causes a current flow in the y-direction. This current hits a sandblasted surface where carriers recombine (recombination velocity $s = 10^6 \text{cm/s}$). The opposite surface is etched ($s = 10^2 \text{cm/s}$). Show that for an ac current of low frequency, rectification occurs. Estimate the upper frequency limit. *Hints:* $s = j_n/e\Delta n = j_p/e\Delta p$. Show that from $dj_n/dy = dj_p/dy = -\Delta n/\tau$ with $L = (D\tau)^{1/2}$ and $L_s = D/s$, the following equations result:

$$d^2 \Delta n/dy^2 + y_d^{-1} d\Delta n/dy = \Delta n/L^2 \tag{5.0.2}$$

and

$$d\Delta n/dy + y_d^{-1}(n_i + \Delta n) = \pm \Delta n/L_s , \tag{5.0.3}$$

where $y_d^{-1} = (e/2k_B T)(\mu_n + \mu_p) E_x B_z$.

5.3. Calculate the differential capacitance for a graded p-n junction: $N_A \propto x$ for $x < 0$ and $N_D \propto x$ for $x > 0$ with the same factor of proportionality on both sides.

5.4. Calculate the frequency dependence of the gain of the bipolar transistor. *Hints:* Solve the time-dependent continuity equation and calculate the complex diffusion length. What is the physical meaning of the maximum frequency in terms of transit time, etc.?

6. Scattering Processes
in a Spherical One-Valley Model

In Chap. 4 we frequently assumed an energy dependence of the momentum relaxation time $\tau_{\mathrm{m}} \propto \varepsilon^r$, where r is a constant, for the calculation of the galvanomagnetic, thermoelectric, thermomagnetic, etc., effects. We will now treat the important scattering mechanisms and find the energy dependence of τ_{m}. For those cases where a power law is found, the magnitude of the exponent r will be determined.

6.1 Neutral-Impurity Scattering

The equation of motion of a carrier in an electric field (4.1.2) contains a *friction term* which is essential for the establishment of a constant drift velocity at a given field intensity. From a microscopical point of view, friction is the interaction of carriers with imperfections of the crystal lattice such as impurities, lattice defects and lattice vibrations. This interaction is called *scattering*. The concept of a *scattering cross section* may be familiar from the theory of transmission of high-energy particles through matter (see, e.g. [Ref. 6.1, p. 110]): the probablity per unit time for a collision $1/\tau_{\mathrm{c}}$ is given by the density of scattering centers N, the cross section of centers σ_{c}, and the velocity v of the particle:

$$1/\tau_{\mathrm{c}} = N \sigma_{\mathrm{c}} v \quad . \tag{6.1.1}$$

τ_{c} is called the *collision time*; it is the mean free time between collisions. For an explanation of this relation, consider N parallel disks of area σ_{c} per unit volume. The particle moves perpendicular to the disks a distance $v\,dt$ in the time interval dt and hits one of the disks with a probability $(N\,\sigma_{\mathrm{c}})\,v\,dt = dt/\tau_{\mathrm{c}}$. σ_{c} is obtained from a *differential cross section* $\sigma(\theta)$ by integrating over the solid angle $d\Omega = 2\pi \sin\theta\,d\theta$, where we assume the scattering center to be spherically symmetric:

$$\sigma_{\mathrm{c}} = 2\pi \int_0^\pi \sigma(\theta) \sin\theta\,d\theta \quad . \tag{6.1.2}$$

θ is the angle of deflection of the particle from its original direction of motion. Hence, after the collision, the component of the particle velocity in the direction of its original motion is $v \cos \theta$. The relative change of this velocity component is therefore

$$\frac{v - v \cos \theta}{v} = 1 - \cos \theta \tag{6.1.3}$$

and this is also the relative change of the corresponding momentum component (the effective mass is assumed to remain constant during the scattering process). Hence, the momentum-transfer cross section σ_m, is given by a modification of (6.1.2), by

$$\sigma_m = 2\pi \int_0^\pi \sigma(\theta)(1 - \cos \theta) \sin \theta \, d\theta \tag{6.1.4}$$

and the momentum relaxation time τ_m is then defined by

$$1/\tau_m = N \sigma_m v = N v 2\pi \int_0^\pi \sigma(\theta)(1 - \cos \theta) \sin \theta \, d\theta \ . \tag{6.1.5}$$

A very fundamental scattering process is the scattering of a conduction electron at a neutral impurity atom in the crystal lattice. A similar process is the scattering of low-energy electrons in a gas. This latter process has been treated quantum mechanically in great detail [6.2]; it had been observed by *Ramsauer*[6.3] before the development of quantum mechanics. The result of the theoretical treatment has been transferred to neutral-impurity scattering in crystals by *Erginsoy* [6.4].

The method applied here is that of *partial waves*: the material wave of the electron is diffracted by the field of the impurity atom in such a way that it fits smoothly to the undistorted wave function outside, which is thought to consist of the partial-wave functions of the plane wave without a scattering center (expansion in Legendre functions of the scattering angle) and the wave function of a scattered wave. Since this is a standard problem in text books on quantum mechanics [Ref. 6.1, Chap. 5], we shall not go into further detail here. The numerical calculation includes both electron exchange effects and the effect of the polarization of the atom by the incident electron. The result can be approximated by a total cross section

$$\sigma = 20 \, a/k \ , \tag{6.1.6}$$

valid for electron energies of up to 25 percent of the ionization energy of the impurity atom; a is the equivalent of the Bohr radius in a hydrogen atom

$$a = \kappa \kappa_0 \hbar^2 / m e^2 = \kappa a_B/(m/m_0) \ , \tag{6.1.7}$$

where $a_B = 0.053$ nm is the Bohr radius, κ is the dielectric constant and m is the *density-of-states effective mass* (Sects. 7.2, 8.1) [6.5]; $k = 2\pi/\lambda = m v/\hbar$

is the absolute magnitude of the electron wave vector. It may be interesting to
note that about the same cross section would have been obtained if the geometrical cross section πa^2 had been multiplied by the ratio λ/a; there is, however,
no simple explanation for this result. In fact, (6.1.6) shows that the cross section varies inversely with the carrier velocity v, while low-energy scattering by a
perfect rigid sphere or by a square-well potential has a cross section that is sub-
stantially independent of velocity [Ref. 6.1, Chap. 5]. For the carrier mobility
μ, we obtain

$$\mu = \frac{e}{20\,a_B\,\hbar}\,\frac{m/m_0}{\kappa\,N^x} \tag{6.1.8}$$

which is independent of temperature. In units of cm^2/Vs, we find

$$\mu = \frac{1.44 \times 10^{22}\,cm^{-3}}{N^x}\,\frac{m/m_0}{\kappa}\ . \tag{6.1.9}$$

For example, for electrons in Ge, where $m/m_0 = 0.12$ and $\kappa = 16$, a mobility of
$1.1 \times 10^3\,cm^2/Vs$ is obtained assuming, e.g., $10^{17}\,cm^{-3}$ neutral impurities. Concerning the dependence on the dielectric constant, neutral impurity scattering
contrasts, as we shall see later on, with ionized-impurity scattering.

Neutral-impurity scattering is always accompanied by other scattering mechanisms such as ionized-impurity scattering and lattice scattering. At very low
temperatures where impurities are neutral due to carrier freeze-out and where
phonons have disappeared to a large extent making ionized impurity scattering
and lattice scattering unimportant, impurity band conduction dominates the
conductivity (Sect. 6.14) and therefore no experimental data on mobility are
available to compare with (6.1.9).

However, both the linewidth of cyclotron resonance and the attenuation of
ultrasonic waves have been shown to be determined at low temperatures by
neutral impurity scattering. Although these effects will be discussed later on
(Sects. 11.11, 7.8 respectively), we consider the results here as far as they can
be explained by neutral-impurity scattering.

Figure 6.1 shows the inverse relaxation time obtained from the linewidth
of cyclotron resonance, plotted vs the impurity concentration [6.6]. The data
obtained for a shallow donor (Sb in Ge) are in good agreement with Erginsoy's
formula (6.1.8) where $\mu = (e/m)\,\tau_m$. However, for shallow acceptors (Ga and
In in Ge), τ_m^{-1} is smaller by an order of magnitude and in agreement with a
theory by *Schwartz* [6.7, 8] which takes account of the fact that by a hydrogenic impurity, a positively charged particle (hole) is scattered differently than
a negatively charged electron.

It may be surprising to find different hydrogenic neutral impurities behaving
so similarly. With respect to effects other than carrier scattering, the situation
can be quite different.

For example, Fig. 6.2 shows [6.9] the attenuation of an ultrasonic wave
of frequency 9 GHz in P-doped, As-doped and Sb-doped Ge as a function of
temperature. The attenuation at temperatures below 30 K is caused by the
immobile electrons bound to the impurities. The dashed curve indicates the

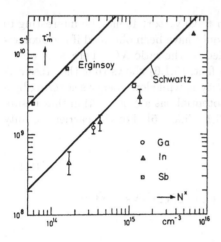

Fig. 6.1: Concentration dependence of the inverse cyclotron relaxation time for neutral impurity scattering in germanium (after [6.6])

attenuation of undoped Ge. The different behavior of the impurities has been ascribed to the fact that the shear strain induced by the acoustic wave acts differently on singlet and triplet states of the electron bound to the impurity (Sect. 3.2). The maximum attenuation has been found theoretically to occur at a temperature which is the ratio of the energy separation of the two states and 1.5 k_B. The known values of the energy separation divided by k_B are 3.7 K for Sb, 33 K for P and 49 K for As. For As, the increasing lattice attenuation masks the decreasing impurity attenuation , and therefore, no maximum is observed.

6.2 Elastic Scattering Processes

While the calculation of the scattering cross section by a superposition of partial waves mentioned above results in series expansions which converge rapidly for $k a \ll 1$, the case of $k a \gg 1$, requires Born's approximation method. This method shall now be applied to the problem of electron scattering by ionized impurity atoms.

The scattering process is considered to be a small *perturbation* of the electron wave by the potential $V(r)$ of the ionized impurity atom. Denoting the crystal volume by V, the ψ-function of the incoming electron having a wave vector k

$$\psi_n = V^{-1/2} \exp[i\,(k_n \cdot r)] \tag{6.2.1}$$

is the solution of the time-independent Schrödinger equation

$$H\psi_n = \hbar\omega_n\psi_n; \quad n = 0, 1, 2, 3, \dots , \tag{6.2.2}$$

where H is the Hamiltonian and $\hbar\omega_n$ are the eigenvalues of the unperturbed

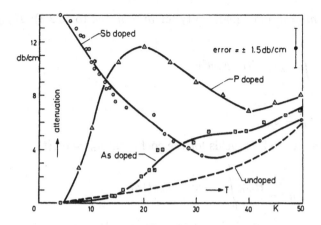

Fig. 6.2: Attenuation of a transverse acoustic wave propagating in the (100) direction of doped and undoped germanium, as a function of temperature. The impurity concentration is about $3{\times}10^{15}$ cm^{-3} and the acoustic frequency is 9 GHz (after [6.9])

problem (i.e., no ionized impurity present) and the normalization of the ψ_n is given by

$$\int |\psi_n|^2 d^3r = 1/V \int d^3r = 1 \ . \tag{6.2.3}$$

For the scattering process [scattering potential $V(r)$], we have to solve the time-dependent Schrödinger equation of the perturbation problem

$$[H + |e|\, V(r)]\,\psi = i\,\hbar\,\partial\psi/\partial t \ , \tag{6.2.4}$$

where ψ is expanded in a series of ψ_n:

$$\psi = \sum_n a_n \psi_n \, \exp(-i\,\omega_n t) \tag{6.2.5}$$

with unknown coefficients $a_n = a_n(t)$. Equation (6.2.4) yields

$$\sum_n a_n \,|e|\, V(r)\,\psi_n \, \exp(-i\,\omega_n t) = i\,\hbar \sum_n (da_n/dt)\,\psi_n \, \exp(-i\,\omega_n t) \ . \tag{6.2.6}$$

If this is multiplied by $\psi_m^* \exp(i\,\omega_m t)$, where m is an integer, and integrated over the crystal volume, we obtain

$$\sum_n a_n \, H_{mn} \, \exp(i\,\omega_{mn} t) = i\,\hbar\, da_m/dt \tag{6.2.7}$$

since the integral

$$\int \psi_m^* \,\psi_n \, d^3r = \frac{1}{V} \int \exp[i(\boldsymbol{k}_n - \boldsymbol{k}_m)\cdot\boldsymbol{r}]\, d^3r \tag{6.2.8}$$

vanishes for $m \neq n$ and equals unity for $m = n$ due to (6.2.3); H_{mn} and ω_{mn} are given by

$$H_{mn} = \int \psi_m^* \, |e| \, V(r) \, \psi_n \, d^3 r \qquad (6.2.9)$$

and

$$\omega_{mn} = \omega_m - \omega_n \ . \qquad (6.2.10)$$

The integration of (6.2.7) yields for the coefficient $a_m = a_m(t)$

$$a_m(t) = -\frac{i}{\hbar} \sum_n H_{mn} \int_0^t a_n \exp(i\,\omega_{mn} t)\, dt \ , \qquad (6.2.11)$$

where we have assumed $V(r)$ and consequently H_{mn} to be time-independent neglecting the thermal motion of the ionized impurity.

The incoming electron is considered to be in an initial state k where $a_k = 1$ and all other a's vanish, and denoting the final state by k', we obtain from (6.2.11)

$$|a_{k'}(t)|^2 = \hbar^{-2} |H_{k'k}|^2 t^2 \sin^2(\omega_{k'k}\, t/2)/(\omega_{k'k}\, t/2)^2 \ . \qquad (6.2.12)$$

For times $t \rightarrow \infty$ (which are long enough for the scattering process to be completed), the function $\sin^2(\omega_{k'k} t/2)/(\omega_{k'k} t/2)^2$ has the properties of a δ-function [1] $2\pi\,\delta(\omega_{k'k} t) = (2\pi\,\hbar/t)\delta(\hbar\,\omega_{k'k})$. The transition probability S per unit time is then given by (*Golden Rule no. 2* [6.10])

$$S(\boldsymbol{k}, \boldsymbol{k}') = |a_{k'}(t)|^2/t = (2\pi/\hbar)\,|H_{k'k}|^2 \delta[\varepsilon(\boldsymbol{k}') - \varepsilon(\boldsymbol{k})] \ . \qquad (6.2.13)$$

For transitions into or within a band, we have to consider not only one single final state k', but a group of possible final states. Therefore, before we integrate over phase space, we have to multiply S (a) by the density of states $g(\varepsilon)\,d\varepsilon$ given by (3.1.7) (neglecting, however, the factor of 2 since the electron spin is *not* changed by the scattering process), (b) by the probability $[1 - f(\boldsymbol{k}')]$ that the final state is not yet occupied, and (c) by the probability $f(\boldsymbol{k})$ that the initial state is occupied. This yields, for the decrease of $f(\boldsymbol{k})$ with time t by the scattering process (*collision*),

$$[-\partial f(\boldsymbol{k})/\partial t]_{\text{coll},k \rightarrow k'} = -V/(2\pi)^3 \int d^3 k'\, S(\boldsymbol{k}, \boldsymbol{k}') f(\boldsymbol{k})[1 - f(\boldsymbol{k}')] \ . \quad (6.2.14)$$

We still have to take into account scattering processes which go in the reverse direction:

$$[-\partial f(\boldsymbol{k})/\partial t]_{\text{coll},k' \rightarrow k} = -V/(2\pi)^3 \int d^3 k'\, S(\boldsymbol{k}', \boldsymbol{k})\, f(\boldsymbol{k}')[1 - f(\boldsymbol{k})] \ . \quad (6.2.15)$$

The total scattering rate consists of the sum of both contributions:

$$(-\partial f/\partial t)_{\text{coll}} = (-\partial f/\partial t)_{\text{coll},k \rightarrow k'} + (-\partial f/\partial t)_{\text{coll},k' \rightarrow k} \ . \qquad (6.2.16)$$

---

[1] For information about δ-functions see, e.g., [Ref. 6.1, p. 57].

For a nondegenerate semiconductor where the final states to a good approximation can be assumed unoccupied, we may cancel the factors $[1 - f(k')]$ and $[1 - f(k)]$ since they are approximately unity. For the equilibrium distribution $f_0(k)$, the time derivative $\partial f_0/\partial t = 0$. According to the principle of detailed balance [6.11],

$$S(k, k')f_0(k) = S(k', k)f_0(k') . \qquad (6.2.17)$$

Hence, in (6.2.15) we can eliminate $S(k', k)$ and obtain for (6.2.16)

$$(-\partial f/\partial t)_{\text{coll}} = V(2\pi)^{-3} \int d^3k' S(k, k')[f(k) - f(k') f_0(k)/f_0(k')] .(6.2.18)$$

For the simple model of band structure $[\varepsilon = \hbar^2 k^2/2m; \nabla_k \varepsilon = (\hbar^2/m)k]$, we obtain from (4.2.4)

$$f(k) = f_0(k) - (\hbar/m)(\partial f_0/\partial \varepsilon)(k \cdot G) . \qquad (6.2.19)$$

Introducing polar coordinates relative to the k-direction we have

$$d^3k' = k'^2 dk' \sin(\theta) d\theta d\varphi$$

$$(k \cdot G) = kG \cos\vartheta \qquad (6.2.20)$$

$$(k' \cdot G) = kG(\cos\vartheta \cos\theta + \sin\vartheta \sin\theta \cos\varphi)$$

Since ionized impurity scattering is highly elastic, $k = k'$, and since the equilibrium distribution $f_0(k)$ does not depend on the direction of k, the ratio $f_0(k)/f_0(k') = 1$ for elastic processes. We then find for the factor in brackets in (6.2.18):

$$f(k) - f(k') = -\frac{\hbar}{m}\left(\frac{\partial f_0}{\partial \varepsilon}\right) kG[\cos\vartheta(1 - \cos\theta) - \sin\vartheta \sin\theta \cos\varphi] .(6.2.21)$$

$S(k, k')$ is independent of φ because of spherical symmetry. Therefore, the integration in (6.2.18) eliminates the last term in (6.2.21). From (6.2.18, 19, 21), we obtain

$$(-\partial f/\partial t)_{\text{coll}} = [f(k) - f_0(k)]/\tau_{\text{m}} , \qquad (6.2.22)$$

where

$$1/\tau_{\text{m}} = V(2\pi)^{-2} \int k'^2 dk' \int_0^\pi S(k, k')(1 - \cos\theta) \sin\theta\, d\theta \qquad (6.2.23)$$

is the inverse momentum relaxation time. The integration over k' is done in the Brillouin zone. A comparision with (6.1.5) reveals that the differential cross section $\sigma(\theta)$ is given by

$$\sigma(\theta) = (Vm)^2(2\pi \hbar)^{-3}k^{-1} \int S(k, k') k' d\varepsilon' , \qquad (6.2.24)$$

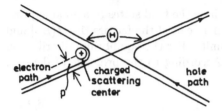

Fig. 6.3: Coulomb scattering of an electron and a hole by a positive ion

where one scattering center in the crystal volume ($N = 1/V$) has been assumed, the electron velocity v has been replaced by $\hbar\, k/m$ and $\varepsilon' = \hbar^2 k'^2/2m$. Due to the δ-function for the elastic scattering process, which according to (6.2.13) is contained in S, the integral is easily solved:

$$\sigma(\theta) = \left(\frac{Vm}{2\pi\hbar^2}|H_{k'k}|\right)^2 \tag{6.2.25}$$

if $|H_{k'k}|^2$ is independent of k'.

6.3 Ionized Impurity Scattering

Let us consider as a scattering center a singly ionized impurity atom of charge $Z\,e$ fixed somewhere inside the crystal. In the classical picture the electron drift orbit is a hyperbola with the ion in one of its focal points depending on the sign of the electronic charge, as shown in Fig. 6.3. The distance p between the ion and the asymptote is called the *impact parameter*. We introduce for convenience the distance

$$K = Z\,e^2/(4\pi\,\kappa\,\kappa_0\,m\,v^2) \tag{6.3.1}$$

for which the potential energy equals twice the kinetic energy: κ is the relative dielectric constant and κ_0 the permittivity of free space. The well-known Rutherford relationship between impact parameter and scattering angle is given by

$$p = K\,\cot\,(\theta/2)\ . \tag{6.3.2}$$

Carriers deflected through an angle between θ and $\theta + d\theta$ into a solid angle $d\Omega$ have an impact parameter with a value between p and $p + dp$ and therefore pass through a ring shaped area $2\pi p\,|dp|$ centered around the ion. The differential cross section is then obtained from

$$\sigma(\theta)\,d\Omega = 2\,\pi p\,|dp| = 2\,\pi p\frac{K\,d\theta/2}{\sin^2\theta/2} = 2\,\pi\,K^2\frac{\cot(\theta/2)\,d\theta/2}{\sin^2\theta/2}\ , \tag{6.3.3}$$

and since $d\Omega = 2\pi \sin\theta \, d\theta = 8\pi \sin^2(\theta/2) \cot(\theta/2) \, d\theta/2$,

$$\sigma(\theta) = \left[\frac{K/2}{\sin^2(\theta/2)} \right]^2 . \tag{6.3.4}$$

The differential cross section has thus been obtained from classical mechanics. It is based on the Coulomb potential of the impurity which has been assumed to extend to infinity. The calculation of τ_m from (6.3.4) runs into difficulties: the integral has no finite value if it begins at a zero scattering angle since $\sigma(0) = \infty$.

In practice, the problem is solved by the fact that the Coulomb potential is not quite correct and a screened potential

$$V(r) = -(Z|e|/4\pi \kappa \kappa_0 r) \exp(-r/L_D) \tag{6.3.5}$$

is more adequate. The idea behind the Debye length L_D is that the electrostatic field of the individual ionized impurity is screened by the surrounding carrier gas. At high impurity concentrations, the ionic space charge will also contribute to screening; we shall consider this contribution later on.

In the vicinity of an ionized impurity, the density of carriers $n(r)$ will be different from the average carrier density n:

$$n(r) = n \, \exp[-|e| \, V(r)/k_B T] \approx n[1 - |e| \, V(r)/k_B T] . \tag{6.3.6}$$

The approximation is valid for a small screening effect. Solving Poisson's equation for spherical symmetry,

$$d^2[rV(r)]/dr^2 = -r \, |e| \, [n(r) - n]/\kappa \, \kappa_0 \tag{6.3.7}$$

with $n(r)$ given by (6.3.6) and assuming that for $r \ll L_D$ the potential $V(r)$ is given by the Coulomb potential, yields (6.3.5) with L_D being the Debye length given by (5.2.22).

The quantum mechanical calculation of the cross section based on the screened potential has an analytical result, in contrast to a classical calculation. It is obvious from (6.2.25) that the first step then will be the calculation of the Hamiltonian matrix element $H_{k'k}$ from (6.2.9):

$$H_{k'k} = \frac{|e|}{V} \int V(r) \, \exp[i \, (k - k') \cdot r] d^3 r . \tag{6.3.8}$$

For the evaluation of the integral we introduce

$$c = |k - k'| \, r \tag{6.3.9}$$

$$\cos\varphi = (k - k')r/c \tag{6.3.10}$$

and

$$z = c \cos\varphi; \quad -c \le z \le c . \tag{6.3.11}$$

Hence, the integral becomes

$$\int_0^\infty V(r)\, 2\pi\, r^2\, dr/c \int_{-c}^{c} \exp(i\,z)\, dz = 4\pi \int_0^\infty V(r)\, r^2 [\sin(c)/c] dr \ . \qquad (6.3.12)$$

From (6.3.5) we now obtain for the matrix element

$$H_{k'k} = -\frac{Z\,e^2}{V\kappa\kappa_0 |k - k'|} \int_0^\infty \exp(-r/L_D) \sin(|k - k'|r) dr$$

$$= -\frac{Ze^2}{V\kappa\,\kappa_0}\frac{1}{|k - k'|^2 + L_D^{-2}} \approx -\frac{Ze^2}{V\kappa\,\kappa_0}\frac{1}{4k^2 \sin^2(\theta/2) + (2\,k\,L_D)^{-2}} \ , (6.3.13)$$

where V is the crystal volume and the approximation is valid for the elastic scattering process considered here where $|k| \approx |k'|$ and

$$|k - k'| \approx 2\,k \sin(\theta/2) \ . \qquad (6.3.14)$$

Equation (6.2.25) yields for the differential cross section

$$\sigma(\theta) = \left[\frac{K/2}{\sin^2(\theta/2) + \beta^{-2}}\right]^2 \ , \qquad (6.3.15)$$

where

$$\beta = 2kL_D \qquad (6.3.16)$$

has been introduced and the distance K is given by (6.3.1). A comparison (6.3.4, 15) reveals that in contrast to the previous calculation , $\sigma(\theta)$ remains finite even for a zero scattering angle.

The calculation of τ_m from (6.1.5) is straightforward with the result

$$\tau_m = \varepsilon^{3/2}\, N_I^{-1} 16\pi\, \sqrt{2m}(\kappa\,\kappa_0/Z\,e^2)^2 [\ln(1 + \beta^2) - \beta^2/(1 + \beta^2)]^{-1} \ , (6.3.17)$$

where N_I is the total concentration of ionized impurities in the crystal. Except for very low values of the carrier velocity $v \propto k \propto \beta$, the term in brackets is nearly constant and τ_m can be said to obey a power law

$$\tau_m \propto \varepsilon^{3/2} \qquad (6.3.18)$$

with an exponent of $+3/2$.

For the averaging procedure, assuming a nondegenerate electron gas,

$$\langle\tau_m\rangle = \frac{4}{3\sqrt{\pi}} \int_0^\infty \tau_m(\varepsilon/k_BT)^{3/2} \exp(-\varepsilon/k_BT) d\varepsilon/k_BT \ , \qquad (6.3.19)$$

we replace ε in β by that value for which the intergrand $(\varepsilon/k_BT)^3 \exp(-\varepsilon/k_BT)$ is a maximum; this is true for $\varepsilon = 3k_BT$. We then denote β by β_{BH} where B and H are the initials of *Brooks* and *Herring* [6.12] to whom this calculation is due:

$$\beta_{BH} = 2\frac{m}{\hbar}\left(\frac{2}{m}3k_B T\right)^{1/2} L_D \tag{6.3.20}$$

or

$$\beta_{BH} = \left(\frac{x}{16}\right)^{1/2}\frac{T}{100\,\text{K}}\left(\frac{m}{m_0}\right)^{1/2}\left(\frac{2.08\times10^{18}\text{cm}^{-3}}{n}\right)^{1/2} \tag{6.3.21}$$

having a numerical value of, e.g., 1 for n-type Ge ($\kappa = 16$; $m/m_0 = 0.12$) with $n = 2.5 \times 10^{17}\,\text{cm}^{-3}$ at 100 K. The Debye length is then 5.5 nm while the average nearest distance between two ionized impurities is 15.9 nm assuming no compensation.

The mobility $\mu = (e/m)\langle\tau_m\rangle$ is given by

$$\mu = \frac{2^{7/2}(4\pi\,\kappa\,\kappa_0)^2(k_B T)^{3/2}}{\pi^{3/2}\,Z^2\,e^3\,m^{1/2}\,N_I[\ln(1+\beta_{BH}^2)-\beta_{BH}^2/(1+\beta_{BH}^2)]} \tag{6.3.22}$$

which in units of cm^2/V s is

$$\mu = \frac{3.68\times10^{20}\text{cm}^{-3}}{N_I}\frac{1}{Z^2}\left(\frac{\kappa}{16}\right)^2\left(\frac{T}{100\,\text{K}}\right)^{1.5}$$

$$\times\frac{1}{(m/m_0)^{1/2}\,[\log(1+\beta_{BH}^2)-0.434\beta_{BH}^2/(1+\beta_{BH}^2)]} \tag{6.3.23}$$

and the log is to the base 10.

Historically, the Brooks–Herring calculation was preceded by a calculation by *Conwell* and *Weisskopf* [6.13] based on (6.3.4) with the requirement of a minimum scattering angle θ_{min}. This angle was obtained from (6.3.2) and a maximum impact parameter p_{max} taken as half the average distance $N_I^{-1/3}$ between adjacent ionized impurity atoms. The calculation arrived at a formula similar to (6.3.23) except that the term in brackets was replaced by $\log(1+\beta_{CW}^2)$ where

$$\beta_{CW} = \frac{1}{Z}\frac{\kappa}{16}\frac{T}{100\text{K}}\left(\frac{2.35\times10^{19}\,\text{cm}^{-3}}{N_I}\right)^{1/3} \tag{6.3.24}$$

does not depend on the carrier concentration but on the ionized impurity concentration. Since the BH and CW results are different only in logarithmic terms, they yield about the same values of the mobility for concentrations up to about 10^{18} cm^{-3}. Beyond this value, most semiconductors become degenerate and the calculations given here are no longer valid. At a constant temperature, the mobility depends on N_I as shown in Fig. 6.4 where $n = N_I$, $Z = 1$, $\kappa = 16$ and $m = m_0$ have been assumed.

At first sight it may be surprising to find that it is difficult to observe experimentally a $\mu \propto T^{3/2}$ behavior over a wide temperature range. However, at

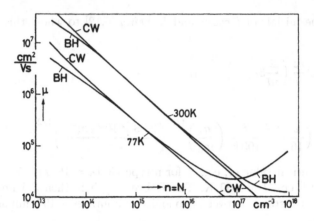

Fig. 6.4: Dependence of the mobility on impurity concentration N_I at 77 and 300 K calculated according to Conwell and Weisskopf (CW) and Brooks and Herring (BH) for a hypothetical uncompensated semiconductor with effective mass equal to the free electron mass, dielectric constant 16, and impurity charge $Ze = e$

high temperatures scattering is predominant while at low temperatures, carriers freeze out at impurity levels thus neutralizing impurities and making N_I a function of T. A discussion of experimental results for $\mu(T)$ will be given after a treatment in which impurity and lattice scattering are taken into account simultaneously (Sect. 6.6).

The problem of shielding of ionized impurities by carriers in the range of carrier freeze-out will now be made plausible in a simplified treatment of the problem. The freeze-out has the effect of an increased density of carriers in the vicinity of an ionized impurity of opposite charge. For simplicity, let us discuss only the combination of electrons and donors. The electron density which enters the Brooks–Herring formula will now be denoted by n'. It is larger than the average electron density n in the semiconductor. The increase $n' - n$ should be proportional to the occupancy $N_{D^\times}/N_D$ of the donors where $N_D = N_{D^+} + N_{D^\times}$ is the total donor concentration, $N_{D^\times}$ the concentration of neutral donors and N_{D^+} that of ionized donors. In a partially compensated n-type semiconductor, there are also N_{A^-} ionized acceptors. Because of charge neutrality,

$$N_{A^-} + n = N_{D^+} + p \approx N_{D^+} \ , \tag{6.3.25}$$

where p is negligibly small. Hence,

$$n' - n \propto (N_D - N_{A^-} - n)/N_D \ . \tag{6.3.26}$$

For the case of a near-complete neutralization of the donors, the increase $n' - n$ becomes small again since there are only few positive ions which disturb the random distribution of electrons in the crystal. Therefore, $n' - n$ should also be proportional to the probability of ionization of donors,

$$N_{D+}/N_D \approx (N_{A-} + n)/N_D .$$

If we divide $n' - n$ by N_D, we may, to a good approximation, assume

$$(n' - n)/N_D = [(N_D - N_{A-} - n)/N_D](N_{A-} + n)/N_D \tag{6.3.27}$$

which yields Brook's formula [6.12]

$$n' = n + (N_D - N_{A-} - n)(N_{A-} + n)/N_D . \tag{6.3.28}$$

The total concentration of ionized impurities N_I is, of course,

$$N_I = N_{D+} + N_{A-} . \tag{6.3.29}$$

For a calculation of n', one has to take into account the fact that the activation energy of donors $\Delta\varepsilon_D$ depends somewhat on N_I. A more refined treatment of scattering reveals that the repulsive scattering process of a carrier at an ionized impurity of the same sign has a cross section different from that of the attractive scattering process where the carrier and ion are oppositely charged. For a discussion on the validity of Born's approximation see, e.g., ([Ref. 6.1, p.325] and [6.14, 15]).[2]

6.4 Acoustic Deformation Potential Scattering of Thermal Carriers

When an acoustic wave propagates in a crystal lattice the atoms oscillate about their equilibrium positions. For small amplitudes A_s this oscillation is harmonic, and for an angular frequency ω_s and wave vector q_s (subscript s for *sound*), can be represented by

$$\delta r = A_s \exp\{\pm i[(q_s \cdot r) - \omega_s t]\} . \tag{6.4.1}$$

At present we shall not discuss the oscillation spectrum of a crystal but consider only long-wavelength acoustic waves where the sound velocity $u_s = \omega_s/q_s$ is a constant. The wavelength $2\pi/q_s$ is much longer than the interatomic distance and the crystal can be treated as a continuous medium. The difference in displacement between two adjacent atoms (average distance a) can be approximated by

$$|\delta r(a) - \delta r(0)| = (\nabla_r \cdot \delta r)a , \tag{6.4.2}$$

where, according to (6.4.1), the periodic dilation $(\nabla_r \cdot \delta r)$ is given by

$$(\nabla_r \cdot \delta r) = \pm i(q_s \cdot \delta r) . \tag{6.4.3}$$

---

[2]For compensated semiconductors, better agreement with experimental data has been obtained with calculations by *Falicov, Cuevas* [6.16, 17]; for many-valley semiconductors [6.18]. A review on ionized impurity scattering was presented by *Chattopadhyay* and *Queisser* [6.19].

Since q_s is a vector in the direction of wave propagation and the product $(q_s \cdot \delta r)$ vanishes for δr transverse to this direction, we shall consider here only longitudinal waves and use a subscript ℓ instead of s:

$$\delta r = A_\ell \exp\{\pm i[(q_\ell \cdot r) - \omega_\ell t]\} \ . \tag{6.4.4}$$

The scattering of conduction electrons by acoustic waves requires the theorem of the deformation potential put forward by *Bardeen* and *Shockley* in 1950 [6.20]. For a principle treatment, consider the variation of the valence band edge with the lattice constant calculated, e.g., for diamond (Fig. 2.8c). This variation can be taken to be linear for a small change in the lattice spacing as it occurs in an acoustic wave. The change in energy of a hole at a transition from one atom to an adjacent atom $\delta\varepsilon_h$ is therefore proportional to $|\delta r(a) - \delta r(0)|$ or, according to (6.4.2), proportional to $(\nabla_r \cdot \delta r)$ in an acoustic wave:

$$\delta\varepsilon_h = \varepsilon_{ac,v}(\nabla_r \cdot \delta r) \ , \tag{6.4.5}$$

where $\varepsilon_{ac,v}$ is a factor of proportionality denoted as the *deformation potential constant* of the valence band. A similar relation applies for the electrons at the conduction band edge:

$$\delta\varepsilon_e = \varepsilon_{ac,c}(\nabla_r \cdot \delta r) \ , \tag{6.4.6}$$

where $\varepsilon_{ac,c}$ is the corresponding constant of the conduction band. Since the energies of electrons and holes are counted positive in opposite directions, Fig. 2.8c shows that both constants have equal sign but, in general, different magnitudes.

Bardeen and *Shockley* proved that for a perturbation treatment of the interaction between the electron and the acoustic wave it is correct to take $\delta\varepsilon = \varepsilon_{ac}(\nabla_r \delta r)$ for the perturbing potential energy [Ref. 6.20, Appendix B]. Since $\delta r = A_\ell \exp[\pm i(q_\ell \cdot r)]$, the matrix element (6.3.8) is given by

$$|H_{k'k}| = \frac{\varepsilon_{ac} \, q_\ell \, A_\ell}{V} \left| \int \exp[i(k - k' \pm q_\ell) \cdot r] \, d^3 r \right| \ . \tag{6.4.7}$$

For the quantum mechanical description, it is more adequate to consider acoustic phonons of energy $\hbar\,\omega_\ell$ and momentum $\hbar\,q_\ell$ rather than acoustic waves. If the condition of momentum conservation

$$k' = k \pm q_\ell \tag{6.4.8}$$

is fulfilled where the plus and minus signs refer to phonon absorption and emission, respectively, the integrand in (6.4.7) is independent of r and the crystal volume V cancels:

$$|H_{k'k}| = \varepsilon_{ac} \, q_\ell \, A_\ell \ . \tag{6.4.9}$$

Equation (6.4.8) is a special case of a more general condition where the rhs also contains a lattice vector of the reciprocal lattice space. Because of the

periodicity of the crystal lattice and the Laue equation, this additional vector
would not impair the result (6.4.9). Such scattering processes are known as
Umklapp processes [6.21, 22]. Since we are dealing with scattering processes
where both k and k' are relatively small, we can neglect the vector of the
reciprocal lattice.

For the vibration amplitude A_ℓ in (6.4.9), we now have to find its quantum
mechanical equivalent. Since we are investigating harmonic oscillations, this is
the matrix element of the space coordinate, say x, for a transition from the Nth
vibrational state either to the $N - 1$ state or to the $N + 1$ state corresponding
to the absorption or emission of a phonon, respectively:

$$A_\ell \rightarrow \left| \int \psi^*_{N\pm1}\, x\, \psi_N\, d^3r \right| = \begin{cases} (N\, \hbar/2M\, \omega_\ell)^{1/2} & \text{for } N \rightarrow N-1 \\ ((N+1)\, \hbar/2M\, \omega_\ell)^{1/2} & \text{for } N \rightarrow N+1 \end{cases}$$

$$(6.4.10)$$

In the Nth state, the oscillatory energy is $(N +1/2)\, \hbar\omega_\ell$ corresponding to N
phonons. Since the crystal contains very many such oscillators, we replace N
by the average number of phonons at a temperature T of the crystal which,
according to Planck, is given by

$$N \rightarrow N_q = [\exp(\hbar\omega_\ell/k_B T) - 1]^{-1} \ . \tag{6.4.11}$$

Taking the volume of a unit cell of the crystal as V, we can replace the oscillator
mass M by the product ϱV where ϱ is the mass density. We finally obtain for
the matrix element of the Hamiltonian

$$|H_{k\pm q,k}| = \varepsilon_{\text{ac}}\, q_\ell\, [(N_q + 1/2 \mp 1/2)\, \hbar/2\varrho\, V\omega_\ell]^{1/2} \ . \tag{6.4.12}$$

The acoustic-phonon energy $\hbar\omega_\ell$ involved here is small compared with the thermal energy $k_B T$. Therefore, N_q can be approximated by $k_B T/\hbar\omega_\ell \gg 1$ and
since $N_q + 1 \approx N_q$ in this approximation, we obtain the same matrix element
for phonon absorption and emission:

$$|H_{k'k}| = \varepsilon_{\text{ac}}\, q_\ell\, [k_B T/2\varrho\, V\omega_\ell^2]^{1/2} = \varepsilon_{\text{ac}}\, [k_B T/2V c_\ell]^{1/2} \ , \tag{6.4.13}$$

where the longitudinal elastic constant $c_\ell = \varrho\omega_\ell^2/q_\ell^2 = \varrho\, u_\ell^2$ has been introduced.
[In Chap. 7 we shall treat the tensor character of ε_{ac}; in brief, we notice that
for a $\langle 100 \rangle$ direction of wave propagation in a cubic lattice, $c_\ell = c_{11}$, for a
$\langle 110 \rangle$ direction $c_\ell = (c_{11} + c_{12} + c_{44})/2$, and for a $\langle 111 \rangle$ direction $c_\ell = (c_{11} + 2c_{12} + 4c_{44})/3$, while for other directions, the waves are not strictly longitudinal
having velocities between the extremes at $\langle 100 \rangle$ and $\langle 111 \rangle$; c_{11}, c_{12}, and c_{44} are
components of the elasticity tensor.]

The matrix element is independent of the electron energy and of the scattering angle. Since it is almost the same for phonon emission and absorption,
we can take care of both processes simply by a factor of 2 in the scattering
probability S, (6.2.13):

$$S \approx \frac{2\pi}{\hbar}\, |H_{k'k}|^2\, [\delta(\varepsilon(k') - \varepsilon(k) + \hbar\, \omega_\ell) + \delta(\varepsilon(k') - \varepsilon(k) - \hbar\omega_\ell)]$$

$$\approx 2\frac{2\pi}{\hbar}|H_{k'k}|^2 \delta[\varepsilon(k') - \varepsilon(k)] \ . \tag{6.4.14}$$

Since $k^2 dk = m^2 v \hbar^{-3} d\varepsilon$, the calculation of the momentum relaxation time according to (6.2.23) yields

$$1/\tau_{\mathrm{m}} = v/l_{\mathrm{ac}} \ , \tag{6.4.15}$$

where the mean free path

$$l_{\mathrm{ac}} = \pi \, \hbar^4 c_\ell/(m^2 \, \varepsilon_{\mathrm{ac}}^2 \, k_{\mathrm{B}} T) \tag{6.4.16}$$

has been introduced which is independent of the carrier velocity. This proves that for the energy dependence of τ_{m},

$$\tau_{\mathrm{m}} \propto \varepsilon^{-1/2} \ , \tag{6.4.17}$$

a power law with an exponent $-1/2$ is valid.

The mean free path decreases with increasing temperature since at higher temperatures more phonons are excited and therefore more *scattering centers* exist. The dependence on the effective mass arises from the density of states, $k^2 dk$. In Chap. 7, a density-of-states effective mass will be defined for a more complex band structure, and it is of course this mass which enters here. For the simple model of band structure ($\varepsilon \propto k^2$) considered at present, there is only one kind of effective mass.

The calculation of the mobility is straightforward:

$$\mu = \frac{2\sqrt{2\pi}}{3} \frac{e \, \hbar^4 \, c_\ell}{m^{5/2} \, (k_{\mathrm{B}})^{3/2} \, \varepsilon_{\mathrm{ac}}^2} T^{-3/2} \tag{6.4.18}$$

which in units of cm^2/V s is given by

$$\mu = 3.06 \times 10^4 \frac{c_\ell/10^{12} \mathrm{dyncm}^{-2}}{(m/m_0)^{5/2}(T/100 \ \mathrm{K})^{3/2}(\varepsilon_{\mathrm{ac}}/eV)^2} \propto T^{-3/2} \ . \tag{6.4.19}$$

For example, for n-type Ge at $T = 100$ K, a mobility of $3 \times 10^4 \mathrm{cm}^2/\mathrm{Vs}$ is calculated ($c_\ell = 1.56 \times 10^{12} \mathrm{dyn/cm}^2$; $m/m_0 = 0.2$; $\varepsilon_{\mathrm{ac}} = 9.5 \mathrm{eV}$). However, at this and higher temperatures, the contribution by optical deformation potential scattering cannot be neglected entirely and modifies the temperature dependence of the mobility : $\mu \propto T^{-1.67}$.

Since the mobility μ is proportional to $m^{-5/2}$, carriers with a small effective mass have a high mobility (e.g., light holes in Ge at low temperatures where acoustic-phonon scattering dominates optical-phonon scattering).

6.5 Acoustic Deformation Potential Scattering of Hot Carriers

We have so far considered acoustic deformation potential scattering as being essentially an elastic process. It may be interesting to see what the average energy loss per unit time of a carrier to the crystal lattice actually is [6.23]. According to definition,

$$\langle -d\varepsilon/dt \rangle_{\text{coll}} = \int \varepsilon(\boldsymbol{k}) \, (\partial f/\partial t)_{\text{coll}} \, d^3k \Big/ \int f \, d^3k \ , \tag{6.5.1}$$

where $(-\partial f/\partial t)_{\text{coll}}$ is given by (6.2.16). By partial integration, (6.5.1) can be manipulated into the form

$$\langle -d\varepsilon/dt \rangle_{\text{coll}} = \int (-d\varepsilon/dt)_{\text{coll}} f(\boldsymbol{k}) \, d^3k \Big/ \int f(\boldsymbol{k}) \, d^3k \ , \tag{6.5.2}$$

where

$$\langle -d\varepsilon/dt \rangle_{\text{coll}} = V(2\pi)^{-3} \int [\varepsilon(\boldsymbol{k}) - \varepsilon(\boldsymbol{k}')] S(\boldsymbol{k}, \boldsymbol{k}') [1 - f(\boldsymbol{k}')] \, d^3k' \tag{6.5.3}$$

is obtained from (6.2.16) and (6.5.1). For the scattering probability $S(\boldsymbol{k}, \boldsymbol{k}')$, we need the exact value

$$S(\boldsymbol{k}, \boldsymbol{k}') = \frac{2\pi}{\hbar} \frac{\varepsilon_{\text{ac}}^2 \, \hbar q}{2\varrho \, V u_\ell} \{ N_q \delta \left[\varepsilon(\boldsymbol{k}') - \varepsilon(\boldsymbol{k}) - \hbar u_\ell q \right]$$

$$+ (N_q + 1) \, \delta[\varepsilon(\boldsymbol{k}') - \varepsilon(\boldsymbol{k}) + \hbar u_\ell q] \} \tag{6.5.4}$$

rather than the approximate value given by (6.4.14); we have replaced ω_ℓ by $u_\ell q_\ell$ and for simplicity, omitted the subscript ℓ of q_ℓ. The integration over d^3k' can be replaced by one over $d^3q = -2\pi q^2 dq \, d(\cos \theta)$. The arguements of the δ-functions are for the simple model of band structure

$$\varepsilon(\boldsymbol{k} + \boldsymbol{q}) - \varepsilon(\boldsymbol{k}) - \hbar u_\ell q$$

$$= \frac{\hbar^2}{2m} (2 \, k q \cos \theta + q^2 - 2m \, u_\ell q/\hbar) = \frac{\hbar^2 q}{2m} (q - q_\beta) \tag{6.5.5}$$

and

$$\varepsilon(\boldsymbol{k} - \boldsymbol{q}) - \varepsilon(\boldsymbol{k}) + \hbar u_\ell q$$

$$= \frac{\hbar^2}{2m} (-2 \, k q \cos \theta + q^2 + 2m \, u_\ell q/\hbar) = \frac{\hbar^2 q}{2m} (q - q_\alpha) \ , \tag{6.5.6}$$

where the constants

$$q_\beta = -2 \, k \cos \theta + 2 \, m \, u_\ell/\hbar; \qquad q_\alpha = 2 \, k \cos \theta - 2 \, m \, u_\ell/\hbar \tag{6.5.7}$$

have been introduced. Let us first integrate over q and afterwards over $\cos\theta$. Due to the δ-function, [3]   the factor $\varepsilon(k) - \varepsilon(k')$ in (6.5.3) becomes $+\hbar u_\ell\, q$ for the case of phonon absorption and $-\hbar u_\ell q$ for emission. Since

$$\delta[(\hbar^2 q/2m)(q - q_{\alpha,\beta})] = (2m/\hbar^2 q)\,\delta\,(q - q_{\alpha,\beta}) \ , \tag{6.5.8}$$

we obtain from (6.5.3)

$$\left(-\frac{d\varepsilon}{dt}\right)_{\text{coll}} = -\frac{m\,\varepsilon_{\text{ac}}^2}{2\pi\,\hbar\,\varrho} \int_{-1}^{+1} d\,(\cos\theta)\,\{q_\beta^3 N_{q_\beta}[1 - f(k + q_\beta)]$$

$$-q_\alpha^3(N_{q_\alpha} + 1)[1 - f(k - q_\alpha)]\}. \tag{6.5.9}$$

The limits of integration are provided by $q_\beta \geq 0$ and $q_\alpha \geq 0$. For the first term we replace $d(\cos\theta)$ by $-(1/2k)dq_\beta$, for the second term by $+(1/2k)dq_\alpha$. In the first integral, the upper limit of integration would result from $\cos\theta = +1$ to be $q_\beta = -2k + 2mu_1/\hbar \approx -2k$ (the approximation is valid since the electron velocity $\hbar k/m$ even at low temperatures is large compared with the sound velocity u_ℓ). Since $q_\beta \geq 0$, we find for the upper limit $q_\beta = 0$. The lower limit (from $\cos\theta = -1$) is denoted by $q_{\beta m}$:

$$q_{\beta m} = 2k + 2m\,u_\ell/\hbar \ . \tag{6.5.10}$$

In the second integral, the upper limit (from $\cos\theta = +1$) is denoted by $q_{\alpha m}$:

$$q_{\alpha m} = 2k - 2m\,u_\ell/\hbar \tag{6.5.11}$$

while the lower limit $-2k - 2mu_\ell/\hbar$ would be negative and is therefore replaced by zero. Hence, (6.5.9) yields

$$\left(-\frac{d\varepsilon}{dt}\right)_{\text{coll}} = \frac{m\,\varepsilon_{\text{ac}}^2}{2\pi\,\hbar\,\varrho\,2k}$$

$$\times\left[\int_{q_{\beta m}}^{0} dq\, q^3 N_q\,[1 - f(k + q)] + \int_{0}^{q_{\alpha m}} dq\, q^3(N_q + 1)\,[1 - f(k - q)]\right]. \tag{6.5.12}$$

This can be manipulated into a more convenient form:

$$\left(-\frac{d\varepsilon}{dt}\right)_{\text{coll}} = -\frac{m\,\varepsilon_{\text{ac}}^2}{4\pi\,\varrho\,k}\left[\int_{q_{\alpha m}}^{q_{\beta m}} dq\, q^3 N_q\,[1 - f(k + q)]\right.$$

$$\left. + \int_{0}^{q_{\alpha m}} dq\, q^3\{N_q[f(k - q) - f(k + q)] - [1 - f(k - q)]\right]. \tag{6.5.13}$$

---

[3] Some properties of the δ-function have been listed, e.g., in [Ref. 6.1, p. 57]. A useful equation is $\delta[f(x)] = \sum_i \delta(x - x_i)/|df/dx|_{x=x_i}$, where $f(x_i) = 0$.

For calculating the difference $f(k - q) - f(k + q)$ we expand

$$f(\varepsilon \pm \hbar u_\ell q) \approx f(\varepsilon) \pm \hbar u_\ell q \, df/d\varepsilon \ . \tag{6.5.14}$$

As in Sect. 4.13, let us assume for $f(\varepsilon)$ a Fermi–Dirac distribution function with an electron temperature T_e:

$$f(\varepsilon) = \{1 + \exp[(\varepsilon - \zeta)/k_B T_e]\}^{-1} \ , \tag{6.5.15}$$

where

$$df/d\varepsilon = f(f - 1)/k_B T_e \tag{6.5.16}$$

is valid. Except for the difference $f(k - q) - f(k + q)$, the functions $f(k \pm q)$ may be approximated by $f(\varepsilon)$; as before, N_q is approximated by $k_B T/\hbar u_\ell q$:

$$\left(-\frac{d\varepsilon}{dt}\right)_{\mathrm{coll}} = -\frac{m\,\varepsilon_{\mathrm{ac}}^2}{4\pi\,\hbar\,\varrho\,k}$$

$$\times \left[\frac{k_B T}{\hbar u_\ell}(1 - f) \int_{q_{\alpha m}}^{q_{\beta m}} q^2 dq + \int_0^{q_{\alpha m}} q^3 dq \left[\frac{2T}{T_e}f(1 - f) - (1 - f)\right]\right] \ . \tag{6.5.17}$$

In the approximation valid for $\hbar k/m \gg u_\ell$, the value of the first integral is $2^4 m u_\ell k^2/\hbar$ while in the second integral the upper limit can be replaced by $2k$:

$$\left(-\frac{d\varepsilon}{dt}\right)_{\mathrm{coll}} = \left(\frac{m\,\varepsilon_{\mathrm{ac}}^2\,k^3}{\pi\,\varrho\,\hbar}\right)(f - 1)\left[\left(\frac{4m\,k_B T}{\hbar^2 k^2}\right) + \left(\frac{2T}{T_e}\right)f - 1\right] \tag{6.5.18}$$

The averaging procedure according to (6.5.2) yields by partial integration

$$\langle -d\varepsilon/dt\rangle_{\mathrm{coll}} = \frac{2m\,\varepsilon_{\mathrm{ac}}^2}{\pi^{3/2}\,\hbar\varrho}\left(\frac{2\,mk_B T_e}{\hbar^2}\right)^{3/2} 2\frac{T_e - T}{T_e}\frac{F_1(\eta)}{F_{1/2}(\eta)} \ , \tag{6.5.19}$$

where the Fermi integrals are given by (3.1.11) and $\eta = \zeta/k_B T_e$ is the reduced Fermi energy.

This expression is simplified to

$$\langle -d\varepsilon/dt\rangle_{\mathrm{coll}} = 4\,m\,u_\ell^2\langle\tau_m^{-1}\rangle(T_e - T)/T \tag{6.5.20}$$

by introducing the average reciprocal momentum relaxation time [4]

$$\langle\tau_m^{-1}\rangle = \langle l_{\mathrm{ac}}^{-1}(2\varepsilon/m)^{1/2}\rangle = \frac{2\,m^2\,\varepsilon_{\mathrm{ac}}^2\,k_B T}{\pi^{3/2}\,\hbar^4 \varrho\,u_\ell^2}\left(\frac{2\,k_B T_e}{m}\right)^{1/2}\frac{F_1(\eta)}{F_{1/2}(\eta)} \ . \tag{6.5.21}$$

Equation (6.5.20) is valid for both degeneracy and nondegeneracy. For the latter case, $\langle\tau_m^{-1}\rangle$ is given by $(2/l_{\mathrm{ac}})(2k_B T_e/\pi m)^{1/2} = 8/(3\pi\langle\tau_m\rangle)$. Therefore,

$$\mu = \frac{8}{3\pi}\frac{e}{m\langle\tau_m^{-1}\rangle} \ . \tag{6.5.22}$$

---

[4]For $\eta \gg 1$ we find $F_1(\eta)/F_{1/2}(\eta) \approx (3\sqrt{\pi}/8)\sqrt{\eta}$ and $\langle\tau_m^{-1}\rangle \propto T$ independent of T_e. The effect of conduction electron screening on the electron–phonon interaction has been treated by *Szymanska, Maneval*: [6.24].

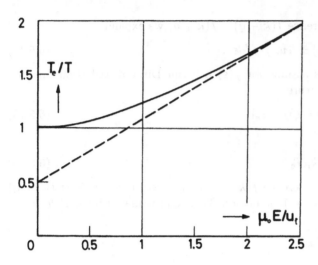

Fig. 6.5: Electron temperature in units of the lattice temperature, as a function of the electric field strength in units of the ratio of the sound velocity to zero-field mobility

With the energy balance equation (4.13.6) and $\mu = \mu_0(T/T_e)^{1/2}$, as is obvious from (6.4.17), we can solve (6.5.20) for T_e/T:

$$T_e/T = \frac{1}{2}\left[1 + \sqrt{1 + \frac{3\pi}{8}\left(\frac{\mu_0 E}{u_\ell}\right)^2}\right] . \tag{6.5.23}$$

For warm carriers ($T_e - T \ll T$), the electron temperature increases with the square of E:

$$T_e/T = 1 + (3\pi\,\mu_0^2/32u_\ell^2)\,E^2 , \tag{6.5.24}$$

while for hot carriers ($T_e \gg T$), the increase is linear:

$$T_e/T = (3\pi/32)^{1/2}\mu_0 E/u_\ell . \tag{6.5.25}$$

In Fig. 6.5 the T_e vs E relationship is plotted. Eliminating T_e from (6.5.23) and $\mu_0/\mu = (T_e/T)^{1/2}$, the field dependence of the mobility is obtained in the form

$$E^2 = \left(\frac{32}{3\pi}u_\ell^2/\mu_0^2\right)(\mu_0/\mu)^2[(\mu_0/\mu)^2 - 1] . \tag{6.5.26}$$

In the warm carrier approximation, the coefficient β is found to be

$$\beta = \frac{\mu - \mu_0}{\mu_0 E^2} = -\frac{3\pi}{64}(\mu_0/u_\ell)^2 = -0.147(\mu_0/u_\ell)^2 , \tag{6.5.27}$$

while for hot carriers, the ratio μ/μ_0 equals [6.25]

$$\mu/\mu_0 = (32/3\pi)^{1/4}(u_\ell/\mu_0 E)^{1/2} = 1.81(u_\ell/\mu_0)^{1/2}E^{-1/2} \tag{6.5.28}$$

and the drift velocity $v_d = \mu E$ increases proportional to $E^{1/2}$. Experimental results are shown in Fig. 4.33, where at $T = 20\,\text{K}$ for field intensities high enough that ionized impurity scattering can be neglected, acoustic-phonon scattering should indeed be predominant.

The field intensity, where deviations from Ohm's law become significant, is u_ℓ/μ_0 according to (6.5.28). Typical values for u_ℓ and μ_0 are 5×10^5 cm/s and 5×10^3 cm^2/Vs, respectively, which yield a *critical* field intensity of 10^2 V/cm.

Let us finally calculate the energy relaxation time of warm carriers. This is defined by (4.13.7) where the lhs for nonequilibrium conditions is $\langle -d\varepsilon/dt \rangle_{\mathrm{coll}}$. For $T_e - T \ll T$, we find for the energy difference by a series expansion

$$\langle \varepsilon(T_e) \rangle - \langle \varepsilon(T) \rangle \approx (T_e - T) \frac{\partial}{\partial T_e} \int \varepsilon f \, d^3k \Big/ \int f \, d^3k$$

$$= \frac{3}{2} k_B (T_e - T) \, \partial \, (T_e F_{3/2}/F_{1/2})/\partial T_e \; . \tag{6.5.29}$$

For the differentiation we take into account that the carrier concentration $n \propto T_e^{3/2} F_{1/2}$ should be independent of T_e and therefore

$$0 = \frac{dn}{dT_e} \propto \frac{3}{2} T_e^{1/2} \, F_{1/2} + T_e^{3/2} \frac{dF_{1/2}}{d\eta} \frac{d\eta}{dT_e} \tag{6.5.30}$$

which yields

$$T_e \, d\eta/dT_e = -(3/2) \, F_{1/2}/F_{-1/2} \; . \tag{6.5.31}$$

Therefore, the differentiation becomes simply

$$\frac{\partial (T_e \, F_{3/2}/F_{1/2})}{\partial T_e} = \frac{F_{3/2}}{F_{1/2}} + \left(1 - \frac{F_{3/2} F_{-1/2}}{F_{1/2}^2} \right) T_e \frac{d\eta}{dT_e} = \frac{5}{2} \frac{F_{3/2}}{F_{1/2}} - \frac{3}{2} \frac{F_{1/2}}{F_{-1/2}}$$

$$\tag{6.5.32}$$

From (4.13.10), (6.5.20, 29) we now obtain for the product $\tau_\varepsilon \langle \tau_m^{-1} \rangle$

$$\tau_\varepsilon \langle \tau_m^{-1} \rangle = \frac{3 \, k_B \, T}{8 \, m \, u_\ell^2} \left(\frac{5}{2} \frac{F_{3/2}}{F_{1/2}} - \frac{3}{2} \frac{F_{1/2}}{F_{-1/2}} \right) \tag{6.5.33}$$

which for the case of nondegeneracy is

$$\tau_\varepsilon \langle \tau_m^{-1} \rangle = \frac{3 \, k_B \, T}{8 \, m \, u_\ell^2} \gg 1 \; . \tag{6.5.34}$$

This product is roughly the number of collisions necessary for the relaxation of energy. For example, for n-Ge at 100 K where $u_\ell = 5.4 \times 10^5$ cm/s, $m/m_0 = 0.2$, the value of the product is about 10^2. Obviously a large number of collisions is necessary to relax the carrier energy while the momentum distribution is already relaxed after one collision, i.e., after a time of the order of magnitude of $\langle \tau_m \rangle$.

In Fig. 6.6 a schematic representation of the relaxation behavior of warm and hot carriers is given [6.26]. Let us assume that an electric field E applied to a semiconductor sample is suddenly increased by a small amount δE such that

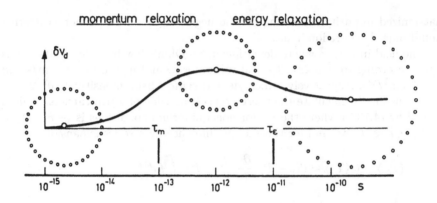

Fig. 6.6: Schematic representation of the relaxation behavior of hot carriers (after [6.26])

the drift velocity of the carriers is increased by an amount δv_d. The change of δv_d with time is represented by the curve. The carrier distribution in k-space at various times is symbolized by dotted circles with diameters which are a measure of the average carrier energy. The initial forward momentum of the carriers caused by the increase in field strength is randomized by collisions in a time of the order of magnitude $\langle \tau_\mathrm{m} \rangle$, while the carrier energy does not yet change appreciably. The increase in carrier energy requires a large time τ_ε. Since the momentum relaxation time $\langle \tau_\mathrm{m} \rangle$ and therefore also the drift velocity decreases with increasing carrier energy, δv_d decreases somewhat during the period of energy relaxation. Now the relaxation is complete and the value of the drift velocity corresponds to the value of the applied field $E + \delta E$.

Since deviations from Ohm's law and effects associated with it such as energy relaxation are best measured at low temperatures where, however, ionized impurity scattering influences the mobility, we will discuss the combined lattice and impurity scattering mechanisms in the following section and consider experimental results in the light of this discussion.

6.6 Combined Ionized Impurity
and Acoustic Deformation Potential Scattering

If there are several scattering mechanisms active, the corresponding scattering rates, which to a good approximation are the inverse momentum relaxation times, have to be added. For the case of combined ionized impurity and acoustic-deformation potential scattering, we have in the relaxation time approximation

$$\frac{1}{\tau_m} = \left(\frac{1}{\tau_m}\right)_{Ion} + \left(\frac{1}{\tau_m}\right)_{ac} . \tag{6.6.1}$$

Since $(1/\tau_m)_{ac} \propto \varepsilon^{1/2}$ and $(1/\tau_m)_{Ion} \propto \varepsilon^{-3/2}$, the ratio

$$(\tau_m)_{ac}/(\tau_m)_{Ion} = q^2(\varepsilon/k_B T)^{-2} , \tag{6.6.2}$$

where q^2 is a factor of proportionality given by

$$q^2 = 6\mu_{ac}/\mu_{Ion} , \tag{6.6.3}$$

and μ_{ac} and μ_{Ion} are the mobilities given by (6.4.18), (6.3.22), respectively. Hence, for τ_m we find

$$\tau_m = (\tau_m)_{ac}(\varepsilon/k_B T)^2/[q^2 + (\varepsilon/k_B T)^2] . \tag{6.6.4}$$

For later convenience, let us calculate the mobility for a Maxwell–Boltzmann distribution with an electron temperature T_e instead of the usual lattice temperature T. According to (4.1.23) for the present case of nondegeneracy, the mobility is

$$\mu = \frac{4}{3\sqrt{\pi}} \frac{e}{m} \int_0^\infty \tau_m \exp(-\varepsilon/k_B T_e) (\varepsilon/k_B T_e)^{3/2} d(\varepsilon/k_B T_e) . \tag{6.6.5}$$

With τ_m given by (6.6.4) and $(\tau_m)_{ac} = \tau_0(\varepsilon/k_B T)^{-1/2}$, according to (6.4.17), we find for the mobility

$$\mu = \mu_{ac} \int_0^\infty \frac{(\varepsilon/k_B T)^{3/2}}{q^2 + (\varepsilon/k_B T)^2} \exp(-\varepsilon/k_B T_e)(\varepsilon/k_B T_e)^{3/2} d(\varepsilon/k_B T_e) , \tag{6.6.6}$$

where $\mu_{ac} = (4/3\sqrt{\pi}) |e| \tau_0/m$ is the zero-field acoustic mobility given by (6.4.18). We introduce for a parameter $\lambda = T/T_e$ and $q' = \lambda q$. Linearizing the denominator of the integrand yields

$$\mu = \mu_{ac}\lambda^{1/2} \left\{ 1 - \frac{q'^2}{2} \left[\int_0^\infty \frac{\exp(-\varepsilon/k_B T_e) \, d(\varepsilon/k_B T_e)}{(\varepsilon/k_B T_e) + i q'} \right. \right.$$

$$\left. \left. + \int_0^\infty \frac{\exp(-\varepsilon/k_B T_e) \, d(\varepsilon/k_B T_e)}{(\varepsilon/k_B T_e) - i q'} \right] \right\} \tag{6.6.7}$$

where i is the imaginary unit. The integrals are evaluated in the complex plane in terms of

$$-si(q) = \int_q^\infty \frac{\sin t}{t} dt ; \qquad -Ci(q) = \int_q^\infty \frac{\cos t}{t} dt \tag{6.6.8}$$

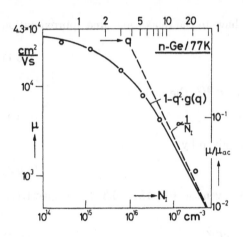

Fig. 6.7: Impurity dependence of the mobility for acoustic deformation potential scattering, according to (6.6.11)

and of the *auxiliary functions* [6.27]

$$f(q) = Ci\,(q)\sin(q) - si\,(q)\cos(q) = 1/q + dg/dq$$

$$g(q) = -Ci\,(q)\cos(q) - si\,(q)\sin(q) = -df/dq \tag{6.6.9}$$

with the result

$$\mu = \mu_{\mathrm{ac}}\lambda^{1/2}[1 - q'^2 g(q')] \ . \tag{6.6.10}$$

For thermal carriers ($\lambda = 1$), we find for the zero-field mobility [6.28]

$$\mu = \mu_{\mathrm{ac}}\,[1 - q^2 g(q)] \ . \tag{6.6.11}$$

The curve in Fig. 6.7 represents the ratio μ/μ_{ac} as a function of q. The data points represent Hall mobility data observed in n-type Ge at 77 K for various impurity concentrations; they have been fitted to the curve by assuming $\mu_{\mathrm{ac}} = 4.3 \times 10^4$ cm^2/V s and a ratio $N_{\mathrm{I}}/q^2 = 6.25 \times 10^{14}$cm^{-3}. For N_{I}, the carrier concentration has been taken which means that impurity compensation has not been taken into account. In addition, optical-phonon scattering known to be present in n-Ge at 77 K has been neglected. Therefore, the values of the fitting parameters should not be taken too seriously.

In the energy balance, ionized impurity scattering may be neglected because it is an elastic process. We shall consider here only the case of warm carriers since for hot carriers ionized impurity scattering [because of $(\tau_{\mathrm{m}})_{\mathrm{Ion}} \propto \varepsilon^{+3/2}$] becomes less important [6.29].

After some algebra we obtain for $\beta/\beta_{\mathrm{ac}}$ [6.30, 31]:

$$\beta/\beta_{\mathrm{ac}} = 1 + 2q^2 - 5q^2 g - 2q^3 f \ . \tag{6.6.12}$$

Figure 6.8 (dashed curve) shows [6.32] the absolute value of β plotted vs $q^2 \propto N_{\mathrm{I}}$. At large values of N_{I}, the curve approaches the dashed straight line which in the log-log plot indicates a proportionality to N_{I}^{-1}. Although

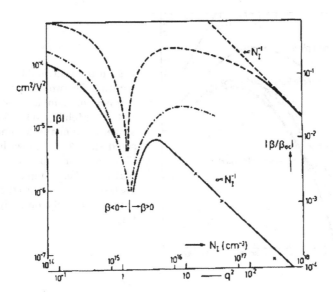

Fig. 6.8: Absolute values of β vs ionized impurity concentration in n-Ge at 77 K. Experimental data in $\langle 100 \rangle$ direction($\times$). Calculation according to (6.6.12 (dashed line) and asymptote N_I^{-1}); by Adawi (6.29) using a variational method (dash - dotted line); acoustic and optical phonon scattering with a ratio of deformation potential constants, $D/\varepsilon_{ac} = 0.4\,\omega_0/u_l$, are taken into account (after [6.32])

experimental data obtained on n-type Ge at 77 K [6.33] (crosses) are much lower, they do show the expected proportionality to N_I^{-1}. Calculations, where optical-phonon scattering (Sect. 6.8) in addition to the mechanisms discussed above, were taken into account, agree qualitatively with observations (dash-dotted curve). In the limit of heavy doping, a Maxwell- Boltzmann distribution function seems to be quite a good approximation because of the strong electron-electron interaction (Sect. 6.11) while the doping is not yet high enough for degeneracy to become important [6.34, 35].

With increasing temperature, the transition of the mobility from predominant ionized impurity scattering to lattice scattering is shown [6.36] in Fig. 6.9. The samples contain doubly-charged zinc atoms which are very efficient in impurity scattering since the charge Ze enters (6.3.22) as $Z^2 e$. For samples 98 C and 104 H the charge of the impurities does not change with temperature. The deviation of the observed mobility from the $T^{3/2}$ dependence in sample 104 H at $T < 30\,\mathrm{K}$ is attributed to impurity band conduction (Sect. 6.14).

A method for obtaining the acceptor and donor concentrations in a semiconductor (actually p-type Si) by conductivity, Hall and magnetoresistance measurements at 77 K in the range of validity of Ohm's law has been presented by *Long* et al. [6.15]. *Brown* and *Bray* [6.37] found that in n-type Ge at impurity

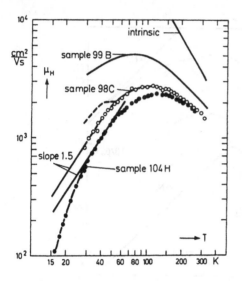

Fig. 6.9: Hall mobility vs temperature for two n-type Zn-doped germanium samples (98 C and 104 H). Mobility data for intrinsic germanium and for uncompensated sample doped with 5×10^{16} Sb atoms/cm^3 are shown for comparison. For sample 98 C, the dashed curve indicates the effect of exposure to light on mobility (after [6.36])

concentrations of less than 10^{15} cm^{-3} between 30 and 300 K, the Brooks–Herring formula describes the ionized impurity scattering very well but overestimates the mobility for higher N_{I} or lower T. *Eagles* and *Edwards* [6.38] in an extension of the calculations by *Herring* and *Vogt* [6.39], investigated the influence of ionized impurity scattering on the galvanomagnetic effects for the case of the many-valley model of band structure to be discussed in Chap. 7.

6.7 Piezoelectric Scattering

If a semiconductor crystal consists of dissimilar atoms such as, e.g., SiC where the bonds are partly ionic (12% in SiC) and the unit cell does not contain a center of symmetry (as in the zincblende lattice or the trigonal lattice), carriers may be scattered by longitudinal acoustic waves due to *piezoelectric scattering* [6.40, 41]. Due to the oscillations of ions, there a dipole moment per unit volume arises known as the *polarization P*. The dielectric displacement D given by

$$D = \kappa_0 E + P \tag{6.7.1}$$

vanishes, since there is no space charge because the spatial displacement of carriers is negligible in comparison to the spatial displacement of the ions. Therefore, there is an ac electric field of intensity

$$E = -P/\kappa_0 . \tag{6.7.2}$$

For a propagating wave of finite wavelength, a strain ($\nabla_r \cdot \delta r$) will exist where δr is the displacement of a lattice atom from its equilibrium position given by

(6.4.1). While for $(\nabla_r \cdot \delta r) = 0$ the dielectric constant κ is defined by the relation $D = \kappa \kappa_0 E$, the dielectric displacement now contains a term which is proportional to $(\nabla_r \cdot \delta r)$:

$$D = \kappa \kappa_0 E + e_{\mathrm{pz}}(\nabla_r \cdot \delta r) . \tag{6.7.3}$$

The factor of proportionality e_{pz} obtained from piezoelectric measurements is of an order of magnitude 10^{-5} As cm^{-2}; this value may be obtained by assuming a typical displacement of the atomic dimension of 10^{-9} cm for about 10^{23} cm^{-3} elementary charges $e = 1.6 \times 10^{-19}$As (one per atom). e_{pz} is called the *piezoelectric constant*.

Equation (6.7.3) yields for $D = 0$

$$E = -(e_{\mathrm{pz}}/\kappa \kappa_0)(\nabla_r \cdot \delta r) . \tag{6.7.4}$$

In an acoustic wave having a propagation vector q, the local variation of $(\nabla_r \cdot \delta r)$ is given by (6.4.3). Since E is proportional to $(\nabla_r \cdot \delta r)$, the potential energy

$$\delta \varepsilon = |e| \int E \, dr = |e| \, E/q \tag{6.7.5}$$

also depends on $(\nabla_r \cdot \delta r)$:

$$\delta \varepsilon = (|e| \, e_{\mathrm{pz}}/\kappa \kappa_0 q)(\nabla_r \cdot \delta r) . \tag{6.7.6}$$

A comparison of this relation with (6.4.6) for nonpolar acoustic scattering reveals that instead of the acoustic deformation potential constant $\varepsilon_{\mathrm{ac}}$, we now have $|e| \, e_{\mathrm{pz}}/\kappa \kappa_0 q$ which in contrast to $\varepsilon_{\mathrm{ac}}$ is not a constant but depends on

$$q = |k' - k| \approx 2|k| \sin(\theta/2) = (2 \, m \, v/\hbar) \sin(\theta/2) . \tag{6.7.7}$$

The absolute magnitude of the matrix element $H_{k'k}$ given by

$$|H_{k'k}| = \frac{|e| \, e_{\mathrm{pz}}}{\kappa \, \kappa_0 \, q} \left(\frac{k_{\mathrm{B}} T}{2 V c_\ell} \right)^{1/2} = \left(\frac{e^2 \, K^2 \, k_{\mathrm{B}} T}{2 V \kappa \kappa_0 \, q^2} \right)^{1/2} \tag{6.7.8}$$

is similar to (6.4.12); here we have introduced the dimensionless *electro-mechanical coupling coefficient* K^2, defined by

$$K^2/(1 - K^2) = e_{\mathrm{pz}}^2/(\kappa \kappa_0 c_\ell) . \tag{6.7.9}$$

The left side for $K^2 \ll 1$ is approximately K^2. For example, in SiC with $e_{\mathrm{pz}} = 10^{-5}$ A s cm^{-2}, $\kappa = 10.2$ and $c_\ell = 1.8 \times 10^7$ N cm^{-2}, it is $K^2 = 6 \times 10^{-4}$. For most polar semiconductors, K^2 is of the order of 10^{-3}. The definition of K^2 given here is related to the power stored in a charged condenser which is proportional to the dielectric constant κ. Let us for simplicity denote the strain $(\nabla_r \cdot \delta r)$ by S. The relationship between tension, strain and electric field strength in one dimension is given by [5]

$$T = c_\ell S - e_{pz} E \ . \tag{6.7.10}$$

In a tension-free crystal $(T = 0)$, the strain induced by the electric field is

$$S = (e_{pz}/c_\ell) E \ . \tag{6.7.11}$$

If we eliminate S from (6.7.3) we obtain for the ratio D/E

$$D/E = \kappa \kappa_0 + e_{pz}^2/c_\ell \ . \tag{6.7.12}$$

We notice that the work necessary for charging the condenser in this case consists of an electric part $\propto \kappa \kappa_0$ and a mechanical part $\propto e_{pz}^2/c_\ell$ (if we can apply a tension so that there is no strain, we have only the electric part). K^2 is now defined by the ratio of the mechanical work to the total work:

$$K^2 = \frac{e_{pz}^2/c_\ell}{\kappa \kappa_0 + e_{pz}^2/c_\ell} \ . \tag{6.7.13}$$

The calculation of the scattering probability from (6.4.14) and of the momentum relaxation time τ_m from (6.2.23) presents no problems:

$$\frac{1}{\tau_m} = \frac{V}{(2\pi)^3} \int 2 \frac{2\pi}{\hbar} \frac{e^2 K^2 k_B T}{2V \kappa \kappa_0 q^2} \delta[\varepsilon(k') - \varepsilon(k)] k^2 \, dk (1 - \cos\theta) \sin\theta \, d\theta \, 2\pi \tag{6.7.14}$$

where q^2 is approximately given by $4k^2 \sin^2(\theta/2)$ and $dk = \hbar^{-1}(m/2\varepsilon)^{1/2} \, d\varepsilon$. The integration yields for τ_m

$$\tau_m = \frac{2^{3/2}\pi \hbar^2 \kappa \kappa_0}{m^{1/2} e^2 K^2 k_B T} \varepsilon^{1/2} \ . \tag{6.7.15}$$

We notice that for the energy dependence of τ_m,

$$\tau_m \propto \varepsilon^{+1/2} \ , \tag{6.7.16}$$

a power law with a positive exponent $+1/2$ is valid. Hence, we expect positive values for the coefficient β for warm carriers.

For nondegenerate thermal carriers, the mobility obtained from (6.7.15) is given by

$$\mu = \frac{16\sqrt{2\pi}}{3} \frac{\hbar^2 \kappa \kappa_0}{m^{3/2} e K^2 (k_B T)^{1/2}} \propto T^{-1/2} \tag{6.7.17}$$

---

[5] Equations (6.7.3, 10) can be written
$D = [\kappa \kappa_0 E/(1 - K^2)] + (e_{pz}/c_\ell)T$,
$S = (e_{pz}/c_\ell)E + c_\ell^{-1}T$,
where K^2 is defined by (6.7.13). Small variations in E and T result in a small change in energy U of $dU = D \, dE + S \, dT$. Now U is an exact integral so that $(\partial D/\partial T)_E = (\partial S/\partial E)_T$. Hence, the same constant e_{pz}/c_ℓ must be used in both equations.

and in units of cm^2/Vs

$$\mu = 2.6 \frac{\kappa}{(m/m_0)^{3/2} K^2 (T/100\,K)^{1/2}} \,. \tag{6.7.18}$$

Assuming typical values $\kappa = 10$, $m/m_0 = 0.1$, $K^2 = 10^{-3}$ and $T = 100$ K, a value of 8.25×10^5 cm^2/Vs is found which is large compared with the mobility due to nonpolar acoustic scattering; the latter is a competing scattering process in semiconductors with partly ionic bonds.

Therefore, it may be interesting to compare the mobility due to piezoelectric scattering (for this purpose denoted as μ_{pz}) with that due to acoustic deformation potential scattering (μ_{ac}) given by (6.4.18). For the ratio μ_{pz}/μ_{ac} we find

$$\frac{\mu_{pz}}{\mu_{ac}} = 0.75 \frac{(\kappa/10)^2 (m/m_0)(T/100\,K)(\varepsilon_{ac}/eV)^2}{(e_{pz}/10^{-5}\,As\,cm^{-2})^2} \,. \tag{6.7.19}$$

At a low enough temperature the mobility in a pure dislocation-free polar semiconductor may be determined by piezoelectric scattering rather than acoustic deformation potential scattering ($\mu_{pz} \ll \mu_{ac}$). A small effective mass m also favors piezoelectric scattering. In a typical case such as $\kappa = 10$, $m/m_0 = 0.1$, $T = 10$ K, $\varepsilon_{ac} = 5$ eV, $e_{pz} = 10^{-5}$ As cm^{-2}, we have $\mu_{pz}/\mu_{ac} \approx 0.2 \ll 1$ and piezoelectric scattering predominates. However, in those polar semiconductors with purities available at present, ionized impurity scattering will probably dominate under these conditions. For this reason there are few practical applications for piezoelectric scattering . The energy transfer rate has been calculated by *Kogan* and others [6.42] and will not be discussed here since it has rarely been observed (see, e.g., Fig. 6.19 inset).

6.8 The Phonon Spectrum of a Crystal

So far we have considered only long-wavelength acoustic phonons in electron scattering processes. They are characterized by a frequency independent sound velocity [6].

In crystals with more than one atom per unit cell, one must also take *optical phonons* into account. In this chapter we will consider the phonon spectra of various semiconductors [6.43, 44].

If there are more than N atoms in a unit cell there are $N - 1$ optical branches, besides the acoustic branch. In a three-dimensional crystal there are two transverse oscillations per longitudinal oscillation which are degenerate with each other in a cubic crystal.

The cubic unit cell of the diamond and zincblende lattices is shown in Fig. 2.11. It contains a central atom and 4 atoms on the cube edges. The atoms

---

[6]The dispersion relation of acoustic lattice vibrations is treated in textbooks on solid state physics

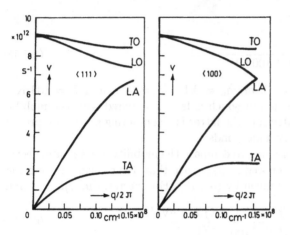

Fig. 6.10: Lattice vibrational spectrum of germanium at 300 K in the $\langle 111 \rangle$ and $\langle 100 \rangle$ directions determined by the scattering of cold neutrons. The transverse branches are doubly degenerate (after [6.51])

on the cube edges are each shared by only 4 adjacent cubes because each atom makes only 4 bonds; therefore, these atoms have to be weighted by a factor of 1/4. The total number of atoms in the unit cell is then given by $1 + 4/4 = 2$.

Besides Raman spectroscopy [6.45, 46], neutron spectroscopy [6.47, 48] provides a very satisfactory experimental method for obtaining the phonon spectrum of a crystal. Results for germanium [6.49] and gallium arsenide [6.50] are illustrated in Figs. 6.10 and 6.11. The optical-phonon temperature is also called the *Debye temperature*[7]. For germanium it has a value of $\hbar\omega_0/k_B = 430$ K and for GaAs a value of 417 K. Due to the ionic binding character in compound semiconductors, atomic oscillations in the optical-phonon mode may be excited by electromagnetic radiation (Sect. 11.7). This explains the name *optical phonon*. For an optical-phonon temperature of, e.g., 400 K, a resonant wavelength of 32 μm in free space is calculated which is in the far-infrared spectrum.

Figure 6.11 shows that for a compound semiconductor with partly ionic bonds, the longitudinal-optical phonon frequency ν_ℓ is somewhat larger than the transverse-optical phonon frequency ν_t at $\mathbf{q} = 0$. In Sect. 11.7 we shall see that the index of refraction which is the square root of the dielectric constant for low frequencies $\sqrt{\kappa}$, is also larger than for optical frequencies where it is $\sqrt{\kappa_{opt}}$; κ is the static dielectric constant. We shall now prove the Lyddane–Sachs–Teller relation [6.53, 54] for polar modes

$$\omega_\ell/\omega_t = \sqrt{\kappa}/\sqrt{\kappa_{opt}} \qquad (6.8.1)$$

which states that the ratio of the two-phonon frequencies at $q = 0$ equals the ratio of the refractive indices.

The mechanical vibration of a polar crystal lattice, the dielectric polarization

---

[7] Actually, this Debye temperature may be somewhat different from the one determined from specific heat measurements.

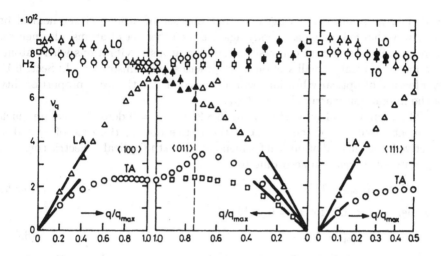

Fig. 6.11: Lattice vibrational spectrum of gallium arsenide at 296 K. The solid points denote frequencies of undetermined polarization. The dashed line indicates the zero boundary in the ⟨011⟩ direction. The solid lines represent the slopes of the corresponding velocities of sound, calculated from the elastic constants (after [6.52])

P, and the electric field E are interrelated by the following set of equations:

$$P = \gamma_{11} E + \gamma_{12} w \tag{6.8.2}$$

$$-d^2 w / dt^2 = \gamma_{22} w - \gamma_{12} E \ , \tag{6.8.3}$$

where the γ_{ik} are coefficients, which will be determined later, and w is a *reduced displacement* given by

$$w = \delta r \sqrt{M \, N_u} \ . \tag{6.8.4}$$

N_u is the number of lattice cells per cm^3 and M is the reduced atomic mass given by

$$1/M = 1/M^+ + 1/M^- \ , \tag{6.8.5}$$

where M^+ and M^- are the masses of the positive and negative ions, respectively. δr is their relative displacement. [8]

In Sect. 7.8 we shall treat the interaction of carriers with coherent acoustic waves in polar lattices with quite similar equations (7.8.17, 18). While for *acoustic* waves and no electric field $E = 0$, the polarization P is proportional to the strain $S = (\boldsymbol{\nabla_r} \cdot \delta r)$, for *optical* modes, P is proportional to δr itself as given by (6.8.2, 24). The essential feature of the optical vibration at q

---

[8]Compare (6.8.23, 24) with (6.7.3, 10), respectively, where $P = D - \kappa_0 E$; $-T \propto d^2 w/dt^2$ (Newton's Second Law); $w \propto S$; the argument for having the same factor e_{pz} in both (6.7.3, 10) also applies to having the same factor γ_{12} in both (6.8.2, 3).

= 0, i.e., at infinite wavelength, is that one set of atoms, either neutral or positively charged, move as a body against the other set of atoms, also neutral or negatively charged, in contrast to the long acoustic vibrations where atoms in a macroscopically small section move practically in unison [9]. In Sect. 6.11 this feature of optical vibrations will also account for the δr – proportionality of the interaction matrix element $H_{k'k}$.

The proportionality of P to E in (6.8.2) for $w = 0$ does not require much comment. Because of the inertia of the lattice atoms, the case of $w \approx 0$ is effective for an ac field at optical frequencies where the optical dielectric constant x_{opt} enters this relation according to

$$P = \kappa_0(\kappa_{\text{opt}} - 1)E \tag{6.8.6}$$

and therefore (6.8.2) for $w = 0$ yields

$$\gamma_{11} = \kappa_0(\kappa_{\text{opt}} - 1) \ . \tag{6.8.7}$$

Equation (6.8.2) for $E = 0$ results in a polarization due to the relative reduced displacement w of the positively charged and negatively charged atoms. In (6.8.3) a static displacement (w for $d^2w/dt^2 = 0$) is shown to be caused by the application of a dc electric field E:

$$w = (\gamma_{11}/\gamma_{22})E \ . \tag{6.8.8}$$

If we eliminate w from (6.8.8), (6.8.2) the polarization P as a function of the applied field E becomes

$$P = (\gamma_{12}^2/\gamma_{22} + \gamma_{11})E \ . \tag{6.8.9}$$

Since at present we deal with static conditions, this relation is usually described by the static relative dielectric constant

$$P = \kappa_0(\kappa - 1)E \ , \tag{6.8.10}$$

where κ_0 is the permittivity of free space. A comparison of these two equations yields

$$\gamma_{12}^2/\gamma_{22} + \gamma_{11} = \kappa_0(\kappa - 1) \ . \tag{6.8.11}$$

If we take γ_{11} as given by (6.8.7) and solve for γ_{12}^2, we find

$$\gamma_{12}^2 = \gamma_{22}\kappa_0(\kappa - \kappa_{\text{opt}}) \ . \tag{6.8.12}$$

Equation (6.8.3) for $E = 0$ is the equation of motion for the atomic lattice vibration which is the well-known result of Hook's law and Newton's second law. The coefficient γ_{22} is the square of the angular frequency of the vibration.

For waves we consider only the field and the polarization in the direction of the wave propagation. Hence, for a transverse mode of oscillation (subscript t), $E = 0$ and

$$\gamma_{22} = \omega_t^2 \ , \tag{6.8.13}$$

---

[9]For example, in Fig. 2.11, atom A and equivalent ones form one set, atom B and equivalent ones the second set.

while for a longitudinal mode of oscillation (subscript ℓ),

$$0 = D = \kappa_0 E + P \ , \tag{6.8.14}$$

where P is given by (6.8.2):

$$0 = \kappa_0 E + \gamma_{11} E + \gamma_{12} w \ . \tag{6.8.15}$$

We solve this equation for E and by eliminating E from (6.8.3), we obtain the equation of motion for the longitudinal mode:

$$-d^2 w/dt^2 = [\gamma_{22} + \gamma_{12}^2/(\kappa_0 + \gamma_{11})] \ w \ . \tag{6.8.16}$$

The factor in brackets is the square of the angular frequency of the vibration:

$$\gamma_{22} + \gamma_{12}^2/(\kappa_0 + \gamma_{11}) = \omega_\ell^2 \ . \tag{6.8.17}$$

From (6.8.13, 17) by taking (6.8.7, 12) into account, we obtain for the ratio ω_ℓ^2/ω_t^2

$$\frac{\omega_\ell^2}{\omega_t^2} = 1 + \frac{\gamma_{12}^2}{\gamma_{22}(\kappa_0 + \gamma_{11})} = 1 + \frac{\kappa_0(\kappa - \kappa_{opt})}{\kappa_0 + \kappa_0(\kappa_{opt} - 1)} = \frac{\kappa}{\kappa_{opt}} \ . \tag{6.8.18}$$

This yields (6.8.1).

By using the electrostatics method, we have tacitly assumed a Coulomb interaction between charges in the lattice although in a lattice wave, of course, the Coulomb interaction is retarded.

Let us finally derive a useful relationship between P and δr for the longitudinal optical mode from (6.8.4, 14, 15):

$$P = -\kappa_0 E = \kappa_0[\gamma_{12}/(\kappa_0 + \gamma_{11})] \delta r \sqrt{M N_u} \ . \tag{6.8.19}$$

Introducing for γ_{12} and γ_{11} their values as given by (6.8.12, 7) and using the Lyddane-Sachs- Teller relation, (6.8.1), we obtain

$$P = \omega_1 [\kappa_0(\kappa_{opt}^{-1} - \kappa^{-1}) M N_u]^{1/2} \delta r \ . \tag{6.8.20}$$

The product $M N_u$ is sometimes approximated by the mass density ϱ. Since the polarization is a dipole moment $e_C \delta r$ per unit volume, which here is the volume of the lattice cell N_u^{-1},

$$P = e_C \delta r/N_u^{-1}, \tag{6.8.21}$$

we find for the *Callen effective charge* [6.55] for longitudinal optical modes

$$e_C = \omega_1 [\kappa_0(\kappa_{opt}^{-1} - \kappa^{-1}) M/N_u]^{1/2} \ . \tag{6.8.22}$$

e_C is usually given in units of the elementary charge. For example, in α-SiC frequencies of $\omega_\ell = (1.82 \pm 0.05) \times 10^{14} \, \text{s}^{-1}$ and $\omega_t = (1.49 \pm 0.01) \times 10^{14} \, \text{s}^{-1}$ have been observed which yield

$$\kappa_{opt}^{-1} - \kappa^{-1} = [(\omega_\ell/\omega_t)^2 - 1]/\kappa = (0.49 \pm 0.15)/\kappa \ . \tag{6.8.23}$$

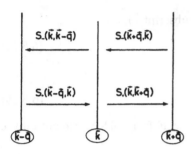

Fig. 6.12: Schematic representation of electron transitions involving phonon emission and phonon absorption

Although the frequencies have been measured quite accurately by Raman spectroscopy, the difference $[(\omega_u/\omega_\ell)^2 - 1]$ is not well known. From the static relative dielectric constant $\kappa = 10.2$, the mass density $\varrho = 3.2\,\mathrm{g\,cm}^{-3}$ and the density of lattice cells $N_u = 1.05 \times 10^{23}\,\mathrm{cm}^{-3}$, a value $e_C/e = 0.40$ has been calculated for SiC. In GaAs, e.g., the value is only 0.17. The corresponding values of the *Szigeti effective charge* e_S [6.56]

$$e_S = e_C\, 3\kappa_{\mathrm{opt}}/(\kappa + 2)\ , \tag{6.8.24}$$

known from the theory of ionic bonds, are 0.67 in SiC and 0.38 in GaAs.

6.9 Inelastic Scattering Processes

So far we have dealt with scattering processes such as impurity scattering and acoustic-phonon scattering which are almost elastic. For optical-phonon scattering processes, however, the phonon energy is of the order of magnitude of the thermal energy of carriers and these processes have to be treated as inelastic. As a consequence, scattering rates from a state k to a state $k' = k + q$ will be very different from those to $k' = k - q$.

In Fig. 6.12, all four possible transitions from and to a state k are shown schematically. For example, $S_-(k, k - q)$ represents the probability for the emission of a phonon of momentum $\hbar\,q$ due to a change of the electron momentum from $\hbar\,k$ to $\hbar\,(k - q)$. Hence (6.2.16) is now given in detail as

$$[-\partial f(k)/\partial t]_{\mathrm{coll}} = V(2\pi)^{-3} \int d^3q \{S_-(k, k - q) f(k)[1 - f(k - q)]$$

$$+ S_+(k, k + q) f(k)[1 - f(k + q)] - S_-(k + q, k) f(k + q)[1 - f(k)]$$

$$- S_+(k - q, k) f(k - q)[1 - f(k)]\} \tag{6.9.1}$$

The subscripts $+$ and $-$ on S indicate phonon absorption and emission, respectively. The usual integration over k' has been replaced by an integration over

q since k' is partly $k + q$ and partly $k - q$. For the case of a non-degenerate electron gas, the factors in brackets can be omitted:

$$[-\partial f(k)/\partial t]_{\text{coll}} = V(2\pi)^{-3} \int d^3q \left\{[S_-(k, k - q) + S_+(k, k + q)] f(k)\right.$$

$$\left. -S_-(k + q, k)f(k + q) - S_+(k - q, k)f(k - q)\right\} . \tag{6.9.2}$$

For inelastic processes we cannot approximate $f(k \pm q)$ by $f(k)$ in order to be able to bring this equation into the form of (6.2.22) and thus define a relaxation time τ_{m}. We shall prove, however, that in the special case where the matrix element $|H_{k'k}|$ is independent of q, the $f(k \pm q)$-factor in the last two terms in (6.9.2) can be transformed into $f_0(k)$ -factors by the integration over q and that even in this inelastic case, τ_{m} can be defined; otherwise a momentum relaxation time τ_{m} does not exist. Since indeed for optical deformation potential scattering the special case is realized, we shall give the proof here.

In the diffusion approximation (6.2.19), $f(k)$ is given by

$$f(k) = f_0(k) + f_1(k) \cos \vartheta , \tag{6.9.3}$$

where, for simplicity, $-(\hbar k\, G/m)\partial f_0/\partial k$ has been denoted by $f_1(k)$. Since without a magnetic field, G is proportional to the electric field E, ϑ is the angle between k and E; the direction of E is the polar axis. The angle between k' and E shall be denoted by ϑ'. Hence,

$$f(k') = f_0(k') + f_1(k') \cos \vartheta' \tag{6.9.4}$$

Let us first prove that [10]

$$\int \cos \vartheta' \delta[\varepsilon(k') - \varepsilon(k) \mp \hbar\omega_0] d^3q = 0 , \tag{6.9.5}$$

where $\hbar\omega_0$ is the optical-phonon energy which is taken as a constant. For the calculation of $\cos \vartheta'$, we introduce polar coordinates for k' with k being the polar axis:

$$k' = k \pm q = (\pm q \sin \theta \cos \varphi, \pm q \sin \theta \sin \varphi, k \pm q \cos \theta) , \tag{6.9.6}$$

where θ is the angle between k and q and φ is the azimuth. In this representation, E is given by

$$E = E(\sin \vartheta \cos \psi, \sin \vartheta \sin \psi, \cos \vartheta) \tag{6.9.7}$$

and $\cos \vartheta'$ by

$$\cos \vartheta' = \frac{(k' \cdot E)}{k' E} = \frac{\pm q \sin \theta \sin \vartheta \cos(\varphi - \psi) \pm q \cos \theta \cos \vartheta + k \cos \vartheta}{\sqrt{q^2 + k^2 \pm 2qk \cos \theta}}$$

$$\tag{6.9.8}$$

---

[10] $k' = \sqrt{k^2 \pm 2m\omega_0/\hbar}$ is independent of q and hence $f_1(k')$, is not subject to integration over q.

Since $d^3q = q^2\,dq\,\sin\theta\,d\theta\,d\varphi$, according to (6.9.5) we first perform the integration over φ which makes the $\cos(\varphi - \psi)$-term vanish and adds a factor of 2π to the other terms; the Dirac δ-function contained in the scattering probabilities S_- and S_+ in (6.9.2)

$$\delta[\varepsilon(\boldsymbol{k} \pm \boldsymbol{q}) - \varepsilon(\boldsymbol{k}) \mp \hbar\omega_0] = \delta[(\hbar^2/2m)(\pm 2kq\cos\theta + q^2) \mp \hbar\omega_0] \qquad (6.9.9)$$

is independent of φ. The integrations over φ and θ are therefore easily performed:

$$\int\limits_0^\pi \int\limits_0^{2\pi} \cos\vartheta'\,d\varphi\,\delta[\varepsilon(\boldsymbol{k} \pm \boldsymbol{q}) - \varepsilon(\boldsymbol{k}) \mp \hbar\omega_0]\sin\theta\,d\theta$$

$$= \frac{2\pi m}{\hbar^2 k q}\cos\vartheta\,\frac{A_\pm - q^2}{\sqrt{k^2 \pm 2m\omega_0/\hbar}} \qquad (6.9.10)$$

where, for simplicity, we have introduced the constant

$$A_\pm = 2k^2 \pm 2m\omega_0/\hbar \ . \qquad (6.9.11)$$

For the final integration over q, we consider only the q-dependent factors

$$\int\limits_{q_1}^{q_2} \frac{A_\pm - q^2}{q}q^2\,dq = \frac{1}{2}(q_2^2 - q_1^2)\left[A_\pm - \frac{1}{2}(q_1^2 + q_2^2)\right] \ . \qquad (6.9.12)$$

Energy and momentum conservation, i.e., the vanishing argument of the δ-function (6.9.9), and $q \geq 0$ yield for the lower limit

$$q_1 = \mp k \pm \sqrt{A_\pm - k^2} \qquad (6.9.13)$$

and for the upper limit

$$q_2 = +k + \sqrt{A_\pm - k^2} \ . \qquad (6.9.14)$$

Figure 6.13 shows these limits in a diagram of q/k vs $\cos\theta$ for $\hbar\omega_0/\varepsilon = 0.5$. With these values the integral given by (6.9.12) vanishes which proves (6.9.5) for both phonon absorption and emission. For (6.9.4) in the form

$$f(\boldsymbol{k} \pm \boldsymbol{q}) = f_0(\boldsymbol{k} \pm \boldsymbol{q}) + f_1(\boldsymbol{k} \pm \boldsymbol{q})\cos\vartheta' \qquad (6.9.15)$$

and (6.2.17) [11] we can manipulate (6.9.2) into a form where we have the distribution functions in the form $f(\boldsymbol{k}) - f_0(\boldsymbol{k})$ on the rhs and τ_m can be defined by (6.2.23). The generalization of (6.9.4) to an arbitrary number of spherical harmonics in the expansion

$$f(\boldsymbol{k}) = \sum_\ell f_\ell(k)\,P_\ell(\cos\vartheta) \qquad (6.9.16)$$

---

[11] $S_-(\boldsymbol{k}+\boldsymbol{q},\boldsymbol{k})f_0(|\boldsymbol{k}+\boldsymbol{q}|) + S_+(\boldsymbol{k}-\boldsymbol{q},\boldsymbol{k})f_0(|\boldsymbol{k}-\boldsymbol{q}|) = S_+(\boldsymbol{k},\boldsymbol{k}+\boldsymbol{q})f_0(\boldsymbol{k}) + S_-(\boldsymbol{k},\boldsymbol{k}-\boldsymbol{q})f_0(\boldsymbol{k})$.

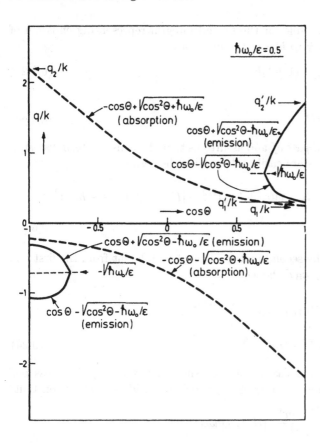

Fig. 6.13: q/k obtained from energy and momentum conservation as a function of the cosine of the scattering angle for phonon absorption and emission

has been given by *Beer* [6.57]. The result for $1/\tau_m$ is obtained from (6.2.22), (6.9.2):

$$\frac{1}{\tau_m} = \frac{V}{(2\pi)^3}\frac{2\pi}{\hbar}\left\{\int_{q_1}^{q_2}|H_{k+q,k}|^2\frac{2\pi m}{\hbar^2 k q}q^2 dq + \int_{q_1'}^{q_2'}|H_{k-q,k}|^2\frac{2\pi m}{\hbar^2 k q}q^2 dq\right\} \quad (6.9.17)$$

where for the first integral representing absorption of $\hbar\omega_0$ in a transition from k to $k+q$, the limits given by (6.9.13, 14) have been written in the form

$$q_1 = -k + \sqrt{k^2 + 2m\omega_0/\hbar} = (a-1)k \ , \tag{6.9.18}$$

where

$$a = (1 + \hbar\omega_0 \, 2m/\hbar^2 k^2)^{1/2} = (1 + \hbar\omega_0/\varepsilon)^{1/2} \tag{6.9.19}$$

corresponding to $\theta = 0$, and

$$q_2 = +k + \sqrt{k^2 + 2m\omega_0/\hbar} = (a+1)k \ , \tag{6.9.20}$$

corresponding to $\theta = \pi$, while for the second integral representing emission of $\hbar\omega_0$ in a transition from $\boldsymbol{k}$ to $\boldsymbol{k} - \boldsymbol{q}$, the limits are

$$q_1' = (1-b)k; \quad q_2' = (1+b)k \ , \tag{6.9.21}$$

where b is given by

$$b = \text{Re}[1 - \hbar\omega_0/\varepsilon]^{1/2} \tag{6.9.22}$$

and Re stands for the *real part of* and assures us that the emission term vanishes for $\varepsilon < \hbar\omega_0$. The integrals are easily solved with the result

$$\frac{1}{\tau_\text{m}} = \frac{2^{1/2}Vm^{3/2}}{\pi\hbar^4}[|H_{k+q,k}|(\varepsilon + \hbar\omega_0)^{1/2} + |H_{k-q,k}|^2 \ \text{Re}\{\varepsilon - \hbar\omega_0\}^{1/2}] \tag{6.9.23}$$

We shall see later that the second matrix element is different from the first one only by a factor $\exp(\hbar\omega_0/k_\text{B}T)$ hence

$$\frac{1}{\tau_\text{m}} = \frac{2^{1/2}Vm^{3/2}}{\pi \ \hbar^4}|H_{k+q,k}|^2 \ [(\varepsilon + \hbar\omega_0)^{1/2}$$

$$+ \exp(\hbar\omega_0/k_\text{B}T)\text{Re}\{\varepsilon - \hbar\omega_0)^{1/2}\}] \ . \tag{6.9.24}$$

The energy loss rate is now easily obtained since the energy gain or loss per collision is the optical-phonon energy $\pm \hbar\omega_0$ which is considered to be a constant:

$$\left(-\frac{d\varepsilon}{dt}\right)_\text{coll} = -\frac{2^{1/2} \ V \ m^{3/2}}{\pi \ \hbar^4}|H_{k+q,k}|\hbar\omega_0$$

$$\times[(\varepsilon + \hbar\omega_0)^{1/2} - \exp(\hbar\omega_0/k_B T) \ \text{Re}\{(\varepsilon - \hbar\omega_0)^{1/2}\}] \ . \tag{6.9.25}$$

In the absorption process the energy change is $+\hbar\omega_0$ while in the emission process it is $-\hbar \ \omega_0$.

6.10 Momentum Balance Equation
and Shifted Maxwellian

A method of calculating the field dependence of the mobility of warm and hot carriers with the simplifying assumption of a Maxwell–Boltzmann distribution with an electron temperature T_e consists of first, averaging τ_m over this distribution and calculating the mobility μ as a function of T_e, second, calculating the relationship between the field strength $\boldsymbol{E}$ and T_e from the energy balance equation

$$\mu \ e \ E^2 = \langle -d\varepsilon/dt \rangle_\text{coll} \ , \tag{6.10.1}$$

and third, eliminating T_e from the two relations.

In another method, the first step is the calculation of an average of τ_m, $\bar{\tau}_m$ from

$$\langle -d(\hbar\, k_E)/dt \rangle_{\text{coll}} = m\, v_d/\bar{\tau}_m \; , \qquad (6.10.2)$$

where the average on the lhs is taken over a *shifted* distribution [see (6.10.12) below]. The quantity $\hbar\, k_E$ is the component of the carrier momentum in the field direction which, on the average, is $m v_d$ where v_d is the drift velocity. The mobility is then approximated by

$$\mu = (|e|/m)\, \bar{\tau}_m \; . \qquad (6.10.3)$$

Since $\mu E = v_d$ and $|e| E$ is the momentum gain per unit time from E, (6.10.3) is obtained from (6.10.2) and the momentum balance equation

$$\langle -d(\hbar\, k_E)/dt \rangle_{\text{coll}} = |e|E \; . \qquad (6.10.4)$$

Otherwise the procedure is the same as above. The results are slightly different from those of the first method. Since the second method has been applied to the case of polar optical scattering [6.58] we shall consider it here briefly.

With k_E being given by

$$k_E = k \cos \vartheta \; , \qquad (6.10.5)$$

we obtain the component of the phonon wave vector in the field direction q_E from (6.9.6, 7):

$$q_E = (\boldsymbol{q} \cdot \boldsymbol{E})/E = q \sin\theta \sin\vartheta \cos(\varphi - \psi) + q \cos\theta \cos\vartheta \; . \qquad (6.10.6)$$

The change of carrier momentum in the field direction is due to the gain or loss of a momentum $\hbar\, q_E$ at an absorption or emission process of a phonon, respectively, multiplied by the scattering probability (Fig. 6.12) and integrated over phase space:

$$[-d(\hbar\, k_E)/dt]_{\text{coll}} = V\,(2\pi)^{-3} \int (\hbar q_E\, S_- - \hbar q_E\, S_+) d^3q \; . \qquad (6.10.7)$$

This is the quantum mechanical equivalent of the classical momentum balance equation.

Since the scattering probability S is independent of the azimuth φ, the integration over φ eliminates the term in (6.10.6) which contains $\cos(\varphi - \psi)$. Hence, from (6.10.7) the momentum loss rate

$$(-dk_E/dt)_{\text{coll}} = \cos \vartheta\, V(2\pi)^{-2} \int \cos\theta (S_- - S_+) q^3 dq \, \sin\theta \; d\theta \qquad (6.10.8)$$

is obtained. Let us denote the rhs by $g(\varepsilon) \cos \vartheta$ where

$$g(\varepsilon) = V(2\pi)^{-2} \int \cos\theta (S_- - S_+) q^3 \, dq \sin\theta \; d\theta \; . \qquad (6.10.9)$$

Because the scattering probabilities S_+ and S_- contain δ-functions, the integration over θ can be done similarly to (6.9.10):

$$\int_0^\pi \cos\theta \, \delta[(\mp\hbar^2 k q/m)\cos\theta + (\hbar^2 q^2/2m) \pm \hbar\omega_0]\sin\theta \, d\theta$$

$$= \pm m/(2\hbar^2 k^2) + m^2 \hbar\omega_0/\hbar^4 k^2 q^2 = \pm(1/4\varepsilon)(1 \pm 2m\, k_B\Theta/\hbar^2 q^2) \ , \tag{6.10.10}$$

where ε and $k_B\Theta$ have been introduced for $\hbar^2 k^2/2m$ and $\hbar\omega_0$, respectively, and the upper signs pertain to phonon emission, the lower ones to absorption. For the function $g(\varepsilon)$ we find

$$g(\varepsilon) = V/,(8\pi\hbar\varepsilon)\left[\int_{q_1}^{q_2} |H_{k+q,k}|^2(1 - 2m\, k_B\,\Theta/\hbar^2 q^2)q^3 dq\right.$$

$$\left.+ \int_{q'_1}^{q'_2} |H_{k-q,k}|^2(1 + 2mk_B\Theta/\hbar^2 q^2)q^3 dq\right] \ , \tag{6.10.11}$$

where the limits are given by (6.9.18, 20, 21).

Next we average (6.10.7) over the shifted Maxwell–Boltzmann distribution function given by

$$f(\boldsymbol{k}) \propto \exp[-(\hbar\boldsymbol{k} - m\,\boldsymbol{v}_d)^2/2m\, k_B T_e] \ , \tag{6.10.12}$$

where the argument for not too large field intensities can be approximated by $-(\hbar^2 k^2 - 2\hbar k\, m\, v_d \cos\vartheta)/2m\, k_B T_e$ and the exponential is approximated by

$$\exp(-\varepsilon/k_B T_e)\exp(\hbar k\, v_d \cos\vartheta/k_B T_e)$$

$$\approx (1 + \hbar k\, v_d \cos\vartheta/k_B T_e)\exp(-\varepsilon/k_B T_e) \ . \tag{6.10.13}$$

The distribution function

$$f(\boldsymbol{k}) = f_0(\varepsilon) + f_1(\varepsilon)\cos\vartheta \tag{6.10.14}$$

therefore consists of a symmetric part $f_0(\varepsilon) \propto \exp(-\varepsilon/k_B T_e)$ and an asymmetric part $f_1(\varepsilon)\cos\vartheta$, where $f_1(\varepsilon)$ is given by

$$f_1(\varepsilon) = f_0(\varepsilon)\, v_d\,(2\, m\,\varepsilon)^{1/2}/k_B T_e \ . \tag{6.10.15}$$

Since the averaging process

$$\langle -dk_E/dt\rangle_{\text{coll}} = \frac{\int \cos\vartheta\, g(\varepsilon)f(\boldsymbol{k})\, k^2\, dk\, \sin\vartheta\, d\vartheta\, d\psi}{\int f(\boldsymbol{k})k^2\, dk\, \sin\vartheta\, d\vartheta\, d\psi} \tag{6.10.16}$$

contains the integrals

$$\int_0^\pi \cos\vartheta\, f(\boldsymbol{k}) \sin\vartheta\, d\vartheta = \int_0^\pi \cos^2\vartheta\, f_1(\varepsilon) \sin\vartheta\, d\vartheta = \frac{2}{3} f_1(\varepsilon) \qquad (6.10.17)$$

and

$$\int_0^\pi f(\boldsymbol{k}) \sin\vartheta d\vartheta = \int_0^\pi f_0(\varepsilon) \sin\vartheta\, d\vartheta = 2 f_0(\varepsilon) \ , \qquad (6.10.18)$$

this is simplified to

$$\left\langle -\frac{dk_E}{dt} \right\rangle_{\mathrm{coll}} = \frac{2^{3/2}}{3\pi^{1/2}} v_{\mathrm{d}} \sqrt{\frac{m}{k_{\mathrm{B}} T_{\mathrm{e}}}} \int_0^\infty g(\varepsilon) \exp\left(-\frac{\varepsilon}{k_{\mathrm{B}} T_{\mathrm{e}}}\right) \frac{\varepsilon}{k_{\mathrm{B}} T_{\mathrm{e}}} d\left(\frac{\varepsilon}{k_{\mathrm{B}} T_{\mathrm{e}}}\right)$$

$$(6.10.19)$$

For $1/\bar{\tau}_{\mathrm{m}}$ we finally obtain from (6.10.2)

$$\frac{1}{\bar{\tau}_{\mathrm{m}}} = \frac{2^{3/2}}{3\pi^{1/2}} \frac{\hbar}{(m\, k_{\mathrm{B}}\, T_{\mathrm{e}})^{1/2}} \int_0^\infty g(\varepsilon) \exp\left(-\frac{\varepsilon}{k_{\mathrm{B}} T_{\mathrm{e}}}\right) \frac{\varepsilon}{k_{\mathrm{B}} T_{\mathrm{e}}} d\left(\frac{\varepsilon}{k_{\mathrm{B}} T_{\mathrm{e}}}\right) \quad (6.10.20)$$

This quantity is somewhat different from $\langle \tau_{\mathrm{m}}^{-1} \rangle$ given by (6.9.23). From (6.10.3) we easily find the mobility as a function of the electron temperature T_{e}.

Next from (6.10.1) we calculate the field strength E as a function of T_{e}:

$$\left(-\frac{d\varepsilon}{dt}\right)_{\mathrm{coll}} = \frac{V}{(2\pi)^3} \int (\hbar\omega\, S_- - \hbar\omega\, S_+) d^3 q \ . \qquad (6.10.21)$$

If, as for optical-phonon scattering, ω does not depend on q, it can be taken outside the integral ($\omega = \omega_0$). Since the remainder on the rhs is just the difference between $1/\tau_{\mathrm{m}}$ for emission and $1/\tau_{\mathrm{m}}$ for absorption, we find

$$\left(-\frac{d\varepsilon}{dt}\right)_{\mathrm{coll}} = \hbar\omega_0 \left[\left(\frac{1}{\tau_{\mathrm{m}}}\right)_- - \left(\frac{1}{\tau_{\mathrm{m}}}\right)_+\right] \qquad (6.10.22)$$

which, for optical deformation potential scattering, yields (6.9.25).

For the average only the symmetric part of the distribution function, $f_0(\varepsilon)$, is relevant.

$$\left\langle -\frac{d\varepsilon}{dt} \right\rangle_{\mathrm{coll}} = \frac{2\hbar\omega_0}{\pi^{1/2}} \int_0^\infty \left[\left(\frac{1}{\tau_{\mathrm{m}}}\right)_- - \left(\frac{1}{\tau_{\mathrm{m}}}\right)_+\right] \qquad (6.10.23)$$

$$\times \exp\left(-\frac{\varepsilon}{k_{\mathrm{B}} T_{\mathrm{e}}}\right) \sqrt{\frac{\varepsilon}{k_{\mathrm{B}} T_{\mathrm{e}}}} d\left(\frac{\varepsilon}{k_{\mathrm{B}} T_{\mathrm{e}}}\right)$$

From(6.10.1, 3) we find for the field intensity

$$E = (\langle -d\varepsilon/dt \rangle_{\text{coll}} m / \bar{\tau}_{\text{m}} e^2)^{1/2} \tag{6.10.24}$$

and for the drift velocity

$$v_{\text{d}} = (\langle -d\varepsilon/dt \rangle_{\text{coll}} \bar{\tau}_{\text{m}} / m)^{1/2} \ , \tag{6.10.25}$$

both as functions of T_{e} where $\bar{\tau}_{\text{m}}$ and $\langle -d\varepsilon/dt \rangle_{\text{coll}}$ are given by (6.10.20, 23), respectively. If both functions of T_{e} are calculated, one can finally obtain a plot of $v_{\text{d}}(E)$.

The main drawback of this method for polar optical scattering is due to the fact that the shifted Maxwell–Boltzmann distribution is a poor approximation to the real distribution: the real one is more spiked in the forward direction and cannot be represented by only one parameter T_{e} [6.59].

6.11 Optical Deformation Potential Scattering

We will now investigate scattering processes of carriers by longitudinal optical phonons in nonpolar crystals. The optical-phonon angular frequency, which in Sect. 6.8 was denoted by ω_ℓ, will now be denoted by ω_0. For the phonon energy $\hbar \omega_0$, a temperature Θ is introduced by the relation

$$\hbar \omega_0 = k_{\text{B}} \Theta \ . \tag{6.11.1}$$

Θ is known as the *Debye temperature* since ω_0 is the highest phonon frequency and in his well-known theory of the specific heat, *Debye* postulated a cutoff of the phonon spectrum at ω_0.

It was mentioned in Sect. 6.8 that in the long-wavelength optical mode of vibration, one set of atoms moves as a body against the second set of atoms with the effect that a carrier in a transition from one atom to an atom of the other set changes its energy by an amount proportional to the atomic displacement:

$$\delta \varepsilon = D \, \delta r \ . \tag{6.11.2}$$

The factor of proportionality D is the *optical deformation potential constant* of the band edge (in units of eV/cm) [6.60, 61]. For simplicity, we have omitted subscripts e or h for electrons or holes, respectively. Equation (6.11.2) is in contrast to the corresponding (6.4.5,6) for the interaction between carriers and acoustic modes of vibration. Otherwise, however, the result of a calculation of the Hamiltonian matrix element [6.25, 6.62]

$$|H_{k \pm q, k}| = D[(N_q + \frac{1}{2} \mp \frac{1}{2}) \, \hbar / 2 \varrho V \omega_0]^{1/2} \tag{6.11.3}$$

is quite similar to (6.4.12). It is independent of q since the optical mode spectrum is stationary at $q = 0$ and N_q depends on ω_0 only:

$$N_q = [\exp(\hbar \omega_0 / k_{\text{B}} T) - 1]^{-1} = [\exp(\Theta / T) - 1]^{-1} \ . \tag{6.11.4}$$

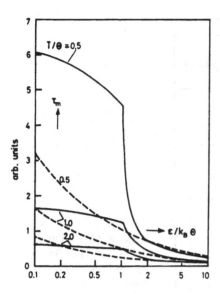

Fig. 6.14: Momentum relaxation time as a function of energy for optical-phonon scattering

Since this is exactly the case treated in Sect. 6.9, we can apply (6.9.24) for a calculation of the momentum relaxation time τ_m:

$$\frac{1}{\tau_m} = \frac{m^{3/2} D^2 N_q}{2^{1/2} \pi \varrho \hbar^2 k_B \Theta} [(\varepsilon + k_B \Theta)^{1/2} + \exp(\Theta/T) \mathrm{Re}\{(\varepsilon - k_B \Theta)^{1/2}\}] \quad (6.11.5)$$

and (6.9.25) for the energy loss rate:

$$-\frac{d\varepsilon}{dt}\Big|_{\mathrm{coll}} = -\frac{m^{3/2} D^2 N_q}{2^{1/2} \pi \varrho \hbar^2} [(\varepsilon + k_B \Theta)^{1/2} - \exp(\frac{\Theta}{T}) \mathrm{Re}\{(\varepsilon - k_B \Theta)^{1/2}\}] \quad (6.11.6)$$

The dependence of τ_m on carrier energy ε is illustrated in Fig. 6.14. At $\varepsilon \geq k_B \Theta$ the emission of optical phonons dominates $\tau_m(\varepsilon)$ which is more pronounced at a lower lattice temperature. A kink similar to the one at $\varepsilon = k_B \Theta$ should also occur at $\varepsilon = 2k_B \Theta, 3k_B \Theta$, etc., when two or three and so on, optical phonons are emitted simultaneously. However, these processes were not incorporated in the theory given above, since the probability for a many-particle process is comparatively small. The dashed curves show τ_m for acoustic-phonon scattering where $\varepsilon_{ac}/u_\ell = D/\omega_0$ and otherwise the same constants were assumed. The high-energy tails of the dashed curves merge with those of the full curves.

The momentum relaxation time decreases with increasing energy, which, for carrier heating, results in deviations from Ohm's law being negative. The zero-field mobility $\mu_0 = (e/m)\langle\tau_m\rangle$, where $\langle\tau_m\rangle$ is given by (6.3.19), is readily evaluated for a nondegenerate carrier gas with the Maxwell–Boltzmann distribution:

$$\mu_0 = \frac{4\sqrt{2\pi}\, e\, \hbar^2 \varrho (k_B \Theta)^{1/2}}{3m^{5/2} D^2} f(T/\Theta) . \quad (6.11.7)$$

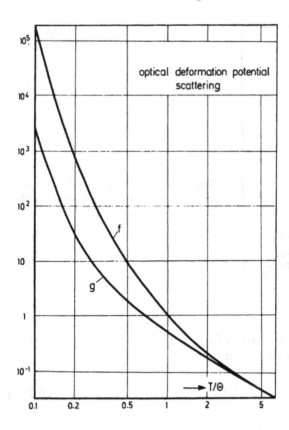

optical deformation potential
scattering

Fig. 6.15: Functions $f(T/\Theta)$ and $g(T/\Theta)$ given by (6.11.8, 19), respectively

The function $f(T/\Theta)$ is given by

$$f(T/\Theta) = (2z)^{5/2}(\exp{(2z)}-1) \int\limits_0^\infty \frac{y^{3/2}\exp{(-2zy)}\,dy}{\sqrt{y+1}+\exp{(2z)}\,\mathrm{Re}\{\sqrt{y-1}\}} \quad .(6.11.8)$$

Here $z = \Theta/2T$ and $y = \varepsilon/k_B\Theta$. The function is shown in Fig. 6.15. The mobility, which is proportional to the function, decreases continuously with increasing temperature. Its numerical value in units of $\mathrm{cm}^2/\mathrm{V\,s}$ is given by

$$\mu = 2.04 \times 10^3 \frac{(\varrho/\mathrm{g\,cm}^{-3})(\Theta/400\,\mathrm{K})^{1/2}}{(m/m_0)^{5/2}(D/10^8\mathrm{eV\,cm}^{-1})^2} f(T/\Theta) \; . \tag{6.11.9}$$

The energy loss rate for hot carriers (6.11.6) averaged over a Maxwell-Boltzmann distribution with an electron temperature T_e is easily evaluated:

$$-\left\langle\frac{d\varepsilon}{dt}\right\rangle_{\mathrm{coll}} = \frac{m^{3/2}D^2(k_B\Theta)^{1/2}}{\pi^{3/2}\,\hbar^2\,\varrho} \frac{(\lambda z)^{1/2}\mathrm{K}_1(\lambda z)}{\sinh(z)} \sinh[(1-\lambda)z] \; , \tag{6.11.10}$$

where $\lambda = T/T_e$ and K_1 is a modified Bessel function.

An energy relaxation time τ_ε which is independent of T_e can be defined rigorously only for the case of warm electrons where the hyperbolic sine may be replaced to a good approximation by its argument

$$\sinh[(1-\lambda)z] \approx (1-\lambda)z \approx z(T_e - T)/T \tag{6.11.11}$$

and $\lambda = 1$ otherwise. From the definition of τ_ε given by (4.13.6, 7), we obtain for the warm electron case

$$\tau_\varepsilon = \frac{3\pi^{3/2}}{4} \frac{\hbar^2 \varrho (k_B \Theta)^{1/2}}{m^{3/2} D^2} \frac{\sinh(z)}{z^{5/2} K_1(z)} , \tag{6.11.12}$$

where $z = \Theta/2T$. For the case of the many-valley model, $m^{3/2}$ has to be replaced by $m_t m_\ell^{1/2}$ where m_t and m_ℓ are the transverse and longitudinal effective masses. The numerical value of τ_ε in the unit s is given by

$$\tau_\varepsilon = 1.6 \times 10^{-12} \frac{(\varrho/\text{g cm}^{-3})(\Theta/400\,\text{K})^{1/2}}{(m/m_0)^{3/2}(D/10^8\,\text{eV cm}^{-1})^2} \frac{\sinh(z)}{z^{5/2}\,K_1(z)} . \tag{6.11.13}$$

We will now apply the method of balance equations (Sect. 6.10) for a calculation of the drift velocity in strong electric fields. Let us first evaluate the function $g(\varepsilon)$ given by (6.10.11). Since the Hamiltonian matrix elements $H_{k\pm q,k}$ do not depend on q, (6.11.3), the integrations are straightforward:

$$g(\varepsilon) = \frac{V}{8\pi\hbar\varepsilon} \frac{\hbar^2 D^2 N_q}{2\varrho V k_B \Theta} \left[\frac{q_2^2 - q_1^2}{2} \left(\frac{q_2^2 + q_1^2}{2} - \frac{2m k_B \Theta}{\hbar^2} \right) \right.$$

$$\left. + \exp\left(\Theta/T\right) \frac{q_2'^2 - q_1'^2}{2} \left(\frac{q_2'^2 + q_1'^2}{2} + \frac{2m k_B \Theta}{\hbar^2} \right) \right] , \tag{6.11.14}$$

$$g(\varepsilon) = \frac{m^2 D^2 N_q}{\pi \hbar^3 \varrho k_B \Theta} [\varepsilon^{1/2}(\varepsilon + k_B \Theta)^{1/2} + \varepsilon^{\Theta/T} \varepsilon^{1/2} \text{Re}\{(\varepsilon - k_B \Theta)^{1/2}\}] . \tag{6.11.15}$$

For the calculation of $1/\bar{\tau}_m$ according to (6.10.20), we introduce the parameter $t = \Theta/2T_e$ and the variable $\xi = \varepsilon/k_B T_e$ for the first term in $g(\varepsilon)$ and $\xi = (\varepsilon - k_B \Theta)/k_B T_e$ for the second term:

$$\frac{1}{\bar{\tau}_m} = \frac{2m^{3/2} D^2 N_q t^{3/2}}{3\pi^{3/2} \hbar^2 \varrho (k_B \Theta)^{1/2}} \{\exp(t)[K_2(t) - K_1(t)]$$

$$+ \exp\left[\left(\frac{\Theta}{T}\right) - t\right] [K_2(t) + K_1(t)]\} \tag{6.11.16}$$

where K_1 and K_2 are modified Bessel functions [6.27]. With N_q given by (6.4.11), we can write

$$\frac{1}{\bar{\tau}_m} = \frac{2m^{3/2}\, D^2(\lambda z)^{3/2}}{3\,\pi^{3/2}\,\hbar^2\varrho(k_B\Theta)^{1/2}\sinh(z)}$$

$$\{\cosh[(1-\lambda)z]\,K_2(\lambda z) + \sinh[(1-\lambda)z]K_1(\lambda z)\} \qquad (6.11.17)$$

where, as usual, $z = \Theta/2T$ and $\lambda = T/T_e$. For the zero-field mobility (6.10.3) for $\lambda = 1$, we find in the present approximation

$$\mu_0 = \frac{4\sqrt{2\pi}\,e\,\hbar^2\varrho(k_B\Theta)^{1/2}}{3\,m^{5/2}\,D^2}g(T/\Theta) \ , \qquad (6.11.18)$$

where the function $g(T/\Theta)$ [not to be confused with $g(\varepsilon)$]

$$g(T/\Theta) = 9\pi\,2^{-7/2}\,z^{-3/2}\sinh(z)/K_2(z) \ , \qquad (6.11.19)$$

for comparison with $f(T/\Theta)$ in (6.11.7), has been plotted in Fig. 6.15. Especially at low temperatures, the values of the mobility obtained by balance equations are somewhat smaller than by the normal procedure.

The calculation of the field strength and the drift velocity, both as functions of λ and according to (6.10.24, 25), where the energy loss rate is given by (6.11.10), yields

$$\frac{v_d}{v_{ds}} = \frac{2^{1/2}\coth^{1/2}(z)}{\lambda^{1/2}z^{1/2}\{1+\coth[(1-\lambda)z]K_2(\lambda z)/K_1(\lambda z)\}^{1/2}} \qquad (6.11.20)$$

and

$$\frac{\mu_0 E}{v_{ds}} = \frac{2^{1/2}\coth^{1/2}(z)\lambda\,\sinh[(1-\lambda)z]K_1(\lambda z)}{z^{1/2}K_2(z)}$$

$$\left\{1 + \coth[(1-\lambda)z]\frac{K_2(\lambda z)}{K_1(\lambda z)}\right\}^{1/2} \qquad (6.11.21)$$

where

$$v_{ds} = \left(\frac{3k_B\Theta}{4\,m\coth(z)}\right)^{1/2} \qquad (6.11.22)$$

is the saturation value of the drift velocity for large field intensities. In the range of saturation, the drift energy of a carrier $m\,v_{ds}^2/2$ is about equal to $k_B\Theta = \hbar\omega_0$, the optical-phonon energy, since at high electron temperatures, phonon emission processes determine the drift velocity. v_{ds} decreases with increasing temperature, but only slowly (13% between $T \ll \Theta$ and $T = \Theta/2$).

Figure 6.16 shows a log-log plot of the calculated drift velocity vs electric field strength for various values of the lattice temperature relative to the Debye temperature T/Θ. Ohm's law is demonstrated by the initial rise at an angle of 45°. At large field strengths the drift velocity saturates; the saturation value,

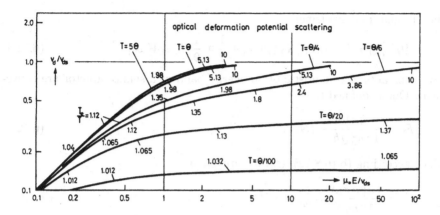

Fig. 6.16: Normalized drift velocity vs normalized electric field intensity, with T_e/T as a parameter, for optical deformation potential scattering. Numbers given along the curves are values of T_e/T (T lattice temperature, T_e carrier temperature, Θ Debye temperature)

which is temperature dependent because of $z = \Theta/(2T)$ in (6.11.22) , has been applied in Fig. 6.16 for the normalization of both v_d and $\mu_0 E$ where μ_0 is the zero-field mobility. The numbers along the curves are the electron temperatures relative to the lattice temperature. At low temperatures, saturation is approached only very gradually while the characteristics deviate from the 45° straight line at already small field intensities.

6.12 Polar Optical Scattering

In polar semiconductors, the interaction of carriers with the optical mode of lattice vibrations is known as *polar optical scattering*. As for piezoelectric scattering, the potential energy of a carrier is given by (6.7.5)

$$\delta\varepsilon = |e|E/q \; , \tag{6.12.1}$$

where the electric field strength E is due to the polarization P of the longitudinal optical lattice vibration given by (6.8.21)

$$E = -P/\kappa_0 = -N_u \, e_C \, \delta r/\kappa_0 \; . \tag{6.12.2}$$

This yields for $\delta\varepsilon$:

$$\delta\varepsilon = (-|e|N_u \, e_C/\kappa_0 \, q) \, \delta r \tag{6.12.3}$$

which, except for the q-dependence of the factor of proportionality between $\delta\varepsilon$ and δr, is very similar to (6.11.2). Hence, in analogy to (6.11.3), we find for the

Hamiltonian matrix element

$$|H_{k\pm q,k}| = (|e|N_u\,e_C/\kappa_0\,q)[(N_q + \frac{1}{2} \mp \frac{1}{2})\,\hbar/2\varrho V\,w_0]^{1/2} \ . \tag{6.12.4}$$

Two constants characteristic of an energy band of a semiconductor are convenient. One is defined by

$$E_0 = \frac{|e|m}{4\pi\,\kappa_0^2\,\varrho\,\hbar\,w_0}(N_u\,e_C)^2 \tag{6.12.5}$$

which according to (6.8.22), can be written as

$$E_0 = \frac{|e|m\,k_B\Theta}{4\pi\,\kappa_0\,\hbar^2}(\kappa_{\text{opt}}^{-1} - \kappa^{-1}) \tag{6.12.6}$$

and in units of kV/cm has a numerical value of

$$E_0 = 16.3\,(m/m_0)\,(\Theta/K)\,(\kappa_{\text{opt}}^{-1} - \kappa^{-1}) \ . \tag{6.12.7}$$

It is called *effective field strength*. The second constant is the dimensionless *polar constant* α defined by

$$\alpha = \frac{\hbar\,|e|\,E_0}{2^{1/2}\,m^{1/2}(\hbar w_0)^{3/2}} = \frac{1}{137}\sqrt{\frac{mc^2}{2\,k_B\Theta}}\left(\frac{1}{\kappa_{\text{opt}}} - \frac{1}{\kappa}\right) \ , \tag{6.12.8}$$

where $1/137 = e^2/(4\pi\,\kappa_0\,\hbar\,c)$ is the fine structure constant and c is the velocity of light. For example, in n-type GaAs, $E_0 = 5.95$kV/cm and $\alpha = 0.067$. With these constants the square of the matrix element (6.12.4) becomes

$$|H_{k\pm q,k}|^2 = \frac{2\pi\,\hbar^2\,|e|E_0}{V\,m\,q^2}\left(N_q + \frac{1}{2} \mp \frac{1}{2}\right)$$

$$= \frac{2^{3/2}\pi\,\hbar\,\alpha(\hbar w_0)^{3/2}}{V\,m^{1/2}\,q^2}\left(N_q + \frac{1}{2} \mp \frac{1}{2}\right) \tag{6.12.9}$$

We have shown in Sect. 6.9 that in the present case, where the matrix element depends on q and the scattering process is inelastic ($\hbar w_0$ is of the order of magnitude of the average carrier energy or even larger), a momentum relaxation time τ_{m} does not exist, strictly speaking. Even so, let us, for an approximation in (6.9.2) replace $f(\mathbf{k}\pm\mathbf{q})$ by $f_0(\mathbf{k})$. Now we obtain for $1/\tau_{\text{m}}$ given by (6.9.17):

$$\frac{1}{\tau_{\text{m}}} = \frac{V}{(2\pi)^3}\,\frac{2\pi}{\hbar}\,\frac{2^{3/2}\,\pi\,\hbar\,\alpha\,(\hbar w_0)^{3/2}\,N_q}{V\,m^{1/2}}$$

$$\times \left[\int_{q_1}^{q_2}\frac{2\pi\,m}{\hbar^2\,k\,q^3}q^2\,dq + \exp(\frac{\Theta}{T})\int_{q_1'}^{q_2'}\frac{2\pi\,m}{\hbar^2\,k\,q^3}q^2\,dq\right] \tag{6.12.10}$$

Solving the integrals yields

$$\frac{1}{\tau_{\mathrm{m}}} = \alpha \, \omega_0 (\hbar \omega_0 / \varepsilon)^{1/2} \, N_q \left[\ln \left| \frac{a+1}{a-1} \right| + \exp(\Theta/T) \ln \left| \frac{1+b}{1-b} \right| \right] \, , \qquad (6.12.11)$$

where a and b are defined by (6.9.19, 22), respectively.

At low temperatures $T \ll \Theta$ where $\varepsilon \ll \hbar \omega_0$, $b \approx 0$ and $N_q \approx \exp(-\Theta/T)$ and $1/\tau_{\mathrm{m}}$ is approximated by:

$$1/\tau_{\mathrm{m}} \approx 2 \alpha \, \omega_0 \, \exp(-\Theta/T) \, . \qquad (6.12.12)$$

The reciprocal momentum relaxation time is then essentially the product of the polar constant and the availability of an optical phonon for absorption. The mobility is simply $(e/m)\tau_{\mathrm{m}}$:

$$\mu = \left[|e| / (2m \, \alpha \, \omega_0) \right] \exp(\Theta/T) \qquad (6.12.13)$$

which in units of cm^2 / V s is given by

$$\mu = 2.6 \times 10^5 \frac{\exp(\Theta/T)}{\alpha(m/m_0) (\Theta/K)} \quad \text{for} \quad T \ll \Theta \, . \qquad (6.12.14)$$

For example, in n-type GaAs where $\Theta = 417\,\mathrm{K}, m/m_0 = 0.072, \alpha = 0.067$, we calculate a mobility of $2.2 \times 10^5 \, \mathrm{cm^2/Vs}$ at 100 K. This is of the order of magnitude of the highest mobilities observed in this material. At this and lower temperatures, the dominant scattering mechanism in compound semiconductors of even the highest purity available at present is impurity scattering. This unfortunately prevents a useful comparison of (6.12.14) with experimental data.

The difficulty in calculating τ_{m} mentioned above can be overcome by, e.g., the use of variational methods [6.63]. The mobility calculated in this way [6.64] for low temperatures $T \ll \Theta$ agrees with the value given by (6.12.14), while for high temperatures, the latter has to be multiplied by $8\sqrt{T/9\pi \Theta}$. In the intermediate range numerical methods have been applied. By variational methods, different momentum relaxation times $\tau_{\mathrm{m}}(\varepsilon)$ are obtained for conductivity, Hall effect and thermoelectric effect. Figure 6.17 shows the exponent r of an assumed power law $\tau_{\mathrm{m}} \propto \varepsilon^r$ as a function of Θ/T for these effects [6.63]. For the case of conductivity, r becomes infinite at $T \approx \Theta$ which demonstrates the difficulty in obtaining a function $\tau_{\mathrm{m}}(\varepsilon)$ valid for all values of ε.

For a calculation of the energy loss rate, we multiply the first term on the rhs of (6.12.11) by $-\hbar \omega_0 = -k_{\mathrm{B}} \Theta$ and the second term by $+k_{\mathrm{B}} \Theta$, since these represent phonon absorption and emission, respectively:

$$\left(-\frac{d\varepsilon}{dt} \right)_{\mathrm{coll}} = \frac{\alpha(k_{\mathrm{B}}\Theta)^{5/2}}{\hbar\sqrt{\varepsilon}} N_q \left(-\ln \left| \frac{a+1}{a-1} \right| + \exp(\Theta/T) \ln \left| \frac{1+b}{1-b} \right| \right)$$

$$(6.12.15)$$

For a nondegenerate electron gas we average the energy loss rate over the Maxwell–Boltzmann distribution with an electron temperature T_{e} :

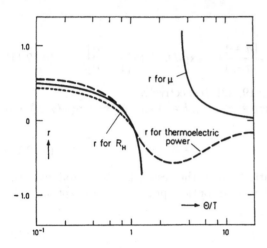

$$\left\langle -\frac{d\varepsilon}{dt} \right\rangle_{\text{coll}} = \frac{2\alpha(\Theta/T_{\rm e})^{3/2} N_q}{\pi^{1/2}\hbar} \left[-\int_0^\infty \ln\left|\frac{a+1}{a-1}\right| \exp(-\varepsilon/k_{\rm B}T_{\rm e}) d\varepsilon \right.$$

$$\left. + \exp(\Theta/T) \int_{k_{\rm B}\Theta}^\infty \ln\left|\frac{1+b}{1-b}\right| \exp(-\varepsilon/k_{\rm B}T_{\rm e}) d\varepsilon \right] k_{\rm B}\Theta \ . \qquad (6.12.16)$$

The first integral on the rhs is easily solved by introducing the variable $\xi = \varepsilon/k_{\rm B}T_{\rm e}$ and the parameter $t = \Theta/(2T_{\rm e})$:

$$k_{\rm B}T_{\rm e} \int_0^\infty \ln\left|\frac{(1+2t/\xi)^{1/2}+1}{(1+2t/\xi)^{1/2}-1}\right| \exp(-\xi)\,d\xi = k_{\rm B}T_{\rm e}\exp(t)\,{\rm K}_0(t) \ , \quad (6.12.17)$$

where K_0 is a modified Bessel function of the second kind [6.27]. For the second integral with the lower limit $k_{\rm B}\Theta$, the variable $\xi = (\varepsilon - k_{\rm B}\Theta)/k_{\rm B}T_{\rm e}$ is introduced:

$$k_{\rm B}T_{\rm e}\exp\left(\frac{\Theta}{T}\right) \int_0^\infty \ln\left|\frac{1+[1-2t/(\xi+2t)]^{1/2}}{1-[1-2t/(\xi+2t)]^{1/2}}\right| \exp-(\xi+2t)\,d\xi$$

$$= k_{\rm B}T_{\rm e}\exp\left(\frac{\Theta}{T}-2t\right)\exp(t){\rm K}_0(t) \ . \qquad (6.12.18)$$

Since $t = \lambda z$ where $\lambda = T/T_{\rm e}$, the average loss rate is finally obtained in the form [6.65]

$$\left\langle -\frac{d\varepsilon}{dt} \right\rangle_{\text{coll}} = \frac{2^{3/2}\alpha}{\pi^{1/2}\hbar}(k_{\rm B}\Theta)^2\,(\lambda z)^{1/2}\,K_0(\lambda z)\frac{\sinh[(1-\lambda)z]}{\sinh(z)} \ . \qquad (6.12.19)$$

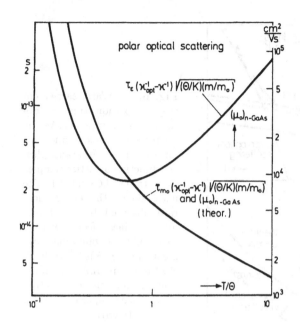

Fig. 6.18: Energy relaxation time τ_ε for warm carriers and momentum relaxation time $\bar{\tau}_m$ as a function of lattice temperature for polar optical scattering

For the energy relaxation time for warm carries, where $\sinh[(1 - \lambda)\, z] \approx (1 - \lambda)z$ and $\lambda \approx 1$ otherwise, we find

$$\tau_\varepsilon = \frac{3\pi^{1/2}}{2^{7/2}\alpha\,\omega_0} \frac{\sinh(z)}{z^{5/2}\,\mathrm{K}_0(z)} = \frac{0.47}{\alpha\,\omega_0} \frac{\sinh(\Theta/2T)}{(\Theta/2T)^{5/2}\mathrm{K}_0(\Theta/2T)} \ . \qquad (6.12.20)$$

In Fig. 6.18 the product

$$\tau_\varepsilon \left(\frac{1}{\kappa_{\mathrm{opt}}} - \frac{1}{\kappa}\right) \left(\frac{\Theta}{K}\right)^{1/2} \left(\frac{m}{m_0}\right)^{1/2}$$

$$= 9.05 \times 10^{-15}\mathrm{s}\,\frac{\sinh(\Theta/2T)}{(\Theta/2T)^{5/2}\mathrm{K}_0(\Theta/2T)} \qquad (6.12.21)$$

has been plotted as a function of T/Θ. There is a minimum at a temperature of about half the Debye temperature Θ (i.e., $\Theta/2T \approx 1$). The minimum, $\tau_\varepsilon = 1.29/\alpha\,\omega_0$, indicates a suitable electromagnetic wavelength for the measurement of τ_ε of $\approx 1/\alpha$ times the optical-phonon wavelength. Since α for most semiconductors is of the order of magnitude 10^{-1} and the optical-phonon wavelength is usually about 30 μm (Sect. 11.7), a suitable wavelength for measurements of τ_ε is about 0.3 mm. Hence, τ_ε can be measured by submillimeter wave methods. Results obtained for n-GaAs and n-InAs are shown in Fig. 6.19 [6.66–68]. The curves calculated according to (6.12.21) have been fitted to the experimental data by assuming suitable values for the polar constant α.

Since the calculation of the drift velocity in strong electric fields by balance equations [6.58] is very similar to the corresponding calculation for optical deformation potential scattering (6.11.18, 31), except that now the matrix elements

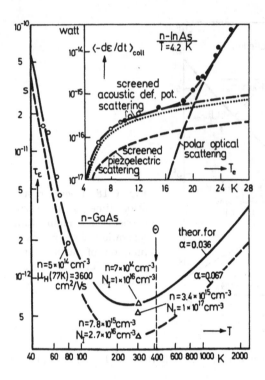

Fig. 6.19: Warm-carrier energy relaxation time τ_ε observed in n-type GaAs by microwave (circles) and far-infrared techniques (triangles) as a function of lattice temperature (after [6.65, 6.66]). The inset shows the energy loss rate observed by the Shubnikov de Haas effect (open circles) and by mobility measurements and $\mu(E)eE^2$ (full circles) in n-type InAs as a function of electron temperature at a lattice temperature of 4.2 K (after (6.68))

are proportional to q^{-2}, we will not go through details of the calculation but only give the results. The reciprocal average momentum relaxation time

$$1/\bar{\tau}_\mathrm{m} = \frac{2^{5/2}\,\alpha\,\omega_0\,(\lambda\,z)^{3/2}}{3\pi^{1/2}\sinh(z)}\{\cosh[(1-\lambda)z]K_1(\lambda z)$$
$$+ \sinh[(1-\lambda)\,z]K_0(\lambda z)\} \tag{6.12.22}$$

where, as usual, $z = \Theta/2T$ and $\lambda = T/T_\mathrm{e}$. The factor in braces is different from the corresponding factor in (6.11.17) in so much as the order of both of the modified Bessel functions is smaller by 1. For the zero-field mobility μ_0, we obtain in the present approximation

$$\mu_0 = \frac{3\,\pi^{1/2}}{2^{5/2}}\,\frac{|e|}{m\alpha\,\omega_0}\,\frac{\sinh(z)}{z^{3/2}\mathrm{K}_1(z)}$$
$$= \frac{31.8\,\mathrm{cm^2/V\,s}}{(\kappa_\mathrm{opt}^{-1} - \kappa^{-1})(\Theta/\mathrm{K})^{1/2}(m/m_0)^{3/2}}\,\frac{\sinh(z)}{z^{3/2}\mathrm{K}_1(z)} \tag{6.12.23}$$

For high temperatures where $z \ll 1$, this is equal to the mobility obtained by variational methods [6.64] times a factor $32/9\,\pi \approx 1.13$. For low temperatures, ($z \gg 1$), it is lower than this value of the mobility by a factor $3T/2\,\Theta$. The mobility calculated from (6.12.23) for electrons in n-type GaAs is shown in Fig.

6.18. At room temperature, a mobility of 7800 cm^2 / V s is found which agrees with the experimental value.

In contrast to optical deformation potential scattering, the drift velocity does not saturate at high field strengths. It is convenient, however, to also introduce here v_{ds} given by (6.11.22). For the drift velocity v_d we find

$$v_d/v_{ds} = \frac{2^{1/2}\coth^{1/2}(z)}{\lambda^{1/2}z^{1/2}\{1 + \coth[(1 - \lambda)z]\,K_1(\lambda z)/K_0(\lambda z)\}^{1/2}} , \quad (6.12.24)$$

where $\lambda = T/T_e$ and $z = \Theta/2T$, which is quite similar to the value for optical deformation potential scattering given by (6.11.20). However, for small values of the argument, i.e., high electron temperatures, the ratio $K_1(\lambda z)/K_0(\lambda z)$ $\approx 1/[\lambda z \ln(2/\lambda z)]$. Therefore,

$$(v_d)_{T_e \to \infty} \approx v_{ds}[2\ln(2/\lambda z)]^{1/2} \to \infty . \quad (6.12.25)$$

Since this is a logarithmic dependence of the drift velocity on carrier temperature, there may be a region in the characteristics where the drift velocity "*seems*" to saturate.

The electric field strength E, in units of the material constant E_0 (6.12.16), is given by

$$\frac{E}{E_0} = \frac{2^{3/2}\lambda z}{(3\pi)^{1/2}} \sinh[(1 - \lambda)z]K_0(\lambda z)$$

$$\frac{\{1 + \coth[(1 - \lambda)z]\,K_1(\lambda z)/K_0(\lambda z)\}^{1/2}}{\sinh(z))} \quad (6.12.26)$$

Figure 6.20 shows a diagram of v_d/v_{ds} vs $\mu_0 E/v_{ds}$ which may be compared with the corresponding diagram for optical deformation potential scattering (Fig. 6.16). At low temperatures ($T/\Theta \ll 1$), there is a plateau where the drift velocity is fairly constant over a range of field intensities, and finally, there is superlinear behavior which has been brought in connection with *dielectric breakdown* [6.59, 69]. At high temperatures ($T \geq \Theta$), no plateau is found and the breakdown characteristics develop right from the straight lines representing Ohm's law. This breakdown has not been observed experimentally in semicondutors since at the high breakdown field intensities, other energy loss mechanisms such as intervalley scattering and impact ionization become important and prevent the high electron temperatures necessary for dielectric breakdown to occur [6.58].

The preceding calculations for all temperatures are, however, based on the assumption of a drifted Maxwell–Boltzmann distribution. For obtaining a better approximation of the true distribution function, the variational method [6.64] and, more recently, the *Monte Carlo procedure*, have been applied [6.70-72]. The latter is a computer simulation of the motion of a carrier in k-space under the influence of both the applied electric field and the collisions. Sets of random numbers are applied for fixing both the scattering rate and the direction of

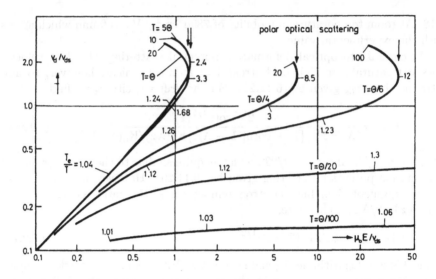

Fig. 6.20: Drift velocity in units of v_{ds} as a function of the electric field intensity (in units of v_{ds}/μ_0) at various lattice temperatures (in units of the Debye temperature Θ) for polar optical scattering. The numbers next to the points are electron temperatures (in units of T). For various lattice temperatures, the dielectric breakdown fields are indicated by arrows

carrier motion after each scattering event. Statistical convergence is obtained after some 10^4 events.

Another recently introduced method is the *iterative method* [6.73] modified by the concept of *self-scattering* [6.74]. Let us assume that after n iterative steps we have arrived at a distribution function f_n. We then use this function to calculate the rhs of the Boltzmann equation which is determined by the scattering process. The distribution function on the lhs, which we denote as f_{n+1}, is then obtained by simple integration. *Self scattering* is an artificial nonphysical process where the carrier in k-space actually remains in position but which considerably simplifies the probability distribution for scattering. This method is well-suited, e.g., for calculations of the ac conductivity in polar semiconductors [6.75].

In alkali halides where $\alpha \geq 1$ (e.g., NaCl: $\alpha = 5.5$, AgBr: $\alpha = 1.6$), the concept of the *polaron* has been very fruitful. Because of the strong polar coupling, the lattice is polarized in the vicinity of a conduction electron and the polarization moves along with the electron. For this reason, the electron effective mass $m_{\rm pol}$ is higher than without the polarization (where it would be m) and the ratio $m_{\rm pol}/m$ is a function of the polar coupling constant α. For the case of weak coupling, $\alpha \ll 1$, the polarization can be considered as a small perturbation and quantum mechanical perturbation theory yields (see, e.g., [6.76]):

$$m_{\rm pol}/m = (1 - \alpha/6)^{-1} \approx 1 + \alpha/6 \ . \tag{6.12.27}$$

The average number of *virtual phonons* carried along by the electron is given by $\alpha/2$. The conduction band edge is depressed by an amount $\alpha\hbar\omega_0$. In a strong magnetic field, where Landau levels are formed in the energy bands (Sect. 9.2), the energy of the first of these levels is given by

$$\varepsilon_{n=1} \approx \frac{3}{2}\hbar\omega_c(1 - \alpha/6) - \alpha\hbar\omega_0 \,, \tag{6.12.28}$$

where ω_c is the cycloton frequency [6.77, 78].

When an electron is strongly interacting with the polar lattice it is called a *polaron* (this should not be confused with the *polariton* mentioned in Sect. 11.7). Hot carriers in polar semiconductors should accordingly be denoted as *hot polarons* [6.79, 80]. At large carrier concentrations n, the carrier–phonon interaction is screened, i.e., $\alpha = \alpha(n)$ [6.81].

The term *small polaron* refers to a polaron where the spatial extent of the lattice distortion is less than or comparable to a lattice constant [6.82]. Since self-trapping of carriers by lattice distortion is to some extent also possible in nonpolar materials, both via acoustic-mode and optical-mode lattice vibrations, small-polaron formation is also possible in, e.g., amorphous silicon. Charge transport by small polarons is via band conduction at low temperatures while at high temperatures, hopping from one lattice site to an equivalent one is predominant. Since the distance of the lattice sites is on an atomic scale, the term *phonon-assisted tunneling* describes the process of small-polaron motion equally well [6.83, 84]. For hopping transport see also Sect. 6.14.

In short-channel GaAs field effect transistors where polar optical scattering is predominant, "ballistic transport" of carriers across the base has been observed [6.85, 86]. As early as 1972 *Ruch* [6.87] showed by Monte Carlo calculations that at a field of 5 kV/cm in GaAs for a time of 10^{-12} s, the drift velocity is three times larger than its low-field (1 kV/cm) saturation value of 10^7 cm/s. During this short time, the electrons travel only 300 nm. Later on base widths of an order of magnitude smaller were produced. During a transit time which is shorter than the energy relaxation time, the drift velocity has not had time to decrease (Fig. 6.6), and the mobility still has its high zero-field value μ_0. At longer times, energy relaxation becomes effective, the distribution function spreads, and the drift velocity decreases from a maximum value of 10^8 cm/s. Although the effect should occur also in silicon, the transient effects in GaAs last an order of magnitude longer than in Si because of the polar optical scattering mechanism. The term "ballistic transport" reminds one of the flight of an electron in a TV tube, with no scattering at all, but since the momentum relaxation is still not small compared with the transit time in the solid state device, it in fact gives a false impression [6.88]. The practical importance of the device for fast computers need hardly be mentioned.

6.13 Carrier–Carrier Scattering

In a process where an electron is scattered by another electron (e-e *scattering*) the total momentum of the electron gas is not changed. Hence, electron–electron scattering as such has little influence on the mobility. However, since it is always combined with another scattering mechanism, which it may enhance, it can have quite an important influence. If, e.g., an electron loses energy $\hbar\omega_0$ by optical-phonon scattering and in **k**-space is replaced due to e-e scattering by another electron with the same energy, the energy loss rate is enhanced.

The change in mobility by this process has been calculated for nonpolar semiconductors by *Appel* [6.89] and for polar semiconductors by *Bate* et al. [6.90]. *Appel* applied the *Kohler* variational method [6.91, 92] which avoids the assumption of a relaxation time (the collision is inelastic!) and instead requires the rate of entropy production caused by all scattering processes involved to be a maximum for the steady state. For dominating ionized impurity scattering in a nondegenerate semiconductor, the mobility $\mu_1(0)$ is reduced to a value μ_1(e–e) by electron–electron scattering by a factor of about 0.6, while for extreme degeneracy, there is no reduction, which may be surprising at first glance. For the intermediate range, *Bate* et al. [6.90] found values for the reduction factor which are plotted vs the reduced Fermi energy, ζ_n/k_BT, in Fig. 6.21. The upper scale gives values of the carrier density valid for n-type InSb at 80 K. There is a marked influence by electron –electron scattering below $10^{17}\,\text{cm}^{-3}$ where $\zeta_n < 7k_BT$. For small electron densities, polar optical scattering will dominate over ionized impurity scattering and we have to consider the change of the polar optical scattering rate by electron–electron scattering. For even smaller electron densities $(n < 10^{14}\,\text{cm}^{-3})$, the influence of electron–electron scattering will vanish since it is proportional to n. [For impurity scattering in addition to electron–electron scattering, this is not quite so (see Fig. 6.21) because the impurity scattering rate is also proportional to n for $N_I = n$ and the ratio $\mu_1(\text{e} - \text{e})/\mu_1(0)$ becomes independent of n.] When acoustic deformation potential scattering is predominant, electron–electron scattering has even less effect. For n-type Ge containing $6.1 \times 10^{14}\,\text{cm}^{-3}$ shallow donors, the maximum change is by a factor of 0.94; this maximum occurs at a temperature of 35 K [6.93].

In an intrinsic semiconductor, electron–hole scattering may influence the mobilities of both types of carriers. In hot carrier experiments, this tends to keep electrons and holes at the same carrier temperature.

6.14 Impurity Conduction and Hopping Processes

In heavily doped semiconductors, one can no longer regard the impurity states as being localized [6.94]. With increasing density of impurities, the average dis-

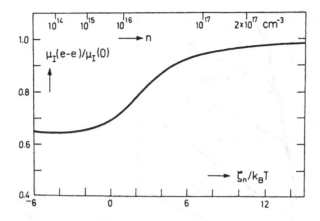

Fig. 6.21: Electron-electron scattering correction for the mobility dominated by ionized impurity scattering, as a function of the reduced Fermi energy. The electron density given at the top scale is valid for the non-parabolic conduction band of InSb at 80 K (after [6.90])

tance between adjacent impurities becomes smaller and the energy levels of the impurities broaden into bands. This is very similar to the process of forming energy bands in a crystal in the tight-binding approximation (Sect. 2.3) except that the impurities are distributed at random (we will not discuss scattering by clusters [6.95]). These bands are at best half filled and therefore, metallic conduction should be expected. However, the bands are narrow since the impurities are still relatively far apart, and therefore a very large effective mass and a low mobility are characteristic of *impurity conduction*. In this respect, impurity conduction is comparable to conduction in poor metals.

At even higher impurity densities, the impurity band merges with the parent band such that a *tail* of localized states is joined onto the parent band of extended states [6.96]. Heavy doping gives rise to structural disorder including vacancies and clusters of impurities and vacancies resulting in, e.g., negative magnetoresistance [6.97]. The effective density of states in the tail can be represented by an exponential with the square of the energy in the argument.

In n-type Ge, impurity conduction prevails only at low temperatures as shown in Fig. 6.22 [6.98]. In the curve labeled $5 \times 10^{14} \, \mathrm{cm}^{-3}$, normal conduction and carrier freeze-out persist up to the highest value of the resistivity. In the more heavily-doped samples, the curve flattens at the onset of impurity conduction. The Hall coefficient then drops markedly unless the doping is extremely high ($> 10^{18} \, \mathrm{cm}^{-3}$) and it remains independent of temperature down to the lowest temperature of about 2 K. Figure 6.23 shows the mobility in heavily-doped boron as a function of temperature [6.99]. The mobility increases with temperature exponentially with an average *activation energy* of 0.1 eV. Such behavior, however, is more characteristic of *hopping processes* than of any type of metallic conduction: carriers are thought to hop from one impurity atom to

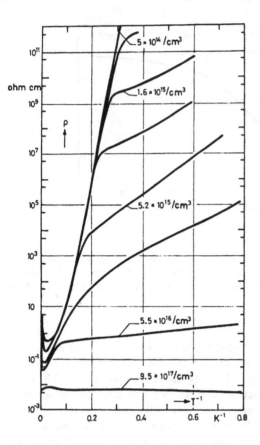

Fig. 6.22: Resistivity of Sb-doped n-type Ge as a function of the reciprocal of the temperature (after [6.98])

the next, and the Coulomb potential around the impurity atom is overcome by means of thermal energy [6.100]. The idea of hopping processes is supplemented in similar cases by the observation that the ac conductivity increases with frequency [6.101]. In heavily-doped n-type Si at frequencies ν between 10^2 and 10^5 Hz and temperatures between 1 and 20 K, an increase proportional to $\nu^{0.74} \ldots \nu^{0.79}$ has been found. Since hopping is a statistical process its probability will be given by $\int G(\tau) \exp(-t/\tau) d\tau$ where τ is the average time between two hops and $G(\tau)$ is a weight factor. The frequency dependence of the conductivity is then obtained by a Laplace transformation: [12]

$$\mathrm{Re}\{\sigma\} \propto \int\limits_0^\infty G(\tau) \frac{\omega^2 \tau^2}{1 + \omega^2 \tau^2} d\tau \quad . \tag{6.14.1}$$

If $G(\tau)$ peaks at a value of $\tau \ll 1/\omega$, a dependence proportional to ω^2 would

_____

[12] An equivalent circuit for this problem is a condenser C in series with a resistor R. The application of a dc voltage results in a current first increasing in a step and then decreasing with time as $\exp(-t/\tau)$ where $\tau = RC$. The ac resistance is $R + 1/(\mathrm{i}\,\omega C)$ and the real part of the conductance R^{-1} is $\omega^2 \tau^2/(1 + \omega^2 \tau^2)$.

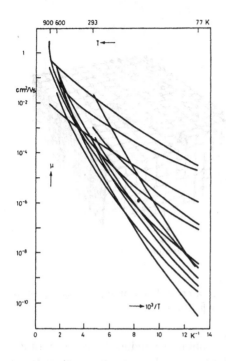

Fig. 6.23: Temperature dependence of the mobility of various carbon doped boron samples (after [6.99])

be expected: at $\tau \gg 1/\omega$ there would be no frequency dependence at all. Thus, by a suitable choice of the function $G(\tau)$, any exponent of ω between 0 and 2 over a certain range of frequencies can be formally explained.

6.15 Dislocation Scattering

Dislocations in a crystal lattice can be considered as belonging to either one of two types shown in Fig. 6.24 or a combination thereof: edge-type or screw-type (spiral) [6.103, 104]. For the formation of the edge-type dislocation, an internal slip of the atomic arrays must occur perpendicular to the dislocation line A–B (slip vector B–C), while for the screw-type, the slip vector is parallel to the dislocation line. The edge-type dislocation has been shown to introduce deep energy levels in germanium and other semiconductors. These levels have been considered as being due to dangling bonds which act as acceptors [6.105]. However, experiments have shown that they can be occupied both by holes and electrons [6.106]. There is still little agreement about the energy of the dislocation states even for the well-known semiconductor n-type Ge. Besides energy levels, dislocation bands have also been discussed [6.107].

It was suggested by *Read* [6.108] that dislocation lines are charged and surrounded by space charge cylinders. Carriers which move at an angle Θ relative

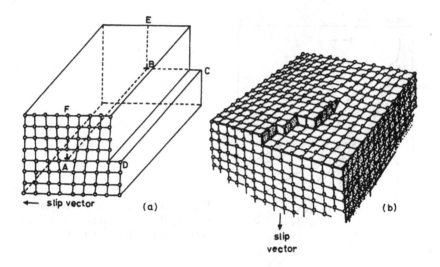

Fig. 6.24: Schematic arrangement of atoms in a crystal containing (**a**) an edge-type dislocation, (**b**) a screw-type dislocation (after [6.102])

to the cylinders are deflected and their mobility is reduced. From a simple mechanical model involving specular reflection at the surface of impenetrable cylinders, a mean free path for the carriers $l = 3/(8\,R\,N)$ was calculated, where R is the cylinder radius and N is the number of dislocation lines per unit area. The scattering probability τ_{m}^{-1} was then assumed to be $(v/l)\sin\Theta$, where v is the carrier velocity, in addition to a term for lattice scattering. A further reduction in the mobility was assumed to be due to the reduction in crystal space available for the conduction electrons by the presence of the impenetrable cylinders.

A more satisfying treatment was given later by *Bonch–Bruevich* and *Kogan* [6.109]. They solved Poisson's equation in cylinder coordinates z, r, φ for a potential $V(r)$:

$$\frac{d^2V}{dr^2} + r^{-1}\frac{dV}{dr} = (n - n_0)\frac{e}{\kappa\,\kappa_0} = n_0[\exp\frac{e\,V}{k_\mathrm{B}T} - 1]\frac{e}{\kappa\,\kappa_0}$$

$$\cong \frac{n_0\,e^2\,V}{\kappa\,\kappa_0\,k_\mathrm{B}T} = \frac{V}{L_\mathrm{D}^2} \tag{6.15.1}$$

where n_0 is the carrier density far away from a space cylinder and L_D is the Debye length given by (5.2.22). The solution is $V = A\,\mathrm{K}_0(r/L_\mathrm{D})$ where A is a constant and K_0 is a modified Bessel function of zero order. If Q is the charge of the dislocation line per unit length and $|\mathbf{E}| = |-\boldsymbol{\nabla}_r V| = A\,\mathrm{K}_1/L_\mathrm{D}$ is the electric field, we have

$$Q/\kappa\,\kappa_0 = \int_0^\infty [r^{-1}\,d(r\,E)/dr]2\,\pi\,r\,dr = 2\pi\,r\,E|_0^\infty = 2\,\pi\,A \ . \tag{6.15.2}$$

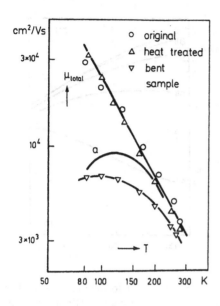

Fig. 6.25: Mobility of electrons in n-type germanium in the same sample before and after plastic bending. Curve (a) has been calculated for the bent sample (after [6.110])

Hence, the potential is given by

$$V(r) = (Q/2\pi\kappa\kappa_0)\,K_0(r/L_D)\ . \tag{6.15.3}$$

By treating the present case of cylindrical symmetry similarly to the previous case of spherical symmetry (Sect. 6.3), Pödör [6.110] was able to calculate the differential cross section

$$\sigma(\Theta) = \frac{2\pi m^2 e^2}{\hbar^4 k_t}\left[\int_0^\infty V(r)I_0(2k_t\, r\sin\theta/2)r\,dr\right]^2$$

$$= \frac{m^2 e^2 f^2}{32\pi\hbar^4\kappa^2\kappa_0^2 k_t^5 a^2}\frac{1}{(\sin^2\theta/2 + \beta^{-2})^2} \tag{6.15.4}$$

and the reciprocal of the momentum relaxation time

$$\tau_m^{-1} = N\,v_t\int_0^{2\pi}(1 - \cos\theta)\sigma(\theta)d\theta = \frac{N e^4 f^2 L_D}{8\kappa^2\kappa_0^2 a^2 m^2}(v_t^2 + \hbar^2/4m^2 L_D^2)^{-3/2}$$

$$\tag{6.15.5}$$

where k_t and v_t are the components of $\boldsymbol{k}$ and v perpendicular to the dislocation lines, $I_0(t)$ is the zero-order Bessel function of the first kind, a is the distance between imperfection centers along the dislocation line and f is their occupation probability $(Q = ef/a)$ [6.111]. Neglecting in the high temperature limit the

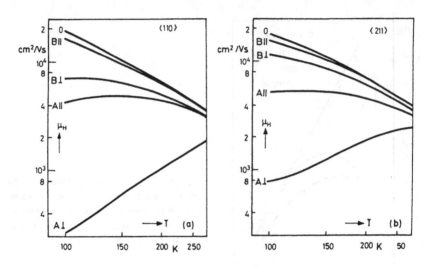

Fig. 6.26: a, b. Hall mobility for n-type germanium in undeformed crystals and in crystals bent at 1103 K about the ⟨110⟩ (left)and the ⟨211⟩ direction (right diagram) (after [6.112])

second term in parenthesis, the mobility is given for a nondegenerate electron gas as

$$\mu = \frac{30\sqrt{2\pi}\,\kappa^2\kappa_0^2 a^2}{e^3 f^2 L_D\, m^{1/2}} \frac{(k_B T)^{3/2}}{N} .$$ (6.15.6)

Figure 6.25 [curve (a)] shows the temperature dependence of $(1/\mu + 1/\mu_1)^{-1}$ where μ_1 is the lattice mobility, for $N = 10^7\,\text{cm}^{-2}$, $\kappa = 16$, $m = 0.3\,m_0$ and $a = 0.35$ nm. The straight line represents $\mu(T)$ for the undeformed sample ($N = 4 \times 10^3\,\text{cm}^{-2}$), the lower curve for the bent sample. N has been calculated from the bending radius. The agreement between the lower curve and curve (a) is considered reasonable. Although the statement has not been made it may be assumed that measurements were made on the sample as a whole. *Van Weeren* et al. [6.112] have shown, however, that in bars cut from the bent sample at the neutral plane (B), the mobility is different from that in bars cut above or below the neutral plane (A), that it is different in crystals bent about the ⟨110⟩ direction from that bent about the ⟨211⟩ direction and, of course, depends upon whether measurements are made with the current parallel or perpendicular to the bending axis. All of these cases are illustrated in Fig. 6.26. As one might suppose, the I–V characteristics for hot carriers are different parallel and perpendicular to the dislocation lines [6.113].

Problems

6.1. For a semiconductor (relative dielectric constant 16) doped with 10^{16} cm^{-3} singly ionized donors, calculate the mobility for temperatures between 30 K and 300 K according to the Conwell–Weisskopf relation assuming an effective mass $m = 0.15\,m_0$.

6.2. Calculate the mobility for scattering at N electric dipoles per cm^3 each with a moment M. For simplicity, neglect screening. Repeat the calculation for quadrupoles.

6.3. Combine scattering at electric dipoles with acoustic deformation potential scattering introducing an appropriate parameter q, in analogy to the one in Sect. 6.6, and the exponential integral $E_1(q)$:

$$\mu/\mu_{\mathrm{ac}} = q^2 \exp(q) E_1(q) - q + 1 \ . \tag{6.0.1}$$

Plot μ/μ_{ac} vs q up to $q = 10$.

6.4. For different scattering mechanisms, none of which is predominant, it is common practice to add the reciprocal relaxation times, as e.g., done in (6.6.1). In fact, it is the scattering rates which should be added. Indicate the problems which arise in doing so.

6.5. Carriers in the inversion layer of a MOSFET can be treated as 2-dimensional. Therefore, the scattering rate is

$$\frac{1}{\tau_{\mathrm{m}}} = \frac{A}{(2\pi)^2} \int k' dk' \int\limits_{0}^{2\pi} S(\boldsymbol{k}, \boldsymbol{k}') d\varphi \ . \tag{6.0.2}$$

Calculate scattering at acoustic phonons (which are 3-dimensional, of course) and show that the carrier energy does not enter the scattering rate. Why is the 2-dimensionality lost at field strengths above 10 kV/cm?

6.6. Calculate the energy loss rate for a degenerate carrier gas for optical deformation potential scattering similar to the calculation that leads to (6.11.10) for a nondegenerate carrier gas. *Hint:* apply a shifted Fermi–Dirac distribution.

6.7. For intrinsic InSb ($n = p = 2 \times 10^{16}$cm^{-3} at 300 K), consider the holes as immobile positive scattering centers for the comparatively light electrons ($m = 0.013\,m_0$) and determine the mobility of the latter taking into account polar optical scattering with a mobility of 4×10^4 cm^2/V s. The dielectric constant is 17.7. *Hint:* add the reciprocal mobilities.

6.8. For small polarons, the frequency dependence of the mobility is given by

$$\mu \propto \omega^{-1} \sinh(\hbar\omega/2k_{\mathrm{B}}T) \exp(-\omega^2\tau^2) \ . \tag{6.0.3}$$

At which value of ω (or mathematically simpler, of τ) is there a maximum? Discuss the physics behind this maximum.

6.9. Calculate the influence of 2-dimensional acoustic deformation potential scattering of Problem 6.5 on the mobility of carriers moving perpendicular to dislocation lines. Neglect the term $\hbar^2/4m^2L_{\mathrm{D}}^2$ for simplicity and assume mean free paths $l_{\mathrm{ac}} \ll l_{\mathrm{disl}}$.

7. Charge Transport and Scattering Processes in the Many-Valley Model

In Sect. 2.4, Figs. 2.26, 27, we saw that the conduction bands of silicon and germanium have constant energy surfaces near the band edge which are either 8 half-ellipsoids or 6 ellipsoids of revolution; these correspond to 4 and 6 energy valleys, respectively. In these and many other semiconductors the *many-valley model* of the energy bands has proved to be a fruitful concept for a description of the observed anisotropy of electrical and optical phenomena. Cyclotron resonance (Sect. 11.11) provides a direct experimental determination of the effective masses in each valley for any crystallographic direction.

In the present chapter we will calculate some important galvanomagnetic effects in the many-valley model of energy bands, deal with intervalley transitions, and finally arrive at the high-field domain and the acoustoelectric instabilities; the latter are also considered for piezoelectric coupling in a one-valley model since the calculation is similar to that for intervalley coupling in nonpolar many-valley semiconductors.

7.1 The Deformation Potential Tensor

In (6.4.5, 6) we defined an *acoustic deformation potential constant* ε_{ac} which in a crystal is actually a tensor. Let us first consider quite generally the six components of a deformation in a crystal [7.1]. In the undeformed crystal we assume a Cartesian coordinate system with unit vectors a, b, c. By a small deformation these vectors are displaced. The displaced unit vectors are called a', b', c' where the origin of the coordinate system is taken at the same lattice point as before the deformation. The displacement is then simply described by the following set of equations:

$$a' = (1 + \varepsilon_{xx})\, a + \varepsilon_{xy}\, b + \varepsilon_{xz}\, c$$

$$b' = \varepsilon_{yx}\, a + (1 + \varepsilon_{yy})\, b + \varepsilon_{yz}\, c \qquad\qquad (7.1.1)$$

$$c' = \varepsilon_{zx}\, a + \varepsilon_{zy}\, b + (1 + \varepsilon_{zz})\, c$$

where ε_{xx}, ε_{xy}, etc., are assumed to be small quantities and their products negligibly small. Therefore, the cosines of angles between a' and b', b' and c', and c' and a' denoted as e_4, e_5, and e_6, respectively, are given by

$$e_4 = (a' \cdot b') = \varepsilon_{yx} + \varepsilon_{xy}$$

$$e_5 = (b' \cdot c') = \varepsilon_{zy} + \varepsilon_{yz} \qquad\qquad (7.1.2)$$

$$e_6 = (c' \cdot a') = \varepsilon_{zx} + \varepsilon_{xz}$$

With the notation

$$e_1 = \varepsilon_{xx}; \quad e_2 = \varepsilon_{yy}; \quad e_3 = \varepsilon_{zz} , \qquad\qquad (7.1.3)$$

we have thus given all six deformation components which have already been applied in (4.12.4). The volume $V = 1$ of a cube having edges a, b, and c is changed by the deformation to

$$1 + \delta V = (a' \cdot [b' \times c']) \approx 1 + \varepsilon_{xx} + \varepsilon_{yy} + \varepsilon_{zz} = 1 + e_1 + e_2 + e_3 . \quad (7.1.4)$$

Hence, the volume dilatation is given by

$$\delta V/V = e_1 + e_2 + e_3 . \qquad\qquad (7.1.5)$$

The shift $r' - r$ of a mass point caused by the deformation can be described in the coordinate system a, b, c by coefficients u, v, w:

$$r' - r = u\,a + v\,b + w\,c . \qquad\qquad (7.1.6)$$

On the other hand, $r' = xa' + yb' + zc'$ and $r = xa + yb + zc$ and therefore, from (7.1.1) we find

$$u = x\,\varepsilon_{xx} + y\,\varepsilon_{yx} + z\,\varepsilon_{zx}$$

$$v = x\,\varepsilon_{xy} + y\,\varepsilon_{yy} + z\,\varepsilon_{zy} \qquad\qquad (7.1.7)$$

$$w = x\,\varepsilon_{xz} + y\,\varepsilon_{yz} + z\,\varepsilon_{zz}$$

Each one of the six coefficients $e_1, e_2, \cdots, e_6$ given by (7.1.2, 3) can now be expressed by derivatives $\partial u/\partial x$, etc.:

$$e_1 = \partial u/\partial x; \quad e_2 = \partial v/\partial y; \quad e_3 = \partial w/\partial z; \qquad\qquad (7.1.8)$$

$$e_4 = \partial v/\partial x + \partial u/\partial y; \quad e_5 = \partial w/\partial y + \partial v/\partial z; \quad e_6 = \partial u/\partial z + \partial w/\partial x$$

The change of, e.g., the conduction band edge $\Delta\varepsilon_c$ of a given valley due to a deformation may be expanded in a Taylor series:

$$\Delta\varepsilon_c = \frac{\partial\varepsilon}{\partial\varepsilon_{xx}}e_1 + \frac{\partial\varepsilon}{\partial\varepsilon_{yy}}e_2 + \frac{\partial\varepsilon}{\partial\varepsilon_{zz}}e_3 + \frac{\partial\varepsilon}{\partial\varepsilon_{xy}}e_4 + \frac{\partial\varepsilon}{\partial\varepsilon_{yz}}e_5 + \frac{\partial\varepsilon}{\partial\varepsilon_{xz}}e_6 + \cdots .(7.1.9)$$

We choose the x-axis of the k-coordinate system as the rotational axis of the ellipsoidal energy surface and introduce two constants [7.2]

$$\Xi_d = \frac{1}{2}\left(\frac{\partial\varepsilon}{\partial\varepsilon'_{yy}} + \frac{\partial\varepsilon}{\partial\varepsilon'_{zz}}\right)$$

$$\Xi_u = \frac{1}{2}\left(2\frac{\partial\varepsilon}{\partial\varepsilon'_{xx}} - \frac{\partial\varepsilon}{\partial\varepsilon'_{yy}} - \frac{\partial\varepsilon}{\partial\varepsilon'_{zz}}\right) = \frac{\partial\varepsilon}{\partial\varepsilon'_{xx}} - \Xi_d \ , \tag{7.1.10}$$

where the deformation relative to the coordinate system of the ellipsoid is denoted by ε'_{ii}. We will show that Ξ_{rmu} is the deformation potential constant for a shear deformation along the symmetry axis of the valley and $\Xi_d + (1/3)\Xi_u$ is the deformation potential constant for a dilatation.

Let us first consider a conduction band valley in germanium located along a $\langle 111\rangle$ axis. Symmetry considerations yield [7.3] $\partial\varepsilon/\partial\varepsilon_{xx} = \partial\varepsilon/\partial\varepsilon_{yy} = \partial\varepsilon/\partial\varepsilon_{zz}$ and $\partial\varepsilon/\partial\varepsilon_{xy} = \partial\varepsilon/\partial\varepsilon_{xz} = \partial\varepsilon/\partial\varepsilon_{yz}$. Hence, we obtain

$$\Delta\varepsilon_c = (\partial\varepsilon/\partial\varepsilon_{xx})(e_1 + e_2 + e_3) + (\partial\varepsilon/\partial\varepsilon_{xy})(e_4 + e_5 + e_6) + \cdots \tag{7.1.11}$$

A rotation of the x-axis into the symmetry axis of the ellipsoid of constant energy in k space may, e.g., be described by the matrix [1]

$$D_1 = \frac{1}{\sqrt{6}}\begin{pmatrix} \sqrt{2} & \sqrt{2} & \sqrt{2} \\ \sqrt{3} & -\sqrt{3} & 0 \\ 1 & 1 & -2 \end{pmatrix} ; \quad D_1^{-1} = \frac{1}{\sqrt{6}}\begin{pmatrix} \sqrt{2} & \sqrt{3} & 1 \\ \sqrt{2} & -\sqrt{3} & 1 \\ \sqrt{2} & 0 & -2 \end{pmatrix} . \tag{7.1.12}$$

The ε-tensor with components ε_{xx}, ε_{xy}, ... is thus related to the ε'-tensor with components $\varepsilon'_{xx}, \varepsilon'_{xy}, \ldots$ by the transformation

$$\varepsilon = D_1^{-1}\varepsilon' D_1 . \tag{7.1.13}$$

A calculation of the sum $e_4 + e_5 + e_6$ from this relation yields

$$e_4 + e_5 + e_6 = 2\varepsilon'_{xx} - \varepsilon'_{yy} - \varepsilon'_{zz} = 2e'_1 - e'_2 - e'_3 . \tag{7.1.14}$$

The sum $e_1 + e_2 + e_3$ is obtained from the fact that the dilatation given by (7.1.4) is invariant to the coordinate transformation:

$$e_1 + e_2 + e_3 = e'_1 + e'_2 + e'_3 . \tag{7.1.15}$$

The last two equations yield for $\Delta\varepsilon_c$:

$$\Delta\varepsilon_c = \left(\frac{\partial\varepsilon}{\partial\varepsilon_{xx}} + 2\frac{\partial\varepsilon}{\partial\varepsilon_{xy}}\right)e'_1 + \left(\frac{\partial\varepsilon}{\partial\varepsilon_{xx}} - \frac{\partial\varepsilon}{\partial\varepsilon_{xy}}\right)(e'_2 + e'_3) . \tag{7.1.16}$$

---

[1] The first column of D_1^{-1} is the $\langle 111\rangle$ unit vector, the second one is the $\langle 1\bar{1}0\rangle$ unit vector, and the third one is the $\langle 11\bar{2}\rangle$ unit vector. If by application of D_1 the cube diagonal ($\langle 111\rangle$) is turned into the x-axis ($\langle 100\rangle$) and at the same time the (x-y) plane diagonal $\langle 1\bar{1}0\rangle$ moves to the y-axis ($\langle 010\rangle$), it is the $\langle 11\bar{2}\rangle$ axis which becomes the z-axis ($\langle 001\rangle$) as shown in Fig. 7.1.

In the principal-axis system, this is by definition

$$\Delta\varepsilon_c = \frac{\partial\varepsilon}{\partial\varepsilon'_{xx}}e'_1 + \frac{\partial\varepsilon}{\partial\varepsilon'_{yy}}e'_2 + \frac{\partial\varepsilon}{\partial\varepsilon'_{zz}}e'_3 \; . \tag{7.1.17}$$

Due to rotational symmetry, $\partial\varepsilon/\partial\varepsilon'_{zz} = \partial\varepsilon/\partial\varepsilon'_{yy}$. A comparison of (7.1.16, 17) yields

$$\frac{\partial\varepsilon}{\partial\varepsilon_{xx}} + 2\frac{\partial\varepsilon}{\partial\varepsilon_{xy}} = \frac{\partial\varepsilon}{\partial\varepsilon'_{xx}} \quad \text{and} \quad \frac{\partial\varepsilon}{\partial\varepsilon_{xx}} - \frac{\partial\varepsilon}{\partial\varepsilon_{xy}} = \frac{\partial\varepsilon}{\partial\varepsilon'_{yy}} \tag{7.1.18}$$

From the definition (7.1.10), we find that

$$\frac{\partial\varepsilon}{\partial\varepsilon'_{xx}} = \Xi_u + \Xi_d \quad \text{and} \qquad \frac{\partial\varepsilon}{\partial\varepsilon'_{yy}} = \Xi_d \; . \tag{7.1.19}$$

Solving (7.1.18) for $\partial\varepsilon/\partial\varepsilon_{xx}$ and $\partial\varepsilon/\partial\varepsilon_{xy}$ thus yields

$$\frac{\partial\varepsilon}{\partial\varepsilon_{xx}} = \Xi_d + \frac{1}{3}\Xi_u \quad \text{and} \quad \frac{\partial\varepsilon}{\partial\varepsilon_{xy}} = \frac{1}{3}\Xi_u \; . \tag{7.1.20}$$

We finally obtain for $\Delta\varepsilon_c$

$$\Delta\varepsilon_c = \left(\Xi_d + \frac{1}{3}\Xi_u\right)(e_1 + e_2 + e_3) + \frac{1}{3}\Xi_u(e_4 + e_5 + e_6) \; . \tag{7.1.21}$$

For the case of hydrostatic pressure, $e_4 = e_5 = e_6 = 0$ and $\Delta\varepsilon_c \propto \delta V/V$ where the factor of proportionality is given by

$$\varepsilon_{ac} = \Xi_d + \frac{1}{3}\,\Xi_u \; . \tag{7.1.22}$$

Experimental values of Ξ_u obtained from piezoresistance or from the acousto-electric effect in n-type Ge are between 15.8 and 19.3 eV. By the positive sign of Ξ_u we emphasize that by a uniaxial compression, the energy of a $\langle 111 \rangle$ valley is lowered relative to the three other valleys. Ξ_u is temperature dependent. Above 60 K it increases with temperature and reaches a value of 21 eV at 100 K [7.4]. At 77 K, the ratio Ξ_d/Ξ_u is about -0.38 according to *Herring* and *Vogt* [7.3] and -0.45 according to *Smith* [7.5]. From these data values of -0.88eV and -1.8 eV are calculated for $\Xi_d + 1/3\,\Xi_u$. A hydrostatic pressure of 3 GPa increases the $\langle 111 \rangle$ minima linearly by 0.03 eV while the $\langle 100 \rangle$ valleys are lowered so that both types of valleys have the same energy. The central valley at $\mathbf{k}=0$, which without pressure is 0.15 eV above the $\langle 111 \rangle$ *satellite valleys*, is raised by 0.21 eV by a pressure of 3 GPa [7.6].

In silicon the lowest conduction band valleys are located on $\langle 100 \rangle$ and equivalent axes. For the $\langle 100 \rangle$ valley, no transformation is required and due to symmetry, $\partial\varepsilon/\partial\varepsilon_{zz} = \partial\varepsilon/\partial\varepsilon_{yy}$, $\partial\varepsilon/\partial\varepsilon_{xy} = 0$, etc.:

$$\Delta\varepsilon_c = \frac{\partial\varepsilon}{\partial\varepsilon_{xx}}e_1 + \frac{\partial\varepsilon}{\partial\varepsilon_{yy}}(e_2 + e_3) \; . \tag{7.1.23}$$

Since $\varepsilon'_{xx} = \varepsilon_{xx}$, etc., (7.1.10) yields

$$\Xi_d = \frac{\partial \varepsilon}{\partial \varepsilon_{yy}}; \quad \Xi_d + \Xi_u = \frac{\partial \varepsilon}{\partial \varepsilon_{xx}} \ . \tag{7.1.24}$$

Therefore, $\Delta \varepsilon_c$ becomes

$$\Delta \varepsilon_c = \Xi_d(e_1 + e_2 + e_3) + \Xi_u e_1$$

$$= \left(\Xi_d + \frac{1}{3}\Xi_u\right)(e_1 + e_2 + e_3) + \frac{1}{3}\Xi_u[(e_1 - e_2) + (e_1 - e_3)] \ . \tag{7.1.25}$$

$\Delta \varepsilon_c$ for the other valleys is easily obtained by a transformation.

Experimental values for Ξ_u of silicon are between 8.5 and 9.6 eV. The deformation potential constant $\Xi_d \ll \Xi_u$ is not well known [7.7, 8].

7.2 Electrical Conductivity

At first we consider valley no. ϱ located in k-space at k_ϱ with its 3 main axes parallel to the coordinate axes. The effective masses in the directions of the axes are denoted by m_1, m_2 and m_3. It is convenient to introduce an *inverse mass tensor* with the diagonal elements $\alpha_{ii} = m_\sigma/m_i$ where the *conductivity-effective mass* m_σ will be defined later [(7.2.19)]. The energy of a carrier is then given by

$$\varepsilon = (\hbar^2/2m_\sigma)(k - k_\varrho)\alpha(k - k_\varrho) \tag{7.2.1}$$

Since α is a diagonal tensor, this relation is simplified to

$$\varepsilon = (\hbar^2/2m_\sigma)w^2 \tag{7.2.2}$$

by introducing a vector **w** with components

$$w_i = \sqrt{\alpha_{ii}}(k_i - k_{\varrho i}) \ . \tag{7.2.3}$$

By this relation an ellipsoidal surface in k-space is transformed into a spherical surface in w-space [7.3]. It is called the *Herring-Vogt transformation*.

In an electric field E and a magnetic field (B is correctly called *induction*), the distribution $f(\mathbf{k})$ of carriers is determined by the Boltzmann equation in the relaxation time approximation (4.2.3)

$$(e/\hbar)(E + \hbar^{-1}[\nabla_k \varepsilon \times B])\nabla_k f = -(f - f_0)/\tau_m \ . \tag{7.2.4}$$

Since for integrals in k-space $dk_i = dw_i/\sqrt{\alpha_{ii}}$ and $f d^3 k = f d^3 w / \sqrt{\alpha_{11}\alpha_{22}\alpha_{33}} = g\,d^3 w$, we define a function

$$g = f/\sqrt{(\alpha_{11}\,\alpha_{22}\,\alpha_{33})} \tag{7.2.5}$$

which for thermal equilibrium is denoted as

$$g_0 = f_0 / \sqrt{(\alpha_{11}\,\alpha_{22}\,\alpha_{33})} \tag{7.2.6}$$

Since $\nabla_k = \sqrt{\alpha}\nabla_w$, we obtain from (7.2.4) by dividing by $\sqrt{\alpha_{11}\alpha_{22}\alpha_{33}}$:

$$(e/\hbar)(\boldsymbol{E} + \hbar^{-1}[\sqrt{\alpha}\nabla_w\varepsilon \times \boldsymbol{B}])\sqrt{\alpha}\nabla_w g = -(g - g_0)/\tau_m \ . \tag{7.2.7}$$

By introducing *effective field intensities* having components

$$E_i^* = \sqrt{\alpha_{ii}}\,E_i \qquad\qquad B_i^* = \sqrt{\alpha_{11}\alpha_{22}\alpha_{33}}\,B_i \tag{7.2.8}$$

we bring (7.2.7) into the form of (7.2.4); the first term of (7.2.7) becomes

$$(\boldsymbol{E}\sqrt{\alpha}\nabla_w g) = \sum_i E_i\sqrt{\alpha_{ii}}\partial g/\partial w_i = \sum_i E_i^*\partial g/\partial w_i = (\boldsymbol{E}^*\nabla_w g) \tag{7.2.9}$$

and the second term

$$[\sqrt{\alpha}\nabla_w\varepsilon \times \boldsymbol{B}]\sqrt{\alpha}\nabla_w g$$

$$= [\sqrt{\alpha_{22}}(\partial\varepsilon/\partial w_2)B_3 - \sqrt{\alpha_{33}}(\partial\varepsilon/\partial w_3)B_2]\sqrt{\alpha_{11}}\partial g/\partial w_1 + \text{cycl.permut.}$$

$$= [(\partial\varepsilon/\partial w_2)\sqrt{\alpha_{11}\alpha_{22}}B_3 - (\partial\varepsilon/\partial w_3)\sqrt{\alpha_{11}\alpha_{33}}B_2]\partial g/\partial w_1 + \text{cycl.permut.}$$

$$= ([\nabla_w\varepsilon \times \boldsymbol{B}^*] \cdot \nabla_w g) \ . \tag{7.2.10}$$

The Joule heat must be invariant to the transformation

$$(\boldsymbol{j} \cdot \boldsymbol{E}) = (\boldsymbol{j}^*\boldsymbol{E}^*) = (\boldsymbol{j}^*\sqrt{\alpha}\boldsymbol{E}) = (\sqrt{\alpha}\boldsymbol{j}^* \ \boldsymbol{E}) \ . \tag{7.2.11}$$

This shows that $\boldsymbol{j}$ is transformed in the following way:

$$\boldsymbol{j} = \sqrt{\alpha}\boldsymbol{j}^* \ . \tag{7.2.12}$$

In w-space the conductivity tensor $\sigma_w = \sigma_w\,(\boldsymbol{B}^*)$ is defined by

$$\boldsymbol{j}^* = \sigma_w\boldsymbol{E}^* \ . \tag{7.2.13}$$

Multiplication of this equation by $\sqrt{\alpha}$ yields

$$\boldsymbol{j} = \sqrt{\alpha}\sigma_w\boldsymbol{E}^* = \sqrt{\alpha}\sigma_w\sqrt{\alpha}\boldsymbol{E} = \sigma\boldsymbol{E} \ . \tag{7.2.14}$$

This shows how σ is transformed:

$$\sigma = \sqrt{\alpha}\sigma_w\sqrt{\alpha} \ . \tag{7.2.15}$$

We will consider here only the case of weak magnetic fields where the conductivity in a spherical valley σ_w is given by (4.2.34) with $\boldsymbol{B}$ replaced by $\boldsymbol{B}^*$ and m by m_σ. The simple case of no magnetic field $\boldsymbol{B}^* = 0$ will be considered first. There we have

$$\sigma = \alpha\sigma_0 = \alpha\,\frac{n\,e^2\langle\tau_m\rangle}{m_\sigma}\begin{pmatrix} 1 & 0 & 0 \\ 0 & 1 & 0 \\ 0 & 0 & 1 \end{pmatrix} \ . \tag{7.2.16}$$

In a crystal lattice of cubic symmetry, the conductivity is isotropic. Due to symmetry at least three valleys form the energy band. If there are three valleys, their long axes are perpendicular to each other. Each one of the other two valleys first has to be transformed into the position of the first valley before the transformation (7.2.3) can be applied. This is done by matrices:

$$D_2 = D_2^{-1} = \begin{pmatrix} 0 & 1 & 0 \\ 1 & 0 & 0 \\ 0 & 0 & -1 \end{pmatrix} \qquad D_3 = D_3^{-1} = \begin{pmatrix} 0 & 0 & 1 \\ 0 & -1 & 0 \\ 1 & 0 & 0 \end{pmatrix} \quad (7.2.17)$$

D_1, of course, is the unit matrix. By summation over all three valleys and taking into account that n, contained in σ_0 given by (4.2.21), is the total carrier concentration assumed to be equally distributed among the three valleys ($n/3$ per valley), we find, assuming an isotropic relaxation time $\langle \tau_m \rangle$

$$\sigma = \frac{1}{3} \sum \alpha_{ii}\, \sigma_0 = \frac{ne^2 \langle \tau_m \rangle}{m_\sigma} \frac{1}{3} \sum \alpha_{ii} = \frac{ne^2 \langle \tau_m \rangle}{m_\sigma} \ . \tag{7.2.18}$$

For six valleys such as shown in Fig. 2.26 for silicon, the same result is obtained as in the three-valley model. Since $\alpha_{ii} = m_\sigma / m_i$, we finally obtain a *conductivity effective mass* m_σ given by

$$\frac{1}{m_\sigma} = \frac{1}{3} \left(\frac{1}{m_1} + \frac{1}{m_2} + \frac{1}{m_3} \right) \tag{7.2.19}$$

if we assume that the ellipsoids are figures of revolution (spheroids), i.e. $m_1 = m_l$; $m_2 = m_3 = m_t$. The conductivity effective mass is also the mass in any particular valley for a direction which is symmetrical to all valleys, e.g., for the three-valley model in $\langle 100 \rangle$ and equivalent directions:

$$\hbar^2 k^2 / 2m_\sigma = \hbar^2 k_1^2 / 2m_1 + \hbar^2 k_2^2 / 2m_2 + \hbar^2 k_3^2 / 2m_3 \ , \tag{7.2.20}$$

where $k_1^2 = k_2^2 = k_3^2 = \frac{1}{3} k^2$.
The *density-of-states effective mass* m_D is defined by

$$m_D = \sqrt[3]{m_1 m_2 m_3} \tag{7.2.21}$$

since the density of states (3.1.10) is proportional to $\int d^3k = \int d^3w / \sqrt{\alpha_{11}\alpha_{22}\alpha_{33}}$ We will now consider the case of germanium where there are four valleys located on the cube diagonals in k-space as shown in Fig. 2.27. The rotation matrix for the $\langle 111 \rangle$-ellipsoid is given by (7.1.12). Figure 7.1 shows the transformation of axes into the $\langle 100 \rangle$, $\langle 010 \rangle$, and $\langle 001 \rangle$-axes of the Cartesian coordinate system. It is important that the rotation does not include a mirror image, i.e., a transformation of a right-hand system into a left-hand system. The rotation matrices for the other three valleys have the same elements as D_1 except for the sign and we give the signs simply as:

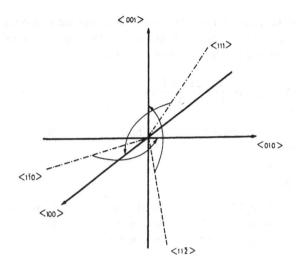

Fig. 7.1: Rotation D_1 of axes of the Cartesian coordinate system

$$
D_2 = \begin{pmatrix} + & + & - \\ + & - & 0 \\ - & - & - \end{pmatrix} ; D_3 = \begin{pmatrix} + & - & + \\ + & + & 0 \\ - & + & + \end{pmatrix} ; D_4 = \begin{pmatrix} - & + & + \\ + & + & 0 \\ - & + & - \end{pmatrix} .
$$

$$(7.2.22)$$

Both the electric field E and the current density j are of course rotated in the same way, i.e., in (7.2.14) we replace j by $D_\varrho j$ and E by $D_\varrho E$:

$$
D_\varrho j = \sqrt{\alpha}\,\sigma_w\sqrt{\alpha}D_\varrho\, E \ . \tag{7.2.23}
$$

Multiplication by D_ϱ^{-1} yields $j = \sigma E$ with σ given by

$$
\sigma = D_\varrho^{-1}\sqrt{\alpha}\,\sigma_w\sqrt{\alpha}\,D_\varrho \ . \tag{7.2.24}
$$

For $B = 0$, σ_w is diagonal and $\sqrt{\alpha}\,\sigma_w\sqrt{\alpha} = \sigma_0\alpha$. Let us introduce the following notations: the effective mass in the direction of the longitudinal axis of the rotational ellipsoid m_l and in the direction of the transverse axis m_t, the anisotropy factor [2] $K = m_l/m_t$, the ratio $(K - 1)/(2K + 1) = \lambda$, and the matrix

$$
D_\varrho^{-1}\alpha\, D_\varrho = \Lambda_\varrho \ , \tag{7.2.25}
$$

where, e.g., Λ_l is given by

$$
\Lambda_l = \frac{m_\sigma}{m_l}\frac{2K + 1}{3}\begin{pmatrix} 1 & -\lambda & -\lambda \\ -\lambda & 1 & -\lambda \\ -\lambda & -\lambda & 1 \end{pmatrix} . \tag{7.2.26}
$$

---

[2] For simplicity, we omit here the usual subscript m to K since the relaxation time has been assumed to be isotropic [see (7.4.20)].

By a summation over all four valleys and taking into account that n, contained in σ_0 as given by (4.2.21), is the total electron concentration which is equally distributed among the 4 valleys, we find

$$\sigma = \sigma_0 \frac{1}{4} \sum_{\varrho=1}^{4} \Lambda_\varrho = \sigma_0 \frac{m_\sigma}{4m_l} \frac{2K+1}{3} 4 = \sigma_0 \; , \tag{7.2.27}$$

where for m_σ its value given by (7.2.19) and $m_1 = m_2 = m_t$, $m_3 = m_l$ have been substituted:

$$m_\sigma = m_l \frac{3}{2K+1} \; . \tag{7.2.28}$$

In a model where 6 or 12 valleys are located on the $\langle 110 \rangle$ and equivalent axes [7.9, 10], the conductivity σ_0 is also isotropic with an effective mass given by (7.2.28), again assuming an isotropic relaxation time τ_m.

7.3 Hall Effect in a Weak Magnetic Field

The terms in σ which are linear in B are given by (4.2.23). As considered above we now have to replace $\boldsymbol{B}$ by $\boldsymbol{B}^*$ and m by m_σ. The Herring–Vogt transformation (7.2.24) yields for $\langle 111 \rangle$ valleys such as in n-Ge:

$$\sigma = \frac{1}{4} \sum_{\varrho=1}^{4} D_\varrho^{-1} \sqrt{\alpha} \begin{pmatrix} 0 & \gamma_0 B_3^* & -\gamma_0 B_2^* \\ -\gamma_0 B_3^* & 0 & \gamma_0 B_1^* \\ \gamma_0 B_2^* & -\gamma_0 B_1^* & 0 \end{pmatrix} \sqrt{\alpha} D_\varrho \; , \tag{7.3.1}$$

where γ_0 is given by (4.2.24) with m replaced by m_σ and $\boldsymbol{B}^*$ in valley no. ϱ has the components

$$B_{i(\varrho)}^* = \sqrt{\alpha_{11}\alpha_{22}\alpha_{33}/\alpha_{ii}} \sum_{k=1}^{3} D_\varrho^{ik} B_k \; . \tag{7.3.2}$$

For example, for valley no.1 which is in the $\langle 111 \rangle$ direction, $B_{(1)}^*$ is given by

$$B_{(1)}^* = \frac{m_\sigma}{m_l} \sqrt{\frac{K}{6}} \begin{pmatrix} \sqrt{2K}(B_1 + B_2 + B_3) \\ \sqrt{3}(B_1 - B_2) \\ B_1 + B_2 - 2B_3. \end{pmatrix} \; . \tag{7.3.3}$$

Let us introduce for a shorthand notation

$$\lambda' = (K+2)/(K-1) \tag{7.3.4}$$

We obtain for the component σ_{12} of the conductivity tensor in the first valley

$$\sigma_{12}^{(1)} = \frac{1}{4} \gamma_0 \frac{m_\sigma}{m_l} \frac{\sqrt{K}}{6} (-2\sqrt{6}B_3^* + 2\sqrt{3K}B_1^*)$$

$$= \frac{1}{4}\gamma_0 \left(\frac{m_\sigma}{m_l}\right)^2 \frac{K}{3}(K-1)(B_1 + B_2 + \lambda'B_3) \ . \tag{7.3.5}$$

$\sigma_{13}^{(1)}$ and $\sigma_{23}^{(1)}$ are similar except that the last factor is replaced by $(-B_1 - \lambda'B_2 - B_3)$ and $(\lambda'B_1 + B_2 + B_3)$, respectively. For valley no. 2 we find

$$\boldsymbol{B}_{(2)}^* = \frac{m_\sigma}{m_l}\sqrt{\frac{K}{6}}\begin{pmatrix} \sqrt{2K}(B_1 + B_2 + B_3) \\ \sqrt{3}(B_1 - B_2) \\ -B_1 - B_2 - 2B_3 \end{pmatrix} \tag{7.3.6}$$

and for the conductivity component σ_{12}

$$\sigma_{12}^{(2)} = \frac{1}{4}\gamma_0\frac{m_\sigma}{m_l}\frac{\sqrt{K}}{6}(-2\sqrt{6}\,B_3^* - 2\sqrt{3K}\,B_1^*)$$

$$= \frac{1}{4}\gamma_0\left(\frac{m_\sigma}{m_l}\right)^2\frac{K}{3}(K-1)(-B_1 - B_2 + \lambda'B_3) \ . \tag{7.3.7}$$

The calculation for the other components and the other valleys is similar. By a summation over all four valleys we obtain

$$\sigma = \frac{1}{4}\gamma_0\left(\frac{m_\sigma}{m_l}\right)^2\frac{K}{3}(K-1)4\lambda'\begin{pmatrix} 0 & B_3 & -B_2 \\ -B_3 & 0 & B_1 \\ B_2 & -B_1 & 0 \end{pmatrix} \ . \tag{7.3.8}$$

The dependence of σ on $\boldsymbol{B}$ is the same as in the spherical one-valley model except that γ_0 is now replaced by a coefficient which we denote as γ_H:

$$\gamma_H = \gamma_0\left(\frac{m_\sigma}{m_l}\right)^2\frac{K}{3}(K-1)\lambda' = \frac{ne^3}{m_H^2}\langle\tau_m^2\rangle \ , \tag{7.3.9}$$

where the *Hall effective mass*

$$m_H = m_l\sqrt{3/[K(K+2)]} \tag{7.3.10}$$

has been introduced and the momentum relaxation time τ_m has been assumed as being isotropic.

For a calculation of the ratio of the Hall and drift mobilities μ_H/μ, we may assume without loss of generality an electric field in the y-direction and a magnetic field in the z-direction. The Hall current density

$$\boldsymbol{j} = \gamma_H\begin{pmatrix} 0 & B_3 & 0 \\ -B_3 & 0 & 0 \\ 0 & 0 & 0 \end{pmatrix}\begin{pmatrix} 0 \\ E_2 \\ 0 \end{pmatrix} = \gamma_H\begin{pmatrix} E_2B_3 \\ 0 \\ 0 \end{pmatrix} \ . \tag{7.3.11}$$

then has only a component in the "1" direction: $j_1 = \gamma_H E_2 B_3$. In addition we have, of course, the longitudinal current density $j_2 = \sigma E_2$ where in the weak-field approximation, $\sigma = \sigma_0$ independent of B. The *Hall field* is given by

$$E_1 = j_1/\sigma = (\gamma_H/\sigma_0^2)j_2 B_3 \tag{7.3.12}$$

and the *Hall coefficient* by

$$R_H = \frac{E_1}{j_2 B_3} = \frac{\gamma_H}{\sigma_0^2} = \frac{1}{ne} \frac{\langle \tau_m^2 \rangle}{\langle \tau_m \rangle^2} \left(\frac{m_\sigma}{m_H} \right)^2 , \qquad (7.3.13)$$

where the *Hall mass factor*

$$\left(\frac{m_\sigma}{m_H} \right)^2 = \frac{3K(K+2)}{(2K+1)^2} . \qquad (7.3.14)$$

This quantity is unity for $K = 1$, 0.78 for $K = 20$ and approaches 0.75 for $K \to \infty$. Since $\mu_H = |R_H| \sigma$ and $\mu = \sigma/(n|e|)$, the Hall factor r_H becomes

$$r_H = \frac{\mu_H}{\mu} = \frac{\langle \tau_m^2 \rangle}{\langle \tau_m \rangle^2} \left(\frac{m_\sigma}{m_H} \right)^2 . \qquad (7.3.15)$$

For silicon the same results are obtained: the weak-field Hall coefficient is isotropic and proportional to a Hall mass factor given by (7.3.14).

7.4 The Weak-Field Magnetoresistance

The terms in σ which are quadratic in $\boldsymbol{B}$ are given by (4.2.31). The Herring–Vogt transformation (7.2.24) yields for the case of n-Ge

$$\sigma = \frac{1}{4} \sum_{\varrho=1}^{4} D_\varrho^{-1} \sqrt{\alpha} \beta_0 \begin{pmatrix} B_2^{*2} + B_3^{*2} & -B_1^* B_2^* & -B_1^* B_3^* \\ -B_1^* B_2^* & B_1^{*2} + B_3^{*2} & -B_2^* B_3^* \\ -B_1^* B_3^* & -B_2^* B_3^* & B_1^{*2} + B_2^{*2} \end{pmatrix} \sqrt{\alpha} D_\varrho , \qquad (7.4.1)$$

where β_0 is given by (4.2.32) with m replaced by m_σ. Depending on the number ϱ of the valley, the $\boldsymbol{B}^*$-components have to be replaced by linear combinations of the B-components as given by (7.3.3, 6), etc. The calculation, although quite tedious, is straightforward and we merely state the result for the component which is quadratic in B (Ref. [7.6] p. 22):

$$\sigma = \beta_M \begin{pmatrix} B_2^2 + B_3^2 & -B_1 B_2 & -B_1 B_3 \\ -B_1 B_2 & B_1^2 + B_3^2 & -B_2 B_3 \\ -B_1 B_3 & -B_2 B_3 & B_1^2 + B_2^2 \end{pmatrix}$$

$$+ \beta_M \frac{2(K-1)^2}{(2K+1)(K+2)} \begin{pmatrix} B_1^2 & 0 & 0 \\ 0 & B_2^2 & 0 \\ 0 & 0 & B_3^2 \end{pmatrix} , \qquad (7.4.2)$$

where

$$\beta_M = -(n e^4/m_l^3) \langle \tau_m^2 \rangle K(K+2)(2K+1)/9 . \qquad (7.4.3)$$

The second term on the rhs of (7.4.2) occurs neither in n-Si where the valleys are located on $\langle 100 \rangle$ and equivalent axes, nor in the spherical band model. For an investigation of this term let us consider the case where j, E and B are parallel to the x-direction. The *longitudinal magnetoresistance* is then given by

$$\frac{\Delta\varrho}{\varrho B^2}\bigg|_{\text{longit.}} = \frac{j(0) - j(B)}{j(0)B^2} = \frac{-\beta_M}{\sigma_0} \frac{2(K-1)^2}{(2K+1)(K+2)}$$

$$= \frac{e^2}{m_l^2} \frac{2K(K-1)^2}{3(2K+1)} \frac{\langle \tau_m^3 \rangle}{\langle \tau_m \rangle} \,, \tag{7.4.4}$$

where in σ both the quadratic and linear terms in B have been considered; the linear term, however, multiplied by $E = (E,0,0)$ vanishes for $B = (B,0,0)$. Introducing a *magnetoresistance mobility*

$$\mu_M = (|e|/m_l)(\langle \tau_m^3 \rangle / \langle \tau_m \rangle)^{1/2} \,,$$

we find

$$(\Delta\varrho/\varrho B^2)_{\text{longit.}} = \mu_M^2 \frac{2K(K-1)^2}{3(2K+1)} \,. \tag{7.4.5}$$

For $K = 20$, which is typical for n-Ge, the K-dependent factor has a value of 118 while it vanishes for $K = 1$; this shows that the effective-mass anisotropy strongly enters the magnetoresistance.

For the case of n-Si, we obtain instead of (7.4.2)

$$\sigma = \begin{pmatrix} \beta'_M(B_2^2 + B_3^2) & -\beta''_M B_1 B_2 & -\beta''_M B_1 B_3 \\ -\beta''_M B_1 B_2 & \beta'_M(B_1^2 + B_3^2) & -\beta''_M B_2 B_3 \\ -\beta''_M B_1 B_3 & -\beta''_M B_2 B_3 & \beta'_M(B_1^2 + B_2^2) \end{pmatrix} \tag{7.4.6}$$

where the coefficients β'_M and β''_M are given by

$$\beta'_M = -(ne^4/m_l^3)\langle \tau_m^3 \rangle K(K^2 + K + 1)/3 \tag{7.4.7}$$

and

$$\beta''_M = -(ne^4/m_l^3)\langle \tau_{rmm}^3 \rangle K^2 \,. \tag{7.4.8}$$

Since σ_{xx} is independent of B_1, there is no longitudinal magnetoresistance in the $\langle 100 \rangle$ direction, in contrast to the case of n-type Ge.

Figure 7.2 shows the observed weak-field magnetoresistance for n-type Ge [7.11] and n-type Si [7.12] as a function of the angle θ between B and the $\langle 100 \rangle$ direction in the [010] plane where j is in the $\langle 100 \rangle$ direction. At $\theta = 0°$, the magnetoresistance is longitudinal; there it is a maximum for n-type Ge and a minimum for n-type Si; in the latter case it should actually vanish, as mentioned above. The case of valleys located on $\langle 110 \rangle$ and equivalent axes will not be considered here [Ref. 7.10, Table VII].

In an experiment the current density j is usually kept constant and voltages are measured from which the Hall field and the longitudinal field intensities are

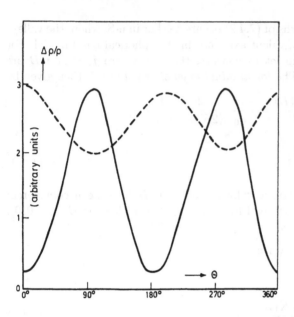

Fig. 7.2: Magnetoresistance of n-type Ge (- - -) and n-type Si (–) as a function of the angle between current (in the (100) direction) and magnetic field in the [010] plane

determined. Instead of σ as a function of E and $\mathbf{B}$, we then need σ as a function of j and $\mathbf{B}$ for a comparision with experimental data. For the case of a weak magnetic field, the magnetoresistance in a cubic crystal is represented by the *inverted Seitz equation* [7.13].

$$\Delta\varrho/(\varrho B^2) = b + c(il + jm + kn)^2 + d(i^2l^2 + j^2m^2 + k^2n^2) , \qquad (7.4.9)$$

where (i, j, k) is a unit vector in the current direction and (l, m, n) is a unit vector in the direction of the magnetic field; b, c, and d are coefficients given by

$$b = -(\beta + \mu_H^2); \quad c = \mu_H^2 - \gamma; \quad d = -\delta , \qquad (7.4.10)$$

where β, γ, and δ are coefficients of the *normal Seitz equation*

$$\frac{j}{\sigma_0} = E + \mu_H[E \times B] + \beta B^2 E + \gamma(E \cdot B)B + \delta \begin{pmatrix} B_1^2 & 0 & 0 \\ 0 & B_2^2 & 0 \\ 0 & 0 & B_3^2 \end{pmatrix} E . (7.4.11)$$

μ_H is the Hall mobility ($= |R_H|\sigma_0$). The reference system for the unit vectors is given by the cubic-lattice vectors. For the longitudinal magnetoresistance, we find from (7.4.9) for the special case that both j and $\mathbf{B}$ are located in the $[1\bar{1}0]$-plane ($i = j = l = m = \sin\theta/\sqrt{2}$; $k = n = \cos\theta$):

$$\Delta\varrho/(\varrho B^2) = (b + c + d) - \frac{1}{2}d\sin^2\theta\,(3\cos^2\theta + 1) . \qquad (7.4.12)$$

The same angular dependence is obtained for $\Delta\varrho/\varrho E^2$ which is characteristic of warm carriers. For any direction of $j \parallel \mathbf{B}$ in a [111]-plane, we find

$$\Delta\varrho/(\varrho B^2) = b + c + d/2 . \qquad (7.4.13)$$

For the transverse magnetoresistance with B also in the $[1\bar{1}0]$ plane but with j in the $\langle 110 \rangle$ direction where $j \perp B$, we obtain

$$\Delta\varrho/(\varrho B^2) = b + \frac{1}{2}d\sin^2\theta \ . \tag{7.4.14}$$

For the B-dependent Hall coefficient R_H, an angular dependence similar to the one given by (4.4.12) has been calculated where the θ-dependent term is proportional to B^2 [7.14].

With the shorthand notations

$$T'_M = \langle \tau_m^3 \rangle \langle \tau_m \rangle / \langle \tau_m^2 \rangle^2 \qquad q = 1 + 3K/(K-1)^2 \ , \tag{7.4.15}$$

we obtain for n-type Si ($\langle 100 \rangle$ valleys) the relations

$$b/\mu_H^2 = T'_M(2K+1)(K^2+K+1)/[K(K+2)^2] - 1$$
$$c/\mu_H^2 = 1 - 3\,T'_M(2K+1)/(K+2)^2 \tag{7.4.16}$$
$$d/\mu_H^2 = -T'_M(2K+1)(K-1)^2/[K(K+2)^2]$$

and

$$b + c = -d > 0 \qquad q = -(\mu_H^2 + b)/d \ . \tag{7.4.17}$$

If the equation for q (7.4.15) is solved for K, we find two solutions indicated by subscripts $+$ and $-$:

$$K_\pm = \frac{1}{2}(2q+1 \pm \sqrt{12q-3})/(q-1) \ . \tag{7.4.18}$$

The solution $K_+ > 1$ yields prolate (cigar-shaped) ellipsoids for the constant-energy surfaces in k-space while $K_- < 1$ yields oblate (disk-shaped) ellipsoids. If the scattering mechanism and therefore T'_M given by (7.4.15) are known, one of the two K values may be excluded. For example, *Stirn* and *Becker* [7.15] found for n-type AlSb, where as in n-type Si, there are $\langle 100 \rangle$ valleys, that impurity scattering, for which (7.4.15) yields $T'_M = 1.58 > 1$, is predominant at 77 K and therefore only the solution $K_+ > 1$ is physically significant.

If the direction of the current density j is the $\langle 100 \rangle$ direction and B is in any direction perpendicular to j, (7.4.9) results in $\Delta\varrho/\varrho^2$ being independent of the direction of B. If this fact cannot be verified experimentally, one may suspect that either a metallic contact at the semiconductor sample acts as a short circuit or that the crystal is inhomogeneous (remember that this is true only for the weak-B case).

Quite often the ratio $\Delta\varrho/\varrho B^2$ is denoted by M with a lower subscript indicating the crystallographic direction of the current density and an upper subscript indicating that of the magnetic field, such as, e.g.,

$$M\,^{001}_{110} = b \qquad M\,^{1\bar{1}0}_{110} = b + \frac{1}{2}d \ .$$

If the second M-value is smaller than the first one, $d < 0$ which is a strong indication for $\langle 100 \rangle$ valleys such as in n-type Si and n-type AlSb.

A criterion for $\langle 111 \rangle$ valleys such as in Ge is given by

$$b = -c \qquad d > 0 \; . \tag{7.4.19}$$

An experimental determination of b, c and d for a single sample with side arms is possible by means of the *planar Hall effect* which was discussed in Sect. 4.7. The coefficient P_H is given by $P_H = c/\sigma_0$ for a current in the $\langle 100 \rangle$ direction and a Hall field $\boldsymbol{E}_p$ in the $\langle 010 \rangle$ direction, while $P_H = (c + d)/\sigma_0$ for $\boldsymbol{j} \parallel \langle 110 \rangle$ and $\boldsymbol{E}_p \parallel \langle \bar{1}10 \rangle$. If we thus determine c for $\boldsymbol{j} \parallel \langle 100 \rangle$ and the coefficient b from the magnetoresistance for the same current direction (but $\boldsymbol{B} \parallel \langle 010 \rangle$), and the quantity $b+c+d$ for $\boldsymbol{B} \parallel \langle 100 \rangle$, we can with only one sample determine b,c, and d. Another possibility is to measure $c+d$ by the planar Hall effect for $\boldsymbol{j} \parallel \langle 110 \rangle$, b by the magnetoresistance for $\boldsymbol{B} \parallel \langle 001 \rangle$, and $b+d/2$ by the magnetoresistance for $\boldsymbol{B} \parallel \langle 110 \rangle$. Because of the finite thickness of the side arms of the sample the experimental data are subject to corrections [7.16].

So far we have assumed the momentum relaxation time τ_m to be isotropic. *Herring* and *Vogt* [7.3], and *Eagles* and *Edwards* [7.17] considered the case of an anisotropic τ_m with components τ_l and τ_t in the longitudinal and the transverse direction, respectively, of a constant - energy ellipsoid of revolution in $\boldsymbol{k}$-space. The ratio $\tau_l/\tau_t = K_\tau$ and the ratio m_l/m_t, now denoted as K_m, yields for K the ratio

$$K = K_m/K_\tau \tag{7.4.20}$$

since the conductivity is determined by the ratio m/τ_m. For example, for b/μ_H^2 for n-type Si, we now obtain instead of (7.4.16)[7.18]

$$\frac{b}{\mu_H^2} = T_M' \frac{(\langle \tau_t^3 K_\tau^2 \rangle/\langle \tau_t^3 \rangle + K_m \langle \tau_t^3 K_\tau \rangle/\langle \tau_t^3 \rangle + K_m^2)(\langle \tau_t K_\tau \rangle/\langle \tau_t \rangle + 2K_m)}{K_m(2\langle \tau_t^2 K_\tau \rangle/\langle \tau_t^2 \rangle + K_m)^2} - 1$$

$$\tag{7.4.21}$$

For lattice scattering, τ_m seems to be fairly isotropic while for impurity scattering in n-type Ge, $K_\tau = 11.2$ has been calculated [7.17]. In n-type Ge with $4 \times 10^{15} \, \mathrm{cm}^{-3}$ impurities, the ratio $K = K_m/K_\tau = 20$ at 4 K and $K = 5$ at 20 K; this temperature dependence has been attributed to a transition from lattice scattering at low temperatures where the impurities are neutral due to the freeze-out of carriers, to a combination of ionized impurity scattering and lattice scattering at somewhat higher temperatures [7.19]. At 77 K, the transition from lattice scattering to ionized impurity scattering may be obtained by a change in the impurity concentration N_I. K_τ as a function of N_I is shown in Fig. 7.3 assuming saturation values of $K_\tau^i = 11$ and 12, for K_τ in the limit of pure ionized impurity scattering, $N_I \to \infty$. The experimental data points have been determined from measurements of the microwave Faraday effect in n-type Ge at 77 K [7.20]; they are consistent with a value of 11 for K_τ^i which agrees

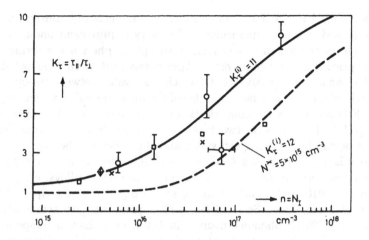

Fig. 7.3: Dependence of the momentum relaxation time anisotropy K_τ on impurity concentration N_I in n-type Ge at 77 K. The full curve has been calculated for combined acoustic deformation potential scattering and ionized impurity scattering. For the dashed curve, isotropic neutral impurity scattering has also been taken into account. Data points: from magnetoresistance measurements: squares [7.21]; diamonds [7.22]; crosses [7.23])

within the accuracy of the experiment with the calculated value of 11.2 given above.

7.5 Equivalent Intervalley Scattering and Valley Repopulation Effects

In Sects. 7.2–7.4 we assumed an equal distribution of carriers in all valleys which are at the same energy and are therefore denoted as *equivalent valleys*. However, any field, pressure or temperature gradient, etc., which introduces a preferential direction in an otherwise isotropic solid may lead to a *repopulation* of valleys since the equilibrium of the valley population is dynamic due to *intervalley scattering*. Scattering *within* a valley is denoted as *intravalley scattering* [7.2]. Before considering repopulation effects, let us investigate the intervalley scattering rate.

In a transition of an electron from one valley in k-space to another one, a large change in momentum, which is of the order of magnitude of the first Brillouin zone radius, is involved. This momentum may be taken up either by an impurity atom or by a phonon near the edge of the phonon Brillouin zone where the acoustic and optical branches are either degenerate or not too far from each other (Fig. 6.10). We will not consider here the case of impurity scattering which is limited to low temperatures and to very impure semiconductors. The phonon case is treated in a way similar to optical-phonon scattering except that the optical-phonon energy $\hbar\omega_0 = k_B\Theta$ is replaced by the somewhat lower *intervalley phonon* energy $\hbar\omega_i = k_B\Theta_i$ which has a value between the optical and the acoustic-phonon energy near the edge of the phonon Brillouin zone ($\Theta_i = 315$ K, $\Theta = 430$ K for Ge). Scattering selection rules have been considered by various authors [7.24]. In n-type Si, two types of processes have been considered: *g scattering* occurs between a given valley and the valley on the opposite side of the same axis, e.g., between a $\langle 100 \rangle$ and a $\langle \bar{1}00 \rangle$ valley, while in the *f process*, a carrier is scattered to one of the remaining equivalent valleys, e.g., between a $\langle 100 \rangle$ and a $\langle 010 \rangle$ valley; both f and g scattering involve a reciprocal lattice vector (*Umklapp process*, Sect. 6.4). g-type phonons have temperatures of about 190 and 720 K; phonons involved in f-scattering have a temperature of 680 K and take part in the repopulation of valleys which will be discussed below [7.25–7.27]. The inverse momentum relaxation time is given by [7.2]

$$1/\tau_i = w_2[\sqrt{(\varepsilon/k_B\Theta_i) + 1} \tag{7.5.1}$$

$$+ \exp(\Theta_i/T)\mathrm{Re}\,\sqrt{(\varepsilon/k_B\Theta_i) - 1}]/[\exp(\Theta_i/T) - 1] \ ,$$

where w_2 is a rate constant. This equation corresponds to (6.11.5) for optical deformation potential scattering. If we average $1/\tau_i$ over a Maxwell–Boltzmann distribution function $\propto \exp(-\lambda\varepsilon/k_BT)$, where $\lambda = T/T_e$, we obtain

$$\langle 1/\tau_i \rangle = w_2\sqrt{2\lambda z/\pi}\cosh[(1 - \lambda)z]K_1(\lambda z)/\sinh(z) \ , \tag{7.5.2}$$

where K_1 is a modified Bessel function of the second kind and $z = \Theta_i/2T$. For thermal carriers $\lambda = 1$. Taking into account acoustic deformation potential scattering in addition to intervalley scattering, the total inverse momentum relaxation time is given by

$$1/\tau_m = 1/\tau_{ac} + 1/\tau_i \ , \tag{7.5.3}$$

where τ_{ac} stands for τ_m given by (6.4.15). The mobility as a function of temperature T has been calculated by *Herring* [7.2]. The ratio μ/μ_0 $vs\,T/\Theta_i$, where μ_0 stands for

$$\mu_0 = \mu_{ac}(T/\Theta_i)^{3/2} \tag{7.5.4}$$

and $k_B\Theta_i = \hbar\omega_i$, is shown in Fig. 7.4 for various values of the parameter w_2/w_1. The rate constant $w_1 = l_{ac}^{-1}(2k_B\Theta_i/m)^{1/2}\Theta_i/T$, where l_{ac} is the acoustic mean free path given by (6.4.16). If an intervalley deformation potential constant D_i is introduced in analogy to the corresponding optical constant, (6.11.2), the ratio $w_2/w_1 = \frac{1}{2}(D_iu_i/\varepsilon_{ac}\omega_i)^2$. It has been shown [7.25–27] that in n-Si, f-type phonon scattering is weak and the observed temperature dependence of the mobility $\mu \propto T^{-2.5}$ is explained by a dependence similar to that shown in Fig. 7.4 with $\Theta_i = 720$ K and $w_2/w_1 = 3$. For n-Ge, w_2 has been determined experimentally from the acousto-electric effect (Sect. 7.8).

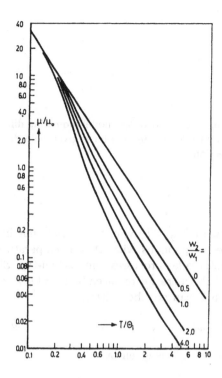

Fig. 7.4: Influence of intervalley scattering on the temperature dependence of the mobility for various values of the coupling constants (after [7.2])

We will now investigate the repopulation of equivalent valleys in a strong electric field. Figure 7.5 shows a two-valley model in k-space with $\boldsymbol{E}$ at an oblique angle relative to the y-axis in *coordinate* space. In both valleys the electron effective mass is given by the direction of $\boldsymbol{E}$. Since the effective masses are proportional to the square of the length of the arrows in Fig. 7.5, they are different for the two valleys except when $\boldsymbol{E}$ is at an angle of 45° relative to both valleys. Because of the different effective masses, electrons are heated by the electrical field intensity $\boldsymbol{E}$ at a different rate. Therefore, in the valley, whose longitudinal axis is at a larger angle relative to $\boldsymbol{E}$, the electron temperature T_e is higher than in the other valley. The transfer rate of carriers from a *hot valley* to a *cool valley* is larger than in the reverse direction according to (7.5.2). Therefore, the equilibrium population of a hot valley is smaller than that of a cool valley. The rate equation for an arbitrary number of valleys and a nondegenerate carrier gas is given by

$$-dn_\varrho/dt = n_\varrho \sum_\sigma{}' \langle 1/\tau_{i,\varrho\to\sigma}\rangle - \sum_\sigma{}' n_\sigma \langle 1/\tau_{i,\sigma\to\varrho}\rangle \;, \qquad (7.5.5)$$

where the first term on the rhs represents the rate at which carriers are scattered *out* of valley no. ϱ, the second term the scattering rate *into* this valley, and the prime at Σ indicates that the sum is taken over all valleys except no. ϱ. Usually, the simplifying assumption is made that the inverse time constant

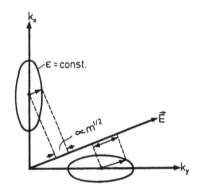

Fig. 7.5: Two-valley model indicating different effective masses for a given field direction

$\langle 1/\tau_{i,\varrho\to\sigma}\rangle$ depends only on the temperature $T_{e\varrho} = T/\lambda_\varrho$ of the carrier emitting valley no.ϱ. Of course, if several types of phonons with Debye temperature $\Theta_{i1}, \Theta_{i2}, \ldots, \Theta_{is}, \ldots$ are involved in the intervalley processes, such as in n-type Si, a sum over all s appears on the rhs of (7.5.2) where z is replaced by $z_s = \Theta_{is}/2T$.

For a total number of valleys N, (7.5.5) can now be written

$$-dn_\varrho/dt = (n_\varrho - n_\varrho^{(0)})(N - 1)\langle 1/\tau_{i\varrho}\rangle \ , \tag{7.5.6}$$

where the equilibrium carrier concentration $n_\varrho^{(0)}$ is given by

$$n_\varrho^{(0)} = {\sum_\sigma}' \frac{n_\sigma\langle 1/\tau_{i\sigma}\rangle}{(N - 1)\langle 1/\tau_{i\varrho}\rangle} \ . \tag{7.5.7}$$

Assuming for simplicity that the field direction is the longitudinal axis of a *cool* valley (subscript c) while all other $N-1$ valleys are equally *hot* (subscript h), we find for (7.5.7)

$$n_c^{(0)}\langle 1/\tau_{ic}\rangle = n_h^{(0)}\langle 1/\tau_{ih}\rangle = n/[\langle 1/\tau_{ic}\rangle^{-1} + (N - 1)\langle 1/\tau_{ih}\rangle^{-1}] \ , \tag{7.5.8}$$

where $n = n_c + (N - 1)n_h$ is the total carrier concentration assumed to be field independent. These rate equations are useful in calculations of not only strong-dc-field effects but also of ac effects such as the acousto-electric effect (Sect. 7.8).

Intervalley scattering also contributes to the energy loss which is treated similarly to optical-phonon scattering except that $\hbar\omega_1$ is taken instead of $\hbar\omega_0$ for the carrier energy change per collision and the deformation potential constant D in nonpolar semiconductors is replaced by a quantity D_{ij} [Ref. 7.28, p. 154]; for the effective mass m, we take the product $m_t^{2/3}m_l^{1/3}$, valid for the valley into which the carrier is scattered. For intravalley scattering (both acoustic and optical), m is also replaced by $m_t^{2/3}m_l^{1/3}$ and, in addition, for acoustic scattering, ε_{ac}^2 by

$$\frac{3}{4}\Xi_d^2[1.31 + 1.61\,\Xi_u/\Xi_d + 1.01\,(\Xi_u/\Xi_d)^2] \tag{7.5.9}$$

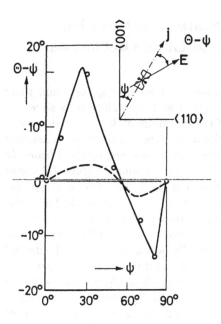

Fig. 7.6: Theoretical $(-\,-\,-)$ and experimental data ($\circ$) of Sasaki angle vs the angle between the current and the (001) direction valid for n-type Ge at 90 K (after [7.29])

in τ_m and by

$$\frac{2}{3}\,\Xi_d^2[1+0.5(m_l/m_t)(1+\Xi_u/\Xi_d)^2] \tag{7.5.10}$$

in $\langle -d\varepsilon/dt\rangle$ (see [7.28] p. 115 and p. 124 for acoustic and p. 152 and p. 155 for optical scattering).

In strong electric fields, even for cubic semiconductors, the current direction may be different from the field direction. We have longitudinal and transverse components of the current density j_l and j_t, respectively. The ratio j_t/j_l yields the *Sasaki angle* $\Theta-\psi$ due to the relation $\tan(\Theta-\psi)=j_t/j_l$. The experimental arrangement used for the measurement of this ratio is similar to the Hall arrangement (Fig. 4.2) except that there is no magnetic field present. Voltages rather than current densities are usually measured and pulse techniques are used to minimize Joule heating effects. The ratio of field intensities $|\,E_t/E_l\,|=|\,j_t/j_l\,|$ is calculated from these data.

Experimental data for n-type Ge at 90 K and 750 V/cm are shown in Fig. 7.6 [7.29]. The effect first published by *Sasaki* et al. [7.29, 30] is known as the *Sasaki–Shibuya effect*. The dashed curve indicates the effect calculated by *Shibuya* for the case of no repopulation. It has been suggested by the authors that the discrepancy is due to repopulation.

7.6 Nonequivalent Intervalley Scattering, Negative Differential Conductivity, and Gunn Oscillations

Besides the usual $\langle 111 \rangle$-minima of the conduction band of germanium, there are upper minima at $k=0$ and along the $\langle 100 \rangle$ and equivalent axes shown in Fig. 2.26. Due to the strong curvature of the minimum at $k = 0$, its density of states is very small and therefore the density of carriers which can be transferred to this valley is negligible. However, carriers are transferred to the *silicon-like* $\langle 100 \rangle$ minima which are 0.18 eV above the $\langle 111 \rangle$ minima. Figure 7.7 shows experimental and theoretical results of the Hall coefficient R_H of n-type Ge at 200 K with j parallel to a $\langle 100 \rangle$ direction which is symmetric to all the normally occupied $\langle 111 \rangle$ valleys [7.31]. R_H increases strongly above 1 kV/cm. This is considered to be due to carrier transfer to the silicon-like valleys where the mobility is much smaller than in the normally occupied valleys. The dashed curves have been calculated for various mobility ratios b. If the carriers in the silicon-like valleys were immobile, the Hall coefficient $R_H \propto 1/n_{\langle 111 \rangle}$ would increase as the number of carriers $n_{\langle 111 \rangle}$ in the normally occupied $\langle 111 \rangle$ valleys decreases. Further observations on n-type Ge will be discussed at the end of this section.

Let us now consider the simpler and more spectacular case of n-type GaAs where the central valley at $k =0$ is lowest in the conduction band and the germanium-like $\langle 111 \rangle$ *satellite* valleys are $\Delta\varepsilon_{\Gamma L} = 0.29$ eV above the former [3] as shown in Fig. 7.8. The effective masses are 0.07 m_0 in the central valley and most likely 0.55 m_0 in each one of the four satellite valleys. Since the effective density of states in *one* valley is proportional to $m^{3/2}$ according to (3.1.10) from now on we will consider only one satellite valley instead of four, with an effective mass of $(0.55^{3/2} \times 4)^{2/3} m_0 = 1.38 \, m_0$.

Because of the small effective mass in the central valley, electrons are heated very effectively by an electric field and gain an energy of magnitude $\Delta\varepsilon_{\Gamma L} = 0.29$ eV at $E \approx 3$ kV/cm. Before reaching this energy, the mobility is not changed considerably due to polar optical mode scattering (Sect. 6.12). At $E \approx 3$ kV/cm, electrons are transferred to the satellite valleys where they are heavy and have a reduced momentum relaxation time because of scattering between these valleys and because of the higher density of states in these valleys. Therefore, the mobility in the satellite valleys is estimated to be only 920 cm^2/V s while the mobility in the central Γ-valley is between 6000 and 7350 cm^2/V s depending on purity; all data are valid for room temperature. As in the case of equivalent-intervalley scattering, the momentum change at carrier transfer is supplied by an intervalley phonon which is either absorbed or emitted (Sect. 7.5 and Fig. 7.8).

These considerations lead to two possible types of drift-velocity-vs-field characteristics plotted in Fig. 7.9. At low field intensities, the slope is the high

---

[3] Until 1976 it was believed that the $\langle 100 \rangle$ valleys were lower than the $\langle 111 \rangle$ valleys. D.E. Aspnes [7.32] first showed that this was not true.

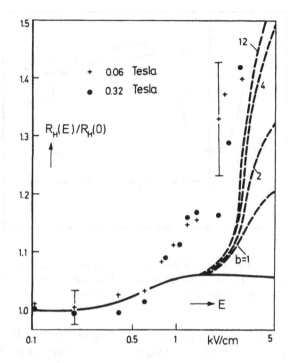

Fig. 7.7: Hall coefficient normalized to unity at zero electric field E = 0 for n-type Ge at a lattice temperature of 200 K and j $\|\langle 111\rangle$ valleys. The dashed curves are valid for electron transfer to $\langle 100\rangle$ valleys and nonequivalent intervalley scattering. The numbers at the dashed curves are the mobility ratios for both types of valleys assumed for the calculation (after [7.31])

mobility μ_1 in the central valley while at high field intensities, the slope is the low mobility μ_2 in the satellite valleys assuming all carriers have been transferred. Curve 2 shows a range of negative differential conductivity (ndc) while curve 1 has a positive slope everywhere. The type of curve observed depends on the ratio of the energy interval $\Delta\varepsilon_{\Gamma L}$ to the thermal energy $k_B T$ which is 0.026 eV at room temperature. Since in GaAs this ratio is very high, a type-2 characteristic is realized in pure GaAs at room temperature. With increased density of impurities, first the threshold field E_t increases and finally the ndc range disappears altogether.

Also in n-type GaSb, the central valley (Γ) is lowest and the next-highest valleys are on the $\langle 111\rangle$ axes (L), $\Delta\varepsilon_{\Gamma L} = 0.075$ eV above the central valley. The energy is only three times the thermal energy at room temperature. Consequently, no ndc is observed. In InSb, the energy $\Delta\varepsilon_{\Gamma L}$ is 0.5 eV but the band gap ε_G is only 0.2 eV and except for very short times $\approx 10^{-9}$s, there is impact ionization across the band gap rather than carrier transfer [7.33].

The idea of an ndc by electron transfer in GaAs and $GaAs_x P_{1-x}$ was first suggested by *Ridley* and *Watkins* [7.34], and *Hilsum* [7.35]. The generation of

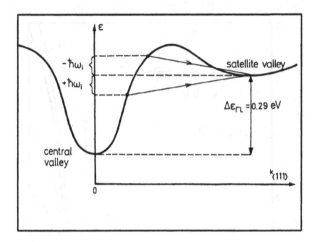

Fig. 7.8: Central and satellite valleys in the conduction band of gallium arsenide [shown schematically]. The full straight lines indicate electron transitions between the valleys with emission or absorption of an intervalley phonon

current oscillations at frequencies up to 10^3 GHz and the application to amplification purposes in this frequency range have been considered by the authors. *Gunn* [7.36] observed oscillations in n-type GaAs and InP at frequencies between 0.47 and 6.5 GHz by applying field intensities E of more than three kV/cm. Further experiments showed that the oscillations were indeed those which had been predicted [7.37].

For a calculation of the ndc, let us follow the simplified treatment given by *Hilsum* [7.35]. He assumes equal electron temperatures in the central valley (subscript 1) and the satellite valley (subscript 2). The ratio of electron densities $\eta = n_1/n_2$ is given by the ratio of the effective densities of states N_{c1}/N_{c2} and the difference in valley energies $\Delta\varepsilon_{\Gamma L}$ by

$$\eta = (N_{c1}/N_{c2})\,\exp(\Delta\varepsilon_{\Gamma L}/k_B T_e) \ . \tag{7.6.1}$$

In later publications [7.28] (p. 254), calculations were performed without making this simplifying assumption resulting in a better reproduction of the experimental results. Equation (7.6.1) yields to a good approximation $n_1 = n\eta/(1+\eta)$ and $n_2 = n/(1+\eta)$ and a conductivity given by

$$\sigma = |e|(n_1\,\mu_1 + n_2\,\mu_2) = |e|\,n\,\mu_2(\eta b + 1)/(\eta + 1) \ , \tag{7.6.2}$$

where $b = \mu_1/\mu_2$ is the mobility ratio. For a voltage V applied to a filamentary sample of length L and area A, the current $I = \sigma AV/L$ vs V has a slope of

$$dI/dV = (A/L)(\sigma + V d\sigma/dV) = (A/L)\sigma[1 + (E/\sigma)d\sigma/dE]$$

$$= \frac{A}{L}n\,|e|\,\mu_2\frac{\eta b + 1}{\eta + 1}\left[1 - \frac{\eta(b-1)}{(\eta+1)(\eta b + 1)}\frac{\Delta\varepsilon_{\Gamma L}E}{k_B T_e^2}\frac{dT_e}{dE}\right] \ . \tag{7.6.3}$$

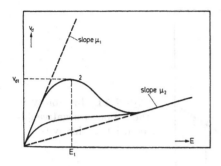

Fig. 7.9: Drift velocity vs field characteristics for transitions from the central valley (mobility μ_1) to the satellite valley (mobility μ_2)

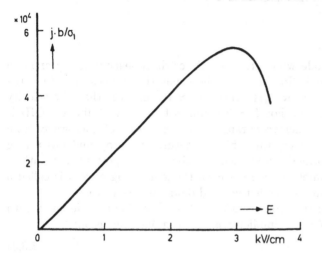

Fig. 7.10: Current density vs field characteristics calculated for n-type gallium arsenide at 373 K assuming polar optical scattering (after [7.35])

The sign of dI/dV depends strongly on dT_e/dE which has been calculated from (6.12.26) and the energy balance (4.13.6). The resulting current-field characteristic is shown in Fig. 7.10. The threshold field strength of three kV/cm calculated from a shifted Maxwell–Boltzmann distribution is not changed appreciably by taking a distribution obtained as a solution of the Boltzmann equation.

We will now consider the generation of oscillations in the range of ndc. From Fig. 7.11 it can be shown that an operating point within the ndc range is not stable [7.34, 38]. At an average field intensity E, two regions of length l_1 and l_2 with field intensities E_1 and E_2 are formed such that the total voltage $V = E_1 l_1 + E_2 l_2$. The high-field *domain* is not fixed locally but drifts with the electrons through the crystal. Inside the domain most of the electrons belong to the satellite valley and the drift velocity of these heavy electrons is the same as that of the light electrons outside the domain. The domain with the field strength E_2 indicated by Fig. 7.11 has a flat top. Such a flat-topped domain has not been observed in GaAs, however [7.38]. Figure 7.12 shows the actual field distribution in the sample taken after successive time intervals. The domain is nucleated at some inhomogeneity of doping close to the cathode and

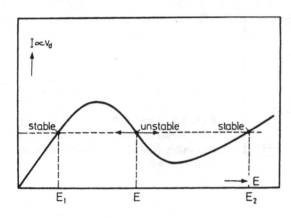

Fig. 7.11: Instability in the region of negative differential conductivity

travels towards the anode while growing. After disappearing in the anode a new domain is generated. In this case, the generation repetition rate is the oscillation frequency given by v_d/L where $v_\mathrm{d} \approx 10^7$ cm/s is the drift velocity and L is the sample length. For $L = 0.1$ mm, a frequency of about 1 GHz is obtained which is in the microwave range. The generation of microwaves from a dc current in a piece of crystal, which is naturally sturdy and has a long lifetime compared to formerly used vacuum tubes, has important applications in RADAR and telecommunication systems; the generating crystal is called a *Gunn diode* (a competing device is the Read diode, see Sect. 10.2).

For a calculation of the growth rate of a high-field domain, we begin with Poisson's equation which, for simplicity, is taken in one dimension:

$$\partial E/\partial x = |e|(n - N_\mathrm{D})/\kappa \kappa_0 \ . \tag{7.6.4}$$

Both the concentration of donors N_D and the dielectric constant κ are assumed as being constant while the carrier concentration n varies with the time t according to

$$n = n_0 + \Delta n \, \exp(\mathrm{i}\omega t) \tag{7.6.5}$$

and similarly, the current density

$$j = j_0 + \Delta j \, \exp(\mathrm{i}\omega t) \tag{7.6.6}$$

and the field strength

$$E = E_0 + \Delta E \, \exp(\mathrm{i}\omega t) \tag{7.6.7}$$

in a *small-signal theory* ($\Delta n \ll n_0; \Delta j \ll j_0; \Delta E \ll E_0$). For the current density we consider the drift term, the diffusion term and the displacement term.

$$j = n|e|v_\mathrm{d} - |e| \, D \, \partial n/\partial x + \kappa \kappa_0 \, \partial E/\partial t \ . \tag{7.6.8}$$

The diffusion constant D is taken as field independent. The small-signal treatment yields

$$\Delta j = n_0|e| \left(\frac{\partial v_\mathrm{d}}{\partial E} \right) \Delta E + |e|v_\mathrm{d}\Delta n - |e| \, D \, \frac{\partial \Delta n}{\partial x} + \mathrm{i}\omega\kappa\kappa_0 \, \Delta E \ . \tag{7.6.9}$$

For Δn and its derivative, we find from (7.6.4)

$$\partial \Delta E/\partial x = (|e|/\kappa\,\kappa_0)\,\Delta n; \qquad \partial^2 \Delta E/\partial x^2 = (|e|/\kappa\,\kappa_0)\,\partial\Delta n/\partial x \ . \quad (7.6.10)$$

The left-hand sides become $-iq\,\Delta E$ and $-q^2\Delta E$, respectively, for a wave propagating as $\Delta E \propto \exp(-iqx)$ with a wave vector q. This yields for (7.6.9)

$$\Delta j = \kappa\,\kappa_0(\tau_d^{-1} + i\omega - iq\,v_d + q^2 D)\,\Delta E \ , \quad (7.6.11)$$

where τ_d is a *differential dielectric relaxation time* given by

$$\tau_d^{-1} = (n_0|e|/\kappa\,\kappa_0)\,\partial v_d/\partial E \ . \quad (7.6.12)$$

Under constant-current conditions $\Delta j = 0$, the electric waves propagating in the crystal have an angular frequency of

$$\omega = q\,v_d + i(\tau_d^{-1} + q^2 D) \ . \quad (7.6.13)$$

The imaginary part of ω represents a damping of the wave:

$$\exp(i\omega t) = \exp(iqv_d t)\,\exp[-(\tau_d^{-1} + q^2 D)\,t] \ . \quad (7.6.14)$$

The real part yields a phase velocity $\mathrm{Re}(\omega)/q = v_d$, the electron drift velocity. Since τ_d^{-1} is proportional to $\partial v_d/\partial E$, it is negative in the range of ndc. If the slope in this range is steep enough so that $-\tau_d^{-1} > q^2 D$, (7.6.14) results in an exponential growth of the amplitude which finally saturates, as shown in Fig. 7.12. The pulse shape of the domain can be investigated by a Fourier analysis where in (7.6.7), a corresponding term has to be taken for each harmonic. Diffusion counteracts the domain formation by its tendency to make the local distribution of electrons uniform.

If the diffusion can be neglected and τ_d and v_d are approximated by their values averaged over the whole sample, we find ΔE by solving the differential equation obtained from (7.6.9, 10):

$$\Delta E = \Delta E(x) = [\Delta j L/(\kappa\,\kappa_0\,v_d\,s)][1 - \exp(-s\,x/L)] \ , \quad (7.6.15)$$

where the dimensionless complex quantity

$$s = (L/v_d)(\tau_d^{-1} + i\omega) \quad (7.6.16)$$

has been introduced. The boundary condition $\Delta E(0) = 0$ is satisfied. The sample impedance $Z = Z(\omega)$ is given by

$$Z(\omega) = \frac{1}{A\,\Delta j}\int_0^L \Delta E dx = \frac{L[L + (L/s)(e^{-s} - 1)]}{A\,\kappa\,\kappa_0\,v_d\,s}$$

$$= \frac{L^2}{A\,\kappa\,\kappa_0\,v_d}\frac{e^{-s} + s - 1}{s^2} \ . \quad (7.6.17)$$

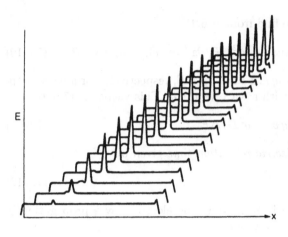

Fig. 7.12: A high-field domain nucleates at an inhomogeneity of doping and moves to the anode of a "long" sample as it grows (after [7.39])

The impedance vanishes at a maximum of $\text{Im}(\omega)$ for $s = -1.77 \pm i\,1.08$, i.e., there is a current instability at a constant applied voltage. Inserting this value into (7.6.16) yields a critical sample length

$$L_{\text{crit}} = 1.77\,v_{\text{d}}(-\tau_{\text{d}}) = \frac{1.77\,\kappa\,\kappa_0\,v_{\text{d}}}{n_0|e|(-\partial v_{\text{d}}/\partial E)} \ . \tag{7.6.18}$$

For domain formation this is the minimum sample length. For n-type GaAs, the product $n_0 L_{\text{crit}}$ is determined by the maximum value of $-\partial v_{\text{d}}/\partial E$ which is 500 cm^2/ Vs, by the average drift velocity of v_{d} of $\approx 10^7$ cm/s, and by the dielectric constant $\kappa = 13.5$. These data give

$$n_0 L_{\text{crit}} = 3 \times 10^{11}\,\text{cm}^{-2} \ . \tag{7.6.19}$$

Assuming, e.g., $L = 0.3$ mm, this yields a miminum carrier concentration of $n_0 = 10^{13}$ cm^{-3}. On the other hand, if the sample is too long the field at the cathode during domain formation is not sufficiently reduced to prevent the launching of a second domain before the first one has reached the anode.

Figure 7.13 shows the shape of a domain moving from left to right. At the leading edge there is a small deficiency of electrons while at the trailing edge there is an excess of electrons: the light Γ-valley electrons are caught by the leading edge where due to the strong electric field, they are converted into heavy X-valley electrons. Because of their low mobility and in spite of the strong electric field, they accumulate at the trailing edge of the pulse. If they should escape from the domain they are immediately reconverted into highly mobile Γ-electrons and catch up with the domain. In this way, the electron accumulation at the trailing edge and the slope $\partial\Delta E/\partial x \propto \Delta n$, (7.6.10), can be understood. A numerical calculation which includes the diffusion of carriers has been published by *Butcher* and *Fawcett* [7.41]. The calculated domain shape has in fact been observed [7.42]; the comparison [7.43] yields an intervalley relaxation time τ_{i} of about 10^{-12}s which is short compared with the drift time of 10^{-9}s.

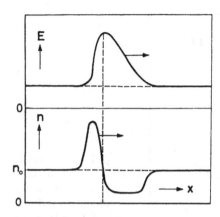

Fig. 7.13: Schematic diagram of the electric field and the electron density vs position for a high-field domain moving towards the right side (after [7.40])

The explanation of the Gunn oscillations by nonequivalent intervalley electron transfer has been supported by two experiments where the energy difference of the valleys was decreased either by hydrostatic pressure [7.44] on GaAs or by varying the composition x of a GaAs$_x$P$_{1-x}$ compound [7.45]. These effects are well known from measurements of the optical absorption edge and will be discussed in Sect. 11.2 where, e.g., Fig. 11.10 shows the shift of the absorption edge in GaAs with hydrostatic pressure. The central valley moves up and the satellite valleys move down with pressure until at 6 GPa they are at the same level of 0.49 eV above the central-valley position at zero pressure. The threshold field for Gunn oscillations decreases with pressure until a value of 1.4 kV/cm at 2.6 GPa is reached. Beyond this pressure there are no more oscillations. With a decreasing energy separation $\Delta\varepsilon_{\Gamma L}$ of the valleys, a lower electron temperature is sufficient for a carrier transfer to the satellite valleys. However, at too low values of $\Delta\varepsilon_{\Gamma L}$ relative to $k_B T$, the ndc vanishes (curve 1 in Fig. 7.9) and the oscillations stop.

Figure 7.14 shows the threshold field as a function of uniaxial pressure in two crystallographic directions [7.46]. The strongest decrease of the threshold field with uniaxial pressure is in the ⟨111⟩ direction, in agreement with the location of the satellite valleys on ⟨111⟩ and equivalent axes: the valley in the direction of the pressure moves down and determines the threshold field, while the other three move up slightly [7.47].

If the Gunn diode is mounted in a resonant circuit, its frequency may be shifted by a factor of two without loss of efficiency. Diodes with nL products of more than a few times 10^{12} cm^{-2}, see (7.6.19), show their largest output at a frequency which is somewhat larger than half the inverse transit time of the domain through the diode. If the nL product is less than 10^{12} cm^{-2}, the diode cannot oscillate but can be used for amplification purposes at a frequency of the inverse transit time [7.48]. In the *limited space charge accumulation oscillator* (LSA), the dc field $\boldsymbol{E}_0$ is supplemented by an ac field $\boldsymbol{E}_1 \cos(2\pi\nu t)$ parallel to $\boldsymbol{E}_0$ of so large an amplitude that for a small part of the period, the differential mobility dv_d/dE is positive [7.49] as shown in Fig. 7.15. At a sufficiently high

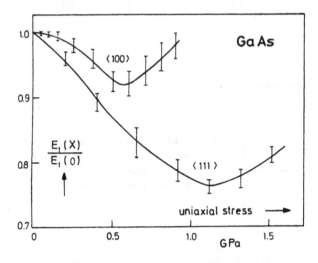

Fig. **7.14**: Normalized threshold field vs uniaxial stress in $\langle 100 \rangle$ and $\langle 111 \rangle$ directions for GaAs (after [7.46])

frequency ν, almost no space charge is generated: $n_0/\nu < 2 \times 10^5 \ \mathrm{cm^{-3} \, Hz^{-1}}$. Due to the decrease of current with increasing voltage it is possible that the power P_{tot}, which an electron gains from the field $E_0 + E_1 \cos(2\pi \nu t)$ during a period $1/\nu$, is smaller than the power P_0 which it would gain from the dc field E_0 alone:

$$P_{\mathrm{tot}} = e\nu \int_0^{1/\nu} E \, v \, dt < P_0 = e\nu E_0 \int_0^{1/\nu} v \, dt \ , \tag{7.6.20}$$

i.e., the yield $(P_0 - P_{\mathrm{tot}})/P_0$ is positive, Maximum yields of 11% at a zero-field mobility μ_0 of 5000 $\mathrm{cm^2/Vs}$ and of 23% at $\mu_0 = 9000 \ \mathrm{cm^2/V \ s}$ have been calculated [7.51].

In another method of suppressing the free formation of Gunn domains, the Gunn diode is coated with an insulating high-dielectric-constant material such as ferroelectric $\mathrm{BaTiO_3}$ [7.52]. In a simplified picture according to (7.6.12), the negative value of τ_{d}^{-1} can be reduced to a magnitude where amplification is possible but free oscillation is not.

After the discovery of the Gunn effect in n-type GaAs and InP, a few more semiconductors with a similar conduction band structure were shown to exhibit this type of oscillation: CdTe [7.53], InSb [7.33], $\mathrm{Ga}_x\mathrm{In}_{1-x}\mathrm{Sb}$ [7.54] and InAs [7.55]. It has been quite surprising to also find oscillations in n-type Ge [7.56], though much weaker, at temperatures below 130 K because there is nearly no difference in mobility between the lowest (L-type) valleys ($3.9 \times 10^4 \ \mathrm{cm^2/V \ s}$ at 77 K) and the upper (X-type) valleys ($5 \times 10^4 \ \mathrm{cm^2/ V \ s}$, observed in n-type Si at 77 K). Nontheless a calculation by *Paige* [7.57] revealed a weak ndc above 2–5 kV/cm at temperatures between 27 K and 77 K where a value of 1×10^8 eV/cm is assumed for the acoustic deformation potential for the transition L $\rightarrow$ X with a phonon temperature $\Theta_{\mathrm{i}} = 320$ K. This accounts for the weak oscillations observed for current directions of e.g., $\langle 100 \rangle$ or $\langle 110 \rangle$, and the fact that there are no oscillations for a $\langle 111 \rangle$ direction; in the latter case there is a repopulation between L-valleys resulting in a lower mobility such that a later transfer to the X-valleys no longer results in an ndc range. For a current in a $\langle 100 \rangle$ direction, the population of the L-type valleys is not changed but the transfer to the $\langle 100 \rangle$ valleys with their low mobility is preferred relative to the $\langle 010 \rangle$ and $\langle 001 \rangle$ valleys, while for a current in a $\langle 110 \rangle$ direction, the populations are not too different from those for a $\langle 100 \rangle$ direction.

After the observation of a Gunn effect in n-type Ge, the question arose as to why a constant-current range instead of an ndc is observed if the Gunn frequency is higher than the maximum frequency of the apparatus, as has been the case in earlier measurements [7.56]. It has been supposed that the field distribution in the sample, although static, may be nonuniform with the formation of a flat-topped domain [7.38]. From Poisson's equation (7.6.4) with $N_D = n_0$ and an average drift velocity $\langle v_d \rangle$ given by

$$n_0 |e| \langle v_d \rangle = n(x) |e| v_d(E) \ , \tag{7.6.21}$$

an explicit relation for $\partial E / \partial x$ is obtained:

$$\partial E / \partial x = (|e| n_0 / \kappa \kappa_0)[\langle v_d \rangle / v_d(E) - 1] \ . \tag{7.6.22}$$

The $v_d(E)$-curve is shown in Fig. 7.16a; $E(x)$ corresponding to two values of $\langle v_d \rangle$ is shown in Fig. 7.16b. For a value of $\langle v_d \rangle$ below the threshold, E is uniform except for a region near the cathode where the material is heavily doped due to the formation of the metallic contact. For a value of $\langle v_d \rangle$ above the threshold, $\max[v_d(E)]$, (7.6.22) yields $\partial E / \partial x > 0$ for all values of x, i.e., throughout the sample, with an inflection point at the threshold field strength E_t. If the product of carrier density and sample length satisfies the condition

$$n_0 L > \kappa \kappa_0 E_t / |e| \ , \tag{7.6.23}$$

the current density j rises only slightly with increasing total voltage across the sample, as shown in Fig. 7.17, once $j > n_0 |e| \max[v_d(E)]$ which is then denoted as the *saturation value* of j. Since small nonuniformities of doping in the sample produce gross nonuniformities of the fields, it is not possible to deduce the sign and magnitude of the differential conductivity near the saturation of j.

Let us finally consider Gunn oscillations in n-type Ge under uniaxial stress in a $\langle 111 \rangle$ direction for a current flowing in the $\langle 11\bar{2} \rangle$ direction which is perpendicular to the stress. The L-type valley whose longitudinal axes is in the stress direction is moved down while the three others are moved up. As shown in Fig. 7.18, the effective masses in the current direction are: m_t in the low valley, 1.3 m_t in two of the upper valleys and 6.4 m_t in the remaining upper valley where $m_t = 0.08 m_0$ is the transverse mass. The threshold field as a function of stress is shown [7.58] in Fig. 7.19 for electron concentrations between 10^{14} and 1.6 $\times 10^{16}$ cm^{-3}. The minimum in this dependence can be explained by assuming that at large stresses and correspondingly large energy differences, high electron temperatures are necessary for electron transfer; at these electron temperatures

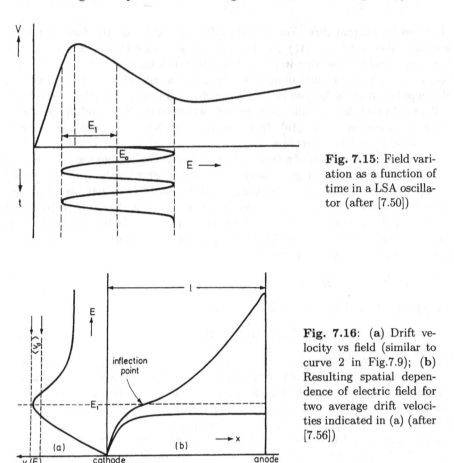

Fig. 7.15: Field variation as a function of time in a LSA oscillator (after [7.50])

Fig. 7.16: (a) Drift velocity vs field (similar to curve 2 in Fig.7.9); (b) Resulting spatial dependence of electric field for two average drift velocities indicated in (a) (after [7.56])

the drift velocity in each valley is kept almost constant due to the emission of energetic phonons [7.59].

In similar experiments with n-type Si, the drift velocity was observed to saturate with $E > 15$ kV/cm at stresses larger than 0.5 GPa but no oscillations have been observed. A Gunn effect bibliography has been compiled in [7.38, 60].

Besides the Gunn effect there are many other effects which show the *voltage controlled* N-shaped characteristic shown in Fig. 7.9. Since these effects produce either amplitudes or oscillation frequencies that are much lower than in the Gunn effect, there are at present no device applications. In some homogeneously doped semiconductors, a *current controlled* S-shaped characteristic is observed which is shown schematically in Fig. 7.20; it leads to a formation of current filaments j_1 and j_2. This behavior has been found, e.g., during impact ionization in compensated n-type Ge (see Chap. 10). For details on effects of this kind see, e.g., [Ref. 7.28, pp. 84–86] and [7.61].

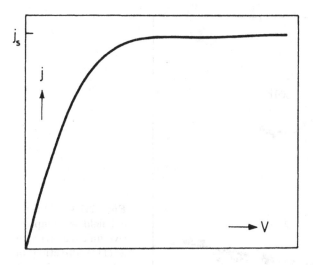

Fig. 7.17: Current voltage characteristic resulting from situation shown in Fig. 7.16, for the case of a "long" sample

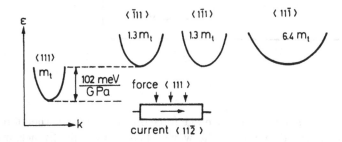

Fig. 7.18: Valleys at the conduction band edge of germanium shifted by uniaxial stress in the ⟨111⟩ direction; effective mass values given in the current direction (after [7.58])

7.7 The Acousto-Electric Effect

The appearance of a dc electric field along the direction of propagation of an acoustic wave in a semiconductor is known as the *acousto-electric effect*. It is due to the drag of carriers by the wave which is similar to the motion of driftwood towards a beach and to charge transport in a linear accelerator. Figure 7.21 shows the sinusoidal variation of the potential energy $\Delta\varepsilon_c$ in a crystal due to an acoustic wave at a certain time. Mobile carriers (represented by open circles) tend to bunch in the potential troughs as indicated by arrows along the curve. However, since the wave moves towards the right with a velocity u_s and a finite time τ_R would be required to reach equilibrium, the carriers can never arrive at the troughs for the case of $\omega\tau_R \approx 1$ where ω is the angular frequency. Therefore, on the front slopes of the wave a higher carrier concentration is found

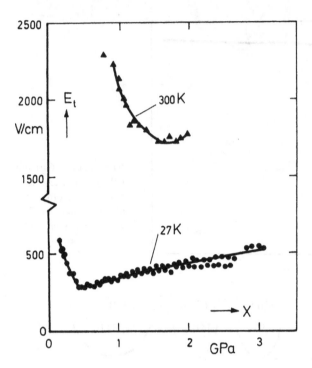

Fig. 7.19: Threshold field vs compressive uniaxial stress in a ⟨111⟩ direction in n-type germanium at 27 K and 300 K (after [7.58])

than on the back slopes. A net average force on the carriers is therefore exerted in the propagation direction of the wave. The carriers finally accumulate on that sample face at which the acoustic wave emerges from the sample. This causes a dc field in the sample. If a dc current is allowed to flow, there is a net energy transfer from the acoustic wave to the electron gas and the wave is attenuated.

On the other hand, an acoustic wave is amplified by a dc current transmitted through the sample if the drift velocity of the carriers is slightly larger than the wave velocity. The carriers will then bunch on the back slopes. It has thus been possible to manufacture acousto-electric amplifiers and also oscillators [7.62]. It may be interesting to note that the *phonon drag* mentioned in Sect. 4.9 is different from the acousto-electric effect; the latter is produced by phase coherent acoustic waves rather than by incoherent phonons.

We will first consider the acousto-electric effect in nonpolar many-valley semiconductors such as n-type Ge. In piezoelectric semiconductors such as CdS where the acousto-electric effect is much stronger, it does not rely upon a many-valley structure; but since the treatment to a large extent is very similar to the effect in n-Ge, we will also consider it here.

A shear wave obtained at frequencies of 20 or 60 MHz from a *Y cut* quartz oscillator is fed via a stopcock lubricant into an n-type Ge crystal in the ⟨100⟩ direction; the wave is polarized in the ⟨010⟩ direction [7.63–65]. Let us denote

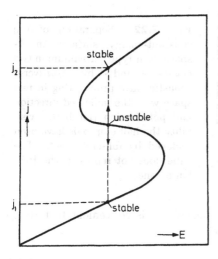

Fig. 7.20: S-shaped current instability with a formation of current filaments

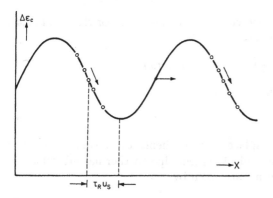

Fig. 7.21: Spatial variation of the potential energy due to propagating acoustic wave and carrier bunching on the front slopes of the wave

these directions by x and y, respectively. For a material displacement

$$\delta y = \delta y_1 \sin(\omega t - qx) \tag{7.7.1}$$

of amplitude δy_1 and angular frequency ω, we obtain a shear strain

$$\varepsilon_{xy} = \partial \delta y / \partial x = -q\, \delta y_1 \, \cos(\omega t - q x) \tag{7.7.2}$$

and a change in potential energy given by the second term in (7.1.21) :

$$\Delta \varepsilon_c = \pm \frac{1}{3}\, \Xi_u\, \varepsilon_{xy} \; , \tag{7.7.3}$$

where the sign depends on the valley in the conduction band of germanium as shown in Fig. 7.22. Hence, there are two classes of conduction band valleys which are indicated by subscripts + and -. Bunching of carriers from different

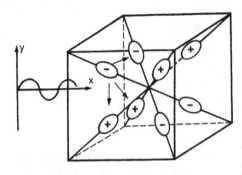

Fig. 7.22: Separation of the constant-energy surfaces in **k**-space for n-type germanium in two classes (+ and -) by a transverse acoustic wave propagating in real space with the indicated direction and polarization (for better visibility the half-ellipsoids have been replaced by full-ellipsoids). The cube does not represent the Brillouin zone

classes of valleys takes place in different phases. It is convenient to formally introduce an *acoustic charge* e_{ac} given by

$$e_{ac} = \frac{1}{3}\Xi_u/\sqrt{c_t} \; , \tag{7.7.4}$$

where c_t is the elastic constant [denoted by c_{44} in elastic-tensor theory (4.12.4)] and an *acoustic potential* φ given by

$$\varphi = -\sqrt{c_t}\, q\, \delta y_1 \cos{(\omega t - qx)} = \varphi_1 \cos{(\omega t - q x)} \; . \tag{7.7.5}$$

In this notation, (7.7.3) simply becomes

$$\Delta\varepsilon_c = \pm e_{ac}\varphi \; . \tag{7.7.6}$$

Let us assume the material displacement and hence the wave amplitude to be small compared to the average displacement due to thermal vibrations so that the carrier densities n_+ and n_- are given by

$$n_\pm = n_0 \exp(\mp e_{ac}\, \varphi/k_B T) \approx n_0(1 \mp e_{ac}\, \varphi/k_B T) \; , \tag{7.7.7}$$

where n_0 is the carrier density in each class of valleys for $\varphi = 0$. For $\varphi \neq 0$ we would have at equilibrium

$$n_+ - n_- = -2n_0\, e_{ac}\, \varphi/k_B T \; . \tag{7.7.8}$$

Naturally, there is no equilibrium and the nonequilibrium transfer rate $R_{+\to-}$ is given by

$$R_{+\to-} = -R_{-\to+} = \frac{2}{3}\langle 1/\tau_i \rangle\, (n_+ - n_- + 2n_0\, e_{ac}\varphi/k_B T) \; , \tag{7.7.9}$$

where τ_i is the intervalley scattering time (Sect. 7.5) and the factor 2/3 arises because in only two out of three intervalley transitions does the electron change the class (indicated by arrows in Fig. 7.22). Particle conservation yields $n_+ + n_- = 2n_0$ and therefore from (7.7.9) we find

$$R_{+\to-} = \frac{4}{3}\langle 1/\tau_i \rangle\, [n_+ - n_0(1 - e_{ac}\varphi/k_B T)] \; . \tag{7.7.10}$$

From the continuity equation

$$R_{+\to-} = \frac{\partial n_-}{\partial t} + \frac{\partial j_-}{\partial x} = -\frac{\partial n_+}{\partial t} - \frac{\partial j_+}{\partial x} \ , \tag{7.7.11}$$

the *particle* current densities given by

$$j_\pm = \mp e_{ac} \frac{\partial \varphi}{\partial x} n_\pm \frac{D_n}{k_B T} - D_n \frac{\partial n_\pm}{\partial x} \tag{7.7.12}$$

and the Einstein relation (between the carrier mobility and the diffusion coefficient D_n), we obtain after some algebra

$$n_+ - n_0 = -n_0 (1 + i \omega \tau_R)^{-1} e_{ac} \varphi / k_B T \ , \tag{7.7.13}$$

where the frequency-dependent relaxation time

$$\tau_R = \left(\frac{4}{3} \langle \frac{1}{\tau_i} \rangle + q^2 D_n \right)^{-1} \approx \left(\frac{4}{3} \langle \frac{1}{\tau_i} \rangle + \frac{\varrho \omega^2}{c_t} D_n \right)^{-1} \tag{7.7.14}$$

has been introduced. Equation (7.7.14) shows that two processes contribute to the relaxation of the distribution, namely, intervalley scattering and diffusion out of the bunches. The average force per particle $\bar{F}$ exerted by the acoustic wave

$$\bar{F} = \langle -e_{ac} \frac{\partial \varphi}{\partial x} (n_+ - n_0) \rangle \langle n_0 \rangle^{-1} \ , \tag{7.7.15}$$

is obtained from (7.7.5, 13):

$$\bar{F} = e E_0 = \frac{\Xi_u^2 J}{9 c_t u_s^2 k_B T} \frac{\omega^2 \tau_R}{1 + \omega^2 \tau_R^2} \ , \tag{7.7.16}$$

where E_0 is the acousto-electric field intensity, $u_s = \omega/q$ is the sound velocity and $J = \frac{1}{2} (\varphi_1^2 u_s)$ is the acoustic energy flux averaged over a period of the wave. For $J = 1 \text{Wcm}^{-2}$ at 60 MHz and 77 K, $\tau_i = 10^{-11}$ s, $\Xi_u = 16$ eV, $c_t = 1.56 \times 10^7$ N cm^{-2}, and $u_s = 5.4 \times 10^5$ cm/s valid for n-type Ge, a field intensity of $1\mu V/$ cm is calculated. Wave amplification by drifting carriers has been treated by, e.g., *Conwell* [Ref. 7.28, p. 142].

From the experimental results [7.63] obtained at temperatures T between 20 and 160 K on five samples of different purity ($10^{14} - 10^{16}$ cm^{-3}), $\langle 1/\tau_i \rangle$ has been determined by means of (7.7.14, 16) and in Fig. 7.23 plotted vs T. At the higher temperatures the data points for all the samples (and especially for the purer ones) fall on the dashed curve which has been calculated from (7.5.2) with $\lambda = 1$ (i.e., $T_e = T$), $\Theta_i = 315$ K and $w_2 = 10^{11}$ s^{-1}. Comparing data on the purest sample at low temperatures and 20 MHz with those at 60 MHz where $\omega^2 \tau_R^2 \approx 1$, the deformation potential constant $\Xi_u = 16$ eV has been determined to an accuracy of 10%. The deviation of the data from the dashed curve have been considered to be due to ionized and neutral impurity scattering.

Let us now investigate the amplification of longitudinal ultrasonic waves in *piezoelectric semiconductors*. This subject has been treated by *White* and

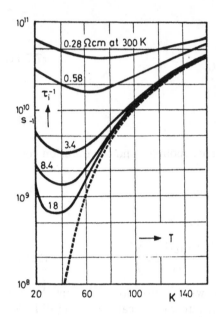

Fig. 7.23: Temperature dependence of the intervalley scattering rate in n-type germanium (after [7.63])

Hutson and *White* [7.66]. The coupling between the electrons and the lattice due to the piezoelectric effect has been discussed in Sect. 6.7. The dielectric displacement is given by (6.7.3) where it is convenient to denote the strain $(\nabla_r \cdot \delta_r)$ by S:

$$D = \kappa \kappa_0 E + e_{pz} S .$$ (7.7.17)

The stress T is given by (6.7.10):

$$T = c_\ell S - e_{pz} E ,$$ (7.7.18)

where the first term on the rhs represents Hook's law while the second term is due to the piezoelectric effect; c_ℓ is the elastic constant at a constant field. Let us denote the mean carrier density by n_0 and the instantaneous local density by $n_0 + f n_s$ where $e n_s$ is the space charge and only a fraction f of the space charge contributes to the conduction process; the rest is trapped at local states in the energy gap. Such traps are more important in piezoelectric semiconductors such as CdS or ZnO than they are in Ge or Si.

The current density is given by

$$j = |e|(n_0 + f n_s) \mu E - e D_n \partial(n_0 + f n_s)/\partial x ,$$ (7.7.19)

where $e < 0$ is valid for electrons. The equation of continuity is

$$e \partial n_s/\partial t + \partial j/\partial x = 0$$ (7.7.20)

and Poisson's equation

$$\partial D/\partial x = e n_s .$$ (7.7.21)

Differentiating the latter equation with respect to t yields

$$\frac{\partial^2 D}{\partial x \partial t} = -\frac{\partial[\sigma + (e/|e|)\mu f \partial D/\partial x]E}{\partial x} + D_n f \frac{\partial^3 D}{\partial x^3} , \qquad (7.7.22)$$

where we have taken (7.7.20) into account. For an *applied* dc field E_0 and an ac field of amplitude $E_1 \ll E_0$ which is due to an ultrasonic wave

$$\delta x = \delta x_1 \exp \text{i}(\omega t - qx) , \qquad (7.7.23)$$

$$E = E_0 + E_1 \exp \text{i}(\omega t - qx) \qquad (7.7.24)$$

and similarly

$$D = D_0 + D_1 \exp \text{i}(\omega t - qx) = D_0 + (\kappa \kappa_0 E_1 - \text{i} q e_{\text{pz}} \delta x_1) \exp \text{i}(\omega t - qx) \qquad (7.7.25)$$

where (7.7.17, 23) have been taken into account, we obtain

$$E_1 = \text{i} \frac{q e_{\text{pz}} \delta x_1}{\kappa \kappa_0} \left[1 - \frac{\text{i}}{\omega \tau_{\text{d}}(\gamma - \text{i}\omega/\omega_{\text{D}})}\right]^{-1} . \qquad (7.7.26)$$

The products of D_1 and E_1 which are second-order terms have been neglected; the dielectric relaxation time τ_{d} given by (5.2.21), a *diffusion frequency* $\omega_{\text{D}} = c_\ell/\varrho D_n$, and a drift parameter $\gamma = 1 - (e/|e|)f \mu E_0/u_s$ have been introduced; μE_0 is the drift velocity and $u_s = \omega/q$ is the sound velocity. If we assume for the wave equation.

$$\partial T/\partial x = \varrho \partial^2 \delta x/\partial t^2 = c_\ell \partial^2 \delta x/\partial x^2 - e_{\text{pz}} \partial E/\partial x \qquad (7.7.27)$$

where (7.7.18) has been taken into account, a solution of the form given by (7.7.23, 24), we find for the relationship between δx_1 and E_1

$$-\varrho\omega^2 \delta x_1 = -c_\ell q^2 \delta x_1 + \text{i} q e_{\text{pz}} E_1 \qquad (7.7.28)$$

This can be written in the usual form

$$-\varrho\omega^2 \delta x_1 = -c'_\ell q^2 \delta x_1 \qquad (7.7.29)$$

by formally introducing a complex elastic constant c'_ℓ. By taking (7.7.26) into account, we obtain for c'_ℓ/c_ℓ

$$\frac{c'_\ell}{c_\ell} = 1 - \text{i} \frac{(e_{\text{pz}}/q) E_1}{\delta x_1} = 1 + \frac{K^2}{1 - \text{i}/[\omega \tau_{\text{d}}(\gamma - \text{i}\omega/\omega_{\text{D}})]} , \qquad (7.7.30)$$

where the electromechanical coupling coefficient $K^2 \ll 1$ given by (6.7.9) has been introduced. The real part of c'_ℓ is simply the product ϱu_s^2 and since $K^2 \ll 1$, we can approximate $u_s = \text{Re}(\sqrt{c'_\ell/\varrho})$ by

$$u_s = u_{s0} \left\{1 + \frac{1}{2} K^2 \left[1 - \frac{1/\omega \tau_{\text{d}} (1/\omega \tau_{\text{d}} + \omega/\omega_{\text{D}})}{\gamma^2 + (1/\omega \tau_{\text{d}} + \omega/\omega_{\text{D}})^2}\right]\right\} , \qquad (7.7.31)$$

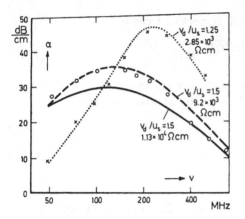

Fig. 7.24: Electron attenuation coefficient vs frequency for shear waves in photoconducting CdS; crosses and dotted curve: observed and calculated data, respectively, for $v_d/u_s = 1.25$ and $\varrho = 1.13 \times 10^4 \Omega$ cm; dashed curve calculated for $v_d/u_s = 1.5$ and $\varrho = 9.2 \times 10^3 \Omega$ cm (after [7.68])

where $\sqrt{c_\ell/\varrho} = u_{s0}$ is the sound velocity for $K^2 = 0$. If we consider q in (7.7.23) as complex with an imaginary part denoted by $-\alpha = \text{Im}(q)$, a positive value of α causes attenuation of the wave. The relation $\sqrt{\varrho}\omega = \sqrt{c_\ell}q$ obtained from (7.7.29) thus yields for α in the same approximation as (7.7.31)

$$\alpha = -\omega \sqrt{\varrho}\,\text{Im}(1/\sqrt{c_\ell}) = \frac{1}{2}\frac{K^2}{\tau_d\, u_{s0}}\frac{\gamma}{\gamma^2 + (1/\omega\,\tau_d + \omega/\omega_D)^2} . \qquad (7.7.32)$$

The attenuation constant is proportional to the drift parameter γ. Assuming $f = 1$ and a drift velocity μE_0 slightly larger than the sound velocity u_s and in the same direction, both γ and α are negative resulting in an amplification of the wave; for electrons (e < 0) the dc field E_0 has to be negative, i.e., opposite to the direction of the wave.

The term $1/\omega\tau_d + \omega/\omega_D$ can be written as $1/\omega\tau_R$ where the frequency dependent relaxation time τ_R has about the same form as τ_R given in (7.7.14):

$$\tau_R = \left(\frac{1}{\tau_d} + \frac{\varrho\,\omega^2}{\text{Re}(c_\ell)}D_n\right)^{-1} \qquad (7.7.33)$$

The maximum relative reduction in the sound velocity occurs at $\gamma = 0$ and amounts to $1/2\,K^2[1 + (u_{s0}/\omega\,L_D)^2]^{-1}$ which may not be appreciable since $K^2 \ll 1$; L_D is the Debye length given by (5.2.22). However, the important fact is the negative attenuation coefficient α. Its maximum value of

$$\frac{1}{8}K^2\sqrt{\omega_D/\tau_d}/u_{s0} = K^2/8\,L_D \qquad (7.7.34)$$

is attained at $\gamma = -(2/u_{s0})\sqrt{D_n/\tau_d}$. For CdS, values of $\gamma = -10.1$ and $-\alpha = 5.6 \times 10^3$ dB/cm have been calculated; even higher values of $-\alpha$, such as 1.3×10^4 dB/cm at $\gamma = -5.1$, are found in ZnO [7.67]. Figures 7.24, 25 show experimental results of α vs frequency and vs γ, respectively, for CdS [7.68]. The curves have been calculated from (7.7.32). The agreement between

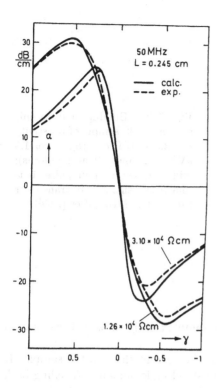

Fig. 7.25: Electronic attenuation coefficient vs drift parameter for 50 MHz shear waves in photoconducting CdS at 300 K (after [7.68])

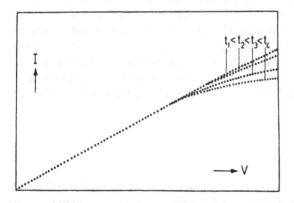

Fig. 7.26: Current voltage characteristics for cadmium sulfide at various time intervals after the application of the voltage (see Fig.7.27)

theoretical and experimental results is very good, although shear waves rather than longitudinal waves have actually been amplified in these experiments.

Figure 7.26 shows the current-voltage characteristics of CdS at various times between $\ll 1\ \mu$s and 80 μs after application of the voltage [7.69]. The deviation from Ohm's law at the larger time intervals is due to the generation of additional acoustic flux (acts by acoustic-phonon scattering) from thermal lattice vibrations due to the acousto-electric effect. The output of a transducer

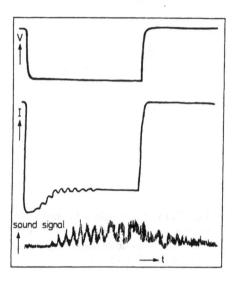

Fig. 7.27: Buildup of ultrasonic flux in a CdS sample of resistivity $\varrho = 3.3 \times 10^4 \Omega$ cm; top trace: 1.5 kV voltage pulse (length: 120 μs); middle trace: current pulse; bottom trace: signal from transducer tuned to 45 MHz (after [7.69])

(= ultrasonic microphone) attached to the sample clearly shows this generation [7.66] (Fig. 7.27, lower trace).

The trace in the middle of Fig. 7.27 is the current through the sample. It shows oscillations which are caused by acousto-electric domains travelling back and forth in the crystal. These domains which are in some respects similar to Gunn domains have been shown to exist by observing the light scattered by them (*Brillouin scattering*)[4]   and by the modulation of microwaves transmitted through the sample which is shown in Fig.7.28 [7.72, 73]. Research has been concerned with the kinetics of domain formation, nonlinear effects such as *parametric up-conversion* of frequencies at high acoustic power levels and electrical instabilities due to the acousto-electric effect (for a review see, e.g., [7.67]).

Problems

7.1. Calculate the Hall coefficient R_{H} for a weak magnetic field for a 2-dimensional carrier gas in a single elliptical valley assuming a current in the direction of one of the two main axes and an isotropic relaxation time.

7.2. From the inverted Seitz equation, show that in a cubic crystal system for any direction within a [111] plane, the longitudinal magnetoresistance in a weak magnetic field is the same and given by $b + c + d/2$.

7.3. For the conduction band, assume two valleys (subscripts 1 and 2) with spheroidal energy surfaces with the axes of rotation being perpendicular to each

---

[4] It is convenient to denote the scattering of visible or infrared light by *acoustic* phonons as *Brillouin scattering*, while light scattering by *optical* phonons (Sect. 11.7) or plasmas in solids is known as *Raman scattering* [7.70, 71].

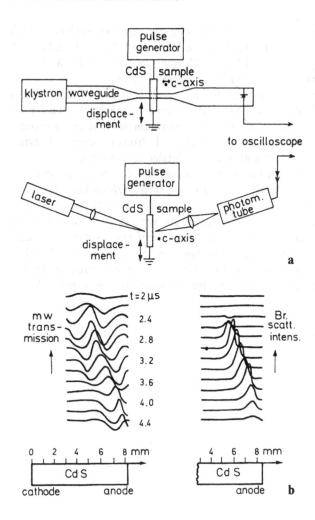

Fig. 7.28: (a) Microwave transmission and Brillouin scattering arrangements for the detection of acoustoelectric domains. (b) Microwave transmission and Brillouin scattering at various times after the application of the voltage. The negative peaks are due to space charge accumulation (after [7.72])

other. For simplicity, assume isotropic energy independent relaxation times for momentum and energy. The carrier concentration in each valley in units of the total one n_0 is assumed:

$$n_1/n_0 = T_2/(T_1 + T_2); \quad n_2/n_0 = T_1/(T_1 + T_2) , \qquad (7.0.1)$$

where T_1 and T_2 are the electron temperatures in the respective valleys. From the energy balance equation, determine the current field characteristics for strong electric fields parallel to the rotational axis of one of the valleys. Show that the limiting conductivity in units of the zero field conductivity is given by $4\, m_1 m_2/(m_1 + m_2)^2$. Discuss the characteristics.

7.4. Calculate the longitudinal weak-field magnetoresistance for n-type GaSb at room temperature for a $\langle 01\bar{1} \rangle$ direction as a function of the electric field strength. Both L-valleys which are 0.1 eV above the Γ-valley, and the Γ-

valley are populated with carrier concentrations n_L and n_Γ , respectively, which in units of 10^{16} cm^{-3} are 3.8, 3.0 for E=0; 4, 2.8 for 0.65 kV/cm; 4.6, 2.2 for 1.5 kV/cm; 5.1, 1.7 for 2 kV/cm; 5.4, 1.4 for 2.5 kV/cm, and 5.7, 1.1 for 3 kV/cm. The mobilities are $\mu_L = 660$ cm^2/Vs and $\mu_\Gamma = 3700$ cm^2/Vs.

7.5. Gunn effect. For a 2-valley conduction band (one L and one Γ) with mobilities $\mu_L = 800$ cm^2/Vs, $\mu_\Gamma = 8000$ cm^2/Vs and an effective- density-of-states ratio $N_{cL}/N_{c\Gamma} = 30$, determine the *I-V* characteristics at fields around 3 kV/cm. Determine the electron temperature at Γ from an energy balance equation assuming $\tau_\varepsilon = 5 \times 10^{-13}$ s at = 300 K lattice temperature.

7.6. Calculate the contribution to the bunching of carriers in acousto- electric domains which is due to the second order terms which have been ne- glected in the small-signal treatment of Sect. 7.7. As *Tien* [7.74] showed, a Fourier series of the carrier concentration converges very slowly. Rather, rapid convergence has been obtained for the Fourier series of the electric potential. For simplicity, drop higher-order terms than the first, i.e., $\varphi = \varphi_0 \cos \theta$ and in [Ref. 7.74, Eq. (12)], consider the second term $\propto n \, d\varphi/d\theta$ as a small perturba- tion. Find $n(\theta)$. Give a numerical example for μE equal to the sound velocity and $e \varphi_0 = k_B T$.

8. Carrier Transport in the Warped-Sphere Model

The valence bands of germanium, silicon and the III–V compounds have an extremum at $k = 0$ and are degenerate there. The constant-energy surfaces for this case are warped spheres which have already been discussed in Sect. 2.4 (Figs. 2.28a-2.28c). In the zincblende lattice typical for III–V compounds, there is no center of inversion, in contrast to the diamond lattice.

8.1 Energy Bands and Density of States

Although in Sect. 2.2 we considered the energy band model in general, we have not treated the warped-sphere model. However, it seems worthwhile to give a brief idea here as to how it is developed from the Schrödinger equation without going into any details of the calculation [1]

If the valence band maximum is at $k = 0$, the so-called $(k \cdot p)$ approximation is a convenient method of calculation. It is a perturbation approach to the problem. Introducing the wave functions $\psi(r) = u(r) \exp[\mathrm{i}(k \cdot r)]$ (2.1.11) in the Schrödinger equation, the form given by (2.1.12) is found. Taking for the momentum operator $p = -\mathrm{i}\hbar \, \nabla_r$ this can be written

$$[p^2/2m + V(r)]u(r) + [\hbar(k \cdot p)/m]u(r) = (\varepsilon - \hbar^2 k^2/2m)\,u(r) \ . \quad (8.1.1)$$

For an energy band calculation at $k = 0$ the term $\hbar \, (k \cdot p) \,/m$ is treated as a small perturbation. For a short-hand notation, let us introduce an eigenvalue

$$\varepsilon' = \varepsilon - \hbar^2 k^2/2m \ . \quad (8.1.2)$$

We will consider here only the diamond lattice where due to crystal symmetry at $k = 0$ (=center of inversion), terms linear in k vanish and ε varies as $|k|^2$ along any direction in k-space. Hence, we have to solve the perturbation equation

$$H_{kp}u(r) = [\hbar(k \cdot p)/m]\,u(r) = \varepsilon' u(r) \quad (8.1.3)$$

to second order where $u(r)$ is given by a linear combination of atomic wave functions

$$u(r) = (ayz + bzx + cxy)/r^2 \ . \quad (8.1.4)$$

---

[1]The following considerations have been adapted from [8.1, 2]

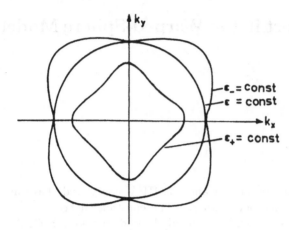

Fig. 8.1: $\langle 100 \rangle$ cross section of the warped–sphere constant–energy surfaces (neglecting spin)

Here, a, b, and c are unknown coefficients; spin–orbit coupling is neglected at present. For a determination of these coefficients $a = a(k_x, k_y, k_z)$, etc., from the set of linear equations (in components) (8.1.3), it is sufficient to consider any physical problem which is proportional to the second order of a vectorial quantity, such as, e.g., the magnetoresistance as a function of the magnetic field, as a model. An inspection of (4.2.34) and neglecting the linear terms (i.e., $\gamma_0 = 0$) shows that the determinant of (8.1.3) must have the form

$$\begin{vmatrix} Ak_x^2 + B(k_y^2 + k_z^2) - \varepsilon & Ck_xk_y & Ck_xk_z \\ Ck_xk_y & Ak_y^2 + B(k_z^2 + k_x^2) - \varepsilon & Ck_yk_z \\ Ck_xk_z & Ck_yk_z & Ak_z^2 + B(k_x^2 + k_y^2) - \varepsilon \end{vmatrix} = 0$$

$$(8.1.5)$$

where A, B, and C are to be calculated by the usual methods of perturbation theory and ε' has been replaced by ε. We will not attempt to do such a calculation but just solve this equation for the $k_x k_y$ plane ($k_z = 0$):

$$[(Ak_x^2 + Bk_y^2 - \varepsilon)(Ak_y^2 + Bk_x^2 - \varepsilon) - C^2 k_x^2 k_y^2](Bk_x^2 + Bk_y^2 - \varepsilon) = 0 \quad (8.1.6)$$

which results in

$$\varepsilon_\pm = \frac{1}{2}(A + B)(k_x^2 + k_y^2) \pm \sqrt{\frac{1}{4}(A + B)^2(k_x^2 + k_y^2)^2 + C^2 k_x^2 k_y^2} \quad (8.1.7)$$

and

$$\varepsilon = B(k_x^2 + k_y^2) . \quad (8.1.8)$$

The curves of constant energy consist of warped circles (8.1.7) and normal circles (8.1.8), a set of which is shown in Fig. 8.1. Including now the spin-orbit interaction at $k = 0$, each one of the three states previously considered is doubled, and to the 6×6 matrix, which in obvious short-hand notation is given by

$$\left| \begin{array}{cc} H_{kp} & 0 \\ 0 & H_{kp} \end{array} \right| ,$$

we have to add a matrix which contains the Pauli spin matrices. In an approximation the total 6×6 matrix can be split into a 2×2 matrix and a 4×4 matrix. The solution of the 2×2 matrix is similar to (8.1.8) except for a shift in energy which we denote by the symbol Δ:

$$\varepsilon = A_0 k^2 - \Delta , \tag{8.1.9}$$

where $k^2 = k_x^2 + k_y^2 + k_z^2$ and A_0 is another coefficient. The 4×4 matrix yields an energy similar to (8.1.7):

$$\varepsilon_\pm = A_0 k^2 \pm \sqrt{(B_0 k^2)^2 + C_0^2(k_x^2 k_y^2 + k_y^2 k_z^2 + k_z^2 k_x^2)} , \tag{8.1.10}$$

where B_0 and C_0 are two more coefficients [8.3]. For the split-off valence band given by (8.1.9), the total angular momentum is $j = 1/2$ while for the heavy and light-hole bands given by the last equation it is $j = 3/2$.

Very often the anisotropy is small, i.e., $|C_0| \ll |B_0|$ and the square root can be expanded. It is then convenient to introduce the coefficients

$$B_0' = \sqrt{B_0^2 + \frac{1}{6} C_0^2} \tag{8.1.11}$$

and

$$\Gamma_\pm = \mp C_0^2/[2B_0'(A_0 \pm B_0')] , \tag{8.1.12}$$

where the upper sign is valid for the light-hole band and the lower sign for the heavy-hole band. With these coefficients we obtain from (8.1.10)

$$\varepsilon_\pm = (A_0 \pm B_0')k^2 \pm B_0' k^2 \left[\sqrt{1 \mp \frac{2(A_0 \pm B_0')\Gamma_\pm}{B_0'} \left(\frac{k_x^2 k_y^2 + \cdots}{k^4} - \frac{1}{6} \right)} - 1 \right] \tag{8.1.13}$$

This is expanded for small values of the second term of the radicand to

$$\varepsilon_\pm = (A_0 \pm B_0')k^2 \left[1 - \Gamma_\pm \left(\frac{k_x^2 k_y^2 + k_y^2 k_z^2 + k_z^2 k_x^2}{k^4} - \frac{1}{6} \right) \right] , \tag{8.1.14}$$

where higher powers of $\Gamma_\pm$ have been neglected; since the second term is $\Gamma_\pm/6$ for $k_x = k_y = k_z = k/\sqrt{3}$, the series converges even for light holes in Si at

4 K where the factor $2(A_0 + B_0')\Gamma_\pm/B_0' \approx 4$; consequently, the approximation is acceptable. In the factor

$$\frac{k_x^2 k_y^2 + k_y^2 k_z^2 + k_z^2 k_x^2}{k^4} - \frac{1}{6} = \frac{1}{2}\left(\frac{2}{3} - \frac{k_x^4 + k_y^4 + k_z^4}{k^4}\right) \tag{8.1.15}$$

which we will denote by $-q$, let us introduce polar coordinates

$$-q = \frac{1}{2}\left[\frac{2}{3} - \sin^4\theta(\cos^4\varphi + \sin^4\varphi) - \cos^4\theta\right] . \tag{8.1.16}$$

With the notation

$$g_\pm = [(A_0 \pm B_0')/k_B T](1 + \Gamma_\pm\, q) \tag{8.1.17}$$

we can write (8.1.14) in the form

$$\varepsilon_\pm/k_B T = -g_\pm k^2 . \tag{8.1.18}$$

Due to the negative sign the energy which is negative for holes yields a positive value for $g_\pm$. Equations (2.1.18) for the heavy and light-hole bands have been obtained from this relation.

The hole concentration p is given by (3.2.14). Let us now calculate the constant N_v by (4.1.29) modified for the negative hole energy:

$$N_v = \frac{1}{4\pi^3}\int \exp(\varepsilon/k_B T)d^3k = \frac{1}{4\pi^3}\int \exp(-gk^2)d^3k , \tag{8.1.19}$$

where d^3k, in polar coordinates, is given by

$$d^3k = k^2 dk \sin\theta d\theta d\varphi = \frac{1}{2}g^{-3/2}\sqrt{gk^2}d(gk^2)\sin\theta d\theta d\varphi . \tag{8.1.20}$$

For simplicity we have omitted the subscripts $\pm$. The integration over gk^2 from 0 to ∞ yields a factor of $\frac{1}{2}\sqrt{\pi}$. For the integration over θ and φ we expand $g^{-3/2}$ where g is given by (8.1.17):

$$g_\pm^{-3/2} = \left(\frac{A_0 \pm B_0'}{k_B T}\right)^{-3/2}\left(1 - \frac{3}{2}\Gamma_\pm q + \frac{15}{8}\Gamma_\pm^2 q^2 - \frac{35}{16}\Gamma_\pm^3 q^3 + - \ldots\right) . \tag{8.1.21}$$

The integration over φ from 0 to 2π and over θ from 0 to π yields for N_v

$$N_v^\pm = 2\left(\frac{m_0 k_B T}{2\pi\hbar^2(A_0 \pm B_0')}\right)^{3/2}$$

$$\times(1 + 0.05\,\Gamma_\pm + 0.0164\,\Gamma_\pm^2 + 0.000908\,\Gamma_\pm^3 \pm \ldots) \tag{8.1.22}$$

which, according to (3.1.10) modified for holes, is given by $2(m_p k_B T/2\pi\hbar^2)^{3/2}$. Hence, the *density-of-states effective mass*, which we denote by m_d instead of m_p, is given by

$$m_\mathrm{d}^\pm = \frac{m_0}{A_0 \pm B_0'}(1 + 0.03333\Gamma_\pm + 0.01057\Gamma_\pm^2 - 1.8 \times 10^{-4}\Gamma_\pm^3 - 3 \times 10^{-5}\Gamma_\pm^4 + \ldots)$$

$$(8.1.23)$$

For a comparison we note the isotropic *conductivity effective mass* m_σ which will be calculated in the following section,

$$m_\sigma^\pm = \frac{m_0}{A_0 \pm B_0'}(1 + 0.3333\Gamma_\pm - 0.02566\Gamma_\pm^2 - 9.5 \times 10^{-4}\Gamma_\pm^3 + \ldots) \ . (8.1.24)$$

In both equations the minus sign in the denominator is valid for heavy holes while the plus sign is valid for light holes. Assuming equal values of the momentum relaxation time for both types of holes, the total conductivity is found

$$\sigma = [(p_\mathrm{h}/m_\mathrm{h}) + (p_\mathrm{l}/m_\mathrm{l})]e^2 \langle \tau_\mathrm{m} \rangle \ , \tag{8.1.25}$$

where the subscripts h and l refer to heavy and light holes, respectively.

The experimental determination of the constants A_0, $|B_0|$, and $|C_0|$ from cyclotron resonance will be discussed in Sect. 11.11. Table 8.1 gives the values of these constants and of the conductivity- effective masses for Si and Ge at 4 K where m_s is the effective mass in the split-off valence band.

According to (8.1.22), the ratio of constants N_v and therefore of the numbers of light and heavy holes is given by

$$p_\mathrm{l}/p_\mathrm{h} = (A_0 - B_0')^{3/2}/(A_0 + B_0')^{3/2}[1 + 0.05(\Gamma_+ - \Gamma_-) + \ldots] \ , \tag{8.1.26}$$

where the light hole band has been assumed as being isotropic. This ratio is 4% for Ge and 16% for Si.

Let us briefly discuss the influence of stress on the valence bands of germanium and silicon [8.5]. With 6 deformation components $e_1 \ldots e_6$ given by (7.1.2, 3) and the deformation potential constants a, b, and d, the change in energy is found:

$$\Delta\varepsilon_\mathrm{v} = a\Delta \pm (\sqrt{\varepsilon_k + \varepsilon_{ek} + \varepsilon_e} - \sqrt{\varepsilon_k}) \quad \text{where} \quad \Delta = e_1 + e_2 + e_3$$

$$\varepsilon_k = (B_0 k^2)^2 + C_0^2(k_x^2 k_y^2 + k_y^2 k_z^2 + k_z^2 k_x^2)$$

$$\varepsilon_e = \frac{b^2}{2}[(e_1 - e_2)^2 + (e_1 - e_3)^2 + (e_2 - e_3)^2]$$

$$\varepsilon_{ek} = B_0 b[3(k_x^2 e_1 + k_y^2 e_2 + k_z^2 e_3) - k^2\Delta] + D_0 d(k_x k_y e_4 + k_y k_z e_5 + k_z k_x e_6)$$

$$D_0 = \sqrt{3B_0^2 + C_0^2} \ , \tag{8.1.27}$$

where the split-off valence band has been neglected in the calculation. For example, Ge, a=3.1 eV , b=2.2 eV , d=4.5 eV while for Si, $a= \Xi_\mathrm{d} + 1/3\ \Xi_\mathrm{u} -$ 1.65 eV, $b = 2.2$ eV and $d=$ 5.3 eV [8.6]. By a dilatation in the $\langle 111 \rangle$ direction, the valence band in Ge is lowered by 40 meV/ GPa given by the first term on the rhs of (8.1.27). At low temperatures ($\approx$ 4 K for $e_1 > 1/450$ in Ge) where the second term is much larger than the thermal energy $k_B T$, only the lower (in terms of hole energy) of the two subbands is populated. The constant-energy surface of this subband is an ellipsoid located at $k = 0$. The semiconductor thus represents the ellipsoidal type and has been labelled *split* p – *germanium*. The transport properties of this semiconductor have been reviewed by Koenig [8.7].

Table 8.1 Coefficients A_0, B_0, C_0 (in units of $\hbar^2/2m_0$), and conductivity effective masses (Si : [8.3]; Ge : [8.4])

	A_0	$|B_0|$	$|C_0|$	m_h/m_0	m_l/m_0	m_s/m_0
Si	4.1	1.6	3.3	0.49	0.16	0.245
Ge	-13.27	8.62	12.4	0.29	0.0437	0.075

8.2 The Electrical Conductivity

If the number of holes in the split-off valence band is negligible, the conductivity is given by (8.1.25) where the effective masses are given by (8.1.24). For a proof of (8.1.24) we consider the conductivity σ given by (4.1.30) where τ_m is given by (4.1.8) and the energy $|\varepsilon|$ by (8.1.18):

$$\tau_m = \tau_0 (g_{\pm} k^2)^r . \tag{8.2.1}$$

$(\partial \varepsilon / \partial k_z)^2$ is calculated from (8.1.13) and expanded for small values of $\Gamma_{\pm}$. The integration is similar to the one that led to (8.1.22) and yields [8.8]

$$\sigma = \frac{4}{3\sqrt{\pi}} \frac{e^2 \tau_0}{m_0} (r + \frac{3}{2})! \, 2 \left(\frac{m_0 k_B T}{2\pi \hbar^2} \right)^{3/2} (A_0 \pm B_0')^{-1/2}$$

$$\times (1 + 0.01667\Gamma + \ldots) \exp[(\varepsilon_v - \zeta)k_B T] . \tag{8.2.2}$$

For $\Gamma=0$, $A_0 \pm B_0' = 1$ and an effective mass m_0, a conductivity $\sigma(m_0)$ would be obtained. Hence:

$$\sigma = \sigma(m_0)(A_0 \pm B_0')^{-1/2}(1 + 0.01667\Gamma + 0.041369\Gamma^2$$

$$+9.0679 \times 10^{-4}\Gamma^3 + 9.1959 \times 10^{-4}\Gamma^4 + \ldots) . \tag{8.2.3}$$

The conductivity effective mass given by (8.1.24) has been obtained from this relation. In this form the result is similar to that for a carrier of isotropic effective mass. In these relations the subscript $\pm$ of Γ has been dropped for simplicity.

8.3 Hall Effect and Magnetoresistance

To obtain a solution of the Boltzmann equation for carriers in electric and magnetic fields, discussed in Sect. 4.2, a quadratic $\varepsilon(\mathbf{k})$-relation has been assumed. Since in the present case (8.1.10) we have a nonquadratic relation, the Boltzmann equation can be solved only for certain approximations. For weak magnetic fields, *Lax* and *Mavroides* [8.8] calculated a series expansion in powers of **B**. Let us briefly consider another method by *McClure* [8.9] valid for arbitrary magnetic field strengths. The carrier velocity given by (4.1.28) is expanded along an orbit in **k**-space (*hodograph*) on a constant-energy surface intersected by a plane normal to **B** in powers of $\omega_c t$ where $\omega_c = \kappa e B(A_0 - B_0')/m_0$ and t is the time period on the hodograph if no electric field is present [2]. For a large deviation of the constant-energy surface from a sphere, quite a few terms of the Fourier expansion have to be retained (up to $5\,\omega_c t$ for heavy holes).

Let us now consider the case of a weak magnetic field. The Boltzmann equation given by (4.2.3) with $\nabla_r f = 0$ and v given by (4.1.28)

$$e\,\hbar^{-1}(\mathbf{E} + \hbar^{-1}[\nabla_{\boldsymbol{k}}\varepsilon \times \mathbf{B}] \cdot \nabla_{\boldsymbol{k}} f) + (f - f_0)/\tau_{\mathrm{m}} = 0 \tag{8.3.1}$$

is solved by assuming

$$f = f_0 - a\,\partial f_0/\partial\varepsilon \tag{8.3.2}$$

which is similar to (4.2.4) where a is a coefficient. The momentum relaxation time τ_{m} can be taken as a scalar quantity since the band extremum is at $\boldsymbol{k} = 0$ and τ_{m} has the symmetry of the cubic lattice. We thus obtain from (8.3.1)

$$\frac{e}{\hbar}\left(\mathbf{E} + \frac{1}{\hbar}[\nabla_{\boldsymbol{k}}\varepsilon \times \mathbf{B}]\right)\left(\frac{\partial f_0}{\partial\varepsilon}\nabla_{\boldsymbol{k}}\varepsilon - \frac{\partial f_0}{\partial\varepsilon}\nabla_{\boldsymbol{k}}a - a\frac{\partial^2 f_0}{\partial\varepsilon^2}\nabla_{\boldsymbol{k}}\varepsilon\right)$$

$$-\frac{a}{\tau_{\mathrm{m}}}\frac{\partial f_0}{\partial\varepsilon} = 0 \tag{8.3.3}$$

For the case of a weak electric field, products of **E** with a and $\nabla_{\boldsymbol{k}}\,a$ can be neglected. The product $([\nabla_{\boldsymbol{k}}\varepsilon \times \mathbf{B}]\nabla_{\boldsymbol{k}}\varepsilon) = 0$ and therefore (8.3.3) is simplified to

$$a = \frac{\tau_{\mathrm{m}}e}{\hbar}(\mathbf{E} \cdot \nabla_{\boldsymbol{k}}\varepsilon) + \frac{1}{\hbar}(\mathbf{B} \cdot [\nabla_{\boldsymbol{k}}\varepsilon \times \nabla_{\boldsymbol{k}}a])$$

$$= \frac{\tau_{\mathrm{m}}e}{\hbar}(\mathbf{E} \cdot \nabla_{\boldsymbol{k}}\varepsilon) + \tau_{\mathrm{m}}e(\mathbf{B} \cdot \boldsymbol{\Omega})a \tag{8.3.4}$$

where an operator

$$\boldsymbol{\Omega} = \hbar^{-2}[\nabla_{\boldsymbol{k}}\varepsilon \times \nabla_{\boldsymbol{k}}] \tag{8.3.5}$$

---

[2]It is assumed that the cyclotron radius is much larger than the lattice constant and therefore the dielectric constant κ is included in the expression for ω_c.

has been introduced. For small values of a we have to a first approximation

$$a_1 = \tau_m e \, \hbar^{-1}(\boldsymbol{E} \cdot \boldsymbol{\nabla}_k \varepsilon) \tag{8.3.6}$$

and to a second approximation

$$a_2 = \tau_m e \, \hbar^{-1}(\boldsymbol{E} \cdot \boldsymbol{\nabla}_k \varepsilon) + \tau_m e^2 \hbar^{-1}(\boldsymbol{B} \cdot \boldsymbol{\Omega})\tau_m(\boldsymbol{E} \cdot \boldsymbol{\nabla}_k \varepsilon) \ , \tag{8.3.7}$$

etc. The current density $\boldsymbol{j}$ is obtained as usual [for the drift velocity see, e.g., (4.1.19)] to any approximation by substituting for a in the relation

$$\boldsymbol{j} = -\frac{e}{4\pi^3\hbar} \int \boldsymbol{\nabla}_k \varepsilon \, a \, \frac{\partial f_0}{\partial \varepsilon} d^3 k \tag{8.3.8}$$

the quantities $a_1, a_2, \ldots$ given by (8.3.6, 7), etc. On the other hand, $\boldsymbol{j}$ is given by its components

$$j_i = \sum_j \sigma_{ij} E_j + \sum_{jk} \sigma_{ijk} E_j B_k + \sum_{jkl} \sigma_{ijkl} E_j B_k B_l + \ldots \ . \tag{8.3.9}$$

A comparison of coefficients yields

$$\sigma_{ij} = -\frac{e^2}{4\pi^3 \, \hbar^2} \int \tau_m \frac{\partial \varepsilon}{\partial k_i} \frac{\partial \varepsilon}{\partial k_j} \frac{\partial f_0}{\partial \varepsilon} d^3 k \ , \tag{8.3.10}$$

$$\sigma_{ijk} = \frac{e^3}{4\pi^3 \hbar^4} \int \tau_m \frac{\partial \varepsilon}{\partial k_i} \frac{\partial f_0}{\partial \varepsilon} \sum_{r,s} \frac{\partial \varepsilon}{\partial k_r} \frac{\partial}{\partial k_s} \left(\tau_m \frac{\partial \varepsilon}{\partial k_j} \right) \delta_{krs} \, d^3 k \ , \tag{8.3.11}$$

where δ_{krs} is the permutation tensor ($= +1$ for a sequence 123 of krs, $= -1$ for a sequence 132, and $= 0$ for two equal subscripts), and σ_{ijkl} is last but not least

$$\sigma_{ijkl} = -\frac{e^4}{4\pi^3 \, \hbar^6} \int \tau_m \frac{\partial \varepsilon}{\partial k_i} \frac{\partial f_0}{\partial \varepsilon}$$

$$\times \sum_{r,s,t,u} \frac{\partial \varepsilon}{\partial k_r} \frac{\partial}{\partial k_s} \left[\tau_m \frac{\partial \varepsilon}{\partial k_t} \frac{\partial}{\partial k_u} \left(\tau_m \frac{\partial \varepsilon}{\partial k_j} \right) \right] \delta_{lrs} \delta_{ktu} d^3 k. \tag{8.3.12}$$

If τ_m is a function of ε independent of the direction of $\boldsymbol{k}$, we can take it before the operator $\boldsymbol{\Omega}$ in (8.3.7) since

$$\boldsymbol{\Omega} \, \tau_m(\boldsymbol{E} \cdot \boldsymbol{\nabla}_k \varepsilon) = (\boldsymbol{E}\boldsymbol{\nabla}_k \varepsilon) \cdot \boldsymbol{\Omega} \, \tau_m + \tau_m \boldsymbol{\Omega}(\boldsymbol{E} \cdot \boldsymbol{\nabla}_k \varepsilon) \tag{8.3.13}$$

where

$$\boldsymbol{\Omega} \, \tau_m = \hbar^{-2}[\boldsymbol{\nabla}_k \varepsilon \times \boldsymbol{\nabla}_k \tau_m] = \hbar^{-2}\frac{d\tau_m}{d\varepsilon}[\boldsymbol{\nabla}_k \varepsilon \times \boldsymbol{\nabla}_k \varepsilon] = 0 \ . \tag{8.3.14}$$

Therefore, in the integrands of (8.3.11, 12) τ_m can be taken outside the parenthesis and combined with the first factor τ_m to yield τ_m^2, τ_m^3, etc. The result of the integrations in the notation of (8.2.3) is given by [8.10].[3]

$$\sigma_{ijk} = \sigma_{ijk}(m_0)(A_0 \pm B_0')^{1/2}$$

$$\times (1 - 0.01667\, \Gamma + 0.017956\, \Gamma^2 - 0.0069857\, \Gamma^3 + \ldots) \tag{8.3.15}$$

---

[3]The value for σ_{xxyy} given by *Mavroides* and *Lax* has been questioned by *Lawaetz*([8.11], p. 875, [8.12] p. 138)

$$\sigma_{xxyy} = \sigma_{xxyy}(m_0)(A_0 \pm B_0')^{3/2}$$

$$\times(1 - 0.2214\,\Gamma + 0.3838\,\Gamma^2 - 0.0167\,\Gamma^3 + 0.00755\,\Gamma^4 + \ldots) \qquad (8.3.16)$$

$$\sigma_{xxxx} = -2\sigma_{xyyx} = \frac{16}{1155}\sigma_{xxyy}(m_0)(A_0 \pm B_0')^{3/2}\,\Gamma^2$$

$$\times(1 - 0.4295\,\Gamma + 0.0188\,\Gamma^2 + 0.0103\,\Gamma^3 + \ldots) \ , \qquad (8.3.17)$$

$$\sigma_{xyxy} = -\sigma_{xxyy}(m_0)(A_0 \pm B_0')^{3/2}$$

$$\times(1 - 0.05\,\Gamma - 0.0469\,\Gamma^2 + 0.004\,\Gamma^3 - 0.00063\,\Gamma^4 + \ldots) \ , \qquad (8.3.18)$$

where $\sigma_{xxyy}(m_0) < 0$ and the existence of a longitudinal magnetoresistance is obvious from $\sigma_{xxxx} \neq 0$; this is a consequence of the warped energy surface. The magnetoresistance in the notation given in the text above (7.4.19) is:

$$M_{100}{}^{100} = -\sigma_{xxxx}/\sigma_0 \ ,$$

$$M_{110}{}^{110} = -(\sigma_{xxxx} + \sigma_{xxyy} + \sigma_{xyyx} + \sigma_{xyxy})/2\sigma_0 \ ,$$

$$M_{100}{}^{010} = -(\sigma_{xxyy}/\sigma_0) - (\sigma_{xyz}/\sigma_0)^2 \ , \qquad (8.3.19)$$

$$M_{110}{}^{\bar{1}10} = -(\sigma_{xxxx} + \sigma_{xxyy} - \sigma_{xyyx} - \sigma_{xyxy})/2\sigma_0 \ .$$

It may be sufficient to give these examples valid for certain principal crystallographic directions only; for more details see [8.10–13]. For a comparison of these calculations with experimental results obtained with p-type Ge [8.14] and p-type Si [8.15], the contributions by light and heavy holes have to be added in the calculation of the conductivity by (8.1.25) and of the magnetoresistance by (4.4.15). In the expression of the conductivity, the light-hole contribution may be neglected due to the small number of these carriers while in the magnetoresistance, the light holes have a very remarkable effect. The assumption of equal relaxation times for light and heavy holes can be easily justified: a given light hole is converted into a heavy hole by lattice scattering for two reasons, namely, (a) energy is carried away by a phonon which makes this process more probable than the inverse process $N_q + 1 > N_q$, [see (6.4.12)], and (b) the heavy-hole density of states and therefore the transition probability are larger than for the inverse process. For the case of ionized impurity scattering, however, the relaxation times for light and heavy holes differ. In a range of temperatures, where light holes are scattered preferably by ionized impurities and heavy holes by the lattice, a satisfactory explanation of experimental results is met with difficulty.

The observed temperature dependence of the Hall mobility in pure p-type Ge (e.g., $N_A = 1.44 \times 10^{13}\,\mathrm{cm}^{-3}$; $N_D = 0.19 \times 10^{13}\,\mathrm{cm}^{-3}$; [8.16]) between 120 K and 300 K,

$$\mu_\mathrm{H} = 2060\,\mathrm{cm}^2/\mathrm{Vs} \times (T/300\,\mathrm{K})^{-2.3} \ , \qquad (8.3.20)$$

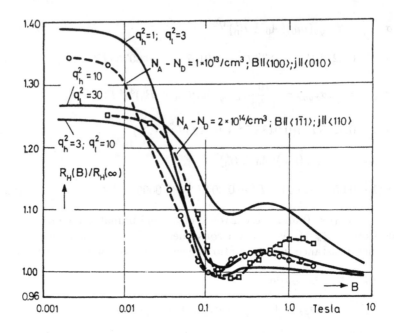

Fig. 8.2: Hall coefficient of p-type germanium vs magnetic field at 77 K observed for two values of the carrier density (dashed curves) and calculated (full curves); for the definition of q^2: (6.6.3). (after [8.18])

has been explained by optical and acoustic–phonon scattering with a value for the ratio of deformation potential constants $D u_1/\varepsilon_{ac}\omega_0$ of $\sqrt{3.8}$ by taking into account effective mass ratios of $m_h/m_0 = 0.35$ and $m_1/m_0 = 0.043$ and a concentration ratio of 23.2. The effective masses have been assumed to be temperature-independent although cyclotron resonance reveals an increase of the light-hole mass by 16% between 10 K and 100 K [8.17]: it does not enter appreciably into the conductivity due to the small relative number of light holes. Between 50 K and 80 K, the mobility is reduced only by 10–20% in comparison with the *acoustic* mobility and is proportional to $T^{-3/2}$. At lower temperatures the influence of ionized impurity scattering increases; very good agreement with the Brooks-Herring formula (6.3.23) is obtained down to 20 K.

The influence of band structure on the galvanomagnetic properties, especially on magnetoresistance and, outside the range $\mu_H B \ll 1$, on the B-dependence of the Hall coefficient R_H, is noteworthy. $R_H(B)$ has a minimum at a value of B which increases with increasing temperature. Experimental data of $R_H(B)$ are shown in Fig. 8.2 [8.18]. The curves are labeled with parameters q_h^2 and q_l^2 which given by (6.6.3), where the subscripts refer to heavy and light holes, respectively. There is qualitative agreement between the calculated and the observed data.

In Table 8.2 the observed magnetoresistance of p-type Ge at 77 K [8.13] is compared with theoretical data [8.10, 11] for various directions of j and B

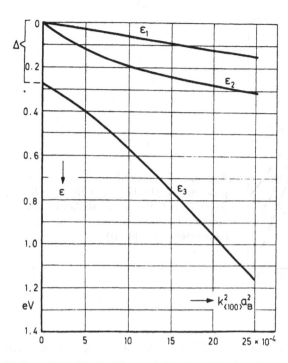

Fig. 8.3: Valence band structure $\varepsilon(k^2)$ for p-type germanium in a $\langle 100\rangle$ direction; $\Delta = 0.29$ eV, similar for p-type silicon where, however, $\Delta = 0.044$ eV (after [8.19])

assuming an isotropic momentum relaxation time of 1.03×10^{-12} s.

The longitudinal magnetoresistance in the $\langle 111\rangle$ direction is smaller by an order of magnitude than in other directions. Therefore, a small misalignment of the sample may cause a considerable error in the measurement, and the agreement between experimental and theoretical data is not as good as in directions where the magnetoresistance is large. In strong magnetic fields (about 10 T), the longitudinal magnetoresistance saturates in a $\langle 111\rangle$ direction but not in a $\langle 100\rangle$ direction.

Galvanomagnetic phenomena in p-type Si are different in some respects from those in p-type Ge. The energy Δ of the split-off valence band is only 0.04 eV in p-type Si; even at room temperature where $k_B T = 0.026\,eV$, it is not appreciably populated because of its small density of states which is in competition with its small effective mass. In band calculations even the small value of Δ given above leads to a nonparabolicity of the other two bands. Especially for the heavy hole band in the $\langle 110\rangle$ direction in k-space, the deviation from the linearity in Fig. 8.3 is remarkable at energies of a few meV; cross sections of the heavy-hole band in a $[1\bar{1}0]$ plane for various energies relative to Δ (0.25, 0.95, and 1.5) were shown in Fig. 2.28b. A calculation of the energy distribution of carriers in a nonparabolic spherical band has been given by *Matz* [8.20]. *Asche* and *Borzeszkowski* [8.21] calculated the temperature dependent mobility of heavy holes in this approximation as shown in Fig. 8.4 for values of 0 and $\sqrt{2.5}$ for $D\,u_1/\varepsilon_{ac}\omega_0$. In a $\langle 110\rangle$ direction there is a swelling of the heavy-hole constant-

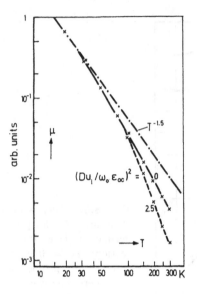

Fig. 8.4: Calculated temperature dependence of the hole mobility in silicon taking band nonparabolicity into account, for two values of the coupling parameter (after [8.21])

energy surface with increasing energy which was shown in Fig. 2.28a. For temperatures $T > 100$ K, the consideration of optical deformation potential scattering in addition to acoustic deformation potential scattering with a value of $\sqrt{2.5}$ for the ratio of the deformation potential constants leads to the observed $T^{-2.9}$ law (Fig. 8.4). Between 50 and 100 K, a $T^{-1.9}$ dependence is calculated for acoustic deformation potential scattering. The deviation of the exponent (-1.9) from the usual value of -1.5 is due to band nonparabolicity.

A consequence of the nonparabolicity of the heavy-hole band is an increase of the conductivity-effective mass with temperature because with increasing thermal energy k_BT the carriers are raised to less parabolic regions of the band. Figure 8.5 shows the conductivity-effective mass in p-type Si vs $10^3/T$ as determined from magneto-Kerr observations [8.22].[4]   This dependence clearly demonstrates the band nonparabolicity [8.23]. Negative differential conductivity in nonparabolic bands has been discussed by *Persky and Bartelink* [8.24] as well as *Fawcett* and *Ruch* [8.25]. The B-dependent Hall coefficient R_H up to 9 T has been found to be anisotropic in p-type Si [8.26]. This and the anisotropy of magnetoresistance which increases with temperature have led to the conclusion that the band is nonparabolic as considered above. The weak-field magnetoresistance in p-type Si observed at 77 K by *Long* [8.27] yields for the coefficients of the inverted Seitz equation (7.4.9) $b = 14.2$, $c = -10.6$ and $d = 0.8$, all in units of 10^7 (cm^2/Vs)$^2 = 10^{-1}$ T^{-2}.

---

[4]The magneto-Kerr effect is a change in ellipticity and polarization of microwave reflection from a sample when exposing it to a longitudinal magnetic field.

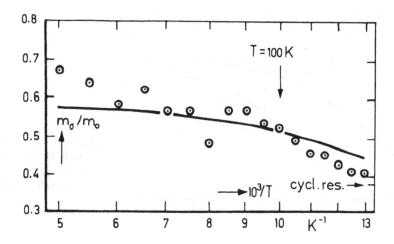

Fig. 8.5: Conductivity-effective mass in p-type silicon obtained from magneto-Kerr effect measurements. The cyclotron resonance value obtained at 4 K is indicated at the vertical scale on the right-hand side of the diagram (after [8.22])

Table 8.2. Experimental and theoretical magnetoresistance of p-Ge at 77 K

$(\Delta\varrho/\varrho B^2)_i^B$	Experim. $\times 10^{-5} T^{-2}$	Theor. $\times 10^{-5} T^{-2}$
$M_{100}^{0\bar{1}0}$	30.4	30.4
$M_{110}^{1\bar{1}0}$	27.0	27.4
M_{100}^{100}	0.14	0.093
M_{110}^{110}	2.0	3.25

8.4 Warm and Hot Holes

While only qualitative agreement between observed and calculated data has been found for galvanomagnetic phenomena in p-type Ge and Si, the situation is even more difficult in treating warm and hot holes. As already mentioned, the time for a conversion of a light hole into a heavy hole is about one collision time, i.e., at a scattering process the conversion takes place with a large probability. Therefore, the electron temperatures of both kinds of carriers may be taken as being equal in the Maxwell–Boltzmann approximation of the hot-carrier distribution function. Let us consider only very strong electric fields. The observed field dependence of the drift velocity of holes in p-type Ge is shown in Fig. 8.6 for two values of the lattice temperature [8.28, 29]. At low temperatures there is a tendency toward saturation. The investigation of the Sasaki–Shibuya effect

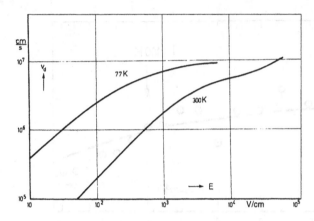

Fig. 8.6: Drift velocity vs field characteristics in p-type germanium (after [8.28])

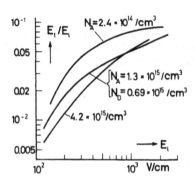

Fig. 8.7: Sasaki-Shibuya effect in p-type germanium for various impurity concentrations at a lattice temperature of 77 K (after [8.30])

(Sect. 7.5, Fig. 7.6) in p-type Ge as shown in Fig. 8.7, yields the same sign as in n-type Ge: the deviation of the field direction from the current direction in both materials is towards the $\langle 111 \rangle$ direction [8.30]. A compensated sample behaves similarly to an uncompensated sample.

Problems

8.1. Show that the conductivity effective mass for mixed conduction by light and heavy holes is given by

$$m_\sigma = (m_{\rm dl}^{3/2} + m_{\rm dh}^{3/2})/(m_{\rm dl}^{3/2}m_{\sigma l}^{-1} + m_{\rm dh}^{3/2}m_{\sigma h}^{-1}) \ . \tag{8.0.1}$$

8.2. In Sect. 8.1, the $\varepsilon(k^2)$ relationship as given by (8.1.10) has been linearized. However, transport in p-type silicon cannot be described adequately, even down to temperatures of 100 K without introducing the nonparabolicity of the valence bands at least to first order. Rather than $\varepsilon(k^2)$, it is more convenient to expand

$k^2(\varepsilon)$:

$$k^2 \approx K_1(\varepsilon + K_2\varepsilon^2) \tag{8.0.2}$$

for $K_2\varepsilon \ll 1$ with two constants K_1 and K_2. For $\tau_m \propto \varepsilon^r$, determine the average $\langle\tau_m\rangle$ of the momentum relaxation time as a function of temperature for a nondegenerate hole gas (one type of holes only).

8.3. Perform the calculation which leads to (8.2.2).

8.4. Calculate the longitudinal and the transverse effective mass of *split p-germanium* parallel and perpendicular to the strain in $\langle100\rangle$ and $\langle111\rangle$ directions.

9. Quantum Effects in Transport Phenomena

In Chap. 2 we learnt how the quantization of the atomic energy levels results in the band structure of the crystalline solid. However, this is not the only domain of quantum mechanics in semiconductivity. Although most transport phenomena can be explained by assuming a classical electron gas, there are some which can be understood only by quantum mechanical arguments. In Sect. 9.1 we will treat phenomena which rely on the quantum mechanical *tunnel effect*, while in Sects. 9.2–9.4 the quantization of electron orbits in a strong magnetic field with the formation of *Landau levels* will be the basis for an understanding of the *oscillatory* behavior of transport phenomena.

9.1 Tunnel Diode, Resonant Quantum Wells, and Superlattices

Although many scientists had observed an *anomalous* current-voltage characteristic in the forward direction of degenerate p-n junctions, it was not until 1958 that *Esaki* [9.1] gave an explanation in terms of a quantum tunneling concept. Since then the *Esaki diode* has become a useful device due to its ultrahigh-speed, low-power and low-noise operation [9.2].

The current-voltage characteristics are shown schematically in Fig. 9.1a. In the reverse direction the current increases as the voltage is increased. In the forward direction it shows a peak at a voltage V_P followed by an interesting region of *negative differential conductivity* (n.d.c.) between V_P and V_V. The various components of the current are plotted in Fig. 9.1b. Let us first consider the tunnel current component at temperatures near absolute zero.

Figure 9.2a shows a simplified diagram of the equilibrium potential distribution in a p-n junction which is degenerate on both sides. The Fermi level ζ is constant across the junction. On the n-side the conduction band is filled with electrons up to ζ, while on the p-side the valence band is filled with holes down to ζ (hatched areas). In Fig. 9.2b the potential in the junction upon application of a *small* forward bias V is given. Electrons from the conduction band of the n-side tunnel through the gap to the empty sites (called *holes*) in the valence band of the p-side.

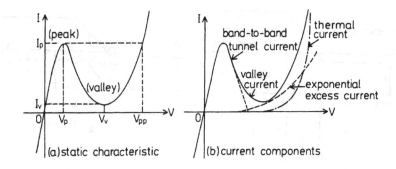

Fig. 9.1: (a) Static current-voltage characteristics of a typical tunnel diode (schematic diagram). (b) The characteristics of Fig. 9.1a decomposed into the current components (after [9.3])

Before we calculate the tunnel current let us consider the case of a *large* forward bias illustrated by Fig. 9.2c. The overlap of the hatched areas in the vertical direction is of interest in the semiconductor laser which will be discussed in Sect. 13.2; there is no horizontal overlap and consequently, no tunneling of electrons. In this case we have only the usual thermal diffusion current indicated in Fig. 9.1b.

The tunneling process is *direct* if in the energy-vs-momentum relationship of the carriers $\varepsilon(k)$, the conduction band edge is located at the same k-value as the valence band edge; otherwise it is *indirect*. In the indirect process, the electron transfer from the conduction band edge to the valence band edge involves a change in electron momentum which may be supplied either by an impurity or a phonon (*impurity-assisted tunneling* and *phonon-assisted tunneling*, respectively). We will consider only direct tunneling here.

The quantum mechanical calculation of the tunnel process through a potential barrier V_0 is similar to the calculations given in (2.1.1–5) [9.4]. The incoming and reflected electron waves are described by

$$\psi = A_1 e^{ikx} + B_1 e^{-ikx} \quad \text{for } x < -b/2 \tag{9.1.1}$$

while the transmitted wave is given by

$$\psi = A_3 e^{ikx} \quad \text{for } x > b/2 \tag{9.1.2}$$

and the wave inside the barrier is a solution of (2.1.3):

$$\psi = A_2 e^{-\beta x} + B_2 e^{\beta x} \quad \text{for } -b/2 \leq x \leq b/2 \ . \tag{9.1.3}$$

The continuity condition for ψ and $d\psi/dx$ at $x = b/2$ and $x = -b/2$ yields for the transmission probability

$$T_t = |A_3/A_1|^2 = \left[1 + \frac{1}{4} \left(\frac{\beta}{k} + \frac{k}{\beta} \right)^2 \sinh^2 \beta b \right]^{-1} \tag{9.1.4}$$

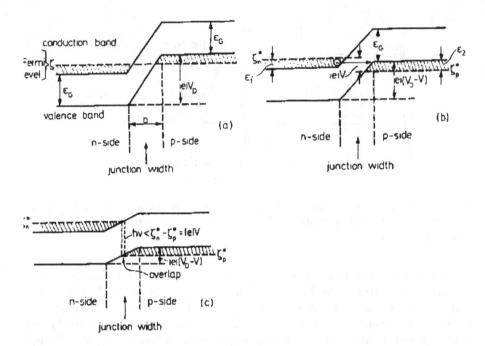

Fig. 9.2: (a) Degenerate p-n junction at thermal equilibrium. (b) Degenerate p-n junction at a small forward bias; tunneling electron is indicated. (c) Degenerate p-n junction at a large forward bias

which for $\beta b \gg 1$ can be approximated by

$$T_t = \left(\frac{4\beta k}{\beta^2 + k^2}\right)^2 \exp\left(-2\beta b\right) \approx \exp\left(-2\beta b\right). \tag{9.1.5}$$

For the second approximation, the prefactor of the exponential has been neglected. So far we have assumed a rectangular potential barrier of width b and height V_0. However, Fig. 9.2b shows that the barrier is triangular and has a height of about ε_G while the energy ε of the electron can be neglected. With β given by $\sqrt{2m\,\hbar^{-2}\varepsilon_G}$, where m is the effective mass and $b \approx \varepsilon_G/|e|E$, where E is the electric field strength in the junction, a calculation modified for the triangular barrier yields an additional factor of 2/3 in the argument of the exponential:

$$T_t = \exp\left(-\frac{4\sqrt{2m}\varepsilon_G^{3/2}}{3|e|E\hbar}\right). \tag{9.1.6}$$

For a parabolic barrier we obtain the same result except that the factor

$4\sqrt{2}/3 = 1.88$ is replaced by $\pi/2^{3/2} = 1.11$. For simplicity we have neglected in this one-dimensional treatment the momentum perpendicular to the direction of tunneling. For phonon-assisted indirect tunneling, ε_G in (9.1.6) has to be replaced by $(\varepsilon_G - \hbar\omega_0)$ where $\hbar\omega_0$ is the phonon energy [9.5]:

$$T_t = \exp\left(-\frac{4\sqrt{2m}(\varepsilon_G - \hbar\omega_0)^{3/2}}{3|e|E\hbar}\right) . \tag{9.1.7}$$

The same expression for photon-assisted tunneling, where $\hbar\omega_0$ is replaced by the photon energy $\hbar\omega$, will be obtained in Sect.11.5 for light absorption in homogeneous semiconductors in an E-field. For a calculation of the current, T_t is treated as a constant in the small voltage range involved.

At thermal equilibrium, the tunneling current $I_{c\to v}$ from the conduction band to the valence band is given by

$$I_{c\to v} = A\,T_t \int\limits_{\varepsilon_c}^{\varepsilon_v} f_c(\varepsilon)g_c(\varepsilon)[1 - f_v(\varepsilon)]g_v(\varepsilon)d\varepsilon , \tag{9.1.8}$$

where $f_c(\varepsilon)$ and $f_v(\varepsilon)$ are the Fermi-Dirac distribution functions of electrons in the conduction band and valence band, respectively, $g_c(\varepsilon)$ and $g_v(\varepsilon)$ are the corresponding densities of states and A is a constant. The reverse current is given by a corresponding expression

$$I_{v\to c} = A\,T_t \int\limits_{\varepsilon_c}^{\varepsilon_v} f_v(\varepsilon)g_v(\varepsilon)[1 - f_c(\varepsilon)]\,g_c(\varepsilon)d\varepsilon , \tag{9.1.9}$$

and with no bias applied due to detailed balance $I_c \to v = I_v \to c$. When the junction is biased, the current can be approximated by [9.1]

$$I = I_{c\to v} - I_{v\to c} = A\,T_t \int\limits_{\varepsilon_c}^{\varepsilon_v} (f_c - f_v)g_c\,g_v\,d\varepsilon , \tag{9.1.10}$$

where in f_c and f_v, the Fermi energies are replaced by the quasi-Fermi energies ζ_n^* and ζ_p^* and the applied voltage $V = (\zeta_n^* - \zeta_p^*)/e$. According to (3.1.7) the densities of states vary as $\sqrt{\varepsilon - \varepsilon_c}$ and $\sqrt{\varepsilon_v - \varepsilon}$, respectively. For $2k_BT \geq \zeta_n^* - \varepsilon_c$ and $\geq \varepsilon_v - \zeta_p^*$, the Fermi-Dirac distribution can be linearized:

$$f_c(\varepsilon) \approx \frac{1}{2} - (\varepsilon - \zeta_n^*)/4k_BT; \quad f_v(\varepsilon) \approx \frac{1}{2} + (\zeta_p^* - \varepsilon)/4k_BT . \tag{9.1.11}$$

Hence, from (9.1.10) we obtain

$$I = A' \frac{|e|V}{4k_BT}(\varepsilon_c - \varepsilon_v)^2 = A' \frac{|e|V}{4k_BT}(\varepsilon_1 + \varepsilon_2 - |e|V)^2 \tag{9.1.12}$$

for $\varepsilon_1 + \varepsilon_2 \geq |e|V$ where A' is another constant and according to Fig. 9.2b, $\varepsilon_1 = \zeta_n^* - \varepsilon_c$ and $\varepsilon_2 = \varepsilon_v - \zeta_p^*$ [9.5].

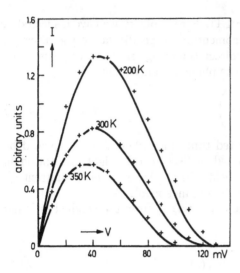

Fig. 9.3: Current-voltage characteristics of a tunnel diode observed (curves) and calculated (crosses) for various temperatures (after [9.6])

Figure 9.3 shows a comparison between experimental curves published by Esaki and points calculated according to (9.1.12) where the constants A' and $\varepsilon_1 + \varepsilon_2$ have been fitted at the peaks of the curves [9.6]. Considering the approximation made in the calculation, the agreement is satisfactory.

For voltages $V \geq (\varepsilon_1 + \varepsilon_2)/|e|$, the tunneling current should decrease to zero and only the normal diode current should flow; in practice, however, there is a current in excess of the latter. This excess current is shown in Fig. 9.1b. It is due to the fact that in a strong electric field the band edges are not as sharp as one might expect from $g(\varepsilon) \propto \sqrt{\varepsilon - \varepsilon_c}$ but have *tails* [9.7]: $g(\varepsilon) \propto \exp(-\text{const} \cdot \varepsilon^2)$ for $\varepsilon < \varepsilon_c$ (see Sect. 11.5). There is tunneling between the tails of the conduction and valence bands. Tunneling also occurs via impurity states in the energy gap.

If the p-n junction is not doped heavily enough so that both sides are just near degeneracy, there is no negative conductivity region but a region of high resistance in the forward direction. The current-voltage characteristics are then opposite to those of a normal diode. Such a diode is called a *backward diode*. It is useful for low-noise rectification of small microwave signals. For further details see, e.g., [9.3].

The scattering of tunneling electrons by impurities in the junction or in a *Schottky barrier* (metal degenerate– semiconductor junction with carrier depletion at the interface), where one or two phonons are emitted, has been investigated by measuring the second derivative of the bias with respect to the current at very low temperatures. Figure 9.4 shows experimental results obtained by *Thomas* and *Queisser* [9.8] on n-type GaAs/Pd Schottky barriers at 2K [9.9]. The type of phonons (transverse-acoustic or longitudinal-optical) which are emitted is indicated in Fig. 9.4. Such measurements are denoted as *tunnel spectroscopy*.

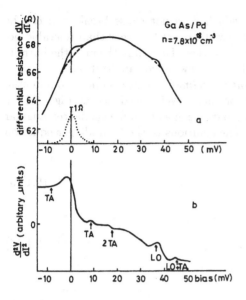

Fig. 9.4: Derivatives of the current-voltage characteristics for a n-GaAs-palladium Schottky barrier at temperatures of 2 K (full line) and 20 K (dashed line); excess differential resistance (dotted line). Arrows indicate phonon energies (per electron charge) in GaAs (after [9.8])

Tunneling through the gap has also been considered by *Zener* [9.10] in reverse biased nondegenerate p-n junctions: if the electric field in the junction is very strong, the probability for tunneling given by (9.1.6) is large enough for this process to lead to a breakdown in the reverse direction of the *I-V* characteristics [9.11]. However, in almost all commercially available *Zener diodes*, breakdown is due to impact ionization [9.12] rather than tunneling (Sect. 10.2).

Twelve years after Esaki's invention of the tunnel diode, *Esaki* and *Tsu* [9.13] proposed a one-dimensional *superlattice*. It consists of two semiconductors which are different in their energy gaps but equal in their crystal structure and which have not too different lattice constants. Two materials which fulfill these conditions, are GaAs and $Ga_{0.7}Al_{0.3}As$. At room temperature ε_G in GaAs is 1.42 eV and at 2K 1.52 eV while for the other compound it is larger by 0.23 eV. By MBE (*molecular beam epitaxy*) [9.14] or MOCVD (*metal organic chemical vapor deposition*) [9.15, 16] about 20 layers of typically 5 nm thickness of alternating GaAs and GaAlAs are deposited on a GaAs substrate. The resulting periodic variation of the band gap ε_G in one dimension simulates the Kronig-Penney periodic-potential-wells model shown in Fig. 2.2. However, now the potential wells and barriers we are considering are, of course, not due to the atoms of the crystal lattice but they are formed by the alternating GaAs and GaAlAs layers with their different energy gaps. It is surprising that the thickness of such a layer may even be only three atomic monolayers because the concept of bands requires many atomic layers of a material. This system is called a one-dimensional *super lattice* (SL). The superlattice constant is usually between 10 and 100 times the atomic lattice constant. Because of the much larger barrier width in the superlattice, a voltage which does not cause breakdown can be applied between two adjacent potential wells which is larger than tolerable between two adjacent atoms in the crystal lattice. Tunneling through the barriers offers the possibility of an n.d.c. and hence new device applications.

Let us first study the effect of *resonant tunneling*[1] between only two potential wells before continuing with the SL problem. With a glance at Fig. 2.1 we reconsider the well-known problem of a particle in a one-dimensional rectangular well that is infinitely deep, $V_0 = \infty$. The solutions of the Schrödinger equation for a time-independent problem

$$\frac{-\hbar^2}{2m}\frac{d^2\psi}{dz^2} = \varepsilon\,\psi \tag{9.1.13}$$

are

$$\varepsilon_n = \frac{\hbar^2}{2m}\left(\frac{n\,\pi}{L_z}\right)^2 \; ; \; n = 1, 2, 3 \ldots \tag{9.1.14}$$

and

$$\psi_n \propto \sin(n\,\pi\,z/L_z) \; , \tag{9.1.15}$$

where m is the carrier effective mass, and L_z is the well width. In three dimensions with unlimited motion in the x- and y-directions, the total energy is

$$\varepsilon = \varepsilon_n + \frac{\hbar^2}{2m}(k_x^2 + k_y^2) \; . \tag{9.1.16}$$

As a result of the quantization of the particle motion normal to the layer, discrete bound states emerge, and the energy of the lowest state increases as L_z decreases. This is called the *quantum size effect* (QSE, see also Sect. 14.3).

Now, we study electrons tunneling between two wells separated by a thin barrier assuming conservation of the electron energy as well as the transverse momentum. The tunnel current integrated over the transverse direction, is given by

$$j = \frac{emk_BT}{2\pi^2\hbar^3}\int\limits_0^\infty |T_t|^2 \ln \frac{1 + \exp[(\zeta - \varepsilon)/k_BT]}{1 + \exp[(\zeta - \varepsilon - eV)/k_BT]}\,d\varepsilon \; , \tag{9.1.17}$$

where $|T_t|^2$ is the transmission coefficient, ζ is the Fermi energy, and ε is the particle energy in the longitudinal direction. This equation can easily be derived from (14.3.3). It is also possible to calculate $|T_t|^2$ as a function of the electron energy similar to the transmission of light through a Fabry-Perot filter [9.17]. When the energy ε of the tunneling electron matches one of the energy levels ε_i in the quantum wells, then the amplitude of the electron de Broglie waves in the well build up due to multiple scattering. The waves leaking in both directions

---

[1]Resonant transmission of electrons by atoms of noble gases is known as the Ramsauer effect (Sect.6.1).

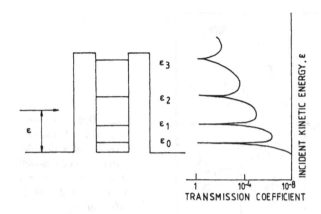

Fig. 9.5: Double – barrier quantum well (lhs) and transmission of an incident electron with energy ε showing resonances at the energy levels of the well.

cancel the reflected waves and enhance the transmitted ones. Near resonances the transmission probability $|T_t|^2$ is given by

$$|T_t|^2 \approx \frac{4 T_1 T_2}{(T_1 + T_2)^2} \frac{(\hbar/\tau)^2}{(\varepsilon - \varepsilon_i)^2 + (\hbar/\tau)^2} , \qquad (9.1.18)$$

where T_1 and T_2 are the transmission amplitudes of the two barriers at $\varepsilon = \varepsilon_i$, and τ is the lifetime width of the resonant state which quasiclassically is $\hbar/[\varepsilon_i(T_1 + T_2)]$. For a system of identical barriers ($T_1 = T_2$) and in the absence of scattering, there is complete transparency for electrons at energies $\varepsilon = \varepsilon_i, (i = 1, 2, 3 \dots)$ which are the resonant energies. Figure 9.5 schematically shows the energy diagram of the well formed by two barriers (*double barrier quantum well*) and the energy ε of the incoming electron, together with the transmission probability, vs ε. The experimental current-voltage characteristics at 77 K and 300 K are depicted in Fig.9.6. The rhs inset shows the structure. The barriers consist of undoped AlAs of 2.5 nm thickness each. The n.d.r. (*negative differential resistance*) exists in both current directions due to the symmetric structure of the device [9.18]. The area is only 2.8×10^{-7} cm^2 and the current density is correspondingly high. Whether the observations can be explained by the Fabry-Perot effect or by a sequential tunneling where the transmission through the two barriers are not correlated depends on the relative value of the elastic-scattering relaxation time called *phase relaxation time* and the tunneling time [9.19, 20]. The latter may be estimated from the barrier height $\Delta\varepsilon$ and the uncertainty relation $\tau \approx h/\Delta\varepsilon$ where h is Planck's constant. For $\Delta\varepsilon \approx 0.2$ eV and a Fermi velocity of order 10^7 cm/s, the transit time is of order 10^{-13}s. Experiments by *Sollner* et al. [9.21] at 25 K on the dc characteristic and the current response under FIR laser illumination ($\lambda = 119\mu$m; $\nu = 2.5$ THz) indicate that the charge transport is very fast, corresponding to $\tau \leq 6 \times 10^{-14}$ s. At 100 K, oscillations at frequencies up to 18 GHz were observed with an output power of 5 μW which are due to the n.d.r. [9.22]. In order to achieve equal transmission for both barriers at the operating point,

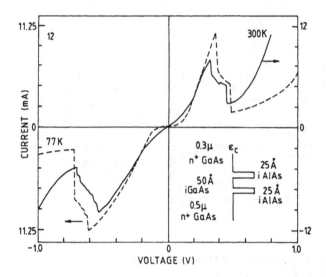

Fig. 9.6: Experimental $I - V$ characteristics of a double-barrier quantum-well structure (see inset) at 77 K and 300 K (after [9.18])

a resonant tunneling structure was proposed [9.23] and built [9.24] in which a symmetric double barrier quantum well was included in the base of a bipolar transistor and shown to operate at room temperature.

The tunneling times were discussed in detail by Jonson [9.25]. Decay rates of electrons trapped in metastable quantum well states of GaAs/GaAlAs double-barrier heterostructures are obtained from time-resolved photoluminescence. For sufficiently thin barriers, tunneling dominates over recombination with holes. Observed lifetimes of about 10^{-10} s are dominated by the time spent in the well, while the barrier traversal time is insignificant. The time it would take an electron at the Fermi level in the conduction band of the doped GaAs to pass the scattering region, if the GaAlAs barriers had not been there, is around 10 fs and should give the approximate scale for the traversal time [9.25]. The optical properties of quantum wells and superlattices will be investigated in Sect.13.3. Optical transitions involve, of course, also those between the valence band and the conduction band of the semiconductor. The alignment of the two types of bands at the interface between the two semiconductors is classified, as shown in Fig.9.7 [9.26]. The various types of band alignment are characterized by the way how the band edges of the smaller-gap semiconductor are located relative to the band gap of the second semiconductor (Fig.9.8). The combination GaAs/GaAlAs is a prototype of the type-I alignment where the band edges of GaAs are both within the band gap of GaAlAs. This is the most intensively studied system. Another system, CdSe/ZnTe, shows the type-II *staggered* alignment. The band gap of CdSe is 1.75eV, that of ZnTe is 2.4 eV, the difference being 0.65 eV. Since the conduction band edge of ZnTe is 1.35 eV above that of CdSe, the valence band edge of ZnTe must be 0.7 eV above that of CdSe. Finally there is the type-II *misaligned* system where both the conduction and

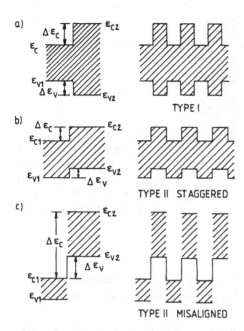

Fig. 9.7: Various types of band alignments at the interface between two semiconductors (lhs) and in a superlattice (rhs).

valence band edges of one material lie entirely within either the conduction band or the valence band of the other. A typical example is InAs/GaSb. A type-III alignment (not shown in Fig. 9.7) occurs when one of the semiconductors has a zero gap or is semimetallic such as HgTe (with CdTe being an appropriate partner). In Fig. 9.8 the band gaps of various III–V compounds vertically arranged according to the band offsets, and their lattice constants are shown. The doping is usually done only in the material with the larger gap, i.e., in the barriers. Most of the carriers drop into the wells where they have a larger mobility far away from the ionized impurities. Strained superlattices consisting of Si/Ge_xSi_{1-x} and Ge/Ge_xSi_{1-x} have found considerable interest because of the possibility of band gap conversion from an indirect gap to a quasi-direct gap due to structurally induced zone-folding along the superlattice growth direction. Electroreflectance spectra (see Sect. 11.5) at 40 K obtained by Pearsall et al. [9.27] from $Ge/Ge_{0.6}Si_{0.4}$- and $Ge/Ge_{0.7}Si_{0.3}$- superlattices show features which do not exist in bulk Ge. These peaks have been identified as being due to transitions involving quasi-direct zone-folded states [9.28]. Another direct semiconductor but without zone-folding should be an alloy of grey tin (α-Sn) and Ge: α-Sn_xGe_{1-x} with $0.24 \leq x \leq 0.7$ [9.29]. Grey tin is a *zero-gap semiconductor*. α-Sn on InSb- or CdTe-substrate does not convert to metallic tin unless the temperature is raised above 130°C. For MBE growth, the substrate is cooled to liquid nitrogen temperature and the surface temperature is briefly raised when the tin atoms hit the surface, by heat pulses from a halogen lamp. Even then only 6% Sn atoms can at most be included in a growing monolayer.

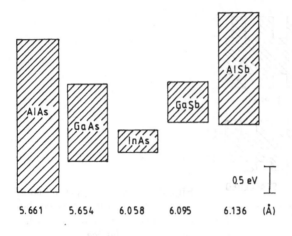

Fig. 9.8: Alignment of band gaps for various III-V semiconductors. The numbers at the bottom are the lattice constants.

For optical experiments see, e.g., Abstreiter et al. [9.30]. Quantum structures of the above mentioned CdTe/Cd$_{1-x}$Zn$_x$Te combination have been reported by *Allegre* et al. [9.31].

All the superlattices mentioned so far are denoted as *compositional superlattices* because the composition varies periodically along the growth direction. Another type are the *doping superlattices* (n-i-p-i) where it is only the doping that varies periodically while the basic material and therefore the band gap remains the same throughout [9.32]. The band-edge variation for 1 1/2 periods is essentially the same as that of the p-n-p transistor shown in Fig. 5.12 but with equal doping concentration on the emitter and the collector side. Unfortunately the advantage of having the same lattice constant throughout the whole sample is more than compensated by the fact that the high doping level required for the tunnel effect drastically reduces the mobility of the carriers. In contrast to the compositional superlattice, it is here not possible to have the doping in the barriers and most of the carriers in the wells. Only optical experiments have been reported.

In superlattices with weakly coupled thick barriers there may, at high applied voltages, be the formation of a high-field domain, (Fig. 9.9). Tunneling from the ground state ε_1 of the well through the barrier to the next well's energy level ε_2 is followed by a transition to ε_1 by emission of a phonon. In this case nearly all of the applied voltage V drops at this barrier, i.e., $eV = \varepsilon_2 - \varepsilon_1$. Such a *high-field domain* is doubled or tripled in length by a corresponding increase of the applied voltage. In a superlattice with 49 periods exactly 48 n.d.c. features have been observed which corroborates this picture [9.33]. From optical absorption and photoluminescence the energy difference of the levels $\varepsilon_2 - \varepsilon_1 = 102$ and 106 meV was determined, in rough agreement with the voltage drop per superlattice period in the transport experiment. Electrically tunable GHz oscillations have been observed in doped GaAs/AlAs superlattices by Kastrup et al. and explained in terms of field domains [9.34].

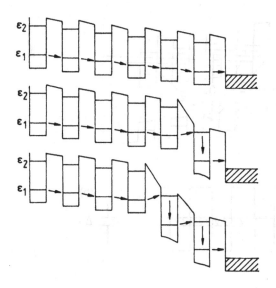

Fig. 9.9: Superlattice under various levels of an applied high voltage may form a high-field domain. The arrows indicate the electron transitions.

An effect of the voltage applied to a crystal lattice first discussed by *Wannier* [9.35] are the *Wannier-Stark ladder* and the *Bloch oscillation*. There is a vast theoretical literature about these effects and only few experiments, in fact mostly optical ones which will be treated in Sect. 13.3. Let us consider an electron in a periodic potential as in Fig. 2.2 but under the action of an externally applied dc field E_0 in the z-direction. In the Schrödinger equation this field adds a potential energy term eE_0z to the zero-field Hamiltonian. Let us treat the problem by solving the equation of motion for an electron of charge $-|e|$ neglecting scattering [9.36]:

$$\hbar\,dk/dt = -|e|E_0 \qquad\qquad (9.1.19)$$

which is, with a constant of integration,

$$k(t) = k_0 - (|e|E_0/\hbar)t \ . \qquad\qquad (9.1.20)$$

The energy $\varepsilon(k)$ as a function of time t is therefore

$$\varepsilon(k) = \varepsilon(k_0 - |e|E_0\,t/\hbar) \qquad\qquad (9.1.21)$$

and for an energy band of lattice period d

$$\varepsilon_\nu(q) = \bar\varepsilon_\nu - \frac{1}{2}\Delta_\nu \cos(q\,d) \qquad\qquad (9.1.22)$$

where according to (2.1.14) $\bar\varepsilon_\nu$ is the middle of the ν-th band, Δ_ν is its width, and q is the lattice wavevector, the velocity $v(t) = \hbar^{-1}d\varepsilon/dk$ is

$$v(t) = \hbar^{-1}\frac{\partial}{\partial k_0}\varepsilon_\nu(k_0 - |e|E_0\,t/\hbar) \ , \qquad\qquad (9.1.23)$$

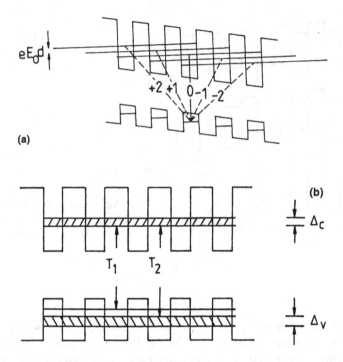

(a)

(b)

Fig. 9.10: (a). Optical transitions between the valence band superlattice levels and those of the conduction band where a Wannier–Stark ladder has emerged under the action of a strong electric field (schematical) (after Mendez [9.37]) (b). Schematic conduction band diagram of a compositional superlattice where electron states extend over all periods of the superlattice at zero applied field.

$$v(t) = (\Delta_\nu d/2\hbar) \sin[(k_0 - |e|E_0\, t/\hbar)d] \ , \tag{9.1.24}$$

where $k(q)$ is the trajectory in q-space . The position $z(t)$ of an electron in coordinate space is found by integration over time t:

$$z(t) = z_0 - (\Delta_\nu/2|e|E_0) \cos[(k_0 - |e|E_0\, t/\hbar)d] \tag{9.1.25}$$

where z_0 is another constant of integration, and thus the energy $\varepsilon_\nu(t) = |e|E_0\, z(t)$ is given by

$$\varepsilon_\nu(t) = |e|E_0\, z_0 - \frac{1}{2}\Delta_\nu \cos[(k_0 - |e|E_0\, t/\hbar)d] \ . \tag{9.1.26}$$

In a constant electric field E_0, the electron oscillates with a frequency $\omega_B/2\pi$ where the *Bloch frequency* ω_B is according to the argument of the cosine in (9.1.26)

$$\omega_B = |e|E_0 d/\hbar \ . \tag{9.1.27}$$

The average energy $\langle \varepsilon_\nu \rangle = eE_0 z_0 + \bar{\varepsilon}_\nu$, according to (9.1.22), is found from a Fourier expansion of the wave function

$$\psi(z,t) \propto \exp \left(-\frac{i}{|e|E_0} \int_{k_0}^{k(t)} \varepsilon_n(k) dk \right) , \qquad (9.1.28)$$

which contains the frequencies

$$\omega_{n,\nu} = \varepsilon_\nu/\hbar + n|e|E_0 d/\hbar; \quad n = \pm 1, \pm 2, \pm 3 \dots \qquad (9.1.29)$$

as Wannier [9.35] first demonstrated. He called the series of energy levels the "Stark ladder" because the action of an electric field on the energy levels in atoms is known as the Stark effect. Today the name *Wannier-Stark ladder* is in common usage. In contrast to the Landau levels caused by magnetic quantization (Sect.9.2) the Wannier-Stark ladder in principle does not have a lowest step. The limits of the index n are determined in a real lattice by its finite length $(2N+1)d$, hence $-N \leq n \leq N$. This calculation will now be applied to a superlattice. The effect of an electric field E_0 on the superlattice bands is usually represented by a tilting of the bands along the field direction as shown in Fig.9.10a. The inclination corresponds to the potential energy $-|e|E_0 z$ in the Schrödinger equation as mentioned above. In a strong field E_0, each of the discrete levels is confined to its particular potential well, as shown for the valence band in Fig.9.10a. At intermediate field strengths, the levels are extended over a number of periods of the order of $\Delta/|e|E_0 d$. Electron excitations from a valence level to conduction levels $n = 0, \pm 1, \pm 2$ are indicated in Fig. 9.10a by dashed lines. At decreasing field intensities E_0, the extension of the levels becomes ever wider until at $E_0 \Delta/N e d$ every level is extended throughout the length of the superlattice and the levels are next to each other as shown in Fig. 9.10b. Carrier scattering at impurities, phonons, and lattice defects with a characteristic time τ will not affect localization as long as the level separation $eE_0 d$ is greater than the collision broadening, i.e.,

$$|e|E_0 d > \hbar/\tau , \qquad (9.1.30)$$

which is equivalent to the condition $\omega_B \tau > 1$, i.e., the completion of at least one full oscillation cycle is required.

We now consider the density-of-states function $g_{2D}(\varepsilon)$ for the miniband split in the Wannier-Stark ladder. For the two-dimensional electron gas in a series of equidistant subbands which are due to e.g. size quantization, (14.3.3) in Sect. 14.3 will help us with the following relation

$$g_{2D}(\varepsilon) = (Vm/2\pi\hbar^2)2 \sum_n \Theta(\varepsilon - \varepsilon_n) , \qquad (9.1.31)$$

where a factor of 2 has been included because of spin degeneracy and the spin summation over the subbands is relieved. Here V is the area of the two-dimensional electron gas, m the effective density-of-states mass of the carriers, and $\Theta(z) = 1$ for $z > 0$ and zero otherwise, is called *Heaviside function*.

We consider a superlattice of length $(2N+1)d$ where d is the superlattice period along the growth direction . The superlattice is clad between infinite potential barriers. For the quasicontinuous spectrum of the conduction miniband we assume (9.1.22) with

$$q\,d = n\,\pi/2(N+1); \quad n = 1, 2, \ldots, 2N + 1 \tag{9.1.32}$$

From (9.1.31) we see that the density of states rises in steps beginning at ε_1 and continuing at ε_2 etc. We assume that the optical interband absorption as a function of the photon energy will likewise go in steps [9.36]. We put in (9.1.28) the value of ε_n given by (9.1.26) and notice that in the imaginary argument of the exponential there is a sine function of t [9.38]. Applying one of the definitions of the Bessel function $J_m(z)$ of the first kind of order m

$$\exp(-iz\sin\varphi) = \sum_{m=-\infty}^{\infty} J_m(z)\,\exp(-im\varphi) \tag{9.1.33}$$

with the argument z being $-2/f$ where

$$f = |e|\,E_0\,d/\frac{1}{2}\Delta \tag{9.1.34}$$

is a reduced field strength, we soon arrive at the final result. For an electron transition from the valence band to the conduction band, we have to take for the miniband width Δ the sum $\Delta_v + \Delta_c$ where the two components refer to valence and conduction band, respectively. Because according to (11.9.11) the square of the absolute value of the momentum matrix element, $|p_{vc}|^2$, enters the oscillator strength, we obtain for this quantity

$$|p_{vc}|^2 \propto \sum_{n=-\infty}^{\infty} J_n^2(-2/f) = \sum_{n=-\infty}^{\infty} J_n^2(+2/f) \ . \tag{9.1.35}$$

The independence of $J_n(z)$ from the sign of z is because $J_n(-z) = -J_n(z)$ is valid. We still have to add as a factor the Heaviside function in order to end up at the equivalent of (9.1.31). Because the superlattice consists of $2N+1$ quantum wells the sum over n runs actually from $-N$ to $+N$ which includes $n = 0$, and with a factor α_0 which is about 0.006 [9.39], and the number $2N + 1$ of wells in one band, the *optical absorption coefficient* of the superlattice subject to an effective field f is [9.40, 41]

$$\alpha(\hbar\omega) = (2N + 1)\alpha_0 \sum_{n=-N}^{N} J_n^2(2/f)\,\Theta(\hbar\omega - (\varepsilon_G + \bar{\varepsilon}_c + \bar{\varepsilon}_v + n|e|E_0\,d))$$

$$\tag{9.1.36}$$

where ε_G is the band gap of the well material. The result of the calculation is shown in Fig. 9.11. The smooth curve, valid for $f = 0$, had been calculated

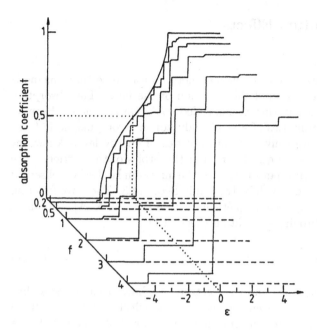

Fig. 9.11:
(a)Theoretical optical absorption coefficient for $N = 20$, normalized to $(2N+1)\alpha_0$ where α_0 is the coefficient for a single quantum well, vs reduced photon energy $[\hbar\omega - (\varepsilon_G + \bar{\varepsilon}_c + \bar{\varepsilon}_v)]/(\Delta_c + \Delta_v)/2$, for increasing values of the reduced field strength f (after [9.40]).

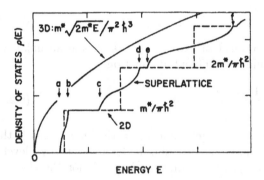

Fig. 9.11: (b) Comparison of the density of states for the three-dimensional and two-dimensional electron systems with that of a superlattice

by an envelope function approximation (Appendix B) from $\pi^{-1} \arccos(-\varepsilon)$ for the density-of-states ratio [9.42] A comparison of this curve and the next curve valid for $f = 0.2$ shows that with increasing number of steps for ever smaller values of f, the smooth curve is the limit of (9.1.36).Experimental results on optical absorption will be discussed in Sect. 13.3.

Unterrainer et al [9.43] have observed in a superlattice sample an *inverse Bloch oscillator* effect: The dc current resulting from an applied dc field is modified by the excitation of intra-miniband transitions caused by a terahertz radiation field from a free-electron laser. The results agree with calculations by *Ignatov* and *Jauho* [9.44], *Winnerl et al.* [9.45] and *Romanov* [9.46].

9.2 Magnetic Quantum Effects

In Sect. 11.11 we will investigate the helical motion of a carrier in a magnetic field. The helix is centered around the direction of the field. The absorption of electromagnetic radiation in this case is called *cyclotron resonance*. Here we consider the transport properties of a semiconductor in strong magnetic fields where the energy bands are converted into a series of energy levels known as *Landau levels* [9.47] due to the quantization of the orbits of the carriers. The most spectacular of magnetic quantum effects, the quantum Hall effect, observed in a 2-dimensional electron gas, will be treated in Sect. 14.3. The wave function $\psi(\mathbf{r})$ of an electron of energy ε in a magnetic induction $\mathbf{B} = [\nabla_r \times \mathbf{A}]$ is obtained as the solution of the Schrödinger equation (Sect. 2.1)

$$\frac{1}{2m}(-i\hbar\nabla_r + |e|\mathbf{A})^2\psi = \varepsilon\,\psi \ , \tag{9.2.1}$$

where the crystal potential $V(r)$ and the electron spin have been neglected but the free electron mass m_0 has been replaced by the effective mass m. If we choose the z-direction of a Cartesian coordinate system as the direction of $\mathbf{B}$, a suitable gauge transformation yields for the vector potential A

$$A = (0, B\,x, 0) \ . \tag{9.2.2}$$

The wave function $\psi(\mathbf{r})$ is the product of $\exp(-ik_y y - ik_z z)$ and an unknown function $\varphi(x)$ for which we obtain the differential equation

$$d^2\varphi/dx'^2 + (2m/\hbar^2)(\varepsilon' - \frac{1}{2}m\omega_c^2 x'^2)\varphi = 0 \ , \tag{9.2.3}$$

where $\varepsilon' = \varepsilon - \hbar^2 k_z^2/2m$, $x' = x - \hbar k_y/|e|B$ and $\omega_c = (|e|/m)B$. This is the Schrödinger equation for a harmonic oscillator having energy levels $\varepsilon' = (n + \frac{1}{2})\hbar\omega_c$ where n is the quantum number. It yields for the electron energy

$$\varepsilon = \left(n + \frac{1}{2}\right)\hbar\,\omega_c + \hbar^2 k_z^2/2m; \quad n = 0, 1, 2, \ldots \tag{9.2.4}$$

The energy in a plane perpendicular to $\mathbf{B}$ is quantized, while the energy of motion in the direction of $\mathbf{B}$, $\varepsilon_{kz} = \hbar^2 k_z^2/2m$, remains unaffected. [2]

Let us now calculate the density of states. The number of allowed values of k_z, which is less than a given $|k_0| = 2\pi/\lambda_0$, is $L_z/(\lambda_0/2) = L_z k_0/\pi = $

---

[2]The orbital radius r of the 1st level is given by $(1/2)(mr^2\omega_c^2) = (1/2)(\hbar\omega_c)$ and turns out to be $\sqrt{\hbar/eB}$ which is typically of the order of 10^{-6}cm, similar to the radius of shallow impurities for which the hydrogen model applies. It is called "magnetic length" l_B. If however, quantization can be neglected, the kinetic energy is $(1/2)m\,r^2\omega_c^2 = \hbar^2 k_F^2/2m$ and the cyclotron radius $\hbar k_F/eB$.

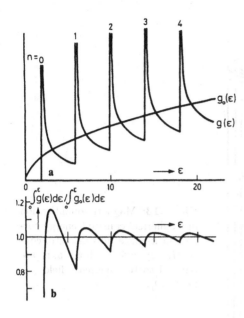

Fig. 9.12: (a) Density of states for no magnetic field g_0 and with field, $g(\varepsilon)$. The number of each Landau level is given in the Figure. [b] Ratio of the number of states below a given energy in a magnetic field relative to the same number without a field (after [9.48]).

$L_z\sqrt{2m\varepsilon_{k_0}}/\pi\hbar$, where L_z is the crystal dimension in the z-direction. By differentiation with respect to ε_{k_0} , the number of allowed k_z, values in the energy range $d\varepsilon_{k_z}$ is found to be

$$g(k_z)d\varepsilon_{kz} = \frac{L_z\sqrt{2m}}{2\pi\hbar\sqrt{\varepsilon_{kz}}}d\varepsilon_{k_z} \ . \tag{9.2.5}$$

Assuming a crystal bounded by planes perpendicular to the x, y and z-axes and having dimensions L_x, L_y and L_z, k_y and k_z due to the periodicity of the Bloch functions have allowed values given by $k_y = n_y 2\pi/L_y$ and $k_z = n_z 2\pi/L_z$, where n_y and n_z are integers. There are $L_y/2\pi$ states per unit extent of k_y. We assume the electron wave function ψ to vanish at the boundaries $x = 0$ and $x = L_x$. x' then has values of between 0 and L_x, and since ψ is centered around $x = \hbar k_y/|e|B$ (according to the definition of x'), k_y extends to $L_x|e|B/\hbar$. The total number of values of k_y is thus given by

$$\mathrm{Max}(n_y) = (L_y/2\pi)\mathrm{Max}(k_y) = (L_y/2\pi)L_x|e|B/\hbar \ . \tag{9.2.6}$$

Multiplication of this expression with the rhs of (9.2.5) and summation over all quantum numbers n yields for the density of states

$$g(\varepsilon)d\varepsilon = \sum_{n=0}^{n_{\max}} \frac{V|e|B\sqrt{2m}}{(2\pi\hbar)^2\sqrt{\varepsilon_{kz}}}d\varepsilon = \frac{V|e|B\sqrt{2m}}{(2\pi\hbar)^2}\sum_{n=0}^{n_{\max}}\frac{d\varepsilon}{\sqrt{\varepsilon - \left(n + \frac{1}{2}\right)\hbar\omega_c}}$$

$$\tag{9.2.7}$$

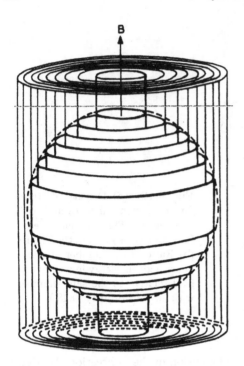

Fig. 9.13: Magnetic quantization of a spherical constant energy surface in k-space into discrete, concentric cylinders whose axis are parallel to the magnetic field (after [9.49])

where the crystal volume $V = L_x L_y L_z$ has been introduced and the sum is to be taken up to a maximum value n_{max} which makes $\varepsilon \geq (n + 1/2)\hbar\omega_c$. The function $g(\varepsilon)$ is shown in Fig. 9.12a together with the density of states for the case of no magnetic field given by (3.1.7) and now denoted by $g_0(\varepsilon)$.

Figure 9.12b shows the ratio

$$\int_0^\varepsilon g(\varepsilon)d\varepsilon \bigg/ \int_0^\varepsilon g_0(\varepsilon)d\varepsilon.$$

For a degenerate semiconductor, which in zero field has a spherical constant-energy surface (*Fermi sphere* $k_x^2 + k_y^2 + k_z^2 = $ const), the allowed k-values in a magnetic field in the z-direction are located on the surfaces of a set of concentric cylinders whose axes are along the k_z-direction (Fig. 9.13). As **B** is increased, the radii of the cylinders become larger and the number of cylinders within a Fermi sphere of an energy which is equal to the Fermi energy ζ becomes smaller. For the case $\hbar\omega_c > \zeta$, only the innermost cylinder with quantum number $n = 0$ is available for accomodation of the carriers; this condition is called the *quantum limit*.

In order to be able to measure quantum effects, $\hbar^2 k_z^2/2m$ in (9.2.4) has to be much smaller than $\hbar\omega_c$, which is equivalent to the condition $k_B T \ll \hbar\omega_c$ or $T/\mathrm{K} \ll B/0.746$ Tesla for $m = m_0$; even for small effective masses, temperatures down to a few K are required in order to observe quantum effects

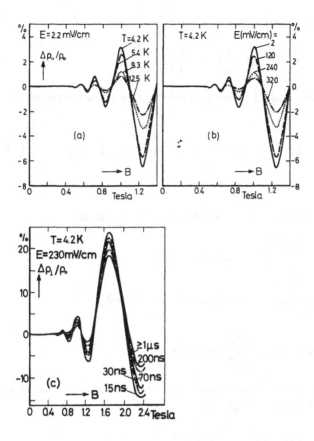

Fig. 9.14: (a) Longitudinal Shubnikov-de Haas effect in n-type InAs at various temperatures for a small electric field strength at long time intervals after the application of the electric field. (b) Same as Fig. 9.14a but for a lattice temperature $T = 4.2$ K and for various electric field strengths. (c) Transverse Shubnikov-de Haas effect in n-type InAs for a lattice temperature $T = 4.2$ K, a strong electric field and various time intervals after the application of the electric field. The non-oscillatory contribution which is negative below 0.5 T has been omitted (after [9.50]).

with available magnetic field intensities. At such low temperatures the carrier gas is degenerate and the average energy $= \zeta_n$ must be $\approx \hbar\omega_c$.

It is clear when the cylinders in Fig. 9.13 move out of the Fermi sphere with increasing magnetic field strength that certain physical properties of the semiconductor show *oscillations* when plotted vs **B**. Figure 9.14a shows [9.46] the longitudinal magnetoresistance $\Delta\varrho/\varrho$ of n-type InAs with a carrier density of $n = 2.5 \times 10^{16}$ cm^{-3} as a function of **B** at various temperatures between 4.2 K and 12.5 K. These oscillations of the magnetoresistance are known as the *Shubnikov - de Haas effect* [9.51]. A calculation for $\Delta\varrho/\varrho$ neglecting the non-oscillatory contribution yields [9.52]

$$\Delta \varrho / \varrho = \sum_{r=1}^{\infty} b_r \cos \left(\frac{2\pi \zeta_n}{\hbar \omega_c} r - \frac{\pi}{4} \right) \, . \tag{9.2.8}$$

To obtain this result, Poisson's transformation equation for an arbitrary function $f(x)$

$$\sum_{n=0}^{n_{\max}} f(x+n) = F(0) + \sum_{r=1}^{\infty} [F(r) \exp(\text{i } 2\pi r x) + F(-r) \exp(-\text{i } 2\pi r x)]$$

$$\tag{9.2.9}$$

has been applied by *Adams* and *Holstein* [9.52] where

$$F(r) = \int_{0}^{n_{\max}+1} f(\xi) \exp(-\text{i} 2\pi r \xi) d\xi \tag{9.2.10}$$

is valid and a Fourier series expansion is involved.

Let us prove

$$\sum_{n=-\infty}^{\infty} \delta(x-n) = \sum_{k=-\infty}^{\infty} \exp(\text{i} 2\pi k x) \tag{9.2.11}$$

where the rhs must be zero except for $x = n$. For $x \neq n$ we replace k by $k + k_1$ with k_1 also being an integer.

$$\sum_{k=-\infty}^{\infty} \exp(\text{i} 2\pi k x) = \sum_{k=-\infty}^{\infty} \exp[\text{i} 2\pi (k+k_1) x]$$

$$= \exp(\text{i} 2\pi k_1) \sum_{k=-\infty}^{\infty} \exp(\text{i} 2\pi k x) \tag{9.2.12}$$

Because $\exp(\text{i} 2\pi k_1) \neq 1$ is valid, the sum for $x \neq n$

$$\sum_{k=-\infty}^{\infty} \exp(\text{i} 2\pi k x) = 0 \tag{9.2.13}$$

vanishes. On the other hand the rhs of (9.2.11) is a periodic function of x with the period being 1. Integration of the rhs from $-\delta$ to $+\delta$ yields

$$\int_{-\delta}^{\delta} \sum_{k=-\infty}^{\infty} \exp(\text{i} 2\pi k x) dx = \sum_{k=-\infty}^{\infty} \frac{\exp(\text{i} 2\pi k \delta) - \exp(-\text{i} 2\pi k \delta)}{\text{i} 2\pi k}$$

$$= \sum_{k=-\infty}^{\infty} \frac{\sin(2\pi k \delta)}{\pi k} \tag{9.2.14}$$

A transition from the sum to an integral is possible in the limit $\delta \to 0$. The integral may be written

$$\int_{-\infty}^{\infty} \frac{\sin(2\pi k\delta)}{\pi k} dk = \frac{2}{\pi} \int_{0}^{\infty} \frac{\sin(2\pi k\delta)}{k} dk = 1. \tag{9.2.15}$$

Hence for x near zero, the rhs of (9.2.11) behaves as if it were a sum of δ - functions. Because of the periodicity and, at non-integer values of x, the zero value of x, (9.2.11) is proven. Taking

$$\sum_{r=0}^{\infty} \int_{0}^{n_{max}} f(x+r) = F(0) + \sum_{r=1}^{\infty} F(r) \cos\left(2\pi rx - \frac{\pi}{4}\right) \tag{9.2.16}$$

into account, where $f(r)$ is any arbitrary function of a variable r, we obtain a periodic function from (9.2.9).

Returning now to (9.2.8) we recall that a calculation of the conductivity from (9.2.7) involves a multiplication with the Fermi-Dirac distribution function and with the (in general) energy-dependent relaxation time, and finally integration. Because of the latter process the result contains the Fermi energy ζ_n instead of the energy ε, and x becomes $\zeta_n/\hbar\omega_c$.

In a plot of $\Delta\varrho/\varrho$ vs $1/B \propto 1/\omega_c$, the period of the oscillation is independent of B; it depends on the carrier concentration via the Fermi energy ζ_n. The coefficients b_r rapidly decrease with increasing values of r and in practice only the $r = 1$ term is retained in the sum:

$$b_r = (-1)^r \sqrt{\frac{\hbar\omega_c}{2\zeta_n r}} \cos\left(r\frac{\pi}{2}\frac{g\,m}{m_0}\right) \frac{r2\pi^2 k_B T/\hbar\omega_c}{\sinh(r2\pi^2 k_B T/\hbar\omega_c)} \exp\left(-\frac{r2\pi^2 k_B T_D}{\hbar\omega_c}\right) \tag{9.2.17}$$

where g is the Landé factor of the carrier [9.53, 54] and T_D is the *Dingle temperature* [9.55] which is a measure of the *natural line width* of the transitions between adjacent Landau levels. For a lifetime τ_c of a carrier in a state of energy ε, the Heisenberg energy uncertainty relation

$$\Delta\varepsilon = h/\tau_c \tag{9.2.18}$$

divided by the level spacing $\hbar\omega_c$ yields an amplitude factor

$$\exp(-\Delta\varepsilon/\hbar\omega_c) = \exp(-2\pi^2 k_B T_D/\hbar\omega_c) , \tag{9.2.19}$$

where

$$k_B T_D = \frac{1}{\pi}\hbar/\tau_c \tag{9.2.20}$$

and since the hyperbolic sin in (9.2.17) may be approximated by an exponential, the combination of this and the exponential given by (9.2.19) yields $\exp[-2\pi^2 k_B(T+T_D)/\hbar\omega_c]$. The temperature T is primarily the carrier temperature T_e. For this reason it is possible to determine T_e for a degenerate electron gas, which is heated by a strong electric field $\mathbf{E}$, by means of the Shubnikov-de Haas effect [9.56]. Figure 9.14b shows oscillations for various values of $\mathbf{E}$ at a constant lattice temperature [9.50]. A comparison of Fig. 9.14a and Fig. 9.14b reveals that the electron gas is heated to a temperature of, e.g., 12.5 K by a field of 320 mV/cm while the lattice is at 4.2 K. Measurements at times between 10^{-8} and 10^{-6}s after the application of the electric field shown in Fig. 9.14c yield a determination of the energy relaxation time which rises from 40 to 70 ns in the electron temperature range from 4.2 to 12.5 K. For the evaluation of the data, the Dingle temperature T_D ($= 7.0\,\mathrm{K}$ for $\tau_c = 3.5 \times 10^{-13}$s in the present case) was assumed to be independent of $\mathbf{E}$ for the range of $\mathbf{E}$ values considered above. Since the mobility variation of the degenerate sample due to electron heating was less than 3%, the influence of T_e on the collision time and therefore on the Dingle temperature was negligibly small. Electron temperatures as a function of the applied electric field strength have been determined by this method, not only for degenerate semiconductors but also for semimetals such as gray tin (α-Sn). At $E = 130$ mV/cm, the electron temperature equals 15 K at a lattice temperature of 4.2 K. No deviation from Ohm's law has been found at field strengths up to this value [9.57].

The lifting of spin degeneracy by a strong magnetic field adds another term to the rhs of (9.2.4):

$$\varepsilon = \left(n + \frac{1}{2}\right)\hbar\omega_c + \hbar^2 k_z^2/2m \pm \frac{1}{2}g\mu_B B \ , \tag{9.2.21}$$

where $\mu_B = e\hbar/2m_0 = 0.577\,\mu\mathrm{eV/T}$ is the Bohr magneton[3] and g is the Landé factor. The ratio $g\mu_B B/\hbar\omega_c = (g/2)m/m_0$ is also a factor in the argument of the cosine in (9.2.12). Values between 0.33 (for n-InSb) and 0.013 (for n-GaAs) have been obtained from Shubnikov-de Haas measurements [9.54]. This yields g values between -44 (for InSb) and 0.32 (for GaAs) for the Landé factor[4] [9.58]. For $m \ll m_0$, spin splitting of levels is negligibly small in general.

An interesting situation in the Shubnikov-de Haas effect arises when under extreme conditions of temperature and field the oscillation amplitude $\Delta\varrho$ in Fig. 9.14 is theoretically larger than ϱ_0. For a two-dimensional electron gas, regions of zero resistance have in fact been observed. In these regions the Hall effect shows plateaus. The phenomenon is known as the *quantum Hall effect* and will be discussed in Sect. 14.4 on surface quantization.

The magneto-optical determination of g will be discussed in Sect. 11.14. Finally, let us briefly consider the *de Haas-van Alphen effect* [9.59]. Oscillations occur in the magnetic susceptibility of metals, semimetals, and degenerate semiconductors. These oscillations are again periodic in $1/B$ and have an amplitude increasing with B. The magnetic moment is given by [9.60]

$$M = -(\partial F/\partial B)_{T=\mathrm{const}} \ , \tag{9.2.22}$$

---

[3] Strictly speaking, the Bohr magneton is $\mu_0\,\mu_B$ where μ_0 is the permeability of free space
[4] The g-factor as calculated from the band structure (2.1.32) reads $g = 2(1 - (m_0/m_n - 1)/(2 + 3\varepsilon_G/\Delta))$ for $\varepsilon_G \ll$ energies of higher bands. For example, n-InSb: $\Delta = 1.5\varepsilon_G$; $m_n/m_0 = 0.012$ yields $g \approx -40$.

where $F = U - TS$ is the free energy, U is the internal energy and S is the entropy of the carriers (Sect. 3.1). The oscillatory part of the magnetic moment is given by [9.57]

$$M_{\text{osc}} = \sum_{r=1}^{\infty} a_r \sin \left(\frac{2\pi\zeta}{\hbar\omega_c} r - \frac{\pi}{4} \right) , \tag{9.2.23}$$

where a_r is similar to b_r except that the factor $\sqrt{\hbar\omega_c/2\zeta}\, r$ contained in b_r given by (9.2.12) has to be replaced by $3n(\hbar\omega_c)^{3/2}/(r^{3/2}4\pi B\sqrt{2\zeta})$ and n is the carrier concentration. This effect has been investigated mostly in metals for a determination of the Fermi surface [9.62].

9.3 Magnetic Freeze-Out of Carriers

Equation (9.2.4) shows that the lowest Landau level (quantum number $n = 0$) with increasing magnetic field intensity rises by $(e\hbar/m)B/2$. Besides the energy bands, the impurity levels are also affected by the magnetic field. The hydrogen model of impurities has been calculated for the case of strong magnetic fields where $\hbar\omega_c/2 \gg$ the Rydberg energy which is Ry ≈ 13.6 eV [9.63]. When the ratio $\gamma = \hbar\omega_c/2$Ry is ≥ 1, the originally spherical wave function of the hydrogen ground state becomes cigar shaped with the longitudinal axis in the direction of $\boldsymbol{B}$. The wave function is compressed, especially in the plane perpendicular to $\boldsymbol{B}$, the average distance between the electrons and the nucleus is reduced and the ionization energy, defined as the difference in energy between the lowest bound state and the lowest state in the conduction band, is increased. For $\gamma \approx 2$ the ionization energy is doubled, but at greater values of γ the increase is sublinear. With increasing ionization energy, the density of conduction electrons n decreases according to (3.2.10) where, however, in the quantum limit for Maxwell-Boltzmann statistics, N_c is given by $\sqrt{2\pi m k_B T e B/h^2}$ according to (9.2.7) for $n_{\text{max}} = 0$ and (3.1.15) (consider that in InSb, e.g., with its large g-factor for conduction electrons, spin-splitting is about as large as Landau splitting and therefore the lowest level will contain electrons of only *one* direction of spin):

$$n = \sqrt{2\pi m k_B T} \frac{|e|B}{h^2} \exp \left[\zeta_n - \frac{1}{2} \frac{\hbar\omega_c}{k_B T} \right] = N_c \exp \left[\zeta_n - \frac{1}{2} \frac{\hbar\omega_c}{k_B T} \right] . \tag{9.3.1}$$

Here B has been assumed as being strong enough that only one spin direction of the carriers is possible. The decrease of n with increasing magnetic field strength is known as the *magnetic freeze-out* of carriers.

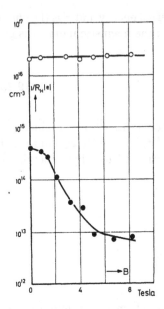

Fig. 9.15: Magnetic freeze-out of carriers in n-type InSb at 4.2 K in a sample with about 10^{14} electrons cm^{-3}(black dots) but not in a sample with about 10^{16} electrons cm^{-3}(circles). (after [9.64])

There are few semiconductors where the condition $\gamma > 1$ is met with available magnetic field strengths. Even these semiconductors have not been obtained pure enough so that the average spacing between neighboring impurity atoms is larger than about ten times the Bohr radius as has been assumed for the calculation (isolated hydrogen model) [9.63]. Hence no quantitative agreement between theory and experiment is to be expected. A semiconductor, which one might expect to show magnetic freeze-out at temperatures between 1 K and 5 K, is n-type InSb. Figure 9.15 shows the results of field-dependent Hall constant measurements obtained at 44.2 K with two samples of different zero-field carrier concentrations, $4 \times 10^{14} cm^{-3}$ and $2 \times 10^{16} cm^{-3}$. For the sample with the higher carrier concentration and consequently higher impurity concentration, the overlap of the impurity wave functions persists to much higher magnetic fields than for the other sample. In Fig. 9.16, the Hall coefficient is plotted vs temperature for various values of B between 0.09 and 8.23×10^{-1} T [9.65]. As should be expected from (3.2.10), freeze-out increases with decreasing temperature. If the Hall coefficient R_H is assumed to be proportional to $\exp(\Delta\varepsilon_D/k_B T)$ with an ionization energy $\Delta\varepsilon_D$, a dependence $\Delta\varepsilon_D \propto B^{1/3}$ on the magnetic field intensity has been found [9.66].

In n-InSb, sub-mm waves ($\lambda = 0.2$mm) are absorbed by the donors in a strong magnetic field (≈ 0.7 T) at 1.8 K (Fig.9.17) and cause photo-conductivity because the magnetically frozen-out electrons are lifted into the conduction band [9.67]. This effect has been applied for infrared detection (Putley detector) [9.68]. At even longer wavelengths (e.g., $\lambda = 1.4$ mm , Fig. 9.17) where free-carrier absorption prevails (Sect.11.10), reverse behavior has been observed and

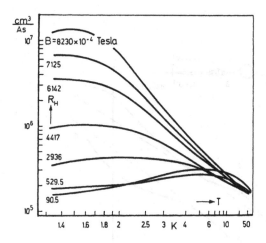

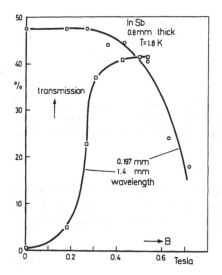

Fig. 9.16: Dependence of the Hall coefficient in n-type InSb on temperature for various values of the magnetic field (after [9.67])

Fig. 9.17: Microwave and far-infrared optical transmission of n-type InSb at 1.8 K (after [9.67])

accounted for by the decreasing density of free carriers with increasing magnetic field.

9.4 The Magnetophonon Effect

In contrast to the Shubnikov-de Haas oscillations which require very low temperatures and a degenerate electron gas [see text after (9.2.7)], there is another

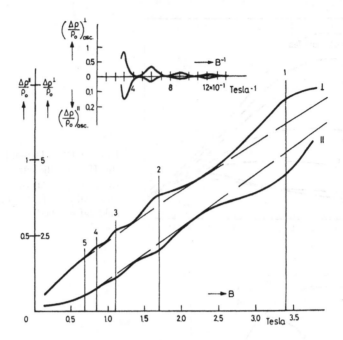

Fig. 9.18: Transverse and longitudinal magnetoresistance for n-type InSb at 90 K. The inset shows the oscillatory part of the curves as a function of the inverse magnetic induction. The numbers at the vertical lines are values of the ratio of phonon frequency and cyclotron resonance frequency (after [9.71])

type of oscillation at higher temperatures which is due to longitudinal optical phonons of energy $\hbar\omega_0$ and which requires no degeneracy. These oscillations occur if the optical-phonon energy is an integer multiple n of the Landau level spacing $\hbar\omega_c$:

$$\hbar\omega_0 = n\hbar\omega_c; \quad n = 1, 2, 3, \dots \ . \tag{9.4.1}$$

By absorption of a phonon, an electron is transferred from a given Landau level no. ϱ to a level no. $\varrho + n$. This effect is known as the *magnetophonon effect*. It was predicted independently by *Gurevich* and *Firsov* [9.69] and by *Klinger* [9.70] and observed by *Firsov* et al. [9.71]. Their experimental data for longitudinal and transverse magnetoresistance obtained on n-type InSb at 90 K are displayed in Fig.9.18. The inset shows the oscillatory part plotted vs $1/B$. The period of the oscillation is constant and given by

$$\Delta(1/B) = e/(m\omega_0) \tag{9.4.2}$$

in agreement with (9.4.1). In n-type InSb at 3.4 T, $m/m_0 = 0.016$ and $\omega_0 = 3.7 \times 10^{13}\,\text{s}^{-1}$, which yields $\Delta(1/B) = 3 \times 10^{-3}/\text{T}$. At resonance the transverse magnetoresistance shows maxima while the longitudinal magnetoresis-

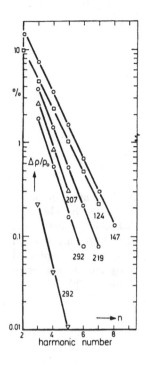

Fig. 9.19: Observed amplitudes of the magnetophonon extrema in n-GaAs in the transverse configuration plotted vs harmonic number. The numbers at the curves are the temperatures in units of K (after [9.75]).

tance shows minima. The period is independent of the carrier density, in contrast to the Shubnikov-de Haas period [see (9.2.8)]. Since for large carrier densities [9.68] and consequently large impurity densities, ionized impurity scattering dominates over optical-phonon scattering, no oscillations have been observed for densities of more than 5×10^{15} cm^{-3}. At very low temperatures, not enough optical phonons are available for the magnetophonon effect to occur, while at too high temperatures in spite of the strong magnetic field, the condition $\omega_c \tau_m \gg 1$ necessary for the observation of Landau levels is not fulfilled. Because of this compromise the effect is quite small and requires refined electronic techniques such as double differentiation [9.73]. The longitudinal magnetophonon effect occurs when two scattering processes such as inelastic scattering by an optical phonon and an elastic transition at an impurity site are operative [9.74].

The amplitudes $\Delta\varrho/\varrho_0$ observed in the transverse configuration with n-type GaAs are plotted vs the harmonic number in Fig.9.19 [9.75]. The straight lines in the semi-log plot suggest a proportionality of $\Delta\varrho$ to $\exp(-\gamma\omega_0/\omega_c)$ with a constant γ which in GaAs is $\gamma = 0.77$ according to observations.

With an optical-phonon frequency ω_0 known, e.g., from Raman scattering, the magnetophonon effect serves for a determination of the effective mass m of the carriers. Figure 9.20 shows the electron and hole effective masses in Ge as a function of the angle between the magnetic field and the $\langle 100 \rangle$ direction in the $[1\bar{1}0]$ plane obtained from magnetophonon measurements. The effective

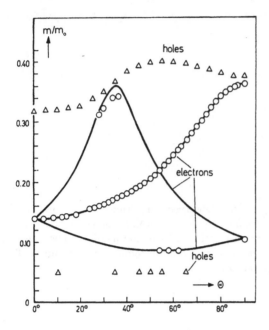

Fig. 9.20: Variation of effective masses in Ge with angle between magnetic field and the (100) direction in the [110] plane, obtained from magneto-phonon resonance (after [9.73]).

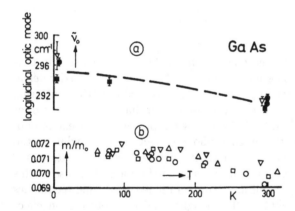

Fig. 9.21: (a) Temperature dependence of the LO phonon frequency at $q=0$ for GaAs. (b) Temperature dependence of the magnetophonon mass for n-type GaAs determined from the LO phonon frequeny and magnetophonon peaks (after [9.75]).

hole masses and the transverse electron mass obtained at 120 K are significantly higher than those obtained from cyclotron resonance at 4 K (Fig.11.52), while the longitudinal electron mass is in agreement with the cyclotron value [9.73]. The discrepancies are explained by the nonparabolicities of the bands: at high temperatures of the magnetophonon measurements, higher parts of the bands are occupied than at 4K. In polar semiconductors due to the polaron effect (Sect. 6.12), the mass derived from magnetophonon oscillations is $1 + \alpha/3$ times the low-frequency mass and $1 + \alpha/2$ times the high-frequency or *bare* mass, where α is the polar coupling constant given by (6.12.8) and has a value of e.g., 0.06 in

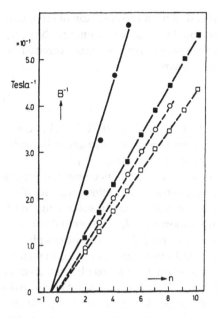

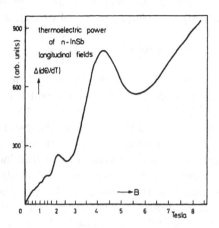

Fig. 9.22: Inverse magnetic fields of the magnetophonon peaks vs harmonic numbers for the series involving the emission of LO-phonons. Open Squares: GaAs at 77 K, ohmic; full squares: GaAs at 20 K, high field; open circles: CdTe at 60 K, ohmic; full circles: CdTe at 14 K, high electric field (after [9.73]).

Fig. 9.23: Variation of the thermoelectric power of n-type InSb with a longitudinal magnetic field at 120 K; peaks caused by magnetophonon resonance (after [9.79]).

n-type GaAs [9.76]. The dependence of the magnetophonon mass m of n-type GaAs on temperature is shown in Fig. 9.21b [9.75]; the variation of the optical-phonon frequency with temperature, which is required for a determination of m, is plotted in Fig. 9.21a. The values of m obtained after correcting for the band nonparabolicity are in agreement with those obtained by cyclotron resonance, Faraday rotation and interband magneto-optic absorption.

While at temperatures below 60 K the magnetophonon effect in n-type GaAs is undetectable, electron heating by electric fields between 1 and 10 V/cm at lattice temperatures between 20 and 50 K results in a reappearance of the oscillation [9.77]. In Fig. 9.22 the $1/B$ -values of the peaks are plotted [9.73] vs harmonic numbers for the series involving the emission of LO-phonons with GaAs and CdTe. For a comparison, the high-temperature thermal-carrier peaks are also plotted. While the extrapolation of thermal peaks to $1/B = 0$ yields $n = 0$, the warm-carrier peaks are extrapolated to a negative value, $-1/2$, of n. Here the dominant energy loss mechanism is associated with the emission of an optical phonon and the capture of the warm electron by a donor of energy $\Delta\varepsilon_D(B)$; at relatively low magnetic fields, the donor energy may be approximated by $\Delta\varepsilon_D(0) + 1/2\,\hbar\omega_c$. Instead of (9.4.1), we now have

$$\hbar\omega_0 = \Delta\varepsilon_D(0) + (n + \frac{1}{2})\hbar\omega_c \ . \tag{9.4.3}$$

A straight line through the $n \geq 7$ peaks of GaAs in Fig. 9.21 fits this equation with $\Delta\varepsilon_D(0) = 5.8$ meV which is consistent with a hydrogenic impurity $\kappa = 12.5$ and $m/m_0 = 0.0675$. The points corresponding to $n \leq 7$ deviate somewhat from this line since the dependence of $\Delta\varepsilon_B$ on B becomes less strong as B is increased. In the warm-electron magnetophonon effect, the mobility and therefore the resistivity is determined by ionized impurity scattering and is very sensitive to a variation of the electron temperature T_e; the oscillation of T_e is caused by the oscillatory B-dependence of the probability of optical-phonon emission from the high-energy tail of the distribution function. Consequently, the energy relaxation time τ_ε should also depend on B in an oscillatory manner under magnetophonon conditions. In CdTe for $E > 7$ V/cm, additional peaks occur which agree with the field position of harmonics at higher temperatures under zero-field conditions and obey (9.4.1). In n-type InSb and GaAs at low temperatures (e.g., 11K) and threshold fields of 0.07 V/cm and 0.8 V/cm, respectively, series of peaks are observed which have been attributed to a simultaneous emission of two oppositely directed transverse acoustic phonons $\hbar\omega_t$ by electrons in the high-energy tail of the distribution:

$$2\hbar\omega_t = n\hbar\omega_c; \qquad n = 1, 2, 3, \ldots \tag{9.4.4}$$

The phonons are thought to be at the X point in the phonon Brillouin zone and have energies of 5.15 meV in InSb and 9.7 meV in GaAs. In infrared lattice absorption (Sect. 11.8), 2-phonon processes have also been observed. The matrix element for the 2-phonon emission has been estimated to be only 10^{-8} of that for the emission of a single LO-phonon [9.73]. A magnetophonon Hall effect has also been measured which is, however, much smaller than the magnetoresistance oscillations [9.78]. Oscillatory terms in ϱ_{xy} arise from the second-order terms of magnitude $1/\omega_c^2\tau_m^2$ which remain in the expression for the Hall coefficient in the high-field limit (4.2.64). A magnetophonon effect in thermoelectric power in n-type InSb at 120 K is shown in Fig. 9.23 [9.79]. The magnetic field is longitudinal. At least five maxima have been identified. The resonance at 4.25 T can be correlated with a transition from the lowest to the next Landau level with the emission of a longitudinal-optical phonon of energy 25 meV.

Problems

9.1. Prove (9.1.4).

9.2 Draw the energy band diagrams of a tunnel diode for the extremes of the current as well as for the ranges below the maximum and above the minimum. How about optical localization of the junction with the Burstein shift (Sect. 11.2)? Discuss the reverse diode current.

9.3. According to quantum mechanics, estimate the frequency limit of a tunnel diode for a 1 eV gap.

9.4. Super Schottky diode. The I-V characteristics of a super Schottky diode consisting of a superconductor on top of a semiconductor is determined by tunneling of electrons. *Hints*: the calculation of the current I is similar to (9.1.10). For the density-of-states function [Ref. [9.80], Eq. (3.26)] is applied. Now the superconductor gap is denoted by 2Δ. With G_{NN} being the normal-state conductance of the junction extrapolated to $T = 0$, the dc current I as a function of the applied voltage is

$$I(V) = \frac{G_{NN}}{e} \int_{\Delta}^{\infty} \varepsilon_k (\varepsilon_k^2 - \Delta^2)^{-1/2} [f(\varepsilon_k - eV) - f(\varepsilon_k + eV)] d\varepsilon_k \ , \quad (9.0.1)$$

where $f(\varepsilon_k) = [1 + \exp(\varepsilon_k/k_B T)]^{-1}$ is the Fermi-Dirac function. Show that for $k_B T \ll eV < \Delta$, the characteristics are simply $I = I_0 \exp(eV/k_B T)$ with $I_0 = G_{NN}(\pi \Delta k_B T/2e^2)^{1/2} \exp(-\Delta/k_B T)$. For ac operation of the diode at frequencies ν where $h\nu$ is of the order of 2Δ, see [9.81].

9.5. Prove (9.3.1). For a compensated semiconductor (N_D and N_A given), calculate the Fermi level ζ_n and apply the rhs of (9.3.1) (e.g., n-InSb at 4.2 K with $m/m_0 = 0.013$; $N_D = 7.8 \times 10^{14}\,\mathrm{cm}^{-3}$; $N_A = 5.4 \times 10^{14}\,\mathrm{cm}^{-3}$; plot $n(B)$).

9.6. Ultrasound absorption by conduction electrons for mean free paths which are large compared to the wavelength is observed only if the carrier velocity approximately equals the sound velocity. For degenerate semiconductors, a strong dependence of sound absorption on an applied longitudinal magnetic field has been observed (*giant quantum oscillations*). Discuss these oscillations in terms of Landau levels and compare with magnetoresistance. At which field strength do you expect ultrasound absorption? *Hint*: draw $\varepsilon(k)$ in the field direction and indicate the sound wave vector at the k axis and the Fermi energy at the ε axis.

9.7. For $m/m_0 = 0.05$ and a phonon energy of 3 meV, determine the magnetophonon maxima for a transverse field.

10. Impact Ionization and Avalanche Breakdown

Some aspects of impact ionization and avalanche breakdown in semiconductors are similar to the corresponding phenomena in gaseous discharges. Semiconductors may serve as model substances for gaseous plasmas since their ionic charges are practically immobile and therefore the interpretation of experimental data is facilitated. Impact ionization has been achieved both in the bulk of homogeneously doped semiconductors at low temperatures and in p-n junctions at room temperature. We will discuss these cases separately.

10.1 Low-Temperature Impact Ionization in Homogeneous Semiconductors

Let us first consider impact ionization of shallow impurities in n-type Ge. Figure 10.1 shows typical I-V characteristics obtained at temperatures between 4.2 and 54.2 K [10.1]. At the lower temperatures most carriers are frozen-out at the impurities. Since the ionization energy is only about 10^{-2} eV, breakdown already occurs at fields of a few V/cm and persists until all impurities are ionized. At the highest temperature in Fig. 10.1, all impurities are thermally ionized and hence there is no breakdown in the range of electric field intensities investigated. (At very high field strengths there is a tunnel effect across the band gap). The onset of breakdown is shown more clearly in Fig. 10.2 together with curves for the reciprocal of the Hall coefficient and the Hall mobility [10.2, 3]. Already in the prebreakdown region there is a gradual increase of the carrier density with field intensity and a maximum in the mobility. At low field intensities, ionized impurity scattering dominates the mobility. As the carrier energy is increased with increasing field intensity, the mobility also increases until lattice scattering becomes dominant and, as a result, from there on the mobility decreases with the field. Since lattice scattering is inelastic, in contrast to ionized impurity scattering, the increase of the carrier concentration with increasing field strength becomes weaker as lattice scattering becomes dominant which is indicated by a bump in the $R_{\mathrm{H}}^{-1}(E)$ curve. The upper part of the characteristics has also been investigated in detail as shown in Fig. 10.3 [10.4]. Curve a is valid for a carrier density of $n = 5.0 \times 10^{14}\,\mathrm{cm}^{-3}$ at complete ionization and has been observed

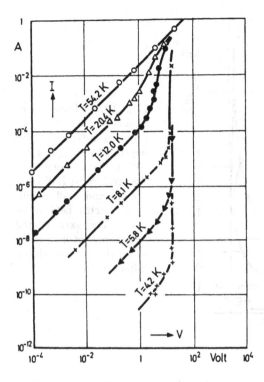

Fig. 10.1: Current-voltage characteristics of n-type Ge at low temperature for various lattice temperatures (after 10.1)

in various samples having cross dimensions ranging from 0.5 to 5 mm. Curve b represents data obtained from a sample with $n = 5.2 \times 10^{15}$ cm^{-3}; both the breakdown field and the maximum conductivity are higher in this case. The decrease in conductivity is due to the above-mentioned decrease in mobility at constant carrier density. The product of the low-field mobility and the breakdown field intensity, which can be approximated by the drift velocity at the onset of breakdown, is about the same for all samples with shallow impurities irrespective of the impurity concentration [10.2, 3].

For the impact ionization of shallow donors, the change of electron concentration n in an n-type semiconductor with time t at a given value of the electric field E and temperature T is given by [10.2].

$$dn/dt = A_T(N_D - N_A) + A_I n[N_D - (N_A + n)]$$

$$-B_T n(N_A + n) - B_I n^2(N_A + n) . \tag{10.1.1}$$

Here $A_T(T)$ and $A_I(E)$ are the coefficients for thermal and impact ionization processes, respectively; $B_T(T,E)$ is the coefficient for thermal recombination of a single electron with an ionized donor; $B_I(T, E)$ is the coefficient for the *Auger process* in which two electrons collide at an ionized donor, one being captured with the other taking off the excess energy. The Auger process is negligible for

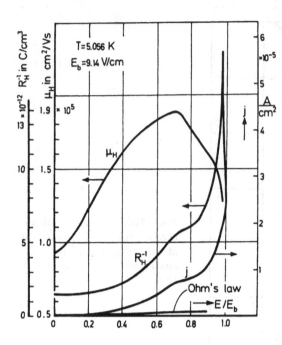

Fig. 10.2: Impact ionization at low temperatures in n-type Ge (Sb doped; $N_D - N_A = 2.2 \times 10^{14} \text{cm}^{-3}$); Current density, reciprocal Hall coefficient, and Hall mobility vs electric field intensity. The extrapolated ohmic current-field characteristic is also shown (after [10.2])

small values of n. N_D-N_A is the concentration of uncompensated donors, N_A+n is the concentration of ionized donors and $N_D - (N_A + n)$ is the concentration of neutral donors. The concentrations of neutral acceptors and holes have been assumed as being negligibly small.

The steady-state value of n which is denoted by n_0 is easily obtained from (10.1.1) for small values of n (we neglect the Auger process and assume $N_A \gg n$):

$$n_0 = A_T(N_D - N_A)/[B_T N_A - A_I(N_D - N_A)]. \tag{10.1.2}$$

Breakdown occurs at a field strength E_b for which the denominator vanishes:

$$B_T(T, E_b)\, N_A - A_I(E_b)\,(N_D - N_A) = 0 \tag{10.1.3}$$

Hence, at a given temperature, the breakdown field strength is a unique function of the compensation ratio N_A/N_D. If we define a time constant

$$\tau = n_0/[A_T(N_D - N_A)] \tag{10.1.4}$$

(10.1.1), subject to the simplifying assumptions that led to (10.1.2), can be written in the form

$$-dn/dt = (n - n_0)/\tau \tag{10.1.5}$$

resulting in an exponential time dependence for a solution. Measurements of the recovery from breakdown yield a variation of τ with E at 4.2 K indicated

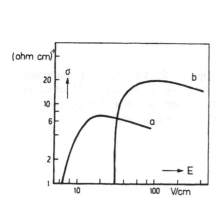

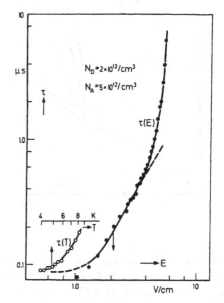

Fig. 10.3: Electrical conductivity of n-type Ge with $N_D - N_A = 5 \times 10^{14} \mathrm{cm}^{-3}$ (curve a) and $5.2 \times 10^{15} \mathrm{cm}^{-3}$ (b) vs electric field strength at 4.2 K (after [10.4])

Fig. 10.4: Variation of recombination time after breakdown with electric field strength at 4.2 K (dots) and with temperature for thermal electrons (circles) in n-type germanium (after [10.5])

by Fig. 10.4 (solid circles) [10.5]. The open circles show the variation of τ with temperature T for thermal electrons. The order of magnitude is $10^{-6} - 10^{-7}$ s for a degree of compensation, $N_A/N_D = 5 \times 10^{12}/2 \times 10^{13} = 0.25$. Equation (10.1.4) shows that τ can be increased by adding compensating impurities. The capture cross section of about 10^{-12} cm^2 is an order of magnitude larger than the geometrical cross section and believed to be due to electron capture in a highly excited orbit with a subsequent cascade of transitions to the ground state [10.5, 6]. In a many-valley semiconductor the breakdown field is anisotropic. The Poole-Frenkel effect discussed at the end of Sect. 5.3 may have to be considered here, too. In small-gap semiconductors such as InSb ($\varepsilon_G = 0.2$ eV), InAs ($\varepsilon_G = 0.4$ eV) and Te ($\varepsilon_G = 0.35$ eV), impact ionization of lattice atoms (i.e., not impurities) with the production of equal numbers of electrons and holes has been observed at temperatures between 4.2 and 300 K. In Fig. 10.5, experimental results obtained in n-type InAs at 77 K at various times after application of the field are shown [10.7]. From the development of the breakdown characteristics with time, a *generation rate*

$$g(E) = (1/n_0) \, dn/dt \tag{10.1.6}$$

has been determined which is shown in Fig. 10.6, together with the theoretical curves obtained by *Dumke* [10.8] and others [10.9] for 0 and 229 K. The recip-

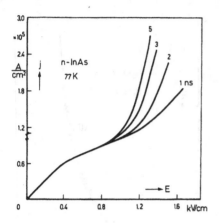

Fig. 10.5: Current-field characteristics for n-type InAs at 77 K for various time intervals after the application of the voltage (after [10.7])

rocal of the generation rate is of the order of magnitude of 10^{-7} to 10^{-8} s. For a transverse magnetic field of 0.6 T at 77 K where $\mu B \approx 2$, the generation rate is decreased by about 40% (Fig. 10.6, dotted curve). This is interpreted by a cooling effect of the magnetic field on the hot carriers. In many semiconductors such as p-type tellurium [10.10] at 77 K and compensated germanium [10.11] at 4.2 K, regions of negative differential conductivity (ndc) have been observed. Figure 10.7 shows I–V characteristics of indium-doped Ge compensated with antimony. Sample B is more heavily doped than sample A. The formation of current filaments in the breakdown region has been observed which is in agreement with the thermodynamic arguments given by *Ridley* [10.12] for S-shaped (current controlled) characteristics [10.13]. The ndc may be due to energy relaxation by deformation potential scattering in the magnetic field of the current [10.14] or due to momentum relaxation by impurity scattering (in compensated material the impurity concentration is higher than normal). For semiconductors with minority carriers having an effective mass much smaller than the majority carriers, one can imagine a process where the minority carriers generated in the avalanche are more easily accelerated by the field and, once they have increased sufficiently in number, dominate the ionization process. In small-gap materials, a *pinch effect* has been observed after the formation of equal numbers of mobile positive and negative charges by the ionization process: the Lorentz force which arises due to the magnetic field of the current drives the carriers to the center of the sample thus forming a filamentary current [Ref. [10.15], p. 379]. The current density in the filament may be so high that the crystal lattice melts locally and after switching off solidifies in polycrystalline form.

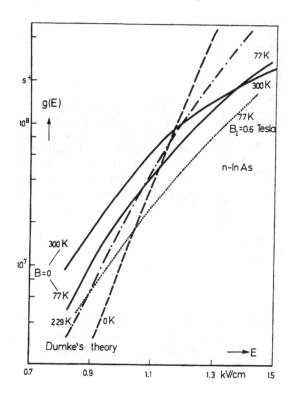

Fig. 10.6: Generation rates: experimental (full curves and, in a transverse magnetic field of 0.6 T, dotted) and calculated according to a theory by Dumke (10.8) dashed and dash-dotted curves (after [10.7])

10.2 Avalanche Breakdown in the p-n Junction

In the depletion layer of a reverse-biased p-n junction, electric field intensities up to 10^6 V/cm can be obtained with a negligible amount of Joule heating. Impact ionization across the band gap is then found even in large-gap semiconductors. In the junction characteristic shown in Fig. 5.8, breakdown occurs at a voltage of 1200 V.

The relative increase in carrier density per unit length is called the *ionization rate*:

$$\alpha = (1/n) \, dn/dx. \qquad (10.2.1)$$

The *multiplication factor* M is determined by the ionization rate

$$1 - M^{-1} = \int_0^d \alpha \, (E) dx \; , \qquad (10.2.2)$$

where the assumption has been made that n_0 carriers have been injected at one end $x = 0$ of the high-field region of width d; $n = M \, n_0$ carriers have been collected at $x = d$ and the ionization rates for electrons and holes are equal. d

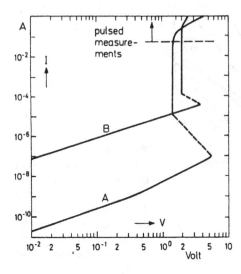

Fig. 10.7: Current-voltage characteristics of indium-doped germanium strongly compensated with antimony at 4.2 K. Sample B is more heavily doped than sample A. Sample thickness is 0.43 mm and contact diameter is 1 mm (after [10.11])

has been assumed to be much larger than the mean free path. Equation (10.2.2) is obtained by considering the increase of the electron current component j_n between x and $x + dx$, which is due to ionization both by electrons and by holes (current component j_p):

$$dj_n/dx = \alpha j_n + \alpha j_p = \alpha j. \tag{10.2.3}$$

The total current j is the same everywhere along the p-n junction. At $x = d$ the current consists only of electrons, i.e., there is no hole injection ($j_n = j$), while at $x = 0$, j_n is the injected electron current which equals $M^{-1}j$ (per definition of the multiplication factor M). With these boundary conditions the integration of (10.2.3) yields (10.2.2). Breakdown occurs when the integral on the rhs of (10.2.2) equals unity, i.e., $M \to \infty$. For a measurement of M as a function of the reverse bias, carriers are usually injected by light [10.16–18].

Figure 10.8 shows experimental results of α vs E for Ge, Si, GaAs, and GaP [10.19]. The curves valid for Ge and Si can be represented by

$$\alpha \propto \exp(-\text{const}/E) \tag{10.2.4}$$

while those for GaAs and GaP are better represented by

$$\alpha \propto \exp(-\text{const}/E^2) \tag{10.2.5}$$

where the constants in the exponentials are fitting parameters. These forms of $\alpha(E)$ shall now be made plausible. A more complete calculation has been given by *Baraff* [10.20].

Let us consider a collision where the initial and final carrier velocities are v_i and v_f, respectively, and the velocities of the pair produced by the collision

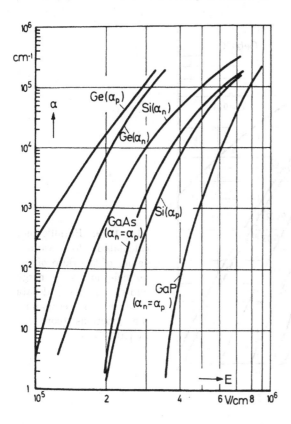

Fig. 10.8: Observed ionization coefficient for avalanche multiplication vs electric field strength for electrons (n) and holes (p) in various semiconductors at room temperature (compiled by [10.19])

are v_n and v_p [Ref. [10.21], p. 171]. For simplicity, we assume a one-dimension model where the incident particle is an electron. Momentum conservation yields

$$m_n v_i = m_n v_f + m_n v_n + m_p v_p \tag{10.2.6}$$

and energy conservation gives

$$m_n \, v_i^2/2 = \varepsilon_G + m_n \, v_f^2/2 + m_n \, v_n^2/2 + m_p \, v_p^2/2 \ . \tag{10.2.7}$$

A minimum of the incident kinetic energy,

$$\varepsilon_i = m_n v_i^2/2 = \varepsilon_G (2 m_n + m_p)/(m_n + m_p) \tag{10.2.8}$$

is obtained for $v_f = v_n = v_p$. Assuming the incident particle to be a hole yields the same value except that m_n and m_p are interchanged. For $m_n = m_p$ we find $\varepsilon_i = 1.5\,\varepsilon_G$. However, depending on the ratio m_n/m_p, the factor following ε_G may have values of between 1 and 2.

The number of carriers with energies $\varepsilon > \varepsilon_i$ is proportional to $\exp(-\varepsilon_i/k_B T_e)$ if the carrier gas is nondegenerate with an electron temperature T_e. The latter has been determined from the energy balance equation

$$e\,E\,v_d = \frac{3}{2}\frac{k_B(T_e - T)}{\tau_\varepsilon} \approx \frac{k_B T_e}{\tau_\varepsilon} \tag{10.2.9}$$

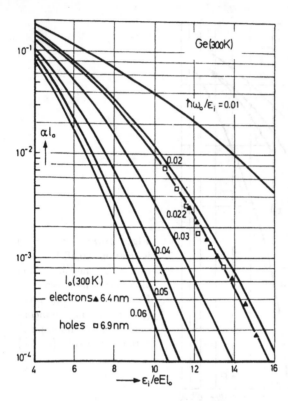

Fig. 10.9: Baraff's plot calculated for various values of the ratio of the optical-phonon energy and $\varepsilon_i = 1.5\,\varepsilon_G$, and data observed in germanium p-n junctions (compiled by [10.19])

neglecting T and the factor $3/2$ for a rough estimate. At the high field intensities of interest here, the drift velocity v_d is practically constant. Assuming a constant energy relaxation time τ_ε and considering the product $v_d \tau_\varepsilon = l_0$ as a carrier mean free path for essentially optical-phonon emission, the ionization rate α is determined by

$$\alpha \propto \exp\left(-\varepsilon_i/k_B T_e\right) = \exp(-\varepsilon_i/e\,E l_0) \ . \tag{10.2.10}$$

Taking the quantity $\varepsilon_i/e l_0$ as a fitting parameter we arrive at (10.2.4). On the other hand, assuming $v_d \tau_\varepsilon$ to be proportional to E yields a dependence $\alpha(E)$ as given by (10.2.5).

Figure 10.9 shows the product $\alpha\, l_0$ as a function of $\varepsilon_i/e\,E\,l_0$ for various values of the ratio of the optical-phonon energy $\hbar\omega_0$ and ε_i [10.19]. Experimental data obtained from Ge p-n-junctions at 300 K have been fitted to the curve for which $\hbar\omega_0/\varepsilon_i = 0.022$, assuming $\varepsilon_i = 1.5\,e_G$, $\varepsilon_G = 1.1$ eV for the direct energy gap at room temperature, and $\hbar\omega_0 = 37$ meV. Values for l_0 of 6.4 nm for electrons and 6.9 nm for holes are obtained. Since the drift velocity is about 6×10^6 cm/s at high field strengths in Ge, values for τ_ε of about 10^{-13}s are found which are, however, more than an order of magnitude lower than those obtained from microwave measurements [10.22]. Even so, the Baraff curve is a good fit to the

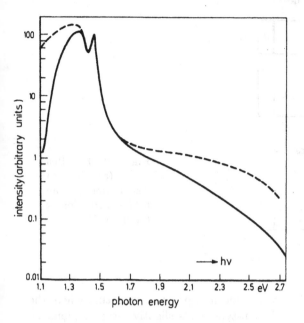

Fig. 10.10: Emission spectrum from a reverse biased GaAs junction at 77 K (original data: full curve; data corrected for spectral sensitivity of the detector: dashed curve (after [10.23]))

experimental data.

In the range of impact ionization in diodes, light is emitted at an efficiency of 10^{-3} to 10^{-5} photons per electron crossing the junction. A typical spectrum obtained from GaAs at 77 K is shown in Fig. 10.10 [10.23]. The maximum of the broad spectrum is near the energy gap. At the onset of breakdown, light is emitted from various spots of the p-n junction rather than uniformly across the junction, and the current is noisy and concentrated on the same sites; these are denoted as *microplasmas*. Dislocation-free junctions do not exhibit the phenomenon of microplasmas.

Impact ionization and a phase shift in ac current due to the carrier transit time through the p-n structure have been utilized for the generation of microwaves. The word IMPATT has been coined from *IMPact ionization Avalanche Transit Time* in order to characterize such devices. A special diode commonly known as the *Read diode* was the first device of this kind [10.24, 25]. We can give only a brief discussion of the physical principles of the Read diode here and for further details, refer the reader to [Ref. [10.26], p. 200], [Ref. [10.21], p. 190] and [10.25].

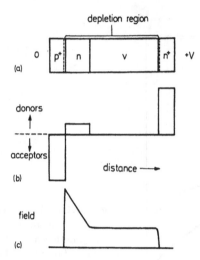

Fig. 10.11: Read diode (a), impurity distribution (b), and field distribution (c) (after [10.21])

Figure 10.11 shows the field distribution in a $p^+n\nu n^+$ structure where the ν region is essentially intrinsic (slightly n-type; for a slightly p-type region, the letter π is used). In the high-field region of the p^+n junction, impact ionization is assumed to occur. In the depletion region the field is assumed as being still high enough that the drift velocity v_d is nearly field-independent. Let us denote the admittance of the depletion region by Y such that the current density injected in the avalanche is given by $j_n = Y V$ where V is the voltage across the whole diode. The charge traveling down the depletion region induces in the external circuit a capacitive current density $\kappa\kappa_0\,\partial E/\partial t = (\kappa\kappa_0/L)\partial V/\partial t$, where L is the length of the depletion region. The external current at time t is induced from carriers generated in the time interval between t and $t - \tau$ where $\tau = L/v_d$ is the carrier transit time. The delay between the injected charge and the induced current is, therefore, on the average $\tau/2$. For a sinusoidal variation of the current at a frequency ω, the induced current j is then given by

$$j \propto Y V \exp\left(-\mathrm{i}\omega\tau/2\right) . \tag{10.2.11}$$

If the phase factor of Y combined with the transit time phase $-\omega\tau/2$ yields a total phase between 90° and 270°, one can show that the current oscillates at the frequency ω even without any applied ac voltage. For a saturated drift velocity of 2×10^7 cm/s (as, e.g., in Si) and a length $L = 10^{-3}$ cm, a frequency of about 5 GHz is obtained. Continuous-operation devices with a power output of several watts have been made. Frequencies as high as 300 GHz have been achieved.

Problems

10.1. A Ge sample at the beginning of low-temperature breakdown conditions is irradiated with chopped light of low intensity and of a photon energy larger than the gap energy. With a chopper frequency of $\omega/2\pi$, a generation rate $G_0 \exp(i\omega t)$ is introduced in (10.1.1). Neglecting terms of third order in n and taking second-order terms as small, is it possible to determine the coefficients or their combinations by a variation of the light intensity and chopper frequency?

10.2. At heavy breakdown, thermal generation and recombination will be less important. Discuss the processes leading to $n(E)$ qualitatively.

11. Optical Absorption and Reflection

The propagation of electromagnetic radiation through a semiconductor depends, in general, on temperature and external pressure and can be modified by electric and magnetic fields. Measurements of these effects provide very useful information about band structure and energy levels in semiconductors. For numerical purposes it may be worth noting that a quantum energy $\hbar\omega$ of 1 eV is equivalent to a wave number ν of 8060 cm^{-1} and to a wavelength λ of 1.24 μm. A frequency of 1 THz corresponds to a wavelength of 299.8 μm [11.1].

11.1 Fundamental Absorption and Band Structure

Many transport properties of semiconductors can be explained by the assumption of an energy gap ε_G between the valence band and the conduction band. In an ideal semiconductor, there are no energy states within this gap. This type of semiconductor should be transparent for light of angular frequencies less than a critical value ω_e given by

$$\hbar\omega_e = \varepsilon_G \tag{11.1.1}$$

if the absorption of light quanta is due only to the transfer of electrons from the valence band to the conduction band. This type of absorption is denoted as *fundamental absorption*, and ω_e is the absorption edge (subscript e).

This is also true for a nonideal semiconductor if the density of conduction electrons and holes (called *free carriers* in optics) is not so large that there is metallic conduction where absorption by the carriers is of the same order of magnitude as the fundamental absorption.

Figure 11.1 shows the extinction coefficient k (imaginary part of the complex refractive index; occasionally the sum of absorption and scattering is also called extinction) and the index of refraction n of germanium [11.2-5]. The absorption coefficient is $4\pi k/\lambda$; the phase shift/length is $2\pi n/\lambda$. The extinction coefficient vanishes at quantum energies less than the energy gap of about 0.7 eV, rises to a peak at energies of a few eV and, as the energy increases to values which are in the soft X-ray range, it decreases again. At the peak, $k \approx n \approx 4$ and the imaginary part of the dielectric constant, $\kappa_i = 2nk$, is about 32.

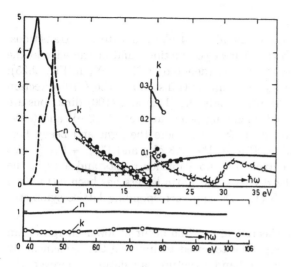

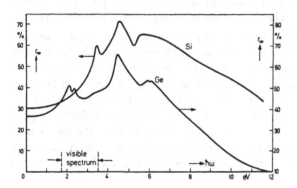

Fig. 11.1: Spectral dependence of the extinction coefficient k for germanium compiled by [11.2]. Refractive index n: experimental (crosses) and calculated by means of a Kramers–Kronig analysis (–)

Fig. 11.2: Room temperature reflectivity of silicon and germanium (after [11.7])

For part of the spectrum the refractive index n has been calculated from $k(\omega)$ by means of the Kramers-Kronig relations [11.6]. This calculation is confirmed by experimental data at energies between 8 and 20 eV, where n is less than 1. At low energies n has a value of about 4 corresponding to a real part of the dielectric constant of 16 according to Maxwell's relation $n^2 = \kappa_r$ valid for $k = 0$. In the X-ray range n approaches unity. Similar results are obtained for other semiconductors.

Figure 11.2 shows the reflection coefficients r_∞ of germanium and silicon up to 12 eV [11.7] and Fig. 11.3 the imaginary part κ_i of the dielectric constant of germanium [11.8]. The reflection coefficient r_∞ and the quantity κ_i have maxima at about the same energies . The maximum reflection is about 70%. κ_i has also been calculated from the band structure shown by Fig. 2.25 [11.9]; the result is indicated by the dashed curve. There is qualitative agreement between the two curves and one can relate maxima in the $\kappa_i(\omega)$-relation to transitions in the energy band structure.

The most prominent maximum at about 4 eV is due to two transitions: one is from a valence band to the nearest conduction band at the edge of the first Brillouin zone in the $\langle 100 \rangle$ direction (denoted as $X_4 \rightarrow X_1$ in Fig. 2.25); the other one, which is $\Sigma_4 \rightarrow \Sigma_1$, is a similar transition in the $\langle 110 \rangle$ direction and is not indicated in Fig. 2.25 where only the $\langle 111 \rangle$ and $\langle 100 \rangle$ directions are shown. All transitions indicated, including peaks at 2 eV, 2.18 eV and 6 eV, originate at the heavy hole band; at $k = 0$, where the light and heavy hole bands are degenerate, transitions $\Gamma'_{25} \rightarrow \Gamma'_2$ (0.8 eV) and $\Gamma'_{25} \rightarrow \Gamma_{15}$ (3 eV) yield shoulders in the $\kappa_i(\omega)$-curve. The notation for points and axes in the first Brillouin zone by capitalized Greek and Latin letters has been introduced along with group theoretical (Sect. 2.4) and corresponding lattice symmetry considerations [11.10].

In the $\kappa_i(\omega)$-diagram, the theoretical maxima occur at lower energies than the experimental ones. One might be tempted to improve the agreement by introducing more parameters in the band structure calculation. However, the fact that the present calculation includes only three adjustable parameters seems to be worth noting in view of the difficulties involved in such calculations.

The reflection spectra of germanium and silicon are very much alike, and so are the gross features of the band structure. Differences exist in the relative position of the conduction subbands: in Si, Γ'_2 is practically identical with Γ'_{15} and therefore Γ'_2 is located below Λ_1 rather than above. The energy differences $L'_3 - L_1$ and $\Lambda_3 - \Lambda_1$ are larger than they are in Ge with the result that the lowest conduction band minimum is not at L_1 but at Δ_1. The reflectivity peak at 3.4 eV in Si is therefore related to a transition different from the one related to the 2.1 eV peak of the Ge curve; to what transition has not yet become clear. It is possibly $\Gamma'_{25} \rightarrow \Gamma_{15}$ which is the direct transition from the top of

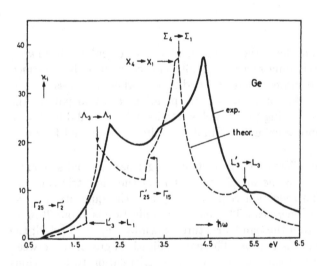

Fig. 11.3: Imaginary part of the dielectric constant for germanium. Solid curve: obtained from observed reflectance data. Dashed curve: calculated (after [11.8])

the valence band to the conduction band at $k = 0$, and also $\Gamma'_{25} \to \Gamma'_2$ and $L'_3 \to L_1$. An optical investigation of germanium -silicon alloys showed that the 3.4 eV peak becomes weaker with decreasing silicon content at about 75 atomic% Si[11.11]. Apparently for larger percentages of germanium, a different transition determines the reflectivity at this energy.

11.2 Absorption Edge: Dependence on Temperature, Pressure, Alloy Composition, and Degeneracy

Let us now consider the edge of the fundamental absorption which, according to (11.1.1), is directly related to the energy gap. Fig . 11.4, 11.5 show the absorption coefficient of GaAs and Ge, respectively, for various temperatures. In both cases, the energy gap decreases with increasing temperature. Figure 11.6 shows the temperature dependence of the gap for germanium. Above 150 K it is linear with a coefficient of - 0.43 meV/K. For semiconductors with energy gaps which are different by an order of magnitude, this coefficient varies only within a factor of two, e.g., GaP. ($\varepsilon_G = 2.24$ eV at 300 K; -0.54 meV/K) and InSb ($\varepsilon_G = 0.167$ eV at 300 K; -0.28 meV/K). Although in PbS, PbSe, and PbTe the gap increases with temperature, the normal behaviour is a decrease.   A comparison of Figs. 11.4, 11.5 reveals a different energy dependence of the absorption coefficients of GaAs and Ge. In Ge we observe a shoulder which is absent in GaAs. This shoulder is due to the fact that in Ge the lowest minimum of the conduction band is at the edge of the first Brillouin zone, while in GaAs it is at $k = 0$. Since the valence band maximum in both cases is at $k = 0$, a transition obeying (11.1.1) in Ge involves a change in momentum while in GaAs it does not. This

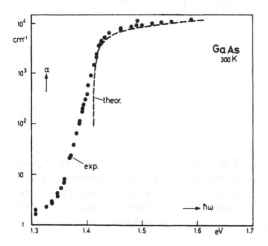

Fig. 11.4: Optical absorption edge of gallium arsenide (after [11.12])

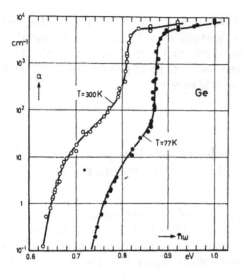

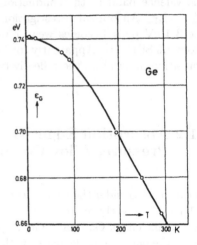

Fig. 11.5: Optical absorption edge of germanium at 77 K and 300 K. The inflection at an absorption coefficient of 10^2 indicates the change from indirect to direct absorption processes (after [11.13])

Fig. 11.6: Temperature dependence of the band gap in germanium (after [11.14])

situation is shown schematically in Fig. 11.7. The transitions are called *indirect* and *direct*, respectively. Both occur in Ge, but the indirect transition, although it requires less energy than the direct transition, has a *smaller* probability and therefore it appears as a shoulder in Fig. 11.5. The conduction band of GaAs is indicated in Fig. 11.7 by a broken curve [7.32] (normalized to the Ge curve at $k = 0$). In this case, the direct transition requires the lowest possible energy

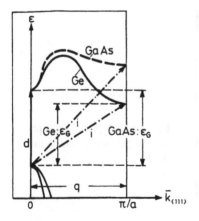

Fig. 11.7: Direct and indirect optical transition in Ge and GaAs (schematic); "i" indicates the indirect transition. The energy scales are normalized for equal direct gap in both semiconductors

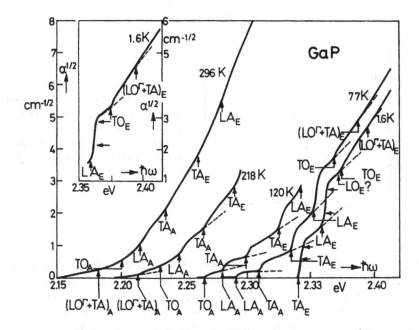

Fig. 11.8: Optical absorption edge of gallium phosphide showing threshold for the formation of free excitons with emission or absorption of phonons (after [11.15])

ε_G. The indirect transition indicated by a dash-dotted arrow may also occur in GaAs but cannot be detected in the absorption spectrum because of its larger energy and smaller probability. Because of these transitions GaAs is called a *direct semiconductor* and Ge an *indirect semiconductor*.

Since the photon momentum is negligibly small [1], momentum conservation in an indirect transition requires the co-operation of another particle. In the course of the transition, this particle changes its momentum by an amount $\hbar q$. The particle may be an intervalley phonon $\hbar\omega_i$ which is either generated or absorbed in the transition. In the case of absorption, energy conservation requires

$$\hbar\omega_e = \varepsilon_G - \hbar\omega_i \ , \tag{11.2.1}$$

instead of (11.1.1). However, in general $\varepsilon_G \gg \hbar\omega_i$. Such a particle may also be an impurity atom or a dislocation. In these cases, scattering processes can be considered as being elastic to a good approximation. The value of q has to be smaller than the reciprocal of the Debye length. In most cases of interest, however, elastic processes are less frequent than phonon interaction.

In Sect. 11.9 a quantum mechanical treatment of optical transitions will be presented. The absorption of light is proportional to the transition probability

---

[1]About 1/2000 of π/a for $\hbar\omega = 1$ eV and a lattice constant $a = 3 \times 10^{-8}$ cm.

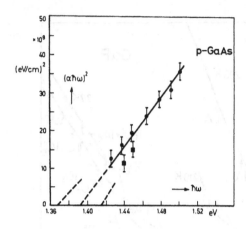

Fig. 11.9: Optical absorption edge of gallium arsenide plotted for a determination of the threshold energy. Square points are valid for a second sample (after [11.16])

which is essentially given by a matrix element times a delta-function integrated over all possible final states in the conduction band. To a first approximation, the matrix element may be considered independent of k and k'. Since $d^3k \propto \sqrt{\varepsilon}\, d\varepsilon$ and $d^3k' \propto \sqrt{-\varepsilon_G - \varepsilon'}\, d\varepsilon'$, where spherical bands with zero energy at the conduction band edge have been assumed, the integration yields

$$\int \sqrt{\varepsilon(-\varepsilon_G - \varepsilon')}\,\delta(\varepsilon - \varepsilon' - \hbar\omega \pm \hbar\omega_i)\, d\varepsilon\, d\varepsilon'$$

$$= \int_0^{\hbar\omega \mp \hbar\omega_i - \varepsilon_G} \sqrt{\varepsilon(-\varepsilon + \hbar\omega \mp \hbar\omega_i - \varepsilon_G)}\, d\varepsilon = \frac{\pi}{8}\left(\hbar\omega \mp \hbar\omega_i - \varepsilon_G\right)^2 \ , \quad (11.2.2)$$

assuming $\hbar\omega \mp \hbar\omega_i > \varepsilon_G$ (otherwise the integral vanishes).

As in any phonon assisted transition process, the matrix element includes the phonon distribution function N_q. Hence, the absorption coefficient α is proportional to

$$\alpha \propto N_q(\hbar\omega + \hbar\omega_i - \varepsilon_G)^2 + (N_q + 1)(\hbar\omega - \hbar\omega_i - \varepsilon_G)^2 \ , \quad (11.2.3)$$

where $N_q = [\exp(\hbar\omega_i/k_B T) - 1]^{-1}$. While a direct transition is weakly temperature dependent due to $\varepsilon_G = \varepsilon_G(T)$, an indirect transition shows, in addition, an exponential dependence because of $N_q(T)$.

At low temperatures and for values of the absorption coefficient of the order of magnitude 10 cm^{-1}, structure is observed in the absorption spectrum which is correlated to longitudinal and transverse-optical and acoustic intervalley phonons. For the case of GaP, where the lowest minimum of the conduction band is in the $\langle 100 \rangle$ direction so that the band structure resembles that of silicon, absorption spectra are shown in Fig. 11.8. The thresholds are sharpened by exciton effects which will be discussed in Sect. 11.3. Except for this structure, the square root of the absorption coefficient is essentially a linear function of $\hbar\omega$ since phonon absorption can be neglected at low temperatures.

The method usually applied for a determination of the absorption edge is illustrated in Fig. 11.9. The square of the product of the absorption coefficient and the quantum energy is plotted vs the quantum energy for GaAs. A straight line can be drawn through the experimental data. Its intersection with the abscissa yields an energy gap of 1.39 eV with an accuracy of ± 0.02 eV. Even better accuracy is obtained by magneto-optical methods, however (see Sect.11.4).

It will be shown in a quantum mechanical treatment (Sect. 11.9) that the absorption coefficient depends on frequency essentially according to a $\sqrt{\hbar\omega - \varepsilon_G}$ law. The reason for plotting $(\alpha\,\hbar\omega)^2$ as the ordinate rather than α^2 can easily be understood: let R/V be the rate per unit volume for transitions of electrons from state $\mathbf{k}$ to state $\mathbf{k}'$, integrated over all final states, by absorption of a photon travelling with a velocity c/n in the medium of refractive index n. Assuming N photons in the volume V, the photon flux $Nc/(nV)$, R/V is then obtained as the product of the photon flux and the absorption coefficient α:

$$R/V = N\frac{c}{nV}\alpha \ . \tag{11.2.4}$$

The energy flux $\bar{S}$ is given by the product of the photon flux and photon energy $\hbar\omega$:

$$\bar{S} = N\frac{c}{nV}\hbar\omega \ . \tag{11.2.5}$$

On the other hand, $\bar{S}$ is the time average of the well known Poynting vector $\mathbf{S}$ as the vector product of the electric field and the magnetic field vectors

$$\mathbf{S} = [\mathbf{E} \times \mathbf{H}] \ . \tag{11.2.6}$$

The vector potential $\mathbf{A}$ in a plane wave of wave vector $\mathbf{q}$ is written as

$$\mathbf{A} = A_0\mathbf{a}\cos(\mathbf{q}\cdot\mathbf{r}-\omega t) = \frac{A_0}{2}\mathbf{a}\left[\exp(i\omega t-i\mathbf{q}\cdot\mathbf{r})+\exp(-i\omega t+i\mathbf{q}\cdot\mathbf{r})\right] \ , \tag{11.2.7}$$

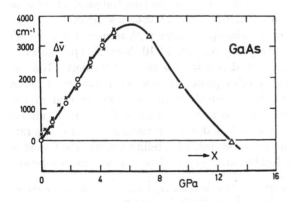

Fig. 11.10: Dependence of the absorption edge of gallium arsenide on hydrostatic pressure (after [11.18])

where A_0 is the amplitude and a is the unit vector in the direction of A and neadless to say that the "$\cdot$" indicates the scalar product. For nonmagnetic media, E and H are given in terms of A by

$$E = -\partial A/\partial t = \omega A_0\, a\, \sin\left(q \cdot r - \omega t\right) \qquad (11.2.8)$$

$$H = [\nabla_r \times A]/\mu_0 = (A_0[a \times q]/\mu_0)\sin(q \cdot r - \omega t) \qquad (11.2.9)$$

neglecting the scalar potential to be discussed below. The polarization vector a is directed in the E-direction. Equation (11.2.6) yields for the time average of $|S|$:

$$\bar{S} = |a(a \cdot q) - q|\,\omega\, A_0^2/2m u_0 \ . \qquad (11.2.10)$$

Assuming for simplicity a transverse electromagnetic wave having a propagation vector of magnitude $|q| = \omega, n/c$, we obtain for $\bar{S}$:

$$\bar{S} = \omega^2 n\, A_0^2/2\mu_0, c \ . \qquad (11.2.11)$$

By eliminating N from (11.2.4, 5), we obtain for the product of the absorption coefficient and the quantum energy

$$\alpha\,\hbar\omega = \frac{377\,\Omega}{n}\,2\,\hbar^2 R/A_0^2 V \ , \qquad (11.2.12)$$

where $\mu_0\, c$ has been replaced by its numerical value of about $377\,\Omega$.

In (11.2.8) we have neglected the scalar potential of the electromagnetic wave. It is well known from the theory of electromagnetic radiation that it can always be made to vanish by means of a gauge transformation.[2]   The Hamiltonian of an electron of momentum $\hbar k$ is then simply given by

$$H(k, r, t) = (\hbar k - eA)^2/2m \ . \qquad (11.2.13)$$

With the effective mass m instead of the free electron mass m_0, the crystal potential is taken into account in a first-order approximation. With this Hamiltonian it will be shown in Sect. 11.9 that the quantum mechanical transition probability R is proportional to A_0^2 which justifies plotting the product $\alpha\,\hbar\omega$, as given by (11.2.12), as a function of the frequency of absorbed light.

We have noticed the shift of the absorption edge with temperature. A shift may also be induced by hydrostatic pressure. Fig. 11.10 shows this phenomenon for the case of GaAs. After an initial rise with a slope of 94 meV/GPa, we find a maximum at 6 GPa and a subsequent decrease with a slope of about -87 meV/GPa. This behaviour is explained by the assumption that at low pressures the conduction band edge is at $k = 0$ (notation: Γ). It rises relative to the valence band maximum, also at $k = 0$, with increasing pressure at a rate of 126 meV/GPa. However, there are two sets of satellite valleys, the lowest of which is 0.29 eV above Γ and located on $k_{\langle 111 \rangle}$ and equivalent axes (notation:

---

[2] $A' = A - \nabla_r f$; $V' = V + \partial f/\partial t$ where $f(r, t)$ is an arbitrary function which can be chosen to yield $V' = 0$. The field intensities E and H are indifferent with respect to f [11.17].

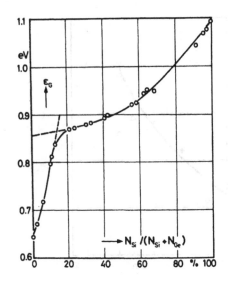

Fig. 11.11: Dependence of the band gap of silicon germanium alloys on composition (after [11.21])

L). These valleys also rise with increasing pressure but at a rate of 55 meV/GPa only.

The second set of valleys on $k_{\langle 100 \rangle}$ and equivalent axes (notation: X) located at 0.48 eV above Γ falls at a rate of 15 meV/GPa. At the meeting point of the L and X valleys at 6 GPa, the X valleys take over the role of forming the conduction band edge. At 6 GPa, the satellite valleys pass the central valley and then form the conduction band edge.

In germanium, pressure coefficients of magnitude 50 meV/GPa for the $\langle 111 \rangle$ minimum (which at low pressure forms the conduction band edge), -20meV/GPa for the $\langle 100 \rangle$ minimum and 120 meV/GPa for the $\langle 000 \rangle$ minimum have been observed. In all semiconductors investigated so far, the pressure coefficient of the $\langle 100 \rangle$ minimum has been found to be negative while those of the X and L-valleys are always positive, the latter with values close to 50 meV/GPa in all semiconductors [11.19, 20].

The absorption edge of alloys as a function of composition has some relation to pressure dependence of its constituents. Fig. 11.11 shows the energy gap obtained from optical absorption in germanium silicon alloys [11.21]. In the germanium-rich alloys, the conduction band edge is at L and rises with increasing silicon content. In the silicon-rich alloys, the band edge rises at a slower pace and is formed by the $\langle 100 \rangle$ minimum. Apparently there is a cross-over of L and X-minima at a silicon content of about 15 atomic %.

Finally we discuss the influence of carrier concentration n on the absorption edge. Such influence has been observed in n-type semiconductors with low effective mass.

Figure 11.12 shows observations in n-InSb [11.22–24]. For n larger than about 10^{18} cm^{-3} where the electron gas becomes degenerate, the optical absorp-

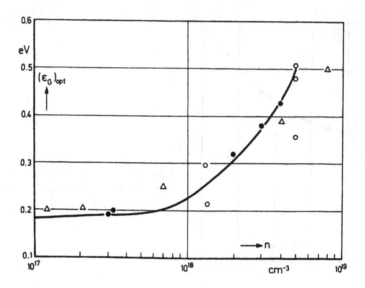

Fig. 11.12: Burstein shift of the apparent optical band gap of indium antimonide with increasing electron concentration n (after [11.22–11.24])

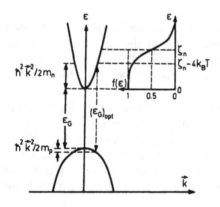

Fig. 11.13: Burstein shift in the energy band model of a direct semiconductor. In the diagram on the rhs the Fermi–Dirac distribution is illustrated

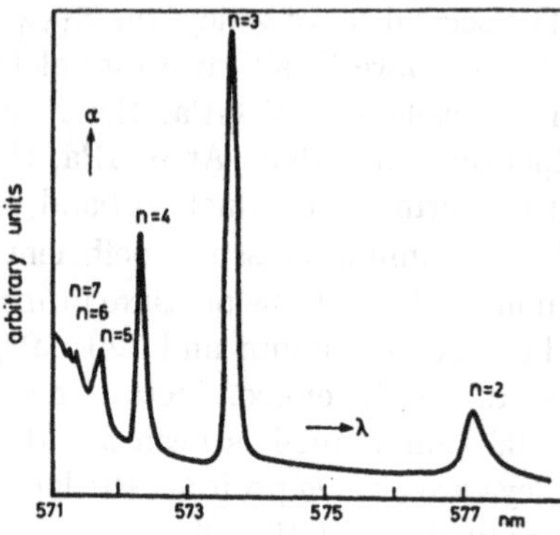

Fig. 11.14: Exciton absorption spectrum of Cu_2O at 4 K (after [11.27])

tion edge $(\varepsilon_G)_{opt}$ rises with increasing values of n. The following equation

$$(\varepsilon_G)_{opt} - \varepsilon_G = (1 + m_n/m_p)(\zeta_n - 4k_BT) \tag{11.2.14}$$

correlates the Burstein shift with the Fermi energy ζ_n [11.25]. Let us assume for simplicity that all energy states in the conduction band up to an energy of

$$\hbar^2|k|^2/2m_n = \zeta_n - 4k_BT \tag{11.2.15}$$

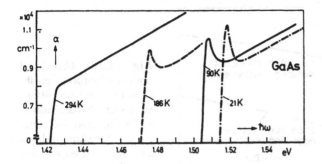

are occupied (since $1 - e^{-4} \approx 99\%$). Because of momentum conservation, the minimum photon energy is not just $\varepsilon_G + \hbar^2 |k|^2 / 2m_n$, but one must also take into account the energy $\hbar^2 |k|^2 / 2m_p$ of the state in the valence band with the same k-vector, as illustrated by Fig. 11.13:

$$\hbar\omega = \varepsilon_G + \frac{\hbar^2 k^2}{2}\left(\frac{1}{m_n} + \frac{1}{m_p}\right) = \varepsilon_G + \frac{\hbar^2 k^2}{2m_n}\left(1 + \frac{m_n}{m_p}\right) . \tag{11.2.16}$$

The electron energy $\hbar^2 k^2 / 2m_n$ is given by (11.2.15). Thus, for the absorption edge $\hbar\omega_e = (\varepsilon_G)_{\text{opt}}$, the value given by (11.2.14) is found. For nonparabolic bands, this relation has to be modified.

11.3 Exciton Absorption and Electron–Hole Droplets

So far we have considered the electron and the hole, which are obtained as a pair by the absorption of a photon, as being completely independent of each other. Actually, this is not always true. From atomic absorption spectra it is well known that besides the ionization continuum there are also discrete absorption lines due to excitations of the atoms. In a semiconductor, such excitations can be described in a simplified manner by an electron and a hole bound to each other by Coulomb interaction. According to the hydrogen model, the binding energy is given by

$$\Delta\varepsilon_{\text{exc}} = -\frac{m_r e^4}{2\hbar^2 (4\pi\,\kappa\kappa_0)^2}\frac{1}{n^2} = -\frac{\Delta\varepsilon_{\text{exc}}^{(1)}}{n^2};\ \ n = 1, 2, 3, \dots\ , \tag{11.3.1}$$

where m_r is the reduced effective mass given by

$$m_r = (m_n^{-1} + m_p^{-1})^{-1} \tag{11.3.2}$$

and κ is the static dielectric constant. These excitations are called *excitons*[11.26]. The ground-state exciton energy $\Delta\varepsilon_{\text{exc}}^{(1)}$ is $(m_r/m_0)\kappa^{-2}\times$ Rydberg energy of 13.6

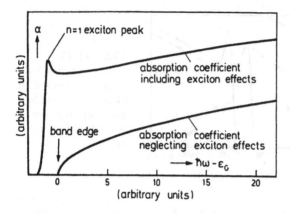

Fig. 11.16: Calculated optical absorption coefficient for direct transitions in the simple band model neglecting and including the n = 1 exciton peak below the band edge (after [11.28])

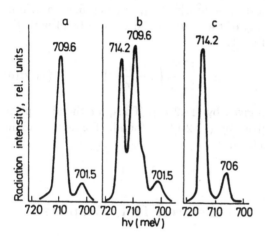

Fig. 11.17: Recombination radiation spectra of pure germanium at various temperatures (**a**) 2.32 K; (**b**) 2.52 K; (**c**) 2.78 K

eV. For values of, e.g., $m_r/m_0 = 0.05$ and $\kappa = 13$, we find 4 meV. Hence, a few meV below the absorption edge a series of discrete absorption lines should be observed. Fig. 11.14 shows such a spectrum found in Cu$_2$O [11.27]. Since the (even parity) exciton is *dipole forbidden* (Sect. 11.9), the line spectrum begins with a quantum number $n = 2$. In allowed transitions observed, e.g., in GaAs (Fig. 11.15), usually only the $n = 1$ peak is observed while the rest of the discrete absorption lines is merged with the absorption edge [11.28]. The absorption spectrum above the edge is the ionization continuum which is different from the spectrum calculated without taking excitons into account by a factor of

$$2\pi \sqrt{\Delta\varepsilon_{\mathrm{exc}}^{(1)}/(\hbar\omega - \varepsilon_{\mathrm{G}})} \left/ \left\{ 1 - \exp\left[2\pi\sqrt{\Delta\varepsilon_{\mathrm{exc}}^{(1)}/(\hbar\omega - \varepsilon_{\mathrm{G}})}\,\right] \right\} \right. .$$

Figure 11.16 shows theoretical absorption spectra both including and neglecting the exciton effects. For $\Delta\varepsilon_{\mathrm{exc}}$ in GaAs, Fig. 11.15 thus yields a value of 3.4

meV at 21 K. At room temperature the exciton peak is completely wiped out since the binding energy is readily supplied by phonons. Also in semiconductors with large carrier concentrations ($n > 2 \times 10^{16}$ cm^{-3} in GaAs), in semimetals and in metals, no excitons exist because free carriers tend to shield the electron hole interaction. Neutral impurities also cause a broadening of the exciton lines and, at large concentrations, cause their disappearance. Since GaAs is a *direct semiconductor* (Sect. 11.2), we call the exciton a *direct exciton*. In GaP, which is an indirect semiconductor, the absorption edge, shown in Fig. 11.8, is determined by *indirect excitons*. The phonon involved in the indirect transition can either be absorbed or emitted. The absorption due to excitons should occur in steps; there should be twice as many steps as there are phonon branches. Actually, shoulders are observed as shown by Fig. 11.8. Similar spectra have been observed in Ge and Si.

At very low temperatures (e.g., 1.6 K in GaP), discrete exciton lines have been observed in indirect semiconductors such as GaP. These lines are interpreted by the assumption of excitons bound to neutral donors (*bound excitons*) [Ref. 11.26, Chap. 3]. For sulfur in GaP, the exciton binding energy of 14 meV is about 10% of the donor ionization energy.

From the hydrogen model discussed above we calculate the Bohr radius of the exciton:

$$a_{\text{exc}} = \frac{\kappa}{m_r/m_0} a_B \ , \tag{11.3.3}$$

where $a_B = \hbar^2/m_0 e^2 \approx 0.053$ nm is the Bohr radius of the hydrogen atom. For example, for $m_r/m_0 = 0.05$ and $\kappa = 13$, we find $a_{\text{exc}} = 13$ nm which is about the same as the radius for a valence electron orbit in a shallow donor or acceptor. Carrier shielding of excitons occurs if the carrier concentration is larger than the concentration $[2/(2\pi)^3]4\pi|k|^3/3$ corresponding to a Fermi wave vector $|k| = 1/a_{\text{exc}}$.

Excitons with the rather large spatial extent given by (11.3.3) are adequately described by the *Wannier model* [11.29]. Frenkel [11.30] considered excitons in a solid consisting of weakly interacting atoms such as rare-gas atoms. In this case, the excitation is essentially that of a single atom or molecule; the interaction of nearest neighbors can be treated as a small perturbation. The radius of the *Frenkel exciton* is, therefore, a few tenths of a nm at most, which is of the order of the lattice constant. Such excitons have been discussed in some alkali halides and organic phosphors.

If a magnetic field is applied perpendicular to the motion of an exciton, the Lorentz force tends to separate the negative electron from the positive hole. An electric field applied along the dipole axis of the exciton would have the same effect on the exciton, and in both cases the optical absorption shows a *Stark effect*. It was observed and measured in CdS by Thomas and Hopfield [11.31]. The experiment demonstrates that the exciton is created with a nonzero velocity [11.32, 33].

At temperatures below about 2.5 K, illumination of pure Ge with the light of a mercury arc produces a 2-line luminescence spectrum; besides the free

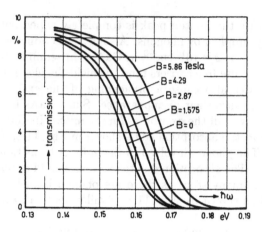

Fig. 11.18: Shift of the optical absorption edge of InSb with magnetic field at room temperature; due to the nonparabolicity of the conduction band the shift is nonlinear in B_z (after [11.39])

exciton line at 714.2 meV which is the only emission at e.g., 2.78 K, there is a second line at 709.6 meV as shown in Fig. 11.17. This lower-energy line is the only emission at, e.g., 2.32 K and lower temperatures [11.34, 35] and has been attributed to the formation of electron-hole droplets [11.36]. Keldysh suggested that the high-density electron-hole conglomerate in such a droplet might have the electrical and thermodynamical properties of a metallic liquid. Scattering of light by these electron-hole droplets in germanium revealed a radius of about 5 μm [11.34]. In a p-n-diode each droplet which diffuses into the region of the built-in electric field yields a pulse of $10^7 - 10^8$ elementary charges; negative electrons and positive holes are separated from each other in the field [11.37]. Hence, the carrier density in a droplet is of the order of 10^{17} cm^{-3} but can be raised by a 1 MW laser pulse up to 10^{19} cm^{-3}. A typical cloud containing 10^5 droplets may have a diameter of 1 mm. A critical temperature of 6.5 K and a critical density representing an inter-particle spacing of 14 nm have been determined for the electron-hole fluid [11.38]. The droplet lifetime at low temperatures is 40 μs; it has been determined from the width of the luminescence spectrum. In contrast, single excitons live only 10 μs before the electron falls into the hole and in the process of annihilation, a photon and a phonon are emitted.

11.4 Interband Transitions in a Magnetic Field

In Chap. 9 it was shown that by a strong magnetic field, Landau levels are formed in the conduction and the valence bands. For a parabolic band (effective mass m), their energies are given by

$$\varepsilon_n = \left(n + \frac{1}{2}\right)\hbar\omega_c + \hbar^2 k_z^2/2m; \quad n = 0, 1, 2, \dots, \tag{11.4.1}$$

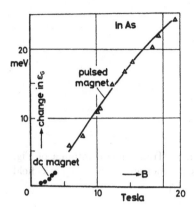

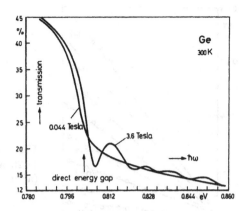

Fig. 11.19: Variation of band gap for indium arsenide with magnetic field (after [11.40])

Fig. 11.20: Oscillatory transmission of germanium at room temperature in magnetic fields of 0.044 Tesla and 3.6 Tesla (after[11.41])

where the magnetic field has been assumed in the z-direction and ω_c is the cyclotron angular frequency

$$\omega_c = (e/m)B_z \tag{11.4.2}$$

For $n = 0$ we find that the band edge moves with B_z such that the energy gap widens:

$$\varepsilon_G(B) = \varepsilon_G(0) + \frac{1}{2}\hbar\left(\frac{e}{m_n} + \frac{e}{m_p}\right)B_z = \varepsilon_G(0) + \frac{e\hbar}{2\,m_r}B_z \ , \tag{11.4.3}$$

where m_r is the reduced mass given by (11.3.2). A corresponding shift of the fundamental absorption edge has been observed. For parabolic bands, this shift should be proportional to B_z.

Figure 11.18 shows the transmission of InSb at the fundamental absorption edge for various values of the magnetic induction [11.39]. With increasing B_z, the edge shifts towards larger energies. In this case the shift is not proportional to B_z.

In Fig. 11.19 the magnetic shift of the conduction band edge in InAs is shown [11.40]. The nonlinear behavior at *small* magnetic fields cannot be explained in the simple Landau model. The nonlinear behavior at *large* magnetic fields is probably due to a nonparabolicity of the conduction band.

Besides the shift of the absorption edge, there is some oscillatory behaviour at somewhat higher photon energies. This is shown in Fig. 11.20 for the case of the direct transition at $k = 0$ in Ge [11.41]. The minima of the transmission curve are plotted as a function of B_z in Fig. 11.21. For each minimum there is a linear relationship between $\hbar\omega$ and B_z. From the slope of line 1, the electron

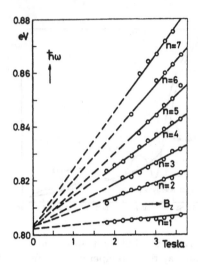

Fig. 11.21: Transmission minima of Fig. 11.20 as a function of the magnetic field (after [11.42])

effective mass in the Γ-valley has been determined as (0.036 ± 0.002) eV at room temperature. The accuracy is better than that obtained by any other method.

At low temperatures, where in (11.4.1) the term $\hbar^2 k_z^2/2m$ is small compared with the first term, transitions between Landau levels of the valence and conduction bands have a fine structure. At large magnetic fields the Landau levels are split because of the two possible electron spins of quantum number $M = \pm 1/2$. The Zeeman splitting energy is given by

$$\Delta\varepsilon_{\pm} = g\, M\, \mu_{\mathrm{B}} B_z \ , \tag{11.4.4}$$

where g is the Landé g-factor of the electrons and $\mu_{\mathrm{B}} = e\hbar/2m_0 = 58\,\mu$ eV/Tesla is the *Bohr magneton*. Fig. 11.22 shows the fine structure of the direct transition in Ge observed at 4.2 K, together with the theoretical line spectrum [11.43]. In the valence band both heavy and light holes have been considered. A rough correlation is found between the theoretical and experimental intensities for all lines except the two lowest; these persist down to zero magnetic field and therefore have been attributed to the direct exciton transition. (An unintentional strain split the valence bands and caused two exciton transitions instead of one.) A value for the g-factor of $g = -2.5$ is found for Ge. For n-InSb a very large negative value of $g = -44$ is obtained in a similar way.

11.5 The Franz-Keldysh Effect

(Electroabsorption and Electroreflectance)

We will now investigate the effect of a strong electric field on light absorption in semiconductors at the fundamental absorption edge. Theoretical treatments

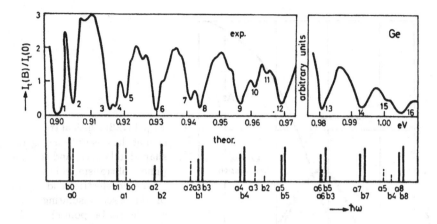

Fig. 11.22: Observed and calculated magnetoabsorption for germanium at 4.2 K and B = 3.89 Tesla parallel to the **E**-vector of the light wave (after [11.43])

of this effect have been given by Franz [11.44] and Keldysh [11.45]. It can best be described as photon-assisted tunneling through the energy barrier of the band gap (for tunneling effects, see also Chap. 9) and exists in insulators as well as semiconductors. In order to distinguish it from hot-electron effects, *semi-insulating semiconductors* are best suited for an experimental investigation. Although the fundamental absorption will only be treated in Sect.11.9, we may consider the effect of an electric field E (parallel to the x-axis) on the transition of an electron from the valence band to the conduction band simply by assuming unperturbed Bloch functions for the band structure. From (11.9.15) we adopt the concept of the reduced effective mass m_r. The potential energy of a tunneling electron moving in the x-direction is given by $-|e|\,E\,x$, where zero energy is taken at $x = 0$. For solving the Schrödinger equation

$$-(\hbar^2/2m_r)\,d^2\psi/dx^2 - |e|E\,x\psi = \varepsilon\,\psi \tag{11.5.1}$$

where ε is the electron energy in the x-direction, we simplify (11.5.1) to

$$d^2\psi/d\xi^2 = -\xi\psi \tag{11.5.2}$$

by introducing the dimensionless coordinate

$$\xi = (x + \varepsilon/|e|\,E)/l \ . \tag{11.5.3}$$

The effective length l is given by

$$l = (\hbar^2/2m_r|e|\,E)^{1/3} \ . \tag{11.5.4}$$

The solution of (11.5.2) is, up to a factor of proportionality, the Airy function Ai($-\xi$) of negative argument [11.46]. It oscillates with amplitudes which decrease with increasing energy. For very large negative values of the argument,

$$\psi \propto |\xi|^{-1/4}\sin(2|\xi|^{3/2}/3 + \pi/4) \ . \tag{11.5.5}$$

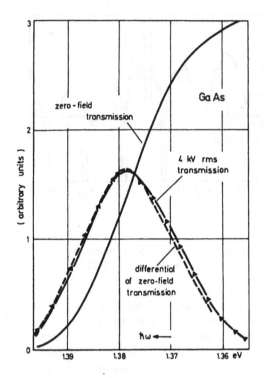

Fig. 11.23: Optical transmission at the absorption edge (S-shaped full curve) and ac component of transmission when a strong ac electric field is applied to a semiinsulating GaAs sample (data points). The differential of the zero-field transmission is shown by the dashed curve (*Franz-Keldysh* effect, after [11.47])

For positive arguments, the Airy function decays exponentially, thus converting the sharp band edge for $E = 0$ to *band tails* extending into the band gap and allowing optical transitions for $\hbar\omega < \varepsilon_\mathrm{G}$. We consider only the tail approximation for the absorption coefficient α:

$$\alpha(\omega) = \frac{\mu_0 c}{\sqrt{\kappa_\mathrm{opt}}} \frac{|e|^3 E f_\mathrm{vc}}{8\pi(\varepsilon_\mathrm{G} - \hbar\omega)} \exp\left[-\frac{4}{3|e|E}\sqrt{\frac{2m_\mathrm{r}}{\hbar^2}}(\varepsilon_\mathrm{G} - \hbar\omega)^{3/2}\right] \quad . \quad (11.5.6)$$

Here, $\sqrt{\kappa_\mathrm{opt}}$ stands for the refractive index. The field dependence is dominated by the exponential function. A corresponding expression for the argument of the exponential function was obtained in the treatment of phonon assisted tunneling (9.1.7). At a given photon energy $\hbar\omega$, the absorption increases with electric field intensity. This can be interpreted as a shift of the absorption edge to lower photon energies. For a shift of 10 meV, e.g., a field intensity E is required which is obtained to a good approximation from

$$3|e|\,E\,\hbar/4\sqrt{2m} = (10^{-2}\,\mathrm{eV})^{3/2} \tag{11.5.7}$$

or, for $m \approx m_0$,

$$E = \frac{4 \times 10^{-3}\,\mathrm{V}^{3/2}}{3\,\hbar}\sqrt{2m_0\,|e|} \approx 5 \times 10^4\,\mathrm{V/cm} \quad . \tag{11.5.8}$$

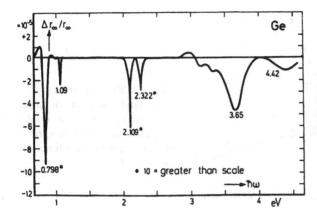

Fig. 11.24: Relative change in reflectivity of germanium upon application of an electric field normal to the reflecting surface (after [11.49])

A convenient method of measurement is to apply an ac voltage of, e.g., 4 kV across a disc-shaped sample of, e.g., 1/3 mm thickness perpendicular to a beam of monochromatic light passing through the sample, and to connect an amplifier to the light detector which is tuned to a frequency of twice the frequency of the ac voltage. The signal as a function of $\hbar\omega$ is proportional to the derivative of the absorption coefficient with respect to $\hbar\omega$. It is obvious that for an arbitrary function f,

$$\alpha = f\left(\frac{\varepsilon_G - \hbar\omega}{E^{2/3}}\right) \tag{11.5.9}$$

yields

$$\frac{d\alpha}{dE} = \frac{2}{3}\frac{\varepsilon_G - \hbar\omega}{E}\frac{d\alpha}{d(\hbar\omega)} . \tag{11.5.10}$$

For not too strong absorption and for short samples, the average transmission $\langle T \rangle$ is essentially proportional to $\exp(-\alpha d)$ where d is the sample length. The change of $\langle T \rangle$ with E is proportional to the change of α with E. The data have to be corrected for Joule heating of the crystal resulting in a change of ε_G with lattice temperature.

Figure 11.23 shows results of $d\alpha/dE$ obtained in semi-insulating GaAs, together with the zero-field transmission [11.47]. The Franz-Keldysh effect has also been observed in other large-gap semiconductors such as CdS, which in the form of pure or compensated crystals, often with deep-level impurities, show little Joule heating at the high voltage applied. Sometimes a large reverse bias applied to a semiconductor diode serves to generate the large electric field with not too much heat generation.

For photon energies larger than the band gap, (11.5.5) predicts oscillatory behavior of the absorption proportional to $E^{2/3}$ observable only for above-band-gap excitation [11.47]. Observations which could be interpreted in this way

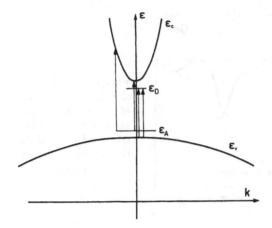

Fig. 11.25: Schematic energy band diagram illustrating spectral spread of impurity absorption due to spread of k of impurity wave function (after [11.50])

can also be explained by the assumption that an electric field destroys exciton absorption [11.48].

So far we have investigated the effect of a strong electric field on absorption of light with frequencies near the fundamental edge. It was shown in Sect. 11.1 that within the range of fundamental absorption, a measurement of reflection is more adequate. The effect of a strong electric field on reflectivity in this range has been observed by the differential method discussed above. The field has been applied perpendicular to the reflecting surface. The relative change in reflectivity $\Delta r_\infty / r_\infty$ for a given magnitude of the electric field intensity observed in germanium at room temperature is shown in Fig. 11.24 in the range of photon energies $\hbar\omega$ between 0.5 and 4.5 eV [11.49]. A comparison of this curve with Fig. 11.3 reveals the increased sensitivity of the electroreflectance method in determining *critical points* (van Hove singularities, see Sect. 11.9). Strongly temperature-dependent dips in the electroreflectance spectrum may indicate excitons at band minima which do not form a band edge. However, exciton Stark effect and exciton quenching [11.7] by the strong electric field will occur and have to be taken into account when evaluating data in terms of band structure.

11.6 Impurity Absorption

The optical absorption spectrum of a semiconductor is modified in two ways by the presence of impurities:

a) There are transitions between the ground state and the excited states of a neutral impurity. The maximum probability for such a transition occurs at an energy $\hbar\omega$ which is of the order of magnitude of the ionization energy $\Delta\varepsilon_A$ for an acceptor and $\Delta\varepsilon_D$ for a donor. This energy is usually much less than the

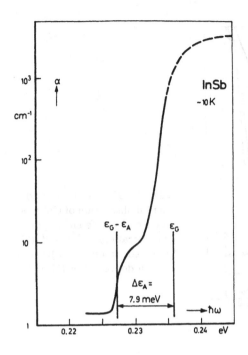

Fig. 11.26: Impurity absorption in InSb at about 10 K (after [11.51])

band gap energy. Such a transition may also occur even though the impurity cannot be ionized thermally (*electrically inactive impurity*).

b) The transition energy may be near the fundamental edge as illustrated by Fig. 11.25. A shallow acceptor level and a shallow donor level are each represented by a horizontal line. The halfwidth of the line is given by $1/a_{imp}$ where the Bohr radius of the impurity atom a_{imp} is given by

$$a_{imp} = \frac{\kappa}{m/m_0} a_B \ . \tag{11.6.1}$$

$a_B \approx 0.05$ nm is the Bohr radius of the hydrogen atom and m is the effective mass of an electron in the conduction band if the impurity is a shallow donor, and of a hole in the valence band if the impurity is a shallow acceptor. Except for the effective mass, the same considerations can be applied here which have been made in connection with the exciton (11.2.3). For deep level impurities and acceptors in group IV and III–V semiconductors, the impurity wave function contains contributions from several bands. In an indirect semiconductor, transitions involve phonons as in band-to-band transitions.

For example, let us consider a transition from a shallow acceptor to the conduction band. The transition probability is proportional to the density N_{A^-} of ionized acceptors which from (3.2.14, 16) is obtained as

$$N_{A^-} = \frac{N_A}{g_A \exp[(\varepsilon_A - \zeta)/k_B T] + 1} \ . \tag{11.6.2}$$

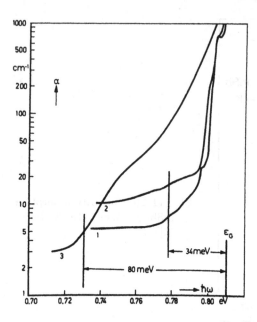

Fig. 11.27: Tail of fundamental absorption of GaSb influenced by different impurity concentrations: (1) and (2): undoped samples, (3): tellurium doped (after [11.51])

For allowed transitions involving acceptors where $m_p \gg m_n$, the absorption coefficient α is given by the proportionality

$$\alpha \, \hbar\omega \propto \frac{N_A \sqrt{\hbar\omega - (\varepsilon_G - \Delta\varepsilon_A)}}{g_A \exp[(\varepsilon_A - \zeta)/k_B T] + 1} \, . \tag{11.6.3}$$

Except for the shift by $\Delta\varepsilon_A$, the spectral variation is the same as for the fundamental absorption. However, the absorption by impurities is temperature dependent and proportional to the impurity concentration.

Experimental results on optical absorption by zinc or cadmium in InSb are shown in Fig. 11.26 [11.51]. Assuming an energy gap ε_G of 235.7 meV at 10 K, an ionization energy $\Delta\varepsilon_A$ of 7.9 meV is found from (11.6.3) (a more accurate value is obtained from magnetoabsorption). The influence of the impurity concentration on the optical absorption of p-GaSb at 10 K is illustrated by Fig. 11.27. The impurity concentrations in units of 10^{17} cm^{-3} measured at 300 K are 1.4 and 2.5 in sample 1 and 2, respectively, while in sample 3, the acceptor impurities are partly compensated by tellurium which is a donor in GaSb. The effect of compensation is to raise the Fermi level to about 80 meV above the valence band edge such that a deep acceptor level, which has not been identified so far, is filled with electrons. Transition of these electrons into the conduction band by absorption of light quanta is indicated by the long tail of the fundamental absorption edge (Fig. 11.27, curve 3). The deep acceptor level is also visible at low photon energies $\hbar\omega \approx 80$ meV (Fig. 11.28) where holes are transferred from the acceptor level to the valence band [11.51]. This corresponds to

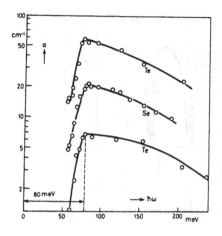

Fig. 11.28: Impurity absorption of GaSb doped with tellurium and selenium of various concentrations (after [11.51])

case (a) discussed above. Since the deep level in GaSb can also be obtained by bombarding the crystal with high-energy electrons, it has been suspected that it is due to lattice defects rather than to impurities [11.52–54].

Low energy photon absorption by shallow donors and acceptors in silicon at 4.2 K is shown in Figs. 11.29 and 11.30, respectively. Sharp bands are obtained in pure samples where line broadening by the interaction of neighboring impurity atoms is negligible. There are several sharp absorption bands at energies which are less than the ionization energy. The absorption band at a wave number of 316.4 cm^{-1} is due to an antimony contamination. The spectra of bismuth and arsenic impurities are very similar; the main difference is a shift in wave number of 140 cm^{-1} which is equivalent to a shift in energy of 17.3 meV. The similarity of the spectra is due to the fact that the valence electron of a shallow donor orbits with a radius which is several times larger than the lattice constant; the electron wave function is determined more by the host lattice than by the substitutional impurity itself. The same is true for holes bound to shallow acceptors. The small difference in the binding energies ε_B between various impurities comes from the difference in the atomic potential of the impurity atom and the replaced host atom. An adequate measure of this atomic potential, which is able to account for the observed chemical trends in ε_B, is provided by the outermost s- and p-orbital energies of the impurity and the host atom [11.55]. At high doping levels in compensated semiconductors there is a formation of ion pairs $N_D^+ N_A^-$.

Optical absorption peaks at 64 and 141 meV in Si are due to an oxygen impurity which is electrically inactive. Even for 10^{18} cm^{-3} oxygen atoms, the absorption coefficient at 4.2 K is only a few cm^{-1} as illustrated by Fig. 11.31. If the more common O^{16} isotope is replaced by the O^{18}, the absorption bands are shifted which proves that the bands are associated with oxygen [11.57].

Valuable information about the assignment of spin states to transitions like those shown in Fig. 11.29, Fig. 10.30 is gained from magneto-optical obser-

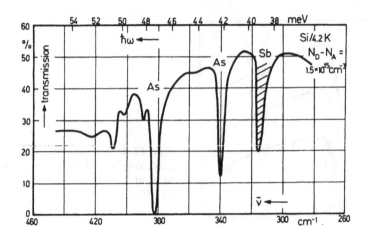

Fig. 11.29: Optical transmission spectra of silicon doped with $1.5 \times 10^{15}\,\mathrm{cm}^{-3}$ arsenic atoms, at 4.2 K; the hatched band arises from antimony contamination (after [11.56])

vations. If a longitudinal magnetic field is applied along a $\langle 100 \rangle$ direction in a germanium crystal at 4.2 K, the Landau levels in the four valleys of the conduction band are left degenerate and Zeeman splitting of the bound states about a phosphorous donor may be investigated. The $1S \rightarrow 2p$, $m = 0$ line at 76.0 cm^{-1} remains unchanged while the $1S \rightarrow 2p$, $m = \pm 1$ line at 91.5 cm^{-1} at low fields splits into a symmetrical doublet as shown in Fig. 11.32 [11.58]. The magnitude of the splitting of this line yields a value for the transverse effective mass in Ge of $(0.077 \pm 0.005)\,m_0$ which agrees well with the value of 0.082 m_0 obtained from cyclotron resonance. In the theory of linear Zeeman splitting [11.59] of donor $2p_\pm$ states in Ge, we have to consider the four ellipsoids of the conduction band and obtain for $\Delta\varepsilon_\pm$ a dependence on the angle θ between $\mathbf{B}$ and the $\langle 100 \rangle$ direction in the $[1\bar{1}0]$ plane, which is shown in Fig. 11.33 [11.60]. The symbol $2\times$ indicates a twofold degeneracy which is due to the fact that two of the four ellipsoids are symmetrical with respect to the $[1\bar{1}0]$ plane. For a $\langle 111 \rangle$ direction, the splitting for the ellipsoid with the longitudinal axis in this direction becomes

$$\Delta\varepsilon_{p\pm} = \pm\frac{e\,\hbar}{2m_t}|B| = \pm\frac{\mu_B|B|}{m_t/m_0} \ , \tag{11.6.4}$$

where m_t is the transverse effective mass and $\mu_B = 57.7\,\mu\mathrm{eV/T}$ is the Bohr magneton. The calculation neglects spin orbit interaction which is justified for strong magnetic fields.

For a spherical band, the linear Zeeman splitting of donor $p_\pm$ states is given by

$$\Delta\varepsilon_{p\pm} = \pm\hbar\omega_c/2 \tag{11.6.5}$$

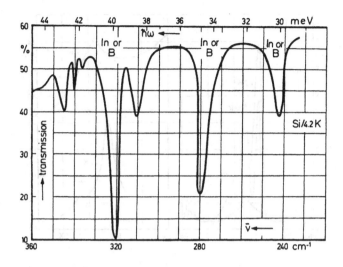

Fig. 11.30: Optical transmission spectra of silicon doped with boron, at 4.2 K (after [11.56])

while for p_0 and s states, there is no splitting but only a quadratic shift. For s states the shift is given by [11.61]

$$\Delta\varepsilon_s = (\hbar\omega_c)^2/(8\Delta\varepsilon_I) \ , \tag{11.6.6}$$

where $\Delta\varepsilon_I$ is the ionization energy and a small magnetic field strength has been assumed ($\hbar\omega_c \ll \Delta\varepsilon_I$).

In a strong magnetic field Landau levels are formed in the conduction band, and the photoionization continuum shows oscillatory behavior. The case of Ge doped with 3×10^{15} cm^{-3} arsenic donors is shown in Fig. 11.34. The correct Landau quantum numbers n assigned to the transmission minima may be obtained by observing the transmission as a function of B at a fixed wave number of, e.g., 137 c m^{-1}. The wave number location of the successive minima in the transmission at two different magnetic fields is indicated in Fig. 11.35. The two straight lines extrapolated to negative values of n intersect at $n = -1/2$. Since the Landau energy is a linear function of $(n + 1/2)\hbar\omega_c$ where $\hbar\omega_c = (e/m)B$, one would expect that for $n = -1/2$, any dependence on B would disappear as is the case. The intersection occurs at a wave number of 108 cm^{-1} which corresponds to an energy of 13.4 meV. It is incorrect to compare this energy with the donor binding energy of 12.8 meV found from electrical measurements at various temperatures [11.62], as pointed out by Howard and Hasegawa [11.63]: In fact, the transitions discussed so far are not transitions to free Landau states but to *bound* states whose energies are lowered relative to the free magnetic states by the Coulomb energy of the impurity (*bound Landau states*). These localized states are found at energies both between the Landau levels and between the zero-Landau level and the zero-field band edge. Later

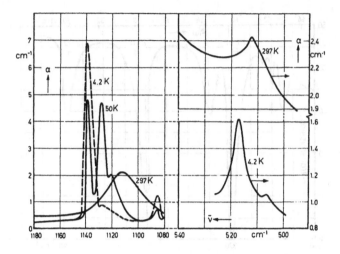

Fig. 11.31: Oxygen absorption bands in silicon at 4.2, 50, and 297 K for a crystal containing oxygen enriched with 12% O^{18} and 1% O^{17}; absorption bands due to O^{16} are at 1106 and 515 cm^{-1}(after [11.57])

experiments by Boyle and Howard [11.64] confirmed this hypothesis. There is no simple way of obtaining the impurity ionization energy from magneto-optical data of this kind [11.65].

11.7 Lattice Reflection in Polar Semiconductors

In polar crystals, transverse-optical lattice vibrations have an electric dipole moment which can be excited by an ac electric field with frequencies in the far-infrared part of the electromagnetic spectrum. Within a small range of frequencies the reflectivity is close to 100%. By multiple reflections this band can therefore be filtered out of a white spectrum. The remaining light beams are called *reststrahlen* (German for residual rays).

Figure 11.36 shows the reflectivity of AlSb observed at room temperature in the spectral range from 18 to 40 μm [11.66]. The reststrahlen band is found between 29 and 31 μm.

We will treat the theoretical problem simply as a classical harmonic oscillator of eigenfrequency ω_e and damping constant γ. In an electric field of amplitude E_1 and angular frequency ω, the classical equation of motion

$$m\, d^2x/dt^2 + m\gamma\, dx/dt + m\omega_e^2\, x = eE_1 \exp(i\omega t) \qquad (11.7.1)$$

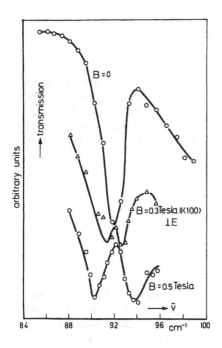

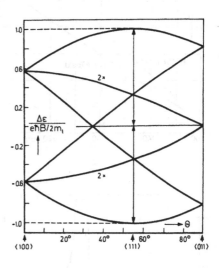

Fig. 11.32: Zeeman splitting of a $2p_\pm$ phosphorous impurity level in germanium at 4.2 K (after [11.58])

Fig. 11.33: Calculated dependence of the linear Zeeman splitting of donor $2p_\pm$ states in germanium on the angle between the magnetic field and the $\langle 100 \rangle$ direction in a $[1\bar{1}0]$ plane (after [11.60])

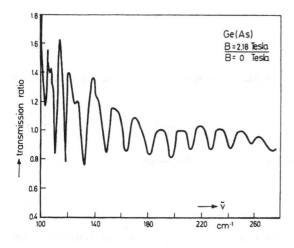

Fig. 11.34: Transmission ratio at 4 K with and without a field of B = 2.18 Tesla in germanium, as a function of wave number. The absorption is due to arsenic impurities (after [11.58])

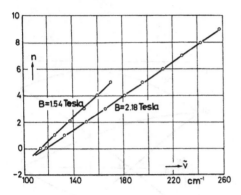

Fig. 11.35: The wave number location of the successive transmission minima of Fig. 11.34 ($B = 2.18$ Tesla) and of a similar plot for $B = 1.54$ Tesla (not shown), (after [11.58])

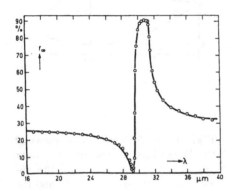

Fig. 11.36: Lattice reflection spectrum of AlSb; the points represent experimental data; the curve represents a fit calculated for a single classical oscillator (after [11.66])

has a solution given by

$$x = \frac{e}{m} \frac{E_1 \exp(i\omega t)}{\omega_e^2 - \omega^2 + i\omega\gamma} \ . \tag{11.7.2}$$

Assuming n oscillators per unit volume, the dipole moment per unit volume is given by

$$\frac{ne^2}{m} \frac{E_1 \exp(i\omega t)}{\omega_e^2 - \omega^2 + i\omega\gamma} = \frac{\omega_p^2}{\omega_e^2 - \omega^2 + i\omega\gamma} \kappa_0 E_1 \exp(i\omega t) \tag{11.7.3}$$

where we have introduced the *plasma angular frequency*

$$\omega_p = \sqrt{\frac{n e^2}{m\kappa_0}} \tag{11.7.4}$$

For the complex relative dielectric constant $\kappa_r - i\kappa_i$, we obtain from (6.8.6, 10)

$$\kappa_r - i\kappa_i = \kappa_{opt} + \frac{\omega_p^2}{\omega_e^2 - \omega^2 + i\omega\gamma} \ , \tag{11.7.5}$$

where κ_{opt} is the dielectric constant at frequencies $\omega, \gg \omega_e$ and ω_p. This equation can be solved for the real and imaginary parts:

$$n^2 - k^2 = \kappa_r = \kappa_{opt} + \omega_p^2 \frac{\omega_e^2 - \omega^2}{(\omega_e^2 - \omega^2)^2 + \omega^2 \gamma^2} \tag{11.7.6}$$

and

$$2nk = \kappa_i = \omega_p^2 \frac{\omega\gamma}{(\omega_e^2 - \omega^2)^2 + \omega^2 \gamma^2} , \tag{11.7.7}$$

where the relationship between κ_r, κ_i, the index of refraction, n (not to be confused with the density of harmonic oscillators) and the extinction coefficient k have been introduced. Solving for n and k we have

$$n = \sqrt{\frac{1}{2}\left(\kappa_r + \sqrt{\kappa_r^2 + \kappa_i^2}\right)} \tag{11.7.8}$$

and

$$k = \sqrt{\frac{1}{2}\left(-\kappa_r + \sqrt{\kappa_r^2 + \kappa_i^2}\right)} . \tag{11.7.9}$$

The reflectivity of an *infinitely thick* plate at normal incidence is given by

$$r_\infty = \frac{(n-1)^2 + k^2}{(n+1)^2 + k^2} . \tag{11.7.10}$$

Figure 11.37 shows κ_r, κ_i, and r_∞ as functions of ω/ω_e with parameters arbitrarily chosen as $\kappa_{opt} = 12$, $(\omega_p/\omega_e)^2 = 3$, $\gamma/\omega_e = 0.05$. Near the resonance frequency ω_e, there is a sharp maximum of κ_i and a broader peak of r_∞ while the frequency behavior of $\kappa_r - \kappa_{opt}$ resembles that of the differential of κ_i with respect to ω/ω_e.

The static dielectric constant κ is obtained from (11.7.6) for $\omega = 0$:

$$\kappa = \kappa_{opt} + (\omega_p/\omega_e)^2 \tag{11.7.11}$$

and in Fig. 11.37, a value of 15 has been chosen. The reflectivity r_∞ has a minimum at a *reduced frequency* given by

$$\frac{\omega}{\omega_e} = \sqrt{\frac{\kappa - 1}{\kappa_{opt} - 1}} \tag{11.7.12}$$

which for values of κ and $\kappa_{opt} \gg 1$ can be approximated by $\sqrt{\kappa/\kappa_{opt}}$. The Lyddane-Sachs-Teller relation (6.8.1) shows that this quantity is given by the ratio ω_0/ω_t where ω_0 and ω_t are the longitudinal and transverse frequencies, respectively, for long-wavelength optical phonons. From (11.7.12), we find that the eigenfrequency ω_e is the transverse frequency ω_t since the transverse oscillations are excited by the transverse-electromagnetic wave. Hence, the reflectivity minimum occurs at the longitudinal-optical phonon frequency ω_0.

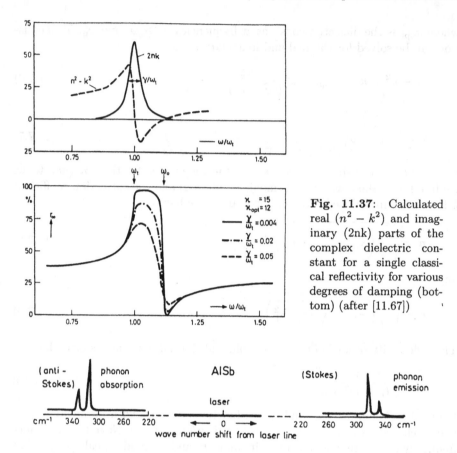

Fig. 11.37: Calculated real $(n^2 - k^2)$ and imaginary $(2nk)$ parts of the complex dielectric constant for a single classical reflectivity for various degrees of damping (bottom) (after [11.67])

Fig. 11.38: Raman spectrum of AlSb at room temperature (after [11.68])

The evaluation of the reflectivity spectrum of AlSb given in Fig. 11.36 yields phonon energies of $\hbar\omega_0 = 42.0$ meV and $\hbar\omega_t = 39.6$ meV. The ratio $\gamma/\omega_t = 5.9(\pm0.5) \times 10^{-3}$, where $\omega_t/2\pi c = \tilde{\nu}_t = 318.8(\pm0.5)$ cm^{-1}. From Raman scattering experiments [11.68], values of $\tilde{\nu}_t = 318.9$ (±0.5)cm^{-1} and $\tilde{\nu}_0 = 339.9$ (±0.5) cm^{-1} have been obtained which are in close agreement with the reflectivity data. The comparatively large inaccuracy in γ is attributed to the difficulty in measuring the absolute magnitude of the maximum reflectivity. The damping constant γ is largely due to lattice anharmonicity.

The Raman spectrum of AlSb is shown in Fig. 11.38. The lines are obtained from the incident laser light either by the absorption or emission of phonons of wave numbers $\tilde{\nu}_t$ (strong lines) or $\tilde{\nu}_0$ (weak lines). (The laser wavelength of 1.06 μm obtained from a YAG:Nd^{3+} laser [3] has been chosen to ensure that the

---

[3]YAG: Yttrium Aluminum Garnet[11.67]

crystal is transparent in this part of the spectrum [11.69, 70]; for a distinction between Raman scattering and Brillouin scattering, see (Sect.7.8)[4].

11.8 Multiphonon Lattice Absorption

In the elemental diamond-type semiconductors, the lattice atoms are not charged and consequently there is no first-order electric moment (*infrared-active fundamental vibration*). However, this and all other semiconductors show a number of weak absorption bands which have been identified as being due to the interaction of the incident photon with several phonons simultaneously (*multiphonon absorption*). Because of their weakness, the lines cannot be found experimentally in the reflectivity spectrum [11.75].

Figure 11.39 shows the absorption spectrum of silicon at various temperatures between 20 and 365 K in the wavelength range from 7 μm (corresponding to about 1500 cm^{-1}) to 20 μm (corresponding to about 300 cm^{-1}) [11.76]. Between 20 K and 77 K there is not much change in the data. The photon energies of the peaks have been found to be combinations of four phonon energies: the transverse (TO) and longitudinal-optical (LO) phonon energies of 59.8 and 51.3 meV, respectively, and the transverse (TA) and longitudinal-acoustic (LA) phonon energies of 15.8 and 41.4 meV, respectively, all at the edge of the phonon Brillouin zone[5] . (For the case of Ge, the phonon spectrum has been shown in Fig. 6.10.) For example, to the highest peak a combination of TO + TA could be assigned, while the small peak at 1450 cm^{-1} could be labeled 3 TO, etc. A total of 10 peaks could be fitted quite well and their dependence on temperature agrees with calculations based on these phonon energies. The peak at 1100 cm^{-1} is, however, due to an oxygen impurity which exists even though the crystal was vacuum grown (see Fig. 11.31).

The phonon energies so obtained compare favorably well with those found from slow-neutron spectroscopy where in a $\langle 100 \rangle$ direction, TO = 58.7 meV, LO = LA = 49.2 meV and TA = 17.9 meV at $q = q_{max}$ [11.77] (Sect. 6.8).

Similar measurements have been made on other group-IV elements, on III–V compounds, on some II–VI compounds and on SiC. An interesting case is that of silicon germanium alloys. The silicon and germanium-like summation bands are not sensitive to the composition which suggests that there are pure Ge and Si conglomerates rather than a truely disordered alloy [11.78]. In addition, two new bands have been observed: one in the Ge-rich alloy and the other in the Si-rich alloy which are impurity bands since they are not present in the pure elements and grow in intensity as the composition is changed toward 50% of

---

[4]In a lattice of strong polarity, an electromagnetic wave is coupled to a sound wave and vice versa. A distinction between phonon and photon can no longer be made and one prefers then to denote the quantum of vibration by the word *polariton*[11.71-74]

[5]An average of the LO and TO frequencies has been denoted as *intervalley phonon frequency* in Sect. 7.5

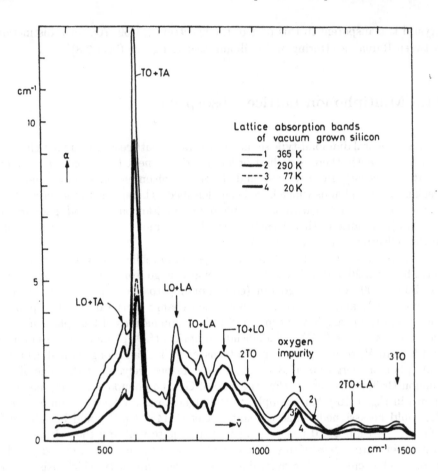

Fig. 11.39: Lattice absorption due to multiphonon processes in vacuum grown silicon (after [11.76])

either constituent. The shift of phonon frequencies with the composition of the alloy is shown in Fig. 11.40.

11.9 Quantum Mechanical Treatment of the Fundamental Optical Absorption Edge

In Sect. 11.2 we have considered the observed optical absorption near the fundamental edge. Now we will treat this problem by means of a quantum mechanical perturbation theory where the electromagnetic wave is considered as a small perturbation of the electron wave [11.7, 79, 80].

The classical Hamiltonian of an electron in an electromagnetic field is given

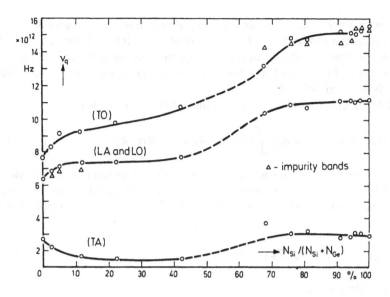

Fig. 11.40: Phonon frequencies in Ge–Si alloys as a function of composition (after [11.78])

by (11.2.13). The quantum mechanical commutation relationship between, e.g., the x-component of the momentum vector $\hbar \mathbf{k} = -i\hbar \, \boldsymbol{\nabla}_r$ and the same component of the vector potential $\mathbf{A}$ is easily found to be [11.80]

$$\hbar k_x A_x - A_x \hbar k_x = -i\hbar \, \partial A_x / \partial x \tag{11.9.1}$$

and therefore in the Hamiltonian $(\hbar k_x - e A_x)^2$ has to be replaced by

$$(\hbar k_x - e A_x)(\hbar k_x - e A_x) = \hbar^2 k_x^2 - 2e A_x \hbar k_x + i e \hbar \, \partial A_x / \partial x + e^2 A_x^2 \ . \tag{11.9.2}$$

For low light intensities the term $\propto A^2$ can be neglected in comparison to the linear terms. The term $\propto (\boldsymbol{\nabla}_r \cdot \mathbf{A})$ can be made to vanish by a suitable gauge transformation (footnote 2, Sect. 11.2). Hence, the time-dependent Schrödinger equation is given by

$$(H_0 + H_1) \, \psi = i\hbar \, \partial \psi / \partial t \ , \tag{11.9.3}$$

where the interaction between the electron and the radiation field is represented by the operator H_1:

$$H_1 = i(e\hbar / m_0)(\mathbf{A} \cdot \boldsymbol{\nabla}_r) \ . \tag{11.9.4}$$

In a perturbation treatment, the matrix element for a transition of an electron from state $\mathbf{k}$ to state $\mathbf{k}'$ (e.g., valence band to conduction band) is given by

$$H_{k'k} = \int \psi_{k'}^* H_1 \psi_k d^3 r \ . \tag{11.9.5}$$

The electron wave functions are the usual Bloch functions $u_x \exp[\mathrm{i}\,(\boldsymbol{k} \cdot \boldsymbol{r})]$ (Sect. 2.1). Since the momentum of the absorbed or emitted photon at optical frequencies is much smaller than the average electron momentum and no phonon is assumed to be involved, the transition in the energy-vs-momentum diagram is nearly vertical ($\boldsymbol{k'} \approx \boldsymbol{k}$).

With $\boldsymbol{A}$ given by (11.2.7), we obtain for the matrix element

$$H_{k'k} = \mathrm{i}(e\hbar A_0/2m_0) \int u_{k'}^* [(\boldsymbol{a} \cdot \boldsymbol{\nabla}_r u_k) + \mathrm{i}(\boldsymbol{a} \cdot \boldsymbol{k})u_k]d^3r \ . \tag{11.9.6}$$

For an allowed transition the second term in the integrand is much smaller than the first term and can be neglected. It would vanish for an exactly vertical transition in the ε-vs-$\boldsymbol{k}$ diagram. With the definition of a momentum matrix element

$$\boldsymbol{p}_{k'k} = -\mathrm{i}\,\hbar \int u_{k'}^* \boldsymbol{\nabla}_r\, u_k\, d^3r \ , \tag{11.9.7}$$

we can write for $H_{k'k}$

$$H_{k'k} = -(e\,A_0/2\,m_0)(\boldsymbol{a} \cdot \boldsymbol{p}_{k'k}) \ . \tag{11.9.8}$$

The transition rate R defined by (11.2.4) is given by the rhs of (6.2.14) with the transition probability $S(k, k')$ given by (6.2.13):

$$R = \frac{2V}{(2\pi)^3} \int d^3k \left(\frac{2\pi}{\hbar}\right) \left(\frac{eA_0}{2m_0}\right)^2 (\boldsymbol{a} \cdot \boldsymbol{p}_{k'k})^2 \delta(\varepsilon' - \varepsilon - \hbar\omega) \ , \tag{11.9.9}$$

where V is the volume of the unit cell, a factor of 2 takes care of the possible change in spin in the photon absorption process, and the initial state has been assumed as being occupied and the final state vacant, i.e., $f(k) = 1$, $f(k') = 0$. Furthermore, induced emission of photons can be neglected because the band gap is usually large compared with $k_B T$.

The argument of the delta function $\varepsilon' - \varepsilon - \hbar\omega$ is a function of $\mathbf{k}$. Because the photon momentum is negligibly small, we have $\mathbf{k}' = \mathbf{k}$ for the transition (assumed to be *direct*) and we write for the delta function

$$\delta(\varepsilon' - \varepsilon - \hbar\omega) = \delta(k' - k)/|\boldsymbol{\nabla}_{\boldsymbol{k}}(\varepsilon' - \varepsilon)|_{\varepsilon'-\varepsilon=\hbar\omega} \ . \tag{11.9.10}$$

Since the initial state is located in the valence band and the final state in the conduction band, it is convenient to introduce subscripts v and c, respectively. Because there are points in the energy band diagram where $\boldsymbol{\nabla}_{\boldsymbol{k}}\varepsilon_c(\mathbf{k}) = \boldsymbol{\nabla}_{\boldsymbol{k}}\varepsilon_v(\mathbf{k})$ and therefore the denominator vanishes, there are large contributions to R at these *critical points*. Compared with the delta function, the factor $(\boldsymbol{a} \cdot \boldsymbol{p}_{vc})^2$ is only a slowly varying function of $\boldsymbol{k}$ and can therefore be taken outside the integral. The product

$$f_{vc} = (2/m_0\,\hbar\omega)(\boldsymbol{a} \cdot \boldsymbol{p}_{vc})^2 \tag{11.9.11}$$

is called the dimensionless *oscillator strength* of the transition and the integral

$$J_{vc}(\omega) = \frac{2}{(2\pi)^3} \int \delta(\varepsilon_c(\mathbf{k}) - \varepsilon_v(\mathbf{k}) - \hbar\omega) d^3 k \tag{11.9.12}$$

the *joint density of states* [dimension: $(eV)^{-1} cm^{-3}$]. Now we obtain for the transition rate R,

$$R = \frac{\pi e^2 A_0^2 \omega V}{4m_0} f_{vc} J_{vc}(\omega) \tag{11.9.13}$$

and from (11.2.12) for the absorption coefficient,

$$\alpha = \frac{\mu_0 c}{n} \frac{\pi e^2 \hbar}{2m_0} f_{vc} J_{vc}(\omega), \tag{11.9.14}$$

where n is the refractive index. The factor $\mu_0 c \pi e^2 \hbar / 2m_0 = 1.098 \times 10^{-16} eVcm^2$.

Now let us consider the joint density of states for the simple case of spherical energy surfaces at $\mathbf{k} = 0$ where

$$\hbar\omega = \varepsilon_G + \hbar^2 k^2 / 2m_n + \hbar^2 k^2 / 2m_p = \varepsilon_G + \hbar^2 k^2 / 2m_r , \tag{11.9.15}$$

where $m_r = (m_n^{-1} + m_p^{-1})^{-1}$ is a reduced effective mass. Let us, for simplicity, denote $\hbar\omega$ by ε. From (11.9.12) we now obtain for J_{vc}

$$J_{vc} = \frac{2}{(2\pi)^3} \frac{d}{d\varepsilon} \left(\frac{4\pi}{3} k^3 \right) = \frac{1}{2\pi^2} \left(\frac{2m_r}{\hbar^2} \right)^{3/2} (\varepsilon - \varepsilon_G)^{1/2} . \tag{11.9.16}$$

The dependence of α on $(\hbar\omega - \varepsilon_G)^{1/2}$ is just the observed frequency dependence for a direct transition. From this value of J_{vc}, we finally obtain for α in units of cm^{-1}

$$\alpha = \frac{2.7 \times 10^5}{n} \left(\frac{2m_r}{m_0} \right)^{3/2} f_{vc} \left[\frac{\hbar\omega - \varepsilon_G}{eV} \right]^{1/2} . \tag{11.9.17}$$

Assuming, e.g., $\hbar\omega - \varepsilon_G = 10$ meV, $m_r = m_0/2$, $f_{vc} = 1$ and $n = 4$, we find $\alpha = 0.67 \times 10^4$ cm^{-1}. This is the observed order of magnitude to which α rises at the fundamental absorption edge.

It has not yet been become clear why the oscillator strength is of the order of magnitude 1, i.e., what $\mathbf{p}_{vc}$ defined by (11.9.7) really means. However, since the Bloch wave functions in the conduction band and the valence band in general are not known explicitly except for their symmetry we will only note that it depends on the polarization of the light and that selection rules for the transitions can be determined from these symmetry properties. As for atomic spectra, an f-sum rule has been shown to exist [11.81] which makes it plausible that $f_{vc} \approx 1$ for allowed dipole transitions.

It is very interesting to investigate the function $J_{vc}(\omega)$ for various types of critical points. Taking (11.9.10) into account, we write for (11.9.12)

$$J_{vc}(\omega) = \frac{2}{(2\pi)^3} \int \frac{dS}{|\nabla_k[\varepsilon_c(\mathbf{k}) - \varepsilon_v(\mathbf{k})]|_{\varepsilon_c - \varepsilon_v = \hbar\omega}} , \tag{11.9.18}$$

where dS is an element of an energy surface $\varepsilon_c(k) - \varepsilon_v(\mathbf{k}) = \hbar\omega$ in k-space. Let us, for simplicity, assume a 2-dimensional k-space and an energy surface [11.82]

$$\varepsilon_c(\mathbf{k}) - \varepsilon_v(\mathbf{k}) = \varepsilon_G - A[\cos(k_x a) + \cos(k_y a) - 2] \ , \tag{11.9.19}$$

where $a > 0$ is a constant and ε_G is the energy gap at $k = 0$. The van Hove singularities are the solutions of

$$\partial(\varepsilon_c - \varepsilon_v)/\partial k_x = A a \sin(k_x a) = 0$$

$$\partial(\varepsilon_c - \varepsilon_v)/\partial k_y = A a \, \sin(k_y a) = 0 \tag{11.9.20}$$

in the first Brillouin zone:

$$k = (0,0), \ (0,\pm\pi/a), \ (\pm\pi/a, 0), \ (\pm\pi/a, \pm\pi/a) \ . \tag{11.9.21}$$

We expand the energy bands around these points according to $\cos x \approx 1 - x^2/2$ and find for $\varepsilon - \varepsilon_G$:

$$\varepsilon - \varepsilon_G = \begin{cases} \frac{1}{2}Aa^2(\Delta k_x^2 + \Delta k_y^2) & \text{minimum at } (0,0) \\ -\frac{1}{2}Aa^2(\Delta k_x^2 + \Delta k_y^2) + 4A & \text{maximum at } (\pi/a, \pi/a) \\ \frac{1}{2}Aa^2(\Delta k_x^2 - \Delta k_y^2) + 2A & \text{saddle point at } (0, \pi/a) \end{cases} \tag{11.9.22}$$

The joint density of states at the extremal points is simply obtained as [6]

$$J_{vc}(\omega) = \frac{2}{(2\pi)^2} \left| \frac{d}{d\varepsilon}[\pi(\Delta k)^2] \right| \frac{1}{\pi A a^2} \tag{11.9.23}$$

both for $\varepsilon > \varepsilon_G$ at $(0,0)$ and for $\varepsilon < \varepsilon_G + 4A$ at $(\pi/a, \pi/a)$ and zero otherwise, while for the saddle point $(0, \pi/a)$ it is

$$J_{vc}(\omega) = \frac{2}{(2\pi)^2} \int_L \frac{ds}{|\nabla_k \varepsilon|} = \frac{2}{(2\pi)^2} 4 \int_{\Delta k_{xm}}^{\Delta k_{xM}} \frac{d(\Delta k_x)}{|\partial\varepsilon/\partial\Delta k_y|}$$

$$= \frac{2}{\pi^2 A a^2} \int_{\Delta k_{xm}}^{\Delta k_{xM}} \frac{d(\Delta k_x)}{\Delta k_y} \tag{11.9.24}$$

where ds is an element of the complete constant-energy line of the two-dimensional $\varepsilon(k_x, k_y)$ surface, and Δk_{xm} and Δk_{xM} are the minimum and the maximum values, respectively, of Δk_x on this line. Solving (11.9.22) (saddle point) for Δk_y and evaluating the integral in (11.9.24), we obtain

$$J_{vc}(\omega) = \frac{2}{\pi^2 A a^2} \{ \ln[\Delta k_{xM} + \sqrt{\Delta k_{xM}^2 - 2(\varepsilon - \varepsilon_G - 2A)/Aa^2}]$$

$$- \ln \sqrt{2(\varepsilon - \varepsilon_G - 2A)/Aa^2}\} \tag{11.9.25}$$

[6]This is a result of the property of the δ function $\int \delta[\varepsilon(k_x, k_y)]dk_x k_y = |(d/d\varepsilon) \int dk_x dk_y|$

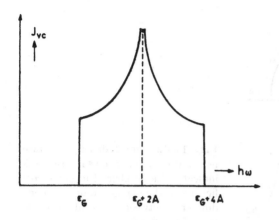

Fig. 11.41: Joint density of states for a two-dimensional surface of constant energy

where $\Delta k_{xm} = \sqrt{2(\varepsilon - \varepsilon_G - 2A)/Aa^2}$ has been taken into account (this is obtained from (11.9.22) for $\Delta k_y = 0$ and valid for a positive value of the radicand) and $\Delta k_{xM} \approx \pi/a$. Equation (11.9.25) shows that $J_{vc}(\omega)$ has a logarithmic singularity at the saddle point, where $\varepsilon = \varepsilon_G + 2A$. The singularity is shown in Fig. 11.41. Such singularities were first analysed by van Hove [11.83] and the points in the Brillouin zone where they occur were identified by Phillips [11.84].

In three dimensions the expansion around a critical point

$$\varepsilon(k) = \varepsilon_c(k) - \varepsilon_v(k) = \varepsilon_0 + \frac{a^2}{2} \sum_{i=1}^{3} A_i (\Delta k_i)^2 \ . \tag{11.9.26}$$

contains three coefficients A_i which may be positive or negative. Saddle points are characterized by either one negative and two positive coefficients (denoted M_1) or two negative and one positive (M_2). For three positive coefficients, we find a minimum (M_0) where J_{vc} varies with $\varepsilon = \hbar\omega$ as given by (11.9.16), while for all coefficients being negative, there is a maximum (M_3) and $J_{vc} \propto (\varepsilon_G - \hbar\omega)^{1/2}$. The functions $J_{vc}(\omega)$ are shown schematically in Fig. 11.42. For a comparison, let us again consider the $\kappa_i(\omega)$ spectrum for Ge shown in Fig. 11.3. It is fairly obvious that, e.g., the edge denoted by $\Lambda_3 \to \Lambda_1$ is of the M_1 type. In Fig. 2.25, the transition has been indicated in the energy band structure. Although it is not possible to see the saddle point in this diagram, one notices that $\nabla_k \varepsilon_c = \nabla_k \varepsilon_v$ for the $k_{\langle 111 \rangle}$ direction indicated there. Another transition of interest to us is the direct transition from the valence band maximum to the conduction band minimum at $k=0$ which is, of course, of the M_0 type (Fig. 11.3)[7]

---

[7]For the correlation between the static dielectric constant and the minimum energy gap see [11.85].

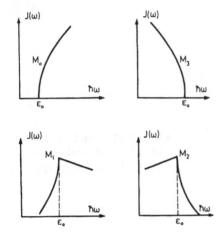

Fig. 11.42: Joint density of states near critical points; the subscript of M indicates the number of negative coefficients in the expansion of the energy difference between the bands as a function of k (after [11.79a])

11.10 Free-Carrier Absorption and Reflection

Light absorption by free carriers (i.e., conduction electrons or holes) shall first be treated by the classical theory of the harmonic oscillator [11.86, 87] in the limit of vanishing binding energy $\hbar\omega_e$. For this case, (11.7.6, 7) yield

$$n^2 - k^2 = \kappa_r = \kappa_{opt} - \omega_p^2/(\omega^2 + \gamma^2) \tag{11.10.1}$$

and

$$2nk = \kappa_i = \omega_p^2\gamma/[\omega(\omega^2 + \gamma^2)] \ . \tag{11.10.2}$$

It will be shown below by a quantum mechanical treatment that the damping constant γ is essentially the inverse momentum relaxation time, τ_m^{-1}. For the case of small damping, we substitute for n in (11.10.2) $\sqrt{\kappa_o}$ which is obtained from (11.10.1) for $\gamma = 0$ and frequencies ω which are well above the plasma frequency ω_p. The extinction coefficient k is then given by

$$k = \omega_p^2\gamma/(2\omega^3\sqrt{\kappa_{opt}}) \tag{11.10.3}$$

and the absorption coefficient α by

$$\alpha = k\frac{4\pi}{\lambda} = \frac{377\Omega}{\sqrt{\kappa_{opt}}}\sigma_0/(\omega\tau_m)^2 \propto \lambda^2 \ , \tag{11.10.4}$$

where σ_0 is the dc conductivity and γ has been replaced by τ_m^{-1}. For example, for $\tau_m = 2 \times 10^{-13}$ s and a wavelength of $\lambda = 1\mu m$, we find $\omega\tau_m \approx 10^2$. A refractive index of, e.g., $\sqrt{\kappa_{opt}} = 4$ and a conductivity of, e.g., $\sigma_0 = 10^2\Omega^{-1}\,\mathrm{cm}^{-1}$ yield $\alpha \approx 1\,\mathrm{cm}^{-1}$ which is small compared with the fundamental absorption where $\alpha \approx 10^4\,\mathrm{cm}^{-1}$. However, in metals where $\sigma_0 \approx 10^6\Omega^{-1}\,\mathrm{cm}^{-1}$, the free-carrier absorption coefficient has the same order of absolute magnitude as the

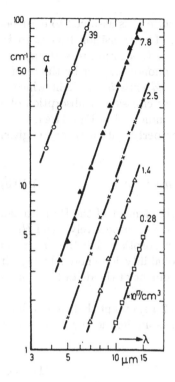

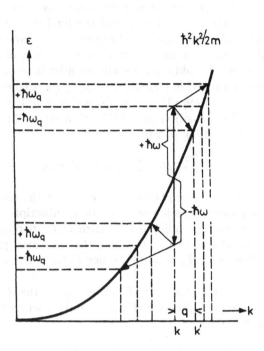

Fig. 11.43: Free–carrier absorption spectrum in n-type indium arsenide for varies carrier densities at room temperature (after [11.88])

Fig. 11.44: Change in carrier momentum as a consequence of a change in energy for intravalley transitions by photon absorption or emission

fundamental absorption in semiconductors, or is even larger (e.g., Ag at $\lambda = 0.6$ μm and a temperature of 300 K: $k = 4.09, n = 0.05$).

Equation (11.10.4) suggests a dependence of α on the second power of the wavelength. Figure 11.43 shows a log-log plot of α vs λ obtained in n-InAs at room temperature for various electron densities. Although a power law is found, the slope of the curves is rather compatible with a λ^3-dependence for α. This discrepancy is resolved by a quantum mechanical treatment of the problem which is different for the various kinds of electron–phonon or electron- -impurity interactions. Figure 11.44 shows that, in fact, an absorption or emission process for a photon of energy $\hbar\omega = \Delta\varepsilon$ must involve a change in momentum $\hbar q$; this is much larger, than the photon momentum itself. The momentum difference may be supplied by a phonon with a momentum $\hbar q = \hbar(k - -k')$ having an energy $\hbar\omega_q$. This phonon may either be absorbed or emitted.

Since 3 particles (electron, photon, and phonon) are involved in the absorption process, we treat the problem with second-order perturbation theory

[11.87, 11.90], where an *intermediate* or *virtual state* k' is assumed to exist between the initial state k and the final state k''. The transition from k to k'' may take place first by absorption of a photon $\hbar\omega$ (transition to state n) and then the subsequent absorption or emission of a phonon $\hbar\omega_q$, or by the reverse process. In addition, we will consider here *induced emission* of light which is not contained in classical theory, again with either emission or absorption of a phonon [11.89]. There are altogether 4 processes indicated in Fig. 11.44.

The transition probability S in second-order perturbation theory is given by [Ref. 11.80, p. 247]

$$S_\pm = \frac{2\pi}{\hbar}(|H_{k'k}|^2_\pm|H_{k''k'}|^2/\hbar^2\omega^2)\delta(\varepsilon' - \varepsilon - \hbar\omega \pm \hbar\omega_q) \ , \tag{11.10.5}$$

where $\hbar\omega$ is the photon energy, $H_{k'k}$ is the matrix element of the Hamiltonian for phonon emission or absorption (subscripts $+$ or $-$, respectively), and $H_{k''k'}$ is the corresponding matrix element for photon absorption. For the process of induced emission of photons, the transition probability is denoted by $S'_\pm$; it differs from $S_\pm$ only by the sign of $\hbar\omega$ in the argument of the delta function.

The transition rate R is given by the rhs of (6.2.14) except that we now have to integrate also over all initial states k where the spin degeneracy introduces a factor of 2:

$$R_\pm = \frac{2V}{(2\pi)^6} \int d^3k \, d^3k' \, S_\pm f(k)[1 - f(k')] \ . \tag{11.10.6}$$

We will limit our treatment to a nondegenerate gas of carriers with an electron temperature T_e where we can neglect $f(k')$, and for $f(k)$ we write

$$f(k) = \frac{n}{N_c} \exp\left(-\frac{\varepsilon}{k_B T_e}\right) = \frac{n}{2}\left(\frac{2\pi\hbar^2}{m\,k_B T_e}\right)^{3/2} \exp\left(-\frac{\varepsilon}{k_B T_e}\right) \ , \tag{11.10.7}$$

where n is the carrier concentration. It is convenient to replace d^3k' in (11.10.6) by d^3q where q is the phonon wave vector.

The matrix element $H_{k''k'}$ is given by the rhs of (11.9.8) except that we now introduce for the momentum matrix element divided by the electron mass:

$$p_{k'k}/m_0 \approx \hbar(k' - k)/m = \hbar\,q/m \tag{11.10.8}$$

neglecting, as in Sect. 11.9, the photon momentum and taking into account the crystal potential [8]   by introducing the carrier effective mass m. Therefore we find for $R_\pm$ from (11.10.6)

---

[8]The effective-mass theory may be applied because the wavelength of both photon and phonon are large compared with atomic dimensions and hence the potential variations over the unit cell are small

$$R_\pm = \frac{Ve^2 A_0^2 n}{2(2\pi)^5 \hbar \omega^2 m^2 N_c} \int |H_{k'k}|_\pm^2 (\mathbf{a} \cdot \mathbf{q})^2$$

$$\times \delta(\varepsilon' - \varepsilon - \hbar\omega \pm \hbar\omega_q) \exp(-\varepsilon/k_B T_e)\, d^3k\, d^3q \ . \tag{11.10.9}$$

In general, the matrix element $H_{k'k}$ depends on q, but not on k. Therefore, the integral over k can be performed right away. The integral of the delta function over θ and φ yields a factor

$$\int \delta[(\hbar^2/2m)(q^2 + 2kq\cos\theta) - \hbar\omega \pm \hbar\omega_q] \sin(\theta) d\theta d\varphi = -2\pi m/\hbar^2 kq$$

$$\tag{11.10.10}$$

and the integral over $k^2 dk$

$$-2\pi m/\hbar^2 q \int\limits_{k_{min}}^{\infty} \exp(-\hbar^2 k^2/2m\, k_B T_e)\, k\, dk$$

$$= (2\pi m^2 k_B T_e/\hbar^4 q) \exp(-\hbar^2 k_{min}^2/2m k_B T_e) \ , \tag{11.10.11}$$

where the lower limit k_{min} is found from energy conservation (argument of the delta function $= 0$) for $\cos\theta = 1$:

$$k_{min}^2 = \left[\frac{m(\hbar\omega \mp \hbar\omega_q)}{\hbar^2 q} - \frac{q}{2} \right]^2 , \tag{11.10.12}$$

while $k_{max}^2 = \infty$ is obtained from $\cos\theta = 0$. [9]   It is convenient to introduce the parameter

$$z_\pm = (\hbar\omega \pm \hbar\omega_q)/2k_B T_e \tag{11.10.13}$$

(subscript $+$ indicates phonon absorption, subscript $-$, emission) and the variable

$$\xi = \hbar^2 q^2/4mk_B T_e \ . \tag{11.10.14}$$

We now obtain for (11.10.11)

$$\int \delta(\varepsilon' - \varepsilon - \hbar\omega \pm \hbar\omega_q) \exp(-\varepsilon/k_B T_e) d^3k$$

$$= \pi \hbar^{-3} m^{3/2} (k_B T_e)^{1/2} \exp(z_\mp) \xi^{-1/2} \exp[-\tfrac{1}{2}(\xi + z_\mp^2/\xi)] \ . \tag{11.10.15}$$

For the integral over $d^3q = q^2\, dq \sin\theta' d\theta' d\varphi'$, we introduce the angle θ' between the unit vector $\mathbf{a}$ of the light polarization and the phonon wave vector $\mathbf{q}$:

---

[9]In a crystal, k_{max} is actually of the order of magnitude π/a where a is the lattice constant; because the integrand depends on k^2 exponentially and is very small for $k = \pi/a$, we replace π/a by ∞.

$$\int\limits_0^\pi (\boldsymbol{a} \cdot \boldsymbol{q})^2 \sin\theta' d\theta' \int\limits_0^{2\pi} d\varphi' = \int\limits_0^\pi q^2 \cos^2\theta' \sin\theta' d\theta' \int\limits_0^{2\pi} d\varphi' = \xi \frac{16\pi m k_B T_e}{3\hbar^2}$$

$$(11.10.16)$$

and finally obtain for the transition rate $R_\pm$

$$R_\pm = \frac{V e^2 A_0^2 n}{2(2\pi)^5 \hbar\omega^2 m^2 N_c} \frac{\pi m^{3/2}(k_B T_e)^{1/2}}{\hbar^3} \frac{16\pi m\, k_B T_e}{3\hbar^2} \frac{(4m\, k_B T_e)^{3/2}}{2\hbar^3}$$

$$\times \int\limits_0^\infty |H_{k'k}|_\pm^2 \exp(z_\mp) \exp\left[-\frac{1}{2}(\xi + z_\mp^2/\xi)\right] \xi d\xi \qquad (11.10.17)$$

and from (11.2.12) for the absorption coefficient

$$\alpha_\pm = R_\pm \frac{2\hbar\mu_0 c}{\sqrt{\kappa_{opt}} A_0^2 \omega} = \frac{\mu_0 c}{\sqrt{\kappa_{opt}}} \frac{2^{3/2}}{3\pi^{3/2}} \frac{n e^2 m^{1/2}(k_B T_e)^{3/2} V}{\hbar^5 \omega^3}$$

$$\times \int\limits_0^\infty |H_{k'k}|_\pm^2 \exp(z_\mp) \exp\left[-\frac{1}{2}(\xi + z_\mp^2/\xi)\right] \xi d\xi \ . \qquad (11.10.18)$$

The limits of integration have been extended to 0 and ∞ in view of the exponential factor in the integrand which vanishes at both limits. Note: $\kappa_{opt} = \kappa_{opt}(\omega)$.

Let us first consider the simple case where the matrix element is independent of q and therefore of ξ so that we can take it outside the integral. The integral is given by [11.46]

$$\int\limits_0^\infty |H_{k'k}|_\pm^2 \exp(z_\mp) \exp\left[-\frac{1}{2}(\xi + z_\mp^2/\xi)\right] \xi d\xi$$

$$= |H_{k'k}|_\pm^2 \exp(z_\mp) 2\, z_\mp^2 K_2(|z_\mp|) \qquad (11.10.19)$$

where K_2 is a modified Bessel function.

Another case of interest is that of $H_{k'k} \propto 1/q$. For this case, we obtain similarly

$$\int\limits_0^\infty q^{-2} \exp(z_\mp) \exp\left[-\frac{1}{2}(\xi + z_\mp^2/\xi)\right] \xi d\xi = \frac{\hbar^2 \exp(z_\mp) 2|z_\mp| K_1(|z_\mp|)}{4 m k_B T_e}$$

$$(11.10.20)$$

The former case is realized for acoustic and optical deformation potential scattering while the latter case pertains to piezoelectric and polar optical scattering.

For ionized impurity scattering no phonon is involved, hence $z_\pm$ becomes $z = \hbar\omega/2k_B T_e$ and since $H_{k'k} \propto 1/q^2$ for neglected screening, we find

$$\int_0^\infty q^{-4} \exp(z) \exp\left[-\frac{1}{2}(\xi + z^2/\xi)\right] \xi d\xi = \left(\frac{\hbar^2}{4mk_B T_e}\right)^2 \exp(z) \, 2\,K_0(|z|)$$

(11.10.21)

Second-order perturbation theory is applicable since the *third particle* is the ionized impurity atom.

Let us first consider the case of *acoustic deformation potential scattering*. The matrix element given by (6.4.13) is the same for phonon emission and absorption. Since the phonon energy is very small $\hbar\omega_q \ll \hbar\omega$, (11.10.13) yields $z_\pm \approx \hbar\omega/2k_B T_e$. Therefore, for photon absorption (subscript a) we find

$$\alpha_a = \frac{\mu_0 c}{\sqrt{\kappa_{opt}}} \frac{2^{5/2}}{3\pi^{3/2}} \frac{ne^2\varepsilon_{ac}^2 m^{1/2} k_B T (k_B T_e)^{3/2}}{\hbar^5 c_l \omega^3} \left(\frac{\hbar\omega}{2k_B T_e}\right)^2$$

$$\times \exp\left(\frac{\hbar\omega}{2k_B T_e}\right) K_2\left(\frac{\hbar\omega}{2k_B T_e}\right)$$

(11.10.22)

where a factor of 2 has been introduced to account for both phonon absorption and emission processes. We still have to treat the induced photon emission (subscript e). This process is obtained from (11.10.5, 9), etc., simply by changing the sign of $\hbar\omega$. Since the argument of K_2 is always positive, we obtain from (11.10.9)

$$\alpha_e \propto |H_{k'k}|_\pm^2 \exp(z_\pm) 2z_\pm^2 K_2(|z_\pm|)$$

$$\propto \left(\frac{\hbar\omega}{2k_B T_e}\right)^2 \exp\left(\frac{\hbar\omega}{2k_B T_e}\right) K_2\left(\frac{\hbar\omega}{2k_B T_e}\right)$$

(11.10.23)

where the rhs is, of course, valid for acoustic scattering. Since the emission term α_e has to be subtracted from α_a, we find for the overall absorption $\alpha = \alpha_a - \alpha_e \propto \exp(z) - \exp(-z) = 2\sinh(z)$:

$$\alpha = \frac{\mu_0 c \, 2^{3/2} ne^2 \varepsilon_{ac}^2 (mk_B T)^{1/2}}{\sqrt{\kappa_{opt}} \, 3\,\pi^{3/2} \, \hbar^3 \, c_l \, \omega} \left(\frac{T}{T_e}\right)^{1/2} \sinh\left(\frac{\hbar\omega}{2k_B T_e}\right) K_2\left(\frac{\hbar\omega}{2k_B T_e}\right)$$

(11.10.24)

Let us next consider the case of thermal equilibrium $T_e = T$ and high temperatures $2k_B T \gg \hbar\omega$. Since for this case $\sinh(z) \approx z$ and $K_2(z) \approx 2/z^2$ we obtain

$$\alpha \approx \frac{\mu_0 \, c \, 2^{7/2} n \, e^2 \varepsilon_{ac}^2 \, m^{1/2} (k_B T)^{3/2}}{\sqrt{\kappa_{opt}} \, 3\pi^{3/2} \hbar^4 \, c_l \, \omega^2} .$$

(11.10.25)

We simplify by introducing the dc mobility given by (6.4.18) and the dc conductivity $\sigma_0 = n e \mu$ neglecting a factor of $32/9\pi \approx 1.13$, we finally obtain (11.10.4). This proves that we were justified in replacing the damping constant γ in the classical model by τ_m^{-1}. Because $\hbar\omega \ll 2k_B T$, it is also correct to neglect the induced emission in the classical treatment. The dependence of α on the carrier temperature T_e in this approximation is just as we expected from (6.4.17):

$$\alpha \propto \gamma = \langle \tau_m^{-1} \rangle \propto T_e^{+1/2} \, , \tag{11.10.26}$$

i.e., there is an increase of the optical absorption with increasing electron temperature. Since for hot carriers according to (6.5.25), T_e increases linearly with the applied dc field E, we find $\alpha \propto E^{1/2}$ for the classical Drude model.

Numerically the absorption cross section is given by

$$\frac{\alpha}{n} = \frac{(0.4693\text{nm})^2 (\lambda/\mu\text{m}) \sinh[7195.3/(\lambda/\mu\text{m})(T_e/\text{K})]}{\sqrt{\kappa_{\text{opt}}} (m/m_0)^2 (\mu/\text{cm}^2\text{V}^{-1}\text{s}^{-1})(T_e/100\text{K})}$$

$$\times K_2[7195.3/(\lambda/\mu\text{m})(T_e/\text{K})] \tag{11.10.27}$$

where $\mu = \mu(T_e) \propto T_e^{-1/2}$ is the acoustic mobility.

For *piezoelectric* scattering, the matrix element is proportional to $1/q$ according to (6.7.8), and (11.10.20) can be applied:

$$\alpha = \frac{\mu_0 c \, 2^{1/2} n \, e^4 \, K^2 (k_B T)^{1/2}}{\sqrt{\kappa_{\text{opt}}} \, 3\pi^{3/2} \hbar^2 \kappa \kappa_0 \, m^{1/2} \omega^2} \left(\frac{T}{T_e} \right)^{1/2} \sinh\left(\frac{\hbar\omega}{2 \, k_B T_e} \right) K_1 \left(\frac{\hbar\omega}{2 \, k_B T_e} \right)$$

$$\tag{11.10.28}$$

In the classical limit, the factor $\sinh(\hbar\omega/2k_B T_e) K_1(\hbar\omega/2k_B T_e) \approx 1$ and $\alpha \propto \lambda^2$ while in the quantum limit it is $\approx (\pi k_B T_e/\hbar\omega)^{1/2}$ and $\alpha \propto \lambda^{2.5}$ independent of the electron temperature T_e. Except for the stronger wavelength dependence, this case is quite similar to acoustic deformation potential scattering. In contrast to the latter, however, α decreases with T_e for $T_e \geq \hbar\omega/3k_B = 4800\text{K}/(\lambda/\mu\text{m})$ while below this range it increases as usual.

For numerical purposes we introduce the mobility μ given by (6.7.17) dependent on $T_e \propto \langle \varepsilon \rangle$ as given by (6.7.16) and obtain for the absorption cross section α/n:

$$\frac{\alpha}{n} = \frac{(0.0771\,\text{nm})^2 (\lambda/\mu\text{m})^2}{\sqrt{\kappa_{\text{opt}}} (\mu/\text{cm}^2\,\text{V}^{-1}\text{s}^{-1})(m/m_0)^2}$$

$$\times \sinh\left[\frac{7195.3}{(\lambda/\mu\text{m})(T/\text{K})} \right] K_1 \left[\frac{7195.3}{(\lambda/\mu\text{m})(T/\text{K})} \right] . \tag{11.10.29}$$

Because of the high mobilities found for piezoelectric scattering in Sect. 6.7, the optical absorption due to this process can usually be neglected.

Let us now consider *optical deformation potential* scattering. The matrix element is given by (6.11.3). It is independent of q and therefore (11.10.19) may be applied. [10] ω_q is now the optical-phonon frequency ω_0. As before we will introduce $z = \hbar\omega_0/2\,k_B\,T$. Since $N_q + 1 = N_q e^{2z}$, we find, except for a common factor, that the four processes yield:

$$\alpha_{a+} \propto \exp(z_+)z_+^2 K_2(z_+); \qquad \alpha_{a-} \propto \exp(2z + z_-)z_-^2 K_2(|z_-|),$$

$$\alpha_{e+} \propto \exp(-z_-)z_-^2 K_2(|z_-|); \qquad \alpha_{e-} \propto \exp(2z - z_+)z_+^2 K_2(z_+),$$

$$(11.10.30)$$

$$\alpha = \alpha_{a+} + \alpha_{a-} - \alpha_{e+} - \alpha_{e-}$$

$$\propto 2\exp(z)[\sinh(z_+ - z)z_+^2 K_2(z_+) + \sinh(z_- + z)z_-^2 K_2(|z_-|)] \ . \ (11.10.31)$$

For $2\exp(z)N_q$ we write $1/\sinh(z)$ and finally obtain from (11.10.18)

$$\alpha = \frac{\mu_0 c}{\sqrt{\kappa_{opt}}} \frac{2^{3/2} n\, e^2 D^2 m^{1/2}}{3\pi^{3/2}\hbar^4 \varrho\, \omega_0\, \omega^3}(k_B T_e)^{3/2}$$

$$\times \frac{\sinh(z_+ - z)z_+^2 K_2(z_+) + \sinh(z_- + z)z_-^2 K_2(|z_-|)}{\sinh(z)} \qquad (11.10.32)$$

For the case of thermal carriers where $T_e = T$, we find $z_+ - z = z_- + z = \hbar\omega/2k_B T$. Introducing the mobility μ given by (6.11.18), α/n becomes

$$\frac{\alpha}{n} = \frac{(4.27 \times 10^{-3}\text{nm})^2}{\sqrt{\kappa_{opt}}}$$

$$\times \frac{(\Theta/100K)(\lambda/\mu m)^3 \sinh(z_0)[z_+^2 K_2(z_+) + z_-^2 K_2(|z_-|)]}{(m/m_0)^2(\mu/\text{cm}^2\,\text{V}^{-1}\text{s}^{-1})(\Theta/2T)^3 K_2(\Theta/2T)} \qquad (11.10.33)$$

where $z_0 = 7195.3/(\lambda/\mu m)(T/K)$ and, at present, $z_\pm = z_0 \pm \Theta/2T$. In Fig. 11.45 the quantity

$$\lambda^3 \sinh(z_0)\frac{z_+^2 K_2(z_+) + z_-^2 K_2(|z_-|)}{(\Theta/2T)^{3/2}\sinh(\Theta/2T)} = \alpha/C_{ODPS} \qquad (11.10.34)$$

which defines a factor of proportionality C_{ODPS}, is plotted vs λ for various values of lattice temperature T and Debye temperature Θ. At long wavelengths, $\alpha \propto \lambda^2$. This is the same as for acoustic scattering. At low temperatures, however, there is a maximum of α at roughly $[7200/(\Theta/2)]\,\mu m$. This corresponds to $z_- = 0$. An inspection of the four components of α, given by (11.10.30), reveals that the maximum is due to α_{a-}; at $\omega = \omega_0$ there is a resonance absorption where a photon is absorbed and subsequently the electron returns to its initial state by emitting an optical phonon of the same frequency. A comparison of

[10]Although for hot carriers the Maxwell-Boltzmann distribution may not be a good approximation of the true distribution, it yields results in analytical form which should, at least qualitatively, be correct.

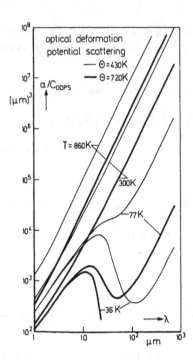

Fig. 11.45: Free–carrier optical absorption coefficient as a function of the wavelength of light for optical deformation potential scattering. C_{ODPS} is a factor of proportionality

(11.10.32) with (11.10.24) for $T_e = T$ and $\hbar\omega = \hbar\omega_0 \gg 2k_B T$ results in a ratio of the absorption coefficients for optical and acoustic deformation potential scattering:

$$\alpha_{opt}/\alpha_{ac} = (4/\sqrt{\pi})(Du_1/\varepsilon_{ac}\omega_0)^2\sqrt{T/\Theta} \ . \qquad (11.10.35)$$

This is of the same order of magnitude as the ratio $(Du_1/\varepsilon_{ac}\omega_0)^2$ which, e.g., for n-Ge is about 0.2. In the quantum limit, $\alpha_{opt} \propto \lambda^{1.5}$ for $\hbar\omega \gg \hbar\omega_0$.

Let us now consider *polar optical* scattering. The Hamiltonian matrix element is given by (6.12.9) and is proportional to $1/q$. Hence, (11.10.20) may be applied.[11]   In order to distinguish between the absorption coefficient α and the polar constant, the latter shall now be denoted as α_{pol}. The result is:

$$\alpha = \frac{\mu_0 c}{\sqrt{\kappa_{opt}}} \frac{2\alpha_{pol}\, ne^2(\hbar\omega_0)^{3/2}}{3\pi^{1/2}\hbar^2\omega^3 m}(k_B T_e)^{1/2}$$

$$\times \frac{\sinh(z_+ - z)z_+ K_1(z_+) + \sinh(z_- + z)|z_-|K_1(|z_-|)}{\sinh(z)} \ . \qquad (11.10.36)$$

The plot of $\alpha = \alpha(\lambda)$ is very similar to that for optical deformation potential scattering. For thermal carriers we introduce the mobility given by (6.12.23) and find for the absorption cross section

---

[11]See previous footnote

$$\frac{\alpha}{n} = \frac{(3.02 \times 10^{-3}\,\mathrm{nm})^2(\Theta/100\mathrm{K})(\lambda/\mu\mathrm{m})^3\sinh(z_0)}{\sqrt{\kappa_{\mathrm{opt}}}(m/m_0)^2(\mu/\mathrm{cm}^2\mathrm{V}^{-1}\mathrm{s}^{-1})(\Theta/2T)^2\mathrm{K}_1(\Theta/2T)}$$

$$\times[z_+\mathrm{K}_1(z_+) + |z_-|\,\mathrm{K}_1(|z_-|)] \tag{11.10.37}$$

The present results for acoustic and optical deformation potential scattering have been extended to the many-valley model by Meyer [11.91]. Gurevich et al. [11.92], König [11.93] and Kranzer and König [11.94] treated the case of polar optical scattering.

Ionized impurity scattering has been treated by Wolfe [11.95]. For $H_{k'k}$ given by (6.3.13) for $L_\mathrm{D}^{-1} \approx 0$ and $|k - k'| = q$, we find from (11.10.18, 21) for the case of absorption

$$\alpha_\mathrm{a} = \frac{\mu_0 c}{\sqrt{\kappa_{\mathrm{opt}}}} \frac{nV^{-1}Z^2e^6}{2^{3/2}\,3\pi^{3/2}\hbar\kappa^2\kappa_0^2 m^{3/2}\omega^3}(k_\mathrm{B}T_\mathrm{e})^{-1/2}\exp(z)\mathrm{K}_0(|z|)\ , \tag{11.10.38}$$

where $z = \hbar\omega/2k_\mathrm{B}T_\mathrm{e}$. The case of induced emission is obtained by changing the sign of $\hbar\omega$. As in the transition from (6.3.13) to (6.3.17), we replace V^{-1} by the impurity concentration N_I. Hence, the total absorption coefficient is given by

$$\alpha = \frac{\mu_0 c}{\sqrt{\kappa_{\mathrm{opt}}}} \frac{nN_\mathrm{I}Z^2e^6(k_\mathrm{B}T_\mathrm{e})^{-1/2}}{2^{1/2}\,3\pi^{3/2}\hbar\kappa^2\kappa_0^2\,m^{3/2}\omega^3}\sinh\left(\frac{\hbar\omega}{2k_\mathrm{B}T_\mathrm{e}}\right)\mathrm{K}_0\left(\frac{\hbar\omega}{2k_\mathrm{B}T_\mathrm{e}}\right) \tag{11.10.39}$$

and the absorption cross section

$$\frac{\alpha}{n} = \frac{(3.88 \times 10^{-1}\mathrm{nm})^2 Z^2 N_\mathrm{I}(\lambda/\mu\mathrm{m})^3}{\sqrt{\kappa_{\mathrm{opt}}}\,10^{17}\,\mathrm{cm}^{-3}\kappa^2(m/m_0)^{3/2}(T_\mathrm{e}/100\mathrm{K})^{1/2}}$$

$$\times \sinh\left[\frac{7195}{(\lambda/\mu\mathrm{m})(T_\mathrm{e}/\mathrm{K})}\right]\mathrm{K}_0\left[\frac{7195}{(\lambda/\mu\mathrm{m})(T_\mathrm{e}/\mathrm{K})}\right] \tag{11.10.40}$$

where the impurity concentration N_I has been related to a typical value of $10^{17}/\mathrm{cm}^3$. Since for an uncompensated semiconductor in the extrinsic range $N_\mathrm{I} = n$, we observe an increase of α with n^2 rather than n, as in the cases treated previously where phonons were involved.

As mentioned by Wolfe [11.95], the Born approximation in the perturbation treatment of ionized impurity scattering may not be applicable for certain cases. Different approximations have been derived from the idea that the process is essentially the inverse process of bremsstrahlung [11.91, 96] for which an exact expression has been given by Sommerfeld [11.97] where, however, the dielectric constant κ and the effective mass ratio m/m_0 have to be introduced.[12]   Denoting the ionization energy of the impurity by $\Delta\varepsilon_\mathrm{I}$, the Born approximation is valid for both $k_\mathrm{B}T_\mathrm{e}$ and $|k_\mathrm{B}T_\mathrm{e} \pm \hbar\omega| \gg \Delta\varepsilon_\mathrm{I}$. In the limit of classical optics where $\hbar\omega \ll \Delta\varepsilon_\mathrm{I}$ but still $k_\mathrm{B}T_\mathrm{e} \gg \Delta\varepsilon_\mathrm{I}$ the Sommerfeld equation yields

---

[12]This has been overlooked in the definition of the interaction parameter in [11.96]; the reason for introducing κ is the same as in the hydrogen model of the shallow impurity.

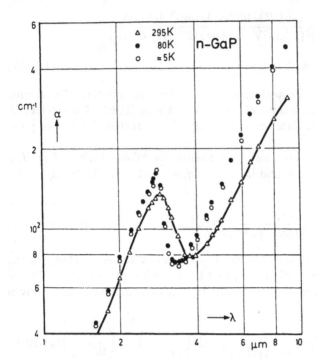

Fig. 11.46: Free–carrier absorption spectrum for n-type gallium phosphide having a carrier density of 1×10^{18} cm^{-3} (after [11.98])

(11.10.39) where, however, the factor $\sinh(z)K_0(z)$ is replaced by the constant $\pi/2\sqrt{3} = 0.907$. For the intermediate case where $\hbar\omega \approx \Delta\varepsilon_{\mathrm{I}}$ and $k_{\mathrm{B}}T_{\mathrm{e}} \ll \Delta\varepsilon_{\mathrm{I}}$, we have to replace this factor by 1 while for $k_{\mathrm{B}}T_{\mathrm{e}} \approx \Delta\varepsilon_{\mathrm{I}}$ and $\hbar\omega \ll \Delta\varepsilon_{\mathrm{I}}$, (11.10.39) is again valid.

In some many-valley semiconductors a structure has been observed in the absorption spectrum which may be due to equivalent intervalley transitions. Fig. 11.46 shows the absorption spectrum of n-type GaP where a maximum at 3 μm is observed [11.98]. In GaP$_x$As$_{1-x}$, its position is nearly independent of the composition x at $x \le 1$. Similar maxima have been found in n-type GaAs [11.99, 100] and in n-type AlSb [11.101]. Free-hole absorption in GaAs shown in Fig. 11.47 has also been explained by interband transitions [11.102, 103].

Finally, let us discuss the case of large carrier concentrations n where the refractive index $\sqrt{\kappa_{\mathrm{opt}}}$ is no longer constant as has been assumed so far. From (11.10.1) we obtain for the free-carrier contribution to the real part of the dielectric constant κ_{r},

$$\Delta\kappa_{\mathrm{r}} = -\omega_{\mathrm{p}}^2/(\omega^2 + \gamma^2) \approx -ne^2/(m\kappa_0\omega^2) \ . \tag{11.10.41}$$

The approximation is valid for small damping. Since the carrier gas is degenerate at the high densities considered here, the ratio n/m should be replaced by $(1/4\pi^3)\langle v\rangle S/3\hbar$, where S is the area of the Fermi surface and $\langle v\rangle$ is the carrier

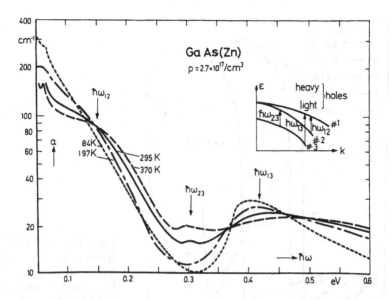

Fig. 11.47: Free–hole absorption in p-type GaAs at various temperatures, and explanation of peaks by the transitions indicated by the inset: upper curve : heavy holes; middle curve: light holes; lower curve: split-off band (after [11.102])

velocity averaged over the Fermi surface [11.104]. For a small value of the extinction coefficient, the reflectivity r_∞ becomes $(\sqrt{\kappa_r} - 1)^2/(\sqrt{\kappa_r} + 1)^2$ and may be nearly unity when the lattice contribution to κ_r is compensated by the negative carrier distribution to yield $\kappa_r \approx 0$. At a slightly different frequency, the carrier contribution may be just sufficient for $\kappa_r = 1$ with the result that the reflectivity nearly vanishes. Observations shown in Fig. 11.48 are in agreement with these predictions [11.104]. The minimum near the *plasma frequency*[13]

$$\bar{\omega}_\mathrm{p} = \omega_\mathrm{p}/\sqrt{\kappa_{\mathrm{opt}}} = \sqrt{ne^2/(m\kappa_0\kappa_{\mathrm{opt}})} \qquad (11.10.42)$$

has been used for a determination of the effective mass m. Values of m increasing with carrier concentration from $0.023\,m_0$ to $0.041\,m_0$ have been determined in n-type InSb from the data in Fig. 11.48. In n-type GaAs, values between $0.078\,m_0$ at $n = 0.49 \times 10^{18}$ cm^{-3} and $0.089\,m_0$ at 5.4×10^{18} cm^{-3} have been found [11.99]. In both cases, the increase of the effective mass with the Fermi energy is due to the nonparabolicity of the conduction band.

The approximation of small damping, where $\omega\tau_\mathrm{m} \gg 1$ [i.e., the free-space wavelength λ/μm $\ll 185.9 \times \tau_\mathrm{m}/10^{-13}$ s] is valid for most semiconductors even in the far-infrared spectrum.

---

[13]In the many-valley model, $1/m$ is to be replaced by $(1/N)\sum 2/m_\varrho$, where $1/m_\varrho$ are the diagonal elements of the effective mass tensor and N is the number of valleys

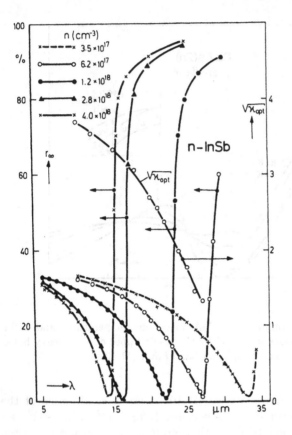

Fig. 11.48: Reflection spectra for five n-type InSb samples at 2995 K; refractive index $\sqrt{\kappa_{opt}}$ valid for the sample with 6.2×10^{17} cm^{-3} electrons (after [11.104])

11.11 Cyclotron Resonance

If a carrier moves in a dc magnetic field B at an oblique angle, its motion is a helix around the direction of the field with an angular frequency known as the *cyclotron frequency* given by

$$\omega_c = (e/m)B = 2\pi\frac{B}{m/m_0} \cdot 28.0\,\mathrm{GHz/T} \qquad (11.11.1)$$

where T stands for the unit Tesla of the magnetic field strength and the projection of the helix on a plain perpendicular to the helix axis is a circle with a radius called cyclotron radius $r_c = \sqrt{\hbar/e\,B}$. The latter is about, e.g., 10 nm in a field of 9 T. The cyclotron wavelength in free space is

$$\lambda_c = 1.07\,\mathrm{cm}\frac{m/m_0}{B/\mathrm{T}} \ , \qquad (11.11.2)$$

If the effective mass $m \approx m_0$ and B is a few 10^{-1} T, the wavelength is in the microwave range. Microwaves incident on the sample and polarized at an angle to B are absorbed; the absorption shows a resonance peak at $\omega = \omega_c$. This phenomenon is called *cyclotron resonance* [11.105]. From a measurement of the resonance frequency, the value of the effective mass m is obtained (11.11.1). As in any resonance observation, a peak is not found for the case of strong damping, i.e., if most carriers make at least one collision before rotating through one radian. Since the number of collisions per unit time is $1/\tau_m$, the condition for a resonance peak is given by

$$\omega_c > 1/\tau_m \ . \tag{11.11.3}$$

This condition poses a severe limit on observations of microwave cyclotron resonance: τ_m should be larger than about 10^{-10} s. This is true only at liquid helium temperatures and even then only in those rare semiconductors which can be highly purified so that at these low temperatures ionized and neutral impurity scattering are negligible.

Another method of observation uses infrared radiation instead of microwaves and strong magnetic fields which are either available in pulsed form (up to about 100 T) or in superconducting hybrid magnets (up to about 30 T). In this case, even at room temperature the resonance condition may be fulfilled ($\tau_m = 10^{-13}$ s corresponds to a wavelength of $\lambda_c = 185.9\mu$m which at ≤ 30 T requires $m/m_0 \leq 0.57$). Since due to induction losses it is difficult to work with pulsed magnetic fields, infrared cyclotron resonance at room temperature is limited to carriers with effective masses of less than about $0.2\, m_0$. At the large fields involved, there is a spin splitting of the levels and a selection rule for the quantum number $\Delta M = 0$ for linearly polarized fields in the *Faraday configuration* (Sect. 11.12). For nonparabolic bands, the effective mass m depends on B in large magnetic fields used in infrared cyclotron resonance. This poses another problem in this type of measurement.

It was shown in Sect. 9.2 that at temperatures T where $k_B T < \hbar\omega_c$, magnetic quantization is observed. For temperatures below 77 K and $\omega_c \approx 10^{13}$ s^{-1}, this condition is fulfilled. Cyclotron resonance may then be considered as a transition between successive Landau levels.

For a simplified treatment of cyclotron resonance, let us consider the equation of motion of a carrier in an ac electric field and a static magnetic field B where we will not include an energy distribution of carriers:

$$d(m\mathbf{v}_d)/dt + m\mathbf{v}_d/\tau_m = e(\mathbf{E} + [\mathbf{v}_d \times \mathbf{B}]) \ . \tag{11.11.4}$$

We introduce a Cartesian coordinate system and choose the direction of $\mathbf{B}$ as the z-axis. For eB/m we introduce the cyclotron frequency ω_c. With $\mathbf{E} \propto \exp(i\omega t)$ and $ne\mathbf{v}_d = \sigma\mathbf{E}$, we find for the nonzero components of the conductivity tensor σ [for $\omega = 0$ see (4.2.21, 24)]

$$\sigma_{xx} = \sigma_{yy} = \sigma_0\tau_m^{-1}\frac{\tau_m^{-1} + i\omega}{(\tau_m^{-1} + i\omega)^2 + \omega_c^2} \ , \tag{11.11.5}$$

$$\sigma_{xy} = -\sigma_{yx} = \sigma_0\tau_m^{-1}\frac{\omega_c}{(\tau_m^{-1} + i\omega)^2 + \omega_c^2} \ , \tag{11.11.6}$$

$$\sigma_{zz} = \sigma_0 \tau_{\mathrm{m}}^{-1} / (\tau_{\mathrm{m}}^{-1} + \mathrm{i}\omega) \ , \tag{11.11.7}$$

where the dc conductivity is $\sigma_0 = n\,e^2\tau_{\mathrm{m}}/m$. We consider a right-hand circularly polarized field where $E_y = -\mathrm{i}\,E_x$ and introduce σ_+ by

$$\sigma_+ = \frac{j_x}{E_x} = \sigma_{xx} + \sigma_{xy}\frac{E_y}{E_x}$$

$$= \sigma_0 \tau_{\mathrm{m}}^{-1}\frac{\tau_{\mathrm{m}}^{-1} + \mathrm{i}\omega - \mathrm{i}\omega_{\mathrm{c}}}{(\tau_{\mathrm{m}}^{-1} + \mathrm{i}\omega)^2 + \omega_{\mathrm{c}}^2} = \frac{\sigma_0 \tau_{\mathrm{m}}^{-1}}{\tau_{\mathrm{m}}^{-1} + \mathrm{i}(\omega + \omega_{\mathrm{c}})} \tag{11.11.8}$$

Similarly, for a left-hand polarization, we have

$$\sigma_- = \frac{\sigma_0 \tau_{\mathrm{m}}^{-1}}{\tau_{\mathrm{m}}^{-1} + \mathrm{i}(\omega - \omega_{\mathrm{c}})} \ . \tag{11.11.9}$$

We will consider here only the *Faraday configuration* where in a transverse electromagnetic wave, $\boldsymbol{E}$ is perpendicular to the dc magnetic field $\boldsymbol{B}$. The power absorbed by the carriers is given by

$$P(\omega) = \frac{1}{2}\mathrm{Re}\{\boldsymbol{j}\,\boldsymbol{E}^*\} \ , \tag{11.11.10}$$

where $\boldsymbol{E}^*$ is the complex conjugate of $\boldsymbol{E}$. For either polarization we have

$$P_\pm = \frac{1}{2}|\boldsymbol{E}|^2\mathrm{Re}\left\{\frac{\sigma_0 \tau_{\mathrm{m}}^{-1}}{\tau_{\mathrm{m}}^{-1} + \mathrm{i}(\omega \pm \omega_{\mathrm{c}})}\right\} = \frac{1}{2}|\boldsymbol{E}|^2\frac{\sigma_0 \tau_{\mathrm{m}}^{-2}}{\tau_{\mathrm{m}}^{-2} + (\omega \pm \omega_{\mathrm{c}})^2} \tag{11.11.11}$$

For a linear polarization which can be considered to be composed of two circular polarizations rotating in opposite directions,

$$P(\omega) = P_+ + P_- = \frac{\sigma_0}{2}|\boldsymbol{E}|^2\left[\frac{1}{1 + (\omega + \omega_{\mathrm{c}})^2\tau_{\mathrm{m}}^2} + \frac{1}{1 + (\omega - \omega_{\mathrm{c}})^2\tau_{\mathrm{m}}^2}\right]$$

$$= \sigma_0|\boldsymbol{E}|^2\frac{1 + (\omega^2 + \omega_{\mathrm{c}}^2)\tau_{\mathrm{m}}^2}{[1 + (\omega^2 - \omega_{\mathrm{c}}^2)\tau_{\mathrm{m}}^2]^2 + 4\omega_{\mathrm{c}}^2\tau_{\mathrm{m}}^2} \tag{11.11.12}$$

At cyclotron resonance $\omega = \omega_{\mathrm{c}}$ and assuming $\omega_{\mathrm{c}}\tau_{\mathrm{m}} \gg 1$, this becomes simply

$$P(\omega_{\mathrm{c}}) = \frac{1}{2}\sigma_0|\boldsymbol{E}|^2 \ . \tag{11.11.13}$$

For this case, the conductivities σ_+ for electrons ($\omega_{\mathrm{c}} < 0$) and σ_- for holes ($\omega_{\mathrm{c}} > 0$) given by (11.11.8, 9) are equal to the dc conductivity σ_0, while outside the resonance they are smaller than σ_0. At low frequencies, the power absorbed by the carriers is given by

$$P(0) = \sigma_0|\boldsymbol{E}|^2/[1 + (\omega_{\mathrm{c}}\tau_{\mathrm{m}})^2] \ . \tag{11.11.14}$$

This equals, of course, $\sigma|\boldsymbol{E}|^2$ with σ given by (4.2.25) neglecting in the present approximation the energy distribution of the carriers.

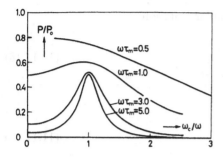

Fig. 11.49: Absorbed microwave power P as a function of $\omega_c/\omega \propto B$ where ω_c is the cyclotron frequecy (after [11.106])

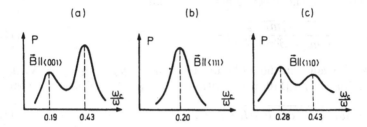

Fig. 11.50: Microwave absorption at 23 GHz in n-type silicon as a function of ω_c/ω for three different directions of the magnetic field (after [11.106])

Figure 11.49 shows the ratio P/P_0 plotted vs ω_c/ω where P_0 stands for $\sigma_0|\boldsymbol{E}|^2$. Since usually in the experiment ω is kept constant and the magnetic field strength is varied, it is quite natural to have ω_c/ω as the independent variable. Obviously the resonance is washed out for $\omega\tau_\mathrm{m} < 1$.

Experimental results obtained on n-type silicon at 4.2 K are shown in Fig.11.50. For a magnetic field in a $\langle 111 \rangle$ crystallographic direction there is a single resonance, while in $\langle 001 \rangle$ and $\langle 110 \rangle$ directions there are two resonances. This can be interpreted by the many-valley model of the conduction band treated in Chap. 7. Since in the $\langle 111 \rangle$ direction there is only one peak, the energy ellipsoids must be located along the $\langle 100 \rangle$ directions which are then equivalent relative to the magnetic field direction.

For an interpretation of Fig. 11.50, we solve (11.11.4) with a mass tensor m in a coordinate system where the tensor is diagonal. For the resonance condition $\omega = \omega_c$, where we may neglect $\boldsymbol{E}$ and τ_m^{-1} for simplicity, we have

$$i\omega_c m_x v_x + e(v_z B_y - v_y B_z) = 0$$

$$i\omega_c m_y v_y + e(v_x B_z - v_z B_x) = 0 \qquad (11.11.15)$$

$$i\omega_c m_z v_z + e(v_y B_x - v_x B_y) = 0$$

Let us denote the direction cosines of $\boldsymbol{B}$ with respect to the three coordinate axes by α, β and γ, respectively. The secular equation

$$\begin{vmatrix} i\omega_c m_x & -eB\gamma & eB\beta \\ eB\gamma & i\omega_c m_y & -eB\alpha \\ -eB\beta & eB\alpha & i\omega_c m_z \end{vmatrix} = 0 \qquad (11.11.16)$$

yields for ω_c:

$$\omega_c = eB\sqrt{\frac{\alpha^2 m_x + \beta^2 m_y + \gamma^2 m_z}{m_x m_y m_z}} . \qquad (11.11.17)$$

For $\omega_c = (e/m)B$, the effective mass m becomes

$$m = \sqrt{\frac{m_x m_y m_z}{\alpha^2 m_x + \beta^2 m_y + \gamma^2 m_z}} . \qquad (11.11.18)$$

If we take the valley in the $\langle 001 \rangle$ direction ($m_x = m_y = m_t = m_l/K; m_z = m_l$), we obtain for B in a $[1\bar{1}0]$ plane with $\alpha = \beta = (\sin\theta)/\sqrt{2}$ and $\gamma = \cos\theta$:

$$m = m_l/\sqrt{K^2 \cos^2\theta + K \sin^2\theta} . \qquad (11.11.19)$$

The valleys in the $\langle 100 \rangle$ and $\langle 010 \rangle$ directions have the same effective mass since they are symmetric to the $[1\bar{1}0]$ plane:

$$m = m_l/\sqrt{K\cos^2\theta + \frac{1}{2}K(K+1)\sin^2\theta} . \qquad (11.11.20)$$

For B in the $\langle 111 \rangle$ direction ($\alpha = \beta = \gamma = 1/\sqrt{3}$), (11.11.19, 20) yield the same value:

$$m = m_l/\sqrt{K(K+2)/3} . \qquad (11.11.21)$$

In Fig. 11.51 the curves calculated from (11.11.19, 20) have been fitted to the data points with effective masses of $m_l/m_0 = 0.90 \pm 0.02$ and $m_t/m_0 = m_l/(Km_0) = 0.192 \pm 0.001$. For any B-direction which is not in the $[1\bar{1}0]$ plane, there should be three resonant frequencies [11.107].

In n-Ge which has four ellipsoids on the $\langle 111 \rangle$ and equivalent axes, there are, in general, 3 resonant frequencies if B is located in the $[1\bar{1}0]$ plane, otherwise there are 4. The analysis of the data yields [11.106] $m_l/m_0 = 1.64 \pm 0.03$ and $m_t/m_0 = (8.19 \pm 0.03) \times 10^{-2}$.

Observations of cyclotron resonance in p-Ge are shown in Fig. 11.52 [11.108]. The two curves indicate two effective masses: the light-hole mass is isotropic in the $[1\bar{1}0]$ plane and has a value of 0.043 m_0, while the heavy-hole mass varies between 0.28 m_0 and 0.38 m_0 depending on the angle θ. In Chap.8 the various effective masses in the warped-sphere model have been treated in detail. The cyclotron effective mass can be calculated by a method proposed by Shockley [11.109]: the equation of motion (neglecting E and $1/\tau_m$ at resonance)

$$d(\hbar k)/dt = (e/\hbar)[\nabla_k \varepsilon \times B] \qquad (11.11.22)$$

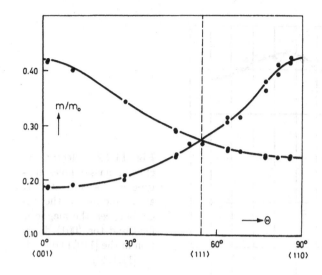

Fig. 11.51: Effective cyclotron mass of electrons in silicon at 4 K as a function of the angle between the magnetic field and the ⟨001⟩ direction in the [1$\bar{1}$0] plane (after [11.107])

is integrated in cyclindrical coordinates where the z-axis is parallel to $\boldsymbol{B}$:

$$\hbar dk_\varphi/dt = \hbar k d\varphi/dt = (e/\hbar)B\,\partial\varepsilon/\partial k \ . \tag{11.11.23}$$

This yields

$$\frac{\hbar^2}{2\pi}\oint\frac{k d\varphi}{\partial\varepsilon/\partial k} = m, \omega_c\frac{1}{2\pi}\oint dt = m \ . \tag{11.11.24}$$

The integration is around a constant ε contour in $\boldsymbol{k}$-space. With $\varepsilon(\boldsymbol{k})$ approximated by (8.1.14) and $\boldsymbol{B}$ located in the [1$\bar{1}$0] plane, we obtain

$$m_\pm/m_0 = (A_0 \pm B_0')^{-1}[1 + \frac{1}{32}(1 - 3\cos^2\theta)^2\Gamma_\pm + \ldots] \ , \tag{11.11.25}$$

where the subscript + refers to light holes and – to heavy holes as in Chap. 8:

$$B_0' = \sqrt{B_0^2 + \frac{1}{4}C_0^2}; \quad \Gamma_\pm = \mp\frac{C_0^2}{2B_0'(A_0 \pm B_0')} \ . \tag{11.11.26}$$

The cyclotron heavy-hole mass in Ge obtained by fitting (11.11.25) (for the case of the minus sign) to the experimental data is 0.3 m_0. For the effective mass in the split-off valence band, 0.075 m_0 is found.

Results of 27.4 μm infrared cyclotron resonance measurements in n-InSb at room temperature are shown in Fig. 11.53 [11.110]. In the upper part of the figure, the variation with $\boldsymbol{B}$ of the spin-split Landau levels is shown. To obtain a good fit of the data it was necessary to adjust ε_G to 0.20 eV while the room temperature *optical gap* is only 0.18 eV. At room temperature, the effective mass is 0.013 m_0. At 77 K, $m = 0.0145\,m_0$ and $\varepsilon_G = 0.225$ eV. The variation of the effective mass with temperature is due to the temperature dependence of

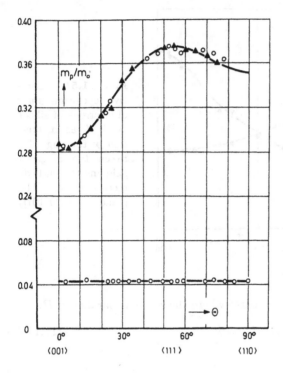

Fig. 11.52: Effective cyclotron mass of holes in p-type germanium at 4 K as a function of the angle between the magnetic field and the ⟨001⟩ direction in the [1\bar{1}0] plane (after [11.108])

the band gap ε_G and has been explained by Kane (Chap. 8). Infrared cyclotron measurements have also been made on n-InAs, n-InP, and n-GaAs [11.110].

Microwave spin resonance of electrons in the conduction band is a magnetic dipole transition and is therefore weak compared with the electric dipole transition at cyclotron resonance. Even so, the large negative g-value in n-InSb has been determined in this way [11.111] (Sect. 11.4).

11.12 Free-Carrier Magneto-Optical Effects

For a plane electromagnetic wave,

$$\boldsymbol{E} = \boldsymbol{E}_1 \exp\left(\mathrm{i}\omega t - \mathrm{i}\,\boldsymbol{q}\cdot\boldsymbol{r}\right); \quad \boldsymbol{B} = \boldsymbol{B}_1\exp(\mathrm{i}\omega\,t - \mathrm{i}\,\boldsymbol{q}\cdot\boldsymbol{r})\ , \tag{11.12.1}$$

in a nonmagnetic conductor, Maxwell's equations are

$$[\boldsymbol{\nabla}\times\boldsymbol{E}] = -\mathrm{i}[\boldsymbol{q}\times\boldsymbol{E}] = -\partial\boldsymbol{B}/\partial t = -\mathrm{i}\omega\boldsymbol{B}\ , \tag{11.12.2}$$

$$[\boldsymbol{\nabla}\times\boldsymbol{B}]/\mu_0 = -\mathrm{i}[\boldsymbol{q}\times\boldsymbol{B}]/\mu_0 = \kappa\kappa_0\partial\boldsymbol{E}/\partial t + \sigma\boldsymbol{E} = \mathrm{i}\omega\kappa\kappa_0\boldsymbol{E} + \sigma\boldsymbol{E} \tag{11.12.3}$$

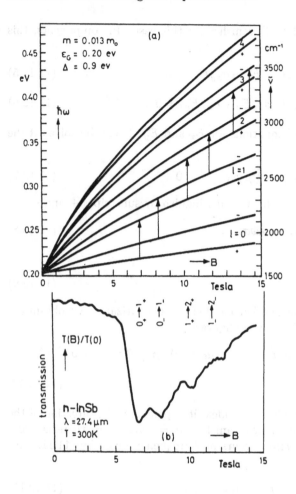

Fig. 11.53: (a) Calculated conduction band Landau levels at $k_z = 0$ for InSb at room temperature, with allowed cyclotron resonance transitions. (b) Observed cyclotron resonance transmission in n-type InSb (after [11.110])

where κ is the relative dielectric constant at the angular frequency ω (i.e., not the static dielectric constant). After eliminating B we obtain

$$(q^2 - \kappa\omega^2/c^2 + i\mu_0\omega\sigma)E = (q \cdot E)q \ , \tag{11.12.4}$$

where c is the velocity of light in free space and σ is a tensor. In the z-direction let us apply a static magnetic field for which we introduce the cyclotron resonance frequency ω_c given by (11.11.1).[14] The conductivity tensor is given by (11.11.5-7). Since the z-components of this tensor do not depend on the static magnetic field, they shall not be of interest to us now; we therefore assume $E_z = 0$.

Let us first consider the *Faraday configuration* where the wave propagates along the direction of the static magnetic field, i.e., $q_x = q_y = 0$ and therefore,

---

[14]The static magnetic field should not be confused with the ac field B given by (11.12.19)

$(\boldsymbol{q} \cdot \boldsymbol{E}) = 0$. The rhs of (11.12.4) vanishes for this case. In components, this equation is given by

$$(q^2 - \kappa\omega^2/c^2)E_x + i\mu_0\omega(\sigma_{xx}E_x + \sigma_{xy}E_y) = 0 \tag{11.12.5}$$

$$(q^2 - \kappa\omega^2/c^2)E_y + i\mu_0\omega(-\sigma_{xy}E_x + \sigma_{xx}E_y) = 0 \ . \tag{11.12.6}$$

This is a homogeneous set of equations for E_x and E_y. It can be solved if the determinant vanishes:

$$(q^2 - \kappa\omega^2/c^2 + i\mu_0\omega\sigma_{xx})^2 + (i\mu_0\omega\sigma_{xy})^2 = 0 \ . \tag{11.12.7}$$

There are two solutions for q^2 which we distinguish by subscripts $+$ and $-$:

$$q^2_{\mp} = \kappa\omega^2/c^2 - i\mu_0\omega\sigma_{xx} \pm \mu_0\omega\sigma_{xy} \ . \tag{11.12.8}$$

Introducing

$$\sigma_{\pm} = \sigma_{xx} \mp i\sigma_{xy} \tag{11.12.9}$$

from (11.11.8, 9) for the right and left-hand circular polarization we obtain the *dispersion relation* for the Faraday configuration:

$$c^2 q^2_{\mp}/\omega^2 = \kappa - i(\mu_0 c^2/\omega)\sigma_{\mp} = \kappa - i(\mu_0 c^2/\omega)\sigma_0\tau_m^{-1}/[\tau_m^{-1} + i(\omega \mp \omega_c)]$$

$$\tag{11.12.10}$$

The lhs is the square of the refractive index; it depends on the direction of the polarization. If the refractive index vanishes, $n = cq/\omega = 0$, there is 100% reflection. This *plasma reflection* occurs at a frequency given by (11.11.8, 9), (11.12.10):

$$0 = \kappa - i(\mu_0 c^2/\omega)\sigma_0\tau_m^{-1}/[\tau_m^{-1} + i(\omega \mp \omega_c)] \ . \tag{11.12.11}$$

We introduce the *plasma frequency* given by (11.10.42):

$$\bar{\omega}_p = \omega_p/\sqrt{\kappa} = \sqrt{ne^2/(m\kappa\kappa_0)} = \sqrt{\mu_0 c^2\sigma_0/(\kappa\tau_m)} \tag{11.12.12}$$

and solve for ω assuming $\tau_m^{-1} \ll |\omega \mp \omega_c|$:

$$\omega = \bar{\omega}_p\{\sqrt{1 + (\omega_c/2\bar{\omega}_p)^2} \pm \omega_c/2\bar{\omega}_p\} \ . \tag{11.12.13}$$

For weak magnetic fields,

$$\sqrt{1 + (\omega_c/2\bar{\omega}_p)^2} \approx 1 + \frac{1}{2}(\omega_c/2\bar{\omega}_p)^2$$

is valid, and $\Delta\omega = \omega - \bar{\omega}_p$ is given by

$$\Delta\omega = \pm\frac{1}{2}\omega_c + \frac{1}{8}\omega_c^2/\bar{\omega}_p \tag{11.12.14}$$

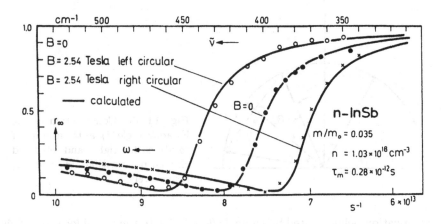

Fig. 11.54: Longitudinal magnetoplasma reflection in n-type InSb at room temperature (after [11.112])

To a first approximation there is a linear shift of the plasma edge with magnetic field intensity. Figure 11.54 shows the observed reflectivity of n-InSb at room temperature for left- and right-hand circular polarization [11.112]. The curve for $B = 0$ is similar to the curves in Fig. 11.48. The electron density is $1.03 \times 10^{18}\,\mathrm{cm^{-3}}$. A magnetic field of 2.54 T causes a shift by $\Delta\omega = 0.65 \times 10^{13}\,\mathrm{s^{-1}}$. In the low-field approximation, the magnetoplasma shift is half the cyclotron frequency. The effective mass relative to the free electron mass is thus obtained from

$$m/m_0 = (e/m_0)B/2\Delta\omega = 1.7 \times 10^{15} \times 2.54 \times 10^{-4}/1.3 \times 10^{13} = 0.035$$

$$(11.12.15)$$

The observed frequency dependence may be compared with (11.12.10). From this comparison a value of 2.8×10^{-13} s is obtained for the momentum relaxation time τ_m. Fig. 11.54 shows for $B = 0$ that $\bar{\omega}_\mathrm{p}$ is $7.5 \times 10^{13}\,\mathrm{s^{-1}}$ and therefore the condition $(\bar{\omega}_\mathrm{p}\tau_\mathrm{m})^2 \gg 1$ is satisfied ($\bar{\omega}_\mathrm{p}\tau_\mathrm{m} = 22.5$). This condition can be written in the form

$$\kappa\varrho/\Omega\,\mathrm{cm} \ll \tau_\mathrm{m}/10^{-13}\,\mathrm{s} \;, \qquad\qquad (11.12.16)$$

where κ is the relative dielectric constant and ϱ is the resistivity. In Sect. 11.15, the case of strong magnetic fields where magnetoplasma waves occur, will be discussed.

Equation (11.12.10) shows that in the presence of a magnetic field, the right- and left-handed circularly-polarized waves have different velocities of propagation $\omega/q_\mp$. A plane-polarized wave can be thought to be composed of two circularly-polarized components. After transmission through a sample of thickness d, by recomposition we again obtain a plane-polarized wave, with the plane

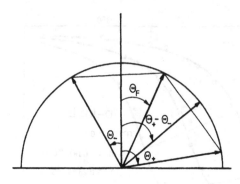

Fig. 11.55: Construction of the Faraday angle Θ_F as the average of rotations for right- and left-hand circular polarizations, Θ_+ and Θ_-, respectively

of polarization being rotated by an angle θ_F relative to the initial location. This rotation is known as the *Faraday effect*. Wave propagation is still along the static magnetic field; this is the *Faraday configuration*. The angle θ_F is the average of the rotations of the two circularly-polarized waves θ_+ and $-\theta_-$ shown in Fig. 11.55:

$$\theta_F = \frac{1}{2}(\theta_+ - \theta_-) = \frac{1}{2}(q_+ - q_-)d \ , \tag{11.12.17}$$

where d is the sample thickness.

From (11.12.10) we obtain for the Faraday rotation

$$\theta_F = \frac{\omega}{2c}\left[\sqrt{\kappa - i(\mu_0 c^2/\omega)\sigma_+} - \sqrt{\kappa - i(\mu_0 c^2/\omega)\sigma_-}\right]d \ . \tag{11.12.18}$$

Introducing the plasma frequency given by (11.12.12) and assuming $\tau_m^{-1} \ll |\omega \pm \omega_c|$, we find

$$\sqrt{\kappa - i\frac{\mu_0 c^2}{\omega}\sigma_\pm} = \sqrt{\kappa}\sqrt{1 - \frac{\bar{\omega}_p^2}{\omega(\omega \pm \omega_c)}} = \sqrt{\kappa}\sqrt{1 - \frac{\bar{\omega}_p^2(\omega \mp \omega_c)}{\omega(\omega^2 - \omega_c^2)}} . \tag{11.12.19}$$

For the case $\bar{\omega}_p^2 \ll \omega(\omega \pm \omega_c)$, the square root can be expanded to

$$\sqrt{\kappa - i\frac{\mu_0 c^2}{\omega}\sigma_\pm} = \sqrt{\kappa}\left[1 - \frac{1}{2}\frac{\bar{\omega}_p^2}{\omega^2 - \omega_c^2} \pm \frac{1}{2}\frac{\bar{\omega}_p^2\omega_c}{\omega(\omega^2 - \omega_c^2)}\right] . \tag{11.12.20}$$

The difference of the square roots in (11.12.18) is $\sqrt{\kappa}\,\bar{\omega}_p^2\omega_c/[\omega(\omega^2 - \omega_c^2)]$. The final result for θ_F, given in degrees instead of radians, is thus

$$\theta_F = \frac{360°}{2\pi}\frac{ne^3 Bd}{m^2\sqrt{\kappa_{opt}}\,\kappa_0\,2c(\omega^2 - \omega_c^2)}$$

$$= \frac{15.1°}{\sqrt{\kappa_{opt}}}\frac{(n/10^{18}\,\text{cm}^{-3})(\lambda/10\,\mu\text{m})^2(B/\text{T})(d/\text{cm})}{(m/m_0)^2(1 - \omega_c^2/\omega^2)} \ , \tag{11.12.21}$$

where for the range of optical frequencies ω given by $\bar{\omega}_p^2 \ll \omega(\omega \pm \omega_c)$, the dielectric constant has been marked by a subscript "opt", and the free-space wavelength λ has been introduced. The sign of θ_F is different for electrons and holes and depends on whether the light propagates in the direction of $\boldsymbol{B}$ or in the opposite direction. θ_F/Bd is called the *Verdet constant*.

In an analysis of data, multiple reflections may have to be taken into account. The true rotation θ_F is then obtained from the observed rotation θ_F' by

$$\theta_F = \theta_F'/[1 + 2r_\infty^2 \exp(-2\alpha d)\cos(4\theta_F')] \; , \tag{11.12.22}$$

where α is the absorption coefficient [11.113].

For the many-valley model, the effective mass m is given by the *Hall effective mass* (7.3.10). For this reason, the Faraday effect may be considered as a *high-frequency Hall effect*. Since the reciprocal relaxation time has been neglected, at small magnetic field intensities the Faraday effect provides a method for determining m which is not obscured by a Hall factor or by the anisotropy of τ_m, in contrast to the usual Hall effect.

For degenerate semiconductors with spherical energy surfaces, the optical effective mass is given by [11.114]

$$\frac{1}{m} = \left(\frac{1}{\hbar^2 k}\frac{d\varepsilon}{dk}\right)_\zeta \; , \tag{11.12.23}$$

where the rhs is evaluated at the Fermi level ζ.

For heavy and light carriers of the same charge, the factor n/m^2 in (11.12.21) is replaced by

$$n/m^2 \to n_h/m_h^2 + n_l/m_l^2 \; , \tag{11.12.24}$$

where the subscripts h and l stand for *heavy* and *light*.

The Faraday ellipticity ε_F is defined by

$$\varepsilon_F = \frac{d}{2}(\text{Im}\{q_+\} - \text{Im}\{q_-\}) \; , \tag{11.12.25}$$

where Im means the imaginary part. In an approximation for small values of τ_m^{-1}, we find from (11.12.10)

$$\text{Im}\{q_\pm\} = -\frac{\sqrt{\kappa}\,\bar{\omega}_p^2(\omega \pm \omega_c)^{-2}\tau_m^{-1}}{2c\sqrt{1 - \bar{\omega}_p^2/[\omega(\omega \pm \omega_c)]}} \; . \tag{11.12.26}$$

For large values of ω, we can neglect the square root in the denominator and obtain

$$\varepsilon_F = \frac{\sqrt{\kappa_{opt}}\,\bar{\omega}_p^2\omega_c\,\omega d}{c(\omega^2 - \omega_c^2)^2\tau_m} = \frac{ne^3 Bd}{m^2\sqrt{\kappa_{opt}}\,\kappa_0\, c\,(1 - \omega_c^2/\omega^2)^2\omega^3\tau_m} \; . \tag{11.12.27}$$

For $\omega_c \ll \omega$ from (11.12.21, 27), a simple expression is found for the ratio ε_F/θ_F:

$$\varepsilon_F/\theta_F = 2/\omega\tau_m \; , \tag{11.12.28}$$

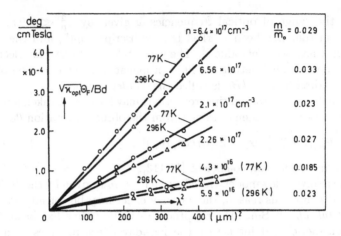

Fig. 11.56: Faraday rotation in n-type InSb. A negative interband rotation (not shown here) causes a deviation from linearity at short wavelengths (after [11.119])

where the unit of θ_F is the radian. A measurement of this ratio directly yields τ_m. However, since it has been assumed that $\tau_m^{-1} \ll \omega$, the ellipticity is small for the range of validity of (11.12.28) [11.115, 116].

At microwave frequencies where $\omega \tau_m \ll 1$, θ_F to a good approximation is obtained from (11.12.21) by replacing $1 - \omega_c^2/\omega^2$ by $(\omega \tau_m)^{-2}$. The ratio ε_F/θ_F becomes $\omega \tau_m$ which is again small [11.117, 118]. The largest value of the ratio is obtained for $\omega \tau_m \approx 1$ and would be most useful for a determination of τ_m.

Experimental data on θ_F vs λ^2 obtained in n-type InSb for various electron densities at 77 K and 296 K are shown in Fig. 11.56. Effective masses obtained from these data increase both with temperature and electron density, in agreement with the well-known nonparabolicity of the conduction band of InSb. In the range of $\lambda = 10$ to $20\mu m$, $\theta_F \propto \lambda^2$, in agreement with (11.12.21).

Figure 11.57 shows data for θ_F vs B at 85 K for a sample with $n \approx 10^{15}\text{cm}^{-3}$ [11.120]. Up to 2 T, $\theta_F \propto B$; above 2 T, however, there is cyclotron resonance where the assumption $\tau_m^{-1} \leq |\omega - \omega_c|$ made in the derivation of (11.12.21) is no longer valid since $\omega = \omega_c$. Two fits made with a more rigorous treatment are given in the figure. At the largest field strengths where $\omega_c \gg \omega$, θ_F is negative and $\propto 1/B$:

$$|\theta_F| = \frac{\mu_0 c}{2\sqrt{\kappa_{opt}}} \, neBd = \frac{0.174°}{\sqrt{\kappa_{opt}}} \frac{(n/10^{10} \text{ cm}^{-3})(d/\text{cm})}{B/\text{T}} \; . \qquad (11.12.29)$$

We will now consider the *transverse* or *Voigt configuration* where $q \perp B$, e.g., $q = (0, q, 0)$ for $E = (E_x, E_y, 0)$ (an electromagnetic wave in a solid may be partly longitudinal). As before, we consider a linearly-polarized wave. Equation (11.12.4) in components is now given by (11.12.5) and by

$$(\kappa \omega^2/c^2 - i\omega \mu_0 \, \sigma_{xx})E_y + i\omega \mu_0 \, \sigma_{xy} E_x = 0 \; . \qquad (11.12.30)$$

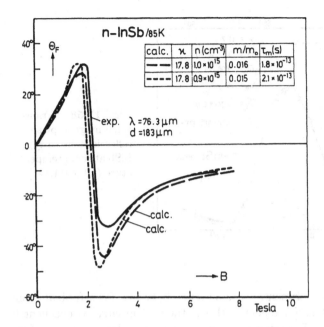

n–InSb/85K

calc.	ϰ	n (cm⁻³)	m/m₀	τ_m(s)
— —	17.8	1.0 × 10¹⁵	0.016	1.8 × 10⁻¹³
- - - -	17.8	0.9 × 10¹⁵	0.015	2.1 × 10⁻¹³

exp. λ = 76.3 μm
d = 183 μm

Fig. 11.57: Observed and calculated Faraday rotation in n-type InSb in a range of magnetic fields which includes cyclotron resonance (after [11.120])

The case $\tau_{\mathrm{m}}^{-1} \ll \omega$ yields for the ratio of the field components

$$E_y/E_x = -i\bar{\omega}_{\mathrm{p}}^2\omega_{\mathrm{c}}/[\omega(\bar{\omega}_{\mathrm{p}}^2 + \omega_{\mathrm{c}}^2 - \omega^2)] \ . \tag{11.12.31}$$

Substituting the ratio E_y/E_x in (11.12.5) yields the *dispersion relation*

$$c^2q^2/(\omega^2\kappa) = 1 + \frac{\bar{\omega}_{\mathrm{p}}^2}{(\omega_{\mathrm{c}}^2 - \omega^2)} - \frac{\bar{\omega}_{\mathrm{p}}^4\omega_{\mathrm{c}}^2}{\omega^2(\omega_{\mathrm{c}}^2 - \omega^2)^2 + \omega^2\bar{\omega}_{\mathrm{p}}^2(\omega_{\mathrm{c}}^2 - \omega^2)} \ . \tag{11.12.32}$$

For magnetic fields of small intensity where $\omega_{\mathrm{c}}^2 \ll \omega^2$, a quadratic equation for ω^2 is obtained:

$$\omega^4 - (2\bar{\omega}_{\mathrm{p}}^2 + c^2q^2/\kappa)\omega^2 + \bar{\omega}_{\mathrm{p}}^2(\bar{\omega}_{\mathrm{p}}^2 - \omega_{\mathrm{c}}^2 + c^2q^2/\kappa) = 0 \ . \tag{11.12.33}$$

For 100% reflectivity where $q = 0$ and for $\omega_{\mathrm{c}} \ll \bar{\omega}_{\mathrm{p}}$, we find a solution

$$\omega_{\pm} = \bar{\omega}_{\mathrm{p}} \pm \frac{1}{2}\omega_{\mathrm{c}} \ , \tag{11.12.34}$$

while for a reflectivity minimum, $c^2q^2 = \omega^2$ and therefore from (11.12.33)

$$\omega'_{\pm} = [\bar{\omega}_{\mathrm{p}}/\sqrt{2(\kappa - 1)}]\sqrt{2\kappa - 1 \pm \sqrt{1 + 4\kappa(\kappa - 1)\omega_{\mathrm{c}}^2/\bar{\omega}_{\mathrm{p}}^2}} \tag{11.12.35}$$

is obtained which for $4\kappa(\kappa - 1)\omega_{\mathrm{c}}^2 \ll \bar{\omega}_{\mathrm{p}}^2$ and $\kappa \gg 1$ can be approximated by

$$\omega'_{\pm} = \bar{\omega}_{\mathrm{p}}[1 \pm \kappa\omega_{\mathrm{c}}^2/(2\bar{\omega}_{\mathrm{p}}^2)] = \bar{\omega}_{\mathrm{p}} \pm \frac{1}{2}\omega_{\mathrm{c}}(\kappa\omega_{\mathrm{c}}/\bar{\omega}_{\mathrm{p}}) \ . \tag{11.12.36}$$

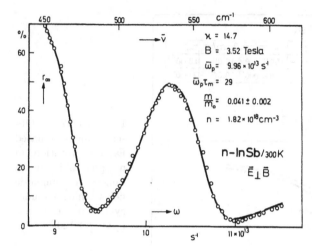

Fig. 11.58: Transverse magnetoplasma reflection in n-type InSb at room temperature (after [11.121])

Figure 11.58 shows experimental data on reflectivity vs ω for n-InSb at room temperature [11.121]. The agreement with the theoretical curve is excellent if τ_m is taken into account. The experiment may serve for a determination of m, τ_m and n.

As in the Faraday experiment, the *transmitted* wave is thought to be decomposed into the two circularly-polarized waves. The *Voigt phase angle* is defined as

$$\theta_V = (q_\perp - q_\parallel)d/2 \ , \tag{11.12.37}$$

where the subscripts refer to the angle between the polarization and the static magnetic field. From the dispersion relation (11.12.32), we obtain for q in the approximation $\omega^2 \gg \omega_c^2$ and $\bar{\omega}_p^2$:

$$q_\perp = \left(\frac{\omega}{c}\right)\sqrt{\kappa}\sqrt{1 - \frac{\bar{\omega}_p^2}{\omega^2} - \frac{\bar{\omega}_p^2\omega_c^2}{\omega^4}} \approx \left(\frac{\omega}{c}\right)\sqrt{\kappa}\left(1 - \frac{\bar{\omega}_p^2}{2\omega^2} - \frac{\bar{\omega}_p^2\omega_c^2}{2\omega^4}\right) \tag{11.12.38}$$

$q_\parallel$ is obtained from this relation by putting $\omega_c = 0$ since the electron motion parallel to the magnetic field is the same as without the field. Hence, we find for θ_V:

$$\theta_V = -\frac{d}{2}\omega\frac{\sqrt{\kappa}}{c}\frac{\bar{\omega}_p^2\omega_c^2}{2\omega^4} = -\frac{\mu_0 c}{4\sqrt{\kappa_{opt}}}\frac{ne^4B^2d}{m^3\omega^3}$$

$$= -\frac{705°}{\sqrt{\kappa_{opt}}}\frac{(n/10^{23}\,\text{cm}^{-3})(\lambda/10\,\mu\text{m})^3(B/\text{T})^2(d/\text{cm})}{(m/m_0)^3} \tag{11.12.39}$$

valid for $\omega_c \ll \omega$. The Voigt phase angle is smaller than the Faraday angle by a factor of $\omega_c/2\omega$ and is therefore more difficult to measure.

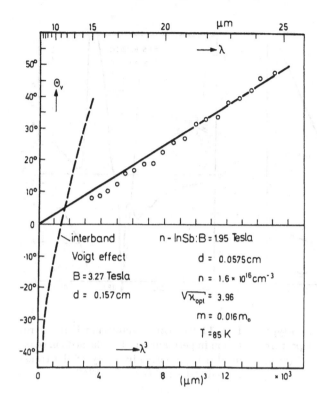

Fig. 11.59: Voigt effect in n-type InSb. An Interband Voigt effect is shown at short wavelengths for somewhat different conditions (dashed curve). The straight line through the origin has been calculated (after [11.110 b])

In the many-valley model, m is replaced by a *magnetoresistance mass* m_M which may be defined from the definition of μ_M given in the text above (7.4.5). For this reason, the Voigt effect may be called the *high-frequency magnetoresistance effect*. Since no τ_m-anisotropy is involved, the Voigt anisotropy is directly related to the band structure, in contrast to the usual magnetoresistance.

For Voigt measurements, the incident beam should be linearly polarized at an angle of 45° relative to the static magnetic field. Although there is no rotation in polarization upon transmission through the sample, the wave becomes elliptically polarized, the ellipticity being [11.122]

$$\varepsilon_V = \tan\theta_V \ . \tag{11.12.40}$$

From measurements of the ellipticity, the Voigt phase angle is thus determined.

Experimental data of θ_V vs λ^3 and B^2 obtained in n-InSb with $n = 1.6 \times 10^{16}$ cm^{-3} at 85 K are shown in Figs. 11.59, 11.60, respectively [11.123]. A slight departure from the θ_V-vs-B^2 linear relationship indicates that the condition $\omega \gg \omega_c$ is not really valid here. At short wavelengths there is an interband transition which will be discussed in Sect. 11.13. As in the case of the Faraday rotation in degenerate semiconductors, the effective mass is the *optical mass* at

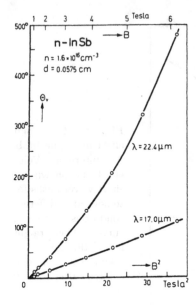

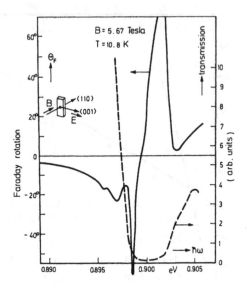

Fig. 11.60: Voigt angle in n-type InSb at 85 K, indicating departure from proportionality to B² (after [11.110 b])

Fig. 11.61: Direct interband Faraday rotation in germanium. The dashed curve indicates the relative intensity of the transmitted radiation (after [11.124])

the Fermi level.

Equation (11.12.32) can be written in the form

$$c^2 q^2/(\omega^2 \kappa) = 1 - (\bar{\omega}_p/\omega)^2 [1 + \omega_c^2/(\omega^2 - \omega_c^2 - \bar{\omega}_p^2)] \; . \tag{11.12.41}$$

This shows that in the Voigt configuration, the cyclotron resonance ($q = \infty$) occurs at

$$\omega = \sqrt{\omega_c^2 + \bar{\omega}_p^2} \; , \tag{11.12.42}$$

while in the configuration usually applied, it is found at $\omega = \omega_c$.

11.13 Interband Magneto-Optical Effects

Magneto-optical effects for free carriers were calculated from the equation of motion (11.11.4). For bound electrons, however, we have to add a term $m\omega_e^2 r$ on the lhs and substitute dr/dt for v_d:

$$m d^2 r/dt^2 + m\omega_e^2 r = e\{E + [(dr/dt) \times B]\} \tag{11.13.1}$$

where the damping term $\propto \tau_m^{-1}$ has been neglected because the eigen-frequency of the harmonic oscillator $\omega_e \gg \tau_m^{-1}$. For the case of right- and left-hand circularly-polarized radiation, the conductivity now becomes instead of (11.11.8, 9):

$$\sigma_{\pm} = i\kappa_0 \omega_p^2 \omega / (\omega_e^2 - \omega^2 \pm \omega\, \omega_c) \;. \tag{11.13.2}$$

For the case of weak fields where $\omega_c \ll (\omega_e^2 - \omega^2)/\omega$, the Faraday rotation becomes

$$\theta_F = \frac{360°}{2\pi} \frac{ne^3\, B\, d}{2m^2 \sqrt{\kappa_{opt}}\, \kappa_o\, c \omega^2 (\omega_e^2/\omega^2 - 1)^2} \tag{11.13.3}$$

which is similar to (11.12.21) except that the factor $(1 - \omega_e^2/\omega^2)$ in the denominator is replaced by $(\omega_e^2/\omega^2 - 1)^2$. The Voigt angle is obtained by a similar calculation:

$$\theta_V = -\frac{\mu_0\, c}{4\sqrt{\kappa_{opt}}} \frac{n\, e^4\, B^2\, d}{m^3 \omega^3 (\omega_e^2/\omega^2 - 1)^3} \;. \tag{11.13.4}$$

Of course, at resonance $\omega = \omega_e$, θ_F and θ_V are determined by τ_m^{-1} which has been neglected here.

Figure 11.61 shows the observed direct interband Faraday rotation in Ge at point Γ in the Brillouin zone as a function of photon energy [11.124]. The reversal in sign cannot be accounted for by (11.13.3). A quantum mechanical treatment with a consideration of the Zeeman splitting of the exciton levels [11.125] is for this case more adequate than the classical oscillator model. In the quantum mechanical treatment, a phenomenological relaxation time τ of the exciton of an order of magnitude of 10^{-13} s and a frequency γB are introduced, where $\gamma = g\,\mu_B/2\hbar$ and g and μ_B are the g-factor and the Bohr magneton, respectively. Sign reversals should occur at frequencies

$$\omega_{\pm} = \omega_e \pm \sqrt{\tau^{-2} + \gamma^2 B^2} \;. \tag{11.13.5}$$

If there are several resonance frequencies ω_e, one resonance may give a contribution at the position of another one and thus prevent one of the two sign reversals in (11.13.5). This would be an explanation for having only one sign reversal in Fig. 11.61.

In the Voigt effect, sign reversals may occur at frequencies

$$\omega_{\pm} = \omega_e \pm \sqrt{(\tau^{-2} + \gamma^2 B^2)/3} \;. \tag{11.13.6}$$

An oscillatory interband Faraday rotation has been observed in Ge at 8 K for magnetic fields of up to 10.3 T [11.126]. The oscillatory effects are of the order of 2% of the total rotation and have been correlated partly with exciton absorption and partly with Landau transitions.

11.14 Magnetoplasma Waves

The dispersion relation for the Faraday configuration (11.12.10) is now approximated for strong magnetic fields where $\omega_c \gg \omega$ [11.127, 128]:

$$c^2 q_\mp^2 / \omega^2 - \kappa = \omega_p^2 / [\omega(\mp\omega_c + i\tau_m^{-1})] \tag{11.14.1}$$

where the plasma frequency

$$\omega_p = \sqrt{ne^2/(m\kappa_0)} \tag{11.14.2}$$

has been introduced for convenience.

For gases where the dielectric constant $\kappa = 1$, (11.14.1) is known as the *Appleton Hartree equation*. For the case of the weak damping $\tau_m^{-1} \ll \omega_c$, we expand (11.14.1) to

$$c^2 q_\mp^2 / \omega^2 - \kappa = \pm\omega_p^2/\omega\omega_c - i\tau_m^{-1}\omega_p^2/\omega\omega_c^2 + \dots \ . \tag{11.14.3}$$

In a typical plasma $\bar{\omega}_p = \omega_p/\sqrt{\kappa} \gg \sqrt{\omega\omega_c}$ and κ is negligible. For a first approximation, we also neglect the damping term; the refractive index

$$cq_\mp/\omega = \sqrt{\pm\omega_p^2/\omega\omega_c} \tag{11.14.4}$$

is real only for the positive sign of the radicand. Depending on the sign of ω_c, i.e., the type of conductivity, either the left- or right-handed circularly polarized wave is transmitted. Its phase velocity

$$\omega/q = c\sqrt{\omega\omega_c/\omega_p^2} = \sqrt{\omega B/ne\mu_0} = 1.77\,\mathrm{cm/s}\sqrt{\frac{(\nu/\mathrm{Hz})(B/\mathrm{T})}{n/10^{23}\,\mathrm{cm}^{-3}}} \tag{11.14.5}$$

can be low. For example, for a semiconductor with 10^{18} carriers/cm^3 in a magnetic field of 10^{-1} T at a frequency of 1 MHz, the velocity is only 1.77×10^5 cm/s and the refractive index is about 10^5. Such a wave is called a *helicon*. The penetration depth δ of the helicon wave is obtained from (11.14.3):

$$\delta = \frac{\lambda}{2\pi} \frac{\omega\omega_c^2}{\tau_m^{-1}\omega_p^2} \sqrt{\frac{\omega_p^2}{\omega\omega_c}} = \frac{c\omega_c^{3/2}\tau_m}{\omega_p\omega^{1/2}} = \sqrt{\frac{1}{e\mu_0 c}} \frac{\mu}{\sqrt{n}} \frac{B^{3/2}\sqrt{\lambda}}{\sqrt{2\pi}}$$

$$= 51.36,\mathrm{cm}/(\mu/\mathrm{cm}^2\mathrm{V}^{-1}\mathrm{s}^{-1})\sqrt{\lambda/\mathrm{cm}}(B/\mathrm{T})^{3/2}/\sqrt{n/\mathrm{cm}^{-3}} \ , \tag{11.14.6}$$

where μ is the carrier mobility and λ the free-space wavelength. For example for n-InSb at 77 K with $n = 1.2 \times 10^{14}$ /cm^3, $\mu = 3.5 \times 10^5$ cm^2/V s at $B = 0.74$T, a 10 GHz wave ($\lambda \approx 3$ cm) has a range of 1.8 cm [11.129] which is larger by a factor of $\sqrt{2}/(\mu B)^{3/2}$ than the range of the order 10^{-2} cm at $B = 0$. At optical wavelengths, a magnetic field which is ten times stronger is required. In microwave experiments there are sometimes *dimensional resonances* which occur

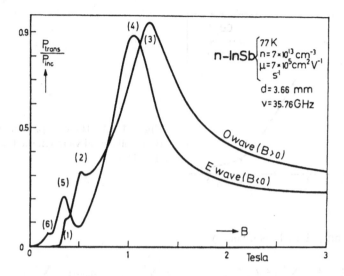

Fig. 11.62: Ratio of transmitted and incident microwave power for a magnetoplasma in n-type InSb at 77 K as a function of the longitudinal magnetic field for right- and left-hand circular polarization of the wave (after [11.130])

if the sample thickness is a multiple integer of half a wavelength in the sample. From measurements of the resonant wavelength, (11.14.5) yields the carrier density n. Figure 11.62 shows resonances measured in n-InSb at a temperature of 77 K and a microwave frequency of 35.76 GHz [11.130; 4.81]. If in a polar crystal the phase velocity is equal to the velocity u_s of transverse sound waves (*shear waves*), an interaction takes place; from (11.14.5) we obtain for this case

$$|u_s| = c\sqrt{\omega\omega_c/\omega_p^2} \ . \tag{11.14.7}$$

Solving for ω and taking $|u_s| \approx 5 \times 10^5$ cm/s, we find

$$\omega = (u_s/c)^2\omega_p^2/\omega_c \approx 5\frac{n/10^{12}\,\mathrm{cm}^{-3}}{B/T}\mathrm{s}^{-1} \ . \tag{11.14.8}$$

In an experiment on Cd_3As_2 with $n = 10^{18}\,\mathrm{cm}^{-3}$ [11.108] shear waves were excited at a frequency $\omega/2\pi$ of 525 kHz by a quartz transducer at one side of a disk-shaped sample. When a longitudinal magnetic field of 1.54 T was applied, helicon waves produced by the coupling were detected by a coil of a few turns around the sample. Figure 11.63 shows the variation of the coil voltage with the applied magnetic field. The experimental data are consistent with (11.14.8).

In intrinsic semiconductors and semimetals, there are equal numbers of electrons and holes. If we can neglect τ_m^{-1} for *both* types of carriers,[15] the disper-

---

[15]This is not the case, e.g., in InSb at room temperature for a range of frequencies where for holes $\tau_m^{-1} > \omega$, while for electrons $\tau_m^{-1} < \omega$.

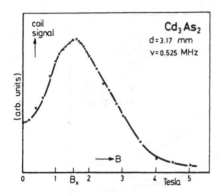

Fig. 11.63: Helicon phonon interaction; coil signal vs magnetic field (after [11.131])

sion relation (11.14.1) is given by

$$c^2 q_{\mp}^2/\omega^2 = \sum_{i=1}^{2} \omega_{pi}^2/[\omega(\mp\omega_{ci} - \omega)] \approx \sum_{i=1}^{2} \omega_{pi}^2/(\mp\omega_{ci}\omega) + \sum_{i=1}^{2} \omega_{pi}^2/\omega_{ci}^2$$

(11.14.9)

The approximation is valid for $\omega_c \gg \omega$. The first term on the rhs vanishes for equal numbers of electrons and holes. The second term yields a phase velocity

$$\omega/q = B/\sqrt{\mu_0 n(m_n + m_p)} = \frac{0.94 \times 10^{15}\,\text{cm/s}(B/\text{T})}{\sqrt{n/\text{cm}^{-3}}\sqrt{(m_n + m_p)/m_0}}$$

(11.14.10)

which is independent of the polarization and therefore also valid for a plane polarized wave. These waves are denoted as *Alfvén waves*. For example in Bi at room temperature, the refractive index cq/ω is 30 for a longitudinal magnetic field of 1.6 T.

If the wave is incident at an oblique angle θ relative to the applied magnetic field, there is an effect of birefringence: one wave which is polarized parallel to the plane of incidence is slower than the perpendicular one by a factor of $\cos\theta$. Alfvén wave measurements yield data for $m_n + m_p$ according to (11.14.10) and therefore for the band structure.

11.15 Nonlinear Optics

The intensity of light emitted by sources other than lasers is low enough to ensure a linear relationship between the polarization P of a dielectric and the electric field strength E:

$$P = \chi E$$

(11.15.1)

where the susceptibility

$$\chi = \kappa_0(\kappa - 1) \tag{11.15.2}$$

and the dielectric constant κ, in general, are second-rank tensors and κ_0 is the permittivity of free space. However, light emitted by a Q-switched ruby laser may have an intensity of up to 10^9 W/cm^2 corresponding to an ac field strength E of about 10^6 V/cm.[16]   Just as in the case of hot carriers where the relationship between current density and E at such high electric field intensities is nonlinear (Fig. 4.31), we have to take into account nonlinear terms in the P-vs-E relationship [11.132, 11.89]:

$$P_i = \sum_k \chi_{ik} E_k + \sum_{kl} \chi_{ikl} E_k E_l + \sum_{klm} \chi_{iklm} E_k E_l E_m + \cdots . \tag{11.15.3}$$

Since in dielectrics the largest contribution to χ comes from bound electrons, this nonlinearity is observed in insulators as well as in semiconductors. However, in semiconductors there is also a free-carrier contribution to the nonlinearity which will be discussed later in this chapter.

In a classical model, the nonlinearity is due to the anharmonicity of the oscillations of the bound electrons in the ac field. In a one-dimensional model, let us assume for the equation of motion of an oscillator

$$d^2x/dt^2 + \omega_0^2 x - \varepsilon x^3 = (e/m)E_1 \cos \omega t , \tag{11.15.4}$$

where damping has been neglected for simplicity and the coefficient of the nonlinear term ε is assumed to be small. To a first approximation we take as a solution $x = x_1 \cos \omega t$, and by applying $4\cos^3 \omega t = \cos(3\omega t) + 3\cos(\omega t)$, we find

$$\frac{3}{4}\varepsilon x_1^3 + (\omega^2 - \omega_0^2)x_1 + (e/m)E_1 = 0 , \tag{11.15.5}$$

where a term $-1/4\varepsilon\, x_1^3 \cos 3\omega t$ has been neglected in this approximation. The second approximation is obtained by integrating

$$d^2x/dt^2 = -\omega^2 x_1 \cos \omega t + (1/4)\,\varepsilon x_1^3 \cos(\omega t) . \tag{11.15.6}$$

which has been obtained from (11.15.4, 5) by eliminating E_1. The solution is nonlinear and given by

$$x = x_1 \cos(\omega t) - \varepsilon(x_1^3/36\omega^2)\cos(3\omega t) . \tag{11.15.7}$$

This yields a dipole moment proportional to e x with frequencies ω and 3ω. Since the polarization P is the dipole moment per unit volume, third harmonic generation is obtained which is adequately described by the third term on the rhs of (11.15.3). Second harmonic generation can be obtained from a term proportional to x^2 in the equation of motion and hence from the second term in (11.15.3). A restoring force proportional to x^2 implies a potential V proportional to x^3, i.e., $V(-x) \neq V(x)$: the dielectric does not possess inversion

---

[16]For comparison: E for sun light on earth is not more than 10 V/cm

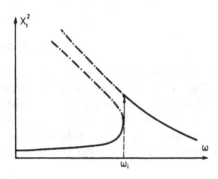

Fig. 11.64: Instability of a classical anharmonic oscillator: the square of the amplitude plotted vs frequency shows instability at a frequency ω_i

symmetry if a second harmonic is generated. Hence, it is the piezoelectric and ferroelectric materials which are of primary interest here.

Equation (11.15.5) which yields the dependence of the oscillation amplitude x_1 on frequency ω, can be written in the form

$$\frac{3}{4}\frac{\varepsilon}{\omega_0^2}x_1^3 = [1 - (\omega/\omega_0)^2]\,x_1 - (e/m\omega_0^2)E_1 \ . \tag{11.15.8}$$

When plotted vs x_1, the lhs is represented by a third-order parabola, while the rhs yields a straight line. For large values of ω there is a unique solution while for small values of ω, the curves cross in three points yielding three possible values of the amplitude. Of course, in the latter case the smallest amplitude is the stable one. At a frequency ω_i where $d\omega/d(x_1^2)$ vanishes, there is an instability which is shown in Fig. 11.64:

$$\frac{\omega_i}{\omega_0} = \sqrt{1 - \left(\frac{3}{2}\right)^{4/3}\varepsilon^{1/3}\left(\frac{eE_1}{m\omega_0^3}\right)^{2/3}} \cong 1 - \frac{1}{2}\left(\frac{3}{2}\right)^{4/3}\varepsilon^{1/3}\left(\frac{eE_1}{m\omega_0^3}\right)^{2/3}$$

$$\tag{11.15.9}$$

Since the series expansion in (11.15.3) is valid only for $\varepsilon E_1^2 \ll (m\omega^3/e)^2$, this shows that the frequency ω_i is smaller than the resonance frequency ω_0 by only a small amount. Hence, in the vicinity of the resonance $\omega = \omega_0$ there is an enhanced multiple-harmonic generation.

The same applies to the generation of beat frequencies $2\omega_1 - \omega_2$ and $2\omega_2 - \omega_1$ which are obtained by applying a field composed of two frequencies ω_1 and ω_2:

$$E = E_1 \exp[\mathrm{i}\omega_1 t - \mathrm{i}(\boldsymbol{q}_1 \cdot \boldsymbol{r})] + E_2 \exp[\mathrm{i}\omega_2 t - \mathrm{i}(\boldsymbol{q}_2 \cdot \boldsymbol{r})] \ . \tag{11.15.10}$$

If the second frequency is twice the first frequency $\omega_2 = 2\omega_1$, the beat frequencies are zero and $3\omega_1$. The former case yields *optical rectification* by *harmonic mixing* (Sect. 4.14).

An important point for the generation of a frequency ω_3 from ω_1 and ω_2 in a dielectric is *phase matching*. For the photons involved in this process, energy conservation yields the condition

$$\omega_3 = \omega_1 + \omega_2 \tag{11.15.11}$$

while momentum conservation yields

$$q_3 = q_1 + q_2 \tag{11.15.12}$$

Assuming q_1 and q_2 to be parallel and introducing an index of refraction $n_i = c|q_i|/\omega_i$, both equations can be fulfilled simultaneously only for $n_1 = n_2 = n_3$. This, however, may not always be true. The power generated at the frequency ω_3 varies as [11.132]

$$P_3 \propto \frac{\sin^2[\frac{1}{2}(q_1 + q_2 - q_3)x]}{(q_1 + q_2 - q_3)^2} \tag{11.15.13}$$

which is a maximum at

$$x = \pi/|q_1 + q_2 - q_3| = \frac{1}{2}|n_1/\lambda_1 + n_2/\lambda_2 - n_3/\lambda_3|^{-1} \tag{11.15.14}$$

where the λ_i are the free-space wavelengths. For example, for frequency doubling this is simply $(\lambda/4)|n_\omega - n_{2\omega}|^{-1}$ which is $\gg \lambda$ for $n_\omega \approx n_{2\omega}$, i.e., for a small dispersion. The drawback of having to use crystals of a length given by (11.15.14) and thus not obtaining a 100% conversion can be overcome by using birefringent crystals, usually uniaxial crystals, if the birefringence exceeds the dispersion. At a certain angle relative to the optical axis, the velocity of the ordinary wave at the frequency ω is the same as the velocity of the extraordinary wave at 2ω. The fundamental and the harmonic are polarized in different directions. Phases are matched over a *coherence length* of several cm, and conversion efficiencies of nearly 100% have been achieved in this way. A prerequisite is, of course, the coherence of the laser beam.

If $\omega_2 = 0$, i.e., a strong (pulsed) dc field is applied in addition to the electromagnetic field, frequency doubling is obtained by an oscillator obeying (11.15.4). Large field amplitudes (of $\approx 10^6$V/cm) may be obtained with small voltages by applying reverse-biased p-n junctions. But even with dc fields of only 10^2 V/cm applied to homogeneously doped n-type InAs at room temperature, the 10.6 μm radiation of a Q-switched CO_2 laser ($\approx 10^5$W/cm^2) could be converted into 5.3 μm radiation. The experimental data displayed in Fig. 11.65 show that the power $P_{2\omega}$ is proportional to the square of the dc current through the sample [11.133]. For electron densities between 1.5×10^{16}cm^{-3} and 3×10^{17} cm^{-3}, $P_{2\omega}$ is proportional to the electron density which proves that this effect is due to the carrier plasma. A generation of beat frequencies from the $\lambda_1 = 10.6\mu m$ and $\lambda_2 = 9.6\mu m$ lines of the CO_2 laser, corresponding to frequencies $\omega_3 = 2\omega_1 - \omega_2$ and $2\omega_2 - \omega_1$ at 8.7μm and 11.8μm, has also been observed [11.135]. Figure 11.66 shows the relative output at 11.8 μm as a function of the electron density (curve a) and of the crystal length x (curve b). The slope of the latter is 2 which indicates a proportionality of the power P_3 to x^2. This dependence is obtained from (11.15.13) for small values of the argument of the sin - function. In an extension of earlier papers, a treatment of optical mixing by mobile carriers in semiconductors has been given by Stenflo [11.135]. Wolff and Pearson [11.136] showed that band nonparabolicity strongly enhances the nonlinearity. Lax et al.

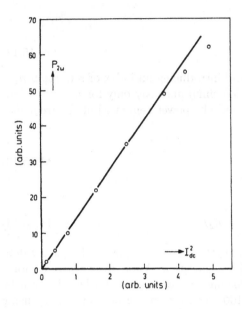

Fig. 11.65: Second-harmonic power (5.3 μm wavelength) as a function of the square of the dc current through an n-type InAs sample [4.4 $\times 10^{16}$ cm^{-3} carriers, conductivity 143 $(\Omega cm)^{-1}$]. At a dc field of 100 V/cm corresponding roughly to $I_{dc}^2 = 1$ arbitrary unit in the figure, the second-harmonic power is about 40 μW for 1 kW input at 10.6 μm (after [11.133])

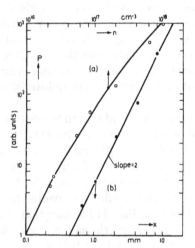

Fig. 11.66: Total mixed signal (11.8 μm) output from n-type InAs as a function of carrier density [open circles: experimental; curve (**a**): calculated] and as a function of sample length [full circles: experimental; straight line (**b**): calculated] (after [11.134])

[11.137] suggested an increase in the output at the beat frequency ω_3 by applying a magnetic field of a magnitude such that ω_3 is near the cyclotron resonance frequency. As shown above, a resonance enhances the effect of nonlinearity.

Problems

11.1. Determine the Burstein shift (if there is one) in n-type GaAs at 77 K and 300 K at carrier concentrations of $n = 10^{17} \text{cm}^{-3}$ and 10^{19}cm^{-3} ($\varepsilon_G = 1.53 \text{ eV} - T \times 5 \times 10^{-4} \text{eV/K}$; $m_n = 0.07 \, m_0$; $m_p = 0.68 \, m_0$).

11.2. Determine the change in the Burstein shift due to nonparabolicity of the conduction band where $k^2 = K_1(\varepsilon + K_2 \, \varepsilon^2)$ with two constants K_1 and K_2.

11.3. For a hot degenerate carrier gas, the Burstein shift depends on the carrier temperature T_e instead of the lattice temperature T. Since degeneracy implies a large number of impurities and ionized impurity scattering is therefore predominant at low temperatures, the relationship between T_e and an applied dc electric field E can be measured by means of the Burstein shift. Make suggestions about the details of the experiment. Calculate how much the Burstein shift is reduced by a field strength of $E = 100 \text{ V/cm}$ at $T = 2 \text{ K}$ in a semiconductor with $n = 10^{16} \text{ cm}^{-3}$; $m_n = 0.04 m_0$; $m_p = 0.4 m_0$; $\mu_n = 2 \times 10^3 \text{ cm}^2/\text{Vs}$ (assumed independent of E); $\tau_\varepsilon = 10^{-10} \text{ s}$.

11.4. How much is the absorption edge of InSb shifted in a magnetic field of 1 T because of the electron spin ($g_n = -50$; $m_n = 0.014 \, m_0$; $m_p = 0.4 \, m_0$; gap energy at $B = 0$ is 0.2 eV)? For simplicity assume $g_p = 0$ for holes.

11.5. Why do $p_\pm$ levels of a shallow impurity show a linear Zeeman effect but not s levels?

11.6. Prove: for a single resonance frequency ω_e assuming $k \ll n$ far away from the resonance, n and k can be approximated by

$$n = 1 + (N e^2/2m\kappa_0)(\omega_e^2 - \omega^2)^{-1}; \qquad (11.0.1)$$

$$k = (N e^2/2m\kappa_0)\gamma\omega/(\omega_e^2 - \omega^2)^2 , \qquad (11.0.2)$$

where N is the number of resonators, m the effective mass and γ a damping constant. This is a model with $\kappa_{\text{opt}} = 1$.

11.7. Assume a semiconductor with scalar hole effective mass m_p and an ellipsoidal conduction band with effective masses m_1, m_2, m_3 in the direction of the main axes. Show that the interband absorption coefficient for a direct transition is proportional to

$$(m_1^{-1} + m_p^{-1})^{-1/2}(m_2^{-1} + m_p^{-1})^{-1/2}(m_3^{-1} + m_p^{-1})^{-1/2} . \qquad (11.0.3)$$

Hint: Apply the transformation given by (7.2.3).

11.8. Tellurium is a direct-gap semiconductor ($\varepsilon_G = 0.335 \text{ eV}$). Both for the valence band and the conduction band the constant-energy surfaces are spheroids with effective masses of $m_{p,t} = 0.1 \, m_0$; $m_{p,l} = 0.3 \, m_0$; and $m_{n,t} = 0.04 \, m_0$; $m_{n,l} = 0.06 \, m_0$, respectively, for small doping concentrations at low temperatures. For a refractive index of $n = 4.8$ and assuming $f_{\text{vc}} = 1 + m_0/m_p$, calculate the absorption edge for a polarization E perpendicular to the c-axis for the photon energies $\hbar\omega = 0.335, 0.340, 0.345,$ and 0.350 eV (the c-axis is the rotational axis of the spheroid!).

11.9. In pure n-Ge at room temperature, both acoustic and optical deformation potential scattering are important. Which one of the two scattering mechanisms dominates the free-carrier absorption at a wavelength of $10\,\mu m$? ($\varepsilon_{ac} = 9.5\,eV$; $c_l = 1.56 \times 10^{12}\,dyn\,cm^{-2}$; $\Theta = 430\,K$; $D = 4.8 \times 10^8\,eV\,cm^{-1}$; density $\varrho = 5.32\,g\,cm^{-3}$).

11.10. By a strong dc field, 10% of the carriers in the Γ valley ($n_\Gamma = 10^{18}\,cm^{-3}$; $m_\Gamma = 0.04\,m_0$) are transferred to four L valleys ($m_l = 1.6\,m_0$; $m_t = 0.08\,m_0$) at 77 K. How does this change the absorption for polar optical scattering at a wavelength of $10\,\mu m$ for polarization (a) parallel to the $\langle 111 \rangle$ axis, (b) parallel to the $\langle 100 \rangle$ axis? (mobility $\mu_\Gamma = 7000\,cm^2/Vs$, $\mu_L = 770\,cm^2/Vs$).

11.11. Neglecting spin, the Landau levels in the nonparabolic conduction band of InSb are described by

$$\varepsilon_n = -\varepsilon_G/2 + [\varepsilon_G^2/4 + \varepsilon_G(n + \tfrac{1}{2})\,\hbar eB/m]^{1/2} \tag{11.0.4}$$

(from Kane's equation for $k_z = 0$), where the gap energy $\varepsilon_G = 0.2\,eV$ and the effective mass $m = 0.014\,m_0$. Calculate the magnetic field strengths for a wavelength of $337\,\mu m$ (HCN laser) where for the first four Landau levels, the transitions $n \to n + 1$ are observed.

11.12. Prove (11.12.22).

11.13. How would you plot the Faraday angle $\theta(\lambda)$ in order to distinguish the free-carrier effect from the interband effect?

11.14. Does the electric or the magnetic field energy dominate in a helicon wave? ($n = 10^{18}\,cm^{-3}$, $B = 0.1\,T$, $\omega/2\pi = 1\,MHz$; $\sqrt{\kappa} = 4$).

11.15. Prove (11.14.6) for weak damping. How deep does a 1 GHz wave in a magnetic field of 0.1 T penetrate in n-InSb? ($m = 0.02\,m_0$, $n = 10^{16}\,cm^{-3}$, $\langle \tau_m \rangle = 10^{-13}s$).

11.16. *Helicon phonon interaction.* For helicon wave propagation in a high-mobility polar semiconductor, we have to consider the rhs of Maxwell's equation $0 = -(cq/\omega)^2\kappa_0\,\partial E/\partial t + j + J$, i.e., besides the electron current contribution j, also the lattice oscillation due to the accompanying acoustic wave represented according to (6.7.3) by $J = iq\,e_{pz}u$ where u is the wave velocity $\partial \delta r/\partial t$ with $\delta r \propto \exp[-(iqz - i\omega t)]$, and e_{pz} is the piezoelectric constant. The propagation direction is, as usual, the z-direction and the oscillation is $E = E_x \pm iE_y$ and $u = u_x \pm iu_y$ where only the minus sign provides a helicon wave. The conductivity which depends on the dc magnetic field, $j/E = \sigma_0/(1 - i\omega_c\tau_m)$ is considered for the case of strong magnetic fields only where $j/E \approx i\,ne/B$ for ω_c being eB/m. Hence, a relationship between E and u is obtained for $E \propto \exp[-(iqz - i\omega t)]$:

$$0 = -i\omega\,(c^2q^2/\omega^2)\,\kappa_0 E + i\,ne\,E/B - iq\,e_{pz}\,u\ .) \tag{11.0.5}$$

A second relationship emerges from the equation of motion taking (7.7.27) into account:

$$\varrho\,\partial^2 r/\partial t^2 = c_l\,\partial^2 r/\partial z^2 - e_{pz}\,\partial E/\partial z\ . \tag{11.0.6}$$

By taking the time derivative:

$$-\omega^2 \varrho u = -q^2 c_l u - \mathrm{i}\,\omega\, e_{pz}\, \partial E/\partial z = -q^2\, c_l\, u - \omega\, q\, e_{pz}\, E \ . \tag{11.0.7}$$

This and the previous equation are a set of homogeneous linear equations for E and u which has a solution only for the vanishing of the determinant. The dispersion relations $\omega(q)$ thus obtained should be plotted and discussed. Indicate where each one of the two branches is helicon-like and where phonon-like. Also discuss the photon - phonon coupling in a polariton [11.138]. Specialize for a transverse wave where the wave vector is perpendicular to the polarization P. Extend the calculations to the influence of free carriers.

12. Photoconductivity

In Sects. 5.8, 5.9 we considered diffusion of carriers which are generated by the absorption of light. In this chapter we will discuss photoconductivity in greater detail with emphasis on trapping processes.

12.1 Photoconduction Dynamics

The experimental arrangement shown in Fig.12.1 allows photoconduction to be observed. Light incident on the semiconductor crystal is absorbed with the effect that additional carriers are produced. If the photon energy is smaller than the band gap, only one kind of carriers may be generated by, e.g., impurity absorption; otherwise, electrons and holes will be generated in pairs. The *dark conductivity*

$$\sigma_0 = |e|(n_0\mu_n + p_0\mu_p) \tag{12.1.1}$$

is thus increased by an amount $\Delta\sigma = |e|(\mu_n\Delta n + \mu_p\Delta p)$. This yields a relative increase of magnitude

$$\frac{\Delta\sigma}{\sigma_0} = \frac{\mu_n\Delta n + \mu_p\Delta p}{\mu_n n_0 + \mu_p p_0} = \frac{b\Delta n + \Delta p}{bn_0 + p_0} , \tag{12.1.2}$$

where the mobility ratio $b = \mu_n/\mu_p$ has been introduced. If in the experiment the sample current is kept constant by a large series resistor, the voltage V across the sample is reduced by an amount $\Delta V = V\Delta\sigma/\sigma_0$. For high sensitivity, a light chopper and a phase sensitive detector are used. For investigations of transient behavior, a flash tube or a Kerr cell chopper combined with a cw light source and an oscilloscope are more appropriate.

If for simplicity we neglect diffusion processes, the continuity equation (5.2.9) becomes

$$\frac{d\Delta n}{dt} = G - \frac{\Delta n}{\tau_n}, \tag{12.1.3}$$

where in the case of uniform light absorption the *generation rate* is given by

$$G = \alpha\eta I/\hbar\omega . \tag{12.1.4}$$

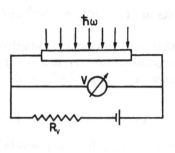

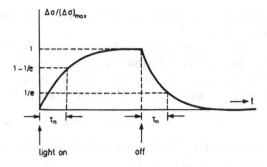

Fig. 12.1: Schematic arrangement for photoconductivity measurements

Fig. 12.2: Dependence of photoconductivity on time t for a relaxation time τ_n

I (in $\mathrm{W\,cm^{-2}}$) is the light intensity, α is the absorption coefficient and η the dimensionless quantum yield. In the case of nonuniform light absorption, the generation rate is a function of the distance x from the surface given by

$$G = \frac{(\alpha\,\eta\,I/\hbar\omega)(1 - r_\infty)[\exp(-\alpha x) + r_\infty \exp(-2\,\alpha\,d + \alpha\,x)]}{[1 - r_\infty^2 \exp(-2\,\alpha\,d)]} \; , \qquad (12.1.5)$$

where r_∞ is the reflectivity of an *infinitely thick* sample [1] and d is the sample thickness.

For the case of photo-ionization of, e.g., acceptor impurities, the product $\alpha\eta$ becomes

$$\alpha\eta = \sigma_\mathrm{A}\,N_\mathrm{A}(1 - f) \; , \qquad (12.1.6)$$

where N_A is the total acceptor concentration, α_A is the photo-ionization cross section and f is the electron occupation probability of the acceptor.

For equilibrium, (12.1.3) yields $\Delta n = G\tau_n$. Assuming pair production $\Delta n = \Delta p$, we obtain from (12.1.2):

$$\frac{\Delta\sigma}{\sigma_0} = G\,\tau_n\,\frac{b+1}{n_0\,b + p_0} \; . \qquad (12.1.7)$$

This shows that it is advantageous for a sensitive photoconductor ($\Delta\sigma \geq \sigma_0$) to have a long time constant τ_n.

The transient behavior, which is observed if the light at low intensity is alternatively switched on and off, is shown in Fig.12.2. The time constant τ_n can easily be determined if the curve for the increase or decrease with time is exponential.

From measurements of τ_n and $\Delta\sigma/\sigma_0$ at equilibrium, the quantum yield η may be determined. For an estimate of Δn let us assume typical values of $\eta = 1, \tau_n = 10^{-4}\,\mathrm{s}, I = 10^{-4}\,\mathrm{W\,cm^{-2}}$, and $\alpha = 10^2\,\mathrm{cm^{-1}}$ at a wavelength of 2

---

[1] *Infinitely thick* means that we can neglect the reflection from the back surface.

μm corresponding to $\hbar\omega = 10^{-19}\,\mathrm{Ws}$. The results are $G = 10^{17}\,cm^{-3}s^{-1}$ and $\Delta n = 10^{13}\,cm^{-3}$.

If the energy of the incident photons is larger than the band gap ε_G, radiative recombination of electron hole pairs by emission of photons $\hbar\omega = \varepsilon_G$ may occur. The recombination rate is proportional to the product $np = (n_0 + \Delta n)(p_0 + \Delta p)$ and hence, with a factor of proportionality C,

$$\Delta n/\tau_n = C(np - n_0 p_0) = C(n_0\,\Delta p + p_0\,\Delta n + \Delta n\,\Delta p) \ . \tag{12.1.8}$$

At equilibrium the lhs is equal to G. For the case $\Delta n = \Delta p$, this is a quadratic equation for Δn. Only at a low light intensity can the product $\Delta n\,\Delta p$ be neglected, and with $G \propto I$, we have $\Delta n \propto I$. At a high light intensity we can, on the other hand, neglect $n_0\,\Delta p + p_0\,\Delta n$ and find $\Delta n \propto \sqrt{I}$.

Besides radiative recombination with the emission of photons $\hbar\omega = \varepsilon_G$, there may also be recombination via *recombination centers* with the emission of photons $\hbar\omega < \varepsilon_G$, or the recombination may be nonradiative. Recombination centers are impurities or imperfections. Typical examples will be studied in Sect. 12.2.

The rate of electron capture by a single type of recombination centers is given by [2]

$$\Delta n/\tau_n = C_n\,n(1 - f) \ , \tag{12.1.9}$$

where $f = N_r^x/N_r$ is the probability that the center is neutral (occupied), $C_n = N_r v \sigma_r$, N_r is the total concentration of recombination centers and σ_r their capture cross section. The rate of thermal emission of electrons from recombination centers is given by

$$G = C_n'\,f \ , \tag{12.1.10}$$

where C_n'/N_r is the thermal ionization probability of a center. The condition for equilibrium in the dark $\Delta n/\tau_n = G$ yields

$$C_n'/C_n = n(1 - f)/f = g^{-1}\,N_c\,\exp(-\Delta\varepsilon_r/k_B T) \tag{12.1.11}$$

assuming nondegeneracy, where the effective density of states N_c is given by (3.1.10) and $\Delta\varepsilon_r = \varepsilon_r - \varepsilon_c$; the spin factor g depends on the kind of impurity. If we denote the electron density in the conduction band by n_1 for the case where the Fermi level coincides with the recombination level, the rhs of (12.1.11) is given by $g_D^{-1} n_1$.

At nonequilibrium conditions which may be due, e.g., to illumination, the net recombination rate of electrons is given by

$$\Delta n/\tau_n = C_n[n(1 - f) - g_D^{-1} n_1 f] \ . \tag{12.1.12}$$

Similarly, the net recombination rate of holes is obtained:

$$\Delta p/\tau_p = C_p[pf - g_A p_1(1 - f)] \ . \tag{12.1.13}$$

---

[2]Compare with (10.1.1) for $A_I = B_I = 0$.

Since electrons and holes recombine in pairs, for a steady state illumination $\Delta n/\tau_n = \Delta p/\tau_p$ which we simply call $\Delta n/\tau$, we find

$$\Delta n/\tau = \frac{C_n C_p(np - n_i^2)}{C_n(n + g_D^{-1}n_1) + C_p(p + g_A p_1)} , \tag{12.1.14}$$

where the intrinsic concentration $n_i = \sqrt{n_1 p_1}$ has been introduced. Since a similar relation $n_i = \sqrt{n_0 p_0}$ holds for the carrier concentrations in the dark, n_0 and p_0, we obtain from (12.1.14) by introducing $n - n_0 = \Delta n$ and $p - p_0 = \Delta p$

$$\Delta n/\tau = \frac{n_0 \Delta p + p_0 \Delta n + \Delta p \Delta n}{C_p^{-1}(n_0 + g_D^{-1}n_1 + \Delta n) + C_n^{-1}(p_0 + g_A p_1 + \Delta p)} . \tag{12.1.15}$$

For brevity we also introduce $\tau_{p0} = C_p^{-1}$, $\tau_{n0} = C_n^{-1}$, $n_1' = g_D^{-1}n_1$, $p_1' = g_A p_1$ and obtain for the case $\Delta n = \Delta p \ll n_0$ and p_0:

$$\tau = \tau_{p0}n_0 + n_1'n_0 + p_0 + \tau_{n0}p_0 + p_1'p_0 + n_0 . \tag{12.1.16}$$

If $p_0 \gg n_0$ and p_1', this yields $\tau = \tau_{n0}$, while for $n_0 \gg p_0$ and n_1', we obtain $\tau = \tau_{p0}$ which is independent of p_0 and n_0, respectively. It is noteworthy that this is in contrast to the behavior in the case of radiative recombination given by (12.1.8) where the lifetime does depend on the carrier concentration. For values of p_0 near the intrinsic value n_i, there is a maximum of the time constant. For the general case where $\Delta n = \Delta p$ and both are not small, (12.1.15) is a second-order equation in Δn with the same kind of behavior as has been discussed in connection with (12.1.8). For $\Delta n \neq \Delta p$ we have a change in f given by

$$\Delta f = (\Delta p - \Delta n)/N_r . \tag{12.1.17}$$

This change may be calculated from $\Delta n/\tau_n = \Delta p/\tau_p$ where both sides are given by (12.1.12, 13), respectively:

$$C_n[\Delta n(1 - f) - n \Delta f - n_1' \Delta f] = C_p(f \Delta p + p \Delta f + p_1' \Delta f) . \tag{12.1.18}$$

For an approximation we replace f, n, and p here by their equilibrium values $n_0/(n_0 + n_1')$, n_0 and p_0, respectively. If we calculate the time constant as a function of n_0, we again find a maximum which, however, in this case is generally not near the value of the intrinsic carrier concentration.

Recombination levels are usually *deep levels*. *Shallow levels*, on the other hand, act as *traps*. For example, a shallow acceptor will trap a hole with a probability, which we denote as $1/\tau_1$, and it will keep it for an average time period of τ_2. The average concentration of trapped holes ΔP is given by the condition of charge neutrality

$$\Delta P = \Delta n - \Delta p . \tag{12.1.19}$$

If $1/\tau_p$ is the probability that a hole recombines with an electron, we have

$$\frac{d \Delta p}{dt} = G - \frac{\Delta p}{\tau_p} - \frac{\Delta p}{\tau_1} + \frac{\Delta P}{\tau_2} . \tag{12.1.20}$$

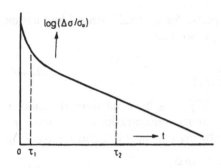

Fig. 12.3: Relaxation of photoconductivity with two recovery time constants for the case of thermal release of carriers from traps

In equilibrium (indicated by a subscript 0) the rate of trapping equals the rate of thermal re-excitation of holes into the valence band:

$$\Delta p_0/\tau_1 = \Delta P_0/\tau_2 \ . \tag{12.1.21}$$

Because phonons are involved in the process of re-excitation, the time constant τ_2 is strongly temperature dependent. The generation rate is then given by

$$G = \Delta p_0/\tau_p \ . \tag{12.1.22}$$

For Δn_0 we obtain from (12.1.19, 21, 22):

$$\Delta n_0 = G\,\tau_p(1 + \tau_2/\tau_1) \ . \tag{12.1.23}$$

The relative conductivity change is now given by (12.1.7) with τ_n being replaced by

$$\tau'_p = \tau_p \frac{1 + b(1 + \tau_2/\tau_1)}{1 + b} \tag{12.1.24}$$

which for $\tau_2 \gg \tau_1$ may be approximated by

$$\tau'_p \approx \tau_p[b/(1 + b)]\tau_2/\tau_1 \ . \tag{12.1.25}$$

This is much larger than the time constant without trapping τ_p. We find as a result that traps considerably increase not only the sensitivity of a photoconductor, but also at the same time, its response time. The reason for this behavior is clear: as more holes are trapped, more electrons stay in the conduction band because of charge neutrality (12.1.19) and the conductivity is larger. The number of trapped holes increases with the average time which a hole spends in a trap τ_2. Of course, the same is true for electrons in a p-type semiconductor.

The conductivity drop after switching off the illumination is illustrated in Fig.12.3 valid for $\tau_2 \gg \tau_1$ and τ_p. For a short period of the order of τ_1 the traps remain at equilibrium and the conductivity drops with the fast time constant τ_p. However, for longer time periods the time constant is given by τ_2. To a first approximation, the sum of two exponentials with time constants τ_p and τ_2 with a smooth transition at $t = \tau_1$ adequately describes the photoconductivity transient behavior. Of course, with a slow apparatus which is not capable of recording the fast time constant τ_p, only the slow drop of $\Delta\sigma$, which at low temperatures may even last hours, is observed.

Illumination of the sample with light with a long wavelength may empty the traps without generating electron-hole pairs, thus considerably decreasing the time constant τ_2 and *quenching* the high photo-sensitivity. If the exciting radiation is *white* and therefore contains a large amount of *quenching* infrared, the photo-sensitivity will be much lower than for monochromatic light.

Surface recombination has been defined by (5.8.1). In a sample of thickness d, a minority carrier has to drift for a distance of less than $d/2$ in order to arrive at the surface. Hence, the rate of recombination is given by

$$d\,\Delta p/dt = -2\,s\,\Delta p/d \qquad (12.1.26)$$

which yields a time constant

$$\tau_p' = d/(2\,s) \ . \qquad (12.1.27)$$

At an etched surface, $s = 10^2$ cm/s, which, e.g., for $d/2 = 1$ mm yields $\tau_p' = 1$ ms. For a sandblasted surface ($s = 10^6$ cm/s), we then have $\tau_p' = 100$ ns. The combination of surface and bulk recombination yields a time constant given by

$$1/\tau_p' = 1/\tau_p + 2\,s/d \ . \qquad (12.1.28)$$

Obviously, the surface condition of a photoconductor affects its overall performance considerably.

Finally we mention that the signal-to-noise ratio of a photoconductor may be determined by generation-recombination noise. These processes (including trapping) are subject to statistics and occur at a random sequence. Therefore, this kind of noise can only be reduced in magnitude by eliminating traps and recombination centers which, however, leads to a decrease in sensitivity as we have seen before. For practical applications see, e.g., [12.1]. Photoconduction in quantum wells and superlattices will be treated in Sect. 13.3.

12.2 Deep Levels in Germanium

Recombination and trapping processes may conveniently be studied with deep levels in germanium which are either *double acceptors* or *double donors*. We will see that the same deep-level impurity atom may act either as a recombination center or a trap depending on the type of conductivity which is determined by the shallow-impurity doping. Therefore, germanium may serve as a model substance for this kind of investigation ([12.2] and [Ref. 12.3, p. 49]).

In Fig.3.7, multiple-acceptor levels in Ge are indicated, together with their *distribution coefficients* (ratios of impurity concentration in the solid phase to the concentration in the liquid phase which is in thermal equilibrium with the

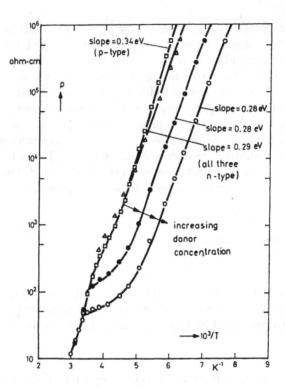

Fig. 12.4: Resistivity as a function of the reciprocal of the temperature for a series of samples from the same Fe doped germanium crystal. With increasing donor concentration both the photosensitivity and the recovery time increase as shown in Fig. 12.5 (after [12.5])

solid phase of germanium [12.4]). Let us, e.g., consider Fe in p-type Ge. Thermal ionization yields a hole and an Fe^- ionized atom:

$$Fe + 0.35\,eV \rightleftharpoons e^+ + Fe^- \ . \tag{12.2.1}$$

In n-type, Ge, however, Fe at low temperatures is present in the form of an Fe^{--} ion, which by thermal ionization yields an electron and also Fe^-:

$$Fe^{--} + 0.27\,eV \rightleftharpoons e^- + Fe^- \ . \tag{12.2.2}$$

Therefore, Fe is an acceptor and Fe^{--} is a donor, and there is usually no Fe in n-type Ge and no Fe^{--} in p-type Ge. The type of conductivity depends on the type of impurity, boron or phosphorus, which is usually present in iron at concentrations of a few ppm depending on the type of purification process. Since boron and phosphorus are so much more easily introduced than iron into a crystal when it grows from the melt, the content of iron is much smaller than that of the shallow impurities.

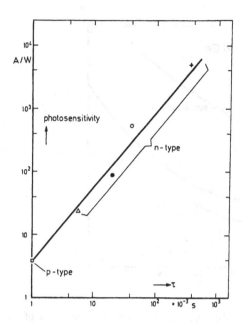

Fig. 12.5: Photo-Hall data for n-type Mn-doped germanium as a function of the inverse temperature at various light intensities. Data at high light levels were obtained using radiation from a tungsten lamp (3000 K) filtered through a 1 mm thick germanium window (after [12.3])

Figure 12.4 shows the temperature dependence of the dark resistivity of Fe-doped n- and p-type Ge. While in n-type samples the slope of the straight lines yields an energy of 0.28 eV, in p-type samples the slope yields 0.34 eV. After correcting for the temperature dependence of the mobility, the temperature dependence of the carrier density is obtained; this is determined by the values of 0.27 and 0.35 eV for the energy indicated in (12.2.1, 2). For example, at a temperature of 150 K, Fig.12.4 shows a very distinct freeze-out of carriers. Illuminating the sample with photons $\hbar\omega \geq \varepsilon_G$ yields electron-hole pairs. After switching off the illumination, the resistivity increases non-exponentially with time. Let us define τ as the time period for an increase of ϱ by 3 orders of magnitude. Fig. 12.5 shows the relationship between the photoconductive sensitivity and τ for p- and n-type samples. Obviously the n-type samples at low temperatures contain traps, while the p-type samples do not.

Equation (12.2.2) shows that at low temperatures in n-type samples there are Fe^{--} ions. Minority carriers are positive holes and these are attracted by the negative ions:

$$Fe^{--} + e^+ \rightleftharpoons Fe^- \ . \tag{12.2.3}$$

The negatively-charged electrons are repelled by negative ions and this explains why Fe^{--} is a hole trap. The fact that an illuminated sample contains Fe^- ions which do not scatter electrons as strongly as the Fe^{--} ions do in the dark, is illustrated by the temperature dependence of the mobility shown in Fig.12.7 (here the deep level is Mn which, however, acts similarly to Fe).

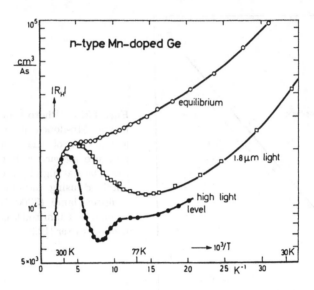

Fig. 12.6: Photo-Hall data for n-type Mn-doped germanium as a function of the inverse temperature at various light intensities. Data at high light levels were obtained using radiation from a tungsten lamp (3000 K) filtered through a 1 mm thick germanium window (after [12.3])

A strong illumination of n-type samples results in the reaction

$$Fe^- + e^+ \rightleftharpoons Fe \tag{12.2.4}$$

with the formation of neutral atoms. These no longer repel electrons and therefore act as recombination centers. The lifetime becomes very short and the sensitivity small. Fig. 12.6 shows the Hall coefficient as a function of temperature at various degrees of illumination for Mn-doped Ge.

In p-type samples we also have recombination centers (neutral Fe atoms) at low illumination levels, (12.2.1). The lifetime and sensitivity are both small.

The dependence of the decay time τ on temperature for n- and p-type Ge doped with deep levels (Mn) is shown in Fig.12.8. The resistivity of the samples had to be chosen low enough so that the dielectric relaxation time, given by (5.2.21), is much smaller than τ. The high-temperature slope of the curve for Mn doped n-Ge is determined by an energy of 0.31 eV. This energy is considered as an activation energy for either the capture of electrons over a repulsive potential barrier or the escape of a hole from a trap. By a *reverse current collection method* by Pell [12.6], it has been shown that in n-type Mn-doped samples, hole escape does take place.

Tellurium in germanium is a double donor with levels 0.1 eV and 0.28 eV below the conduction band:

$$Te + 0.10\,eV \rightleftharpoons e^- + Te^+ , \tag{12.2.5}$$

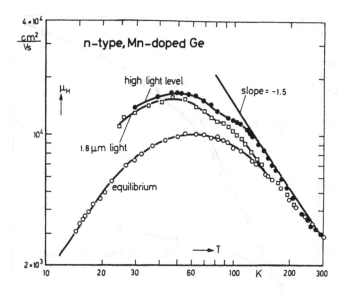

Fig. 12.7: Hall mobility data corresponding to the Hall coefficient data of Fig. 12.6 (after [12.3])

$$\mathrm{Te}^+ + 0.28\,\mathrm{eV} \rightleftharpoons \mathrm{e}^- + \mathrm{Te}^{++} \; . \tag{12.2.6}$$

In p-type samples at low temperatures, we have Te^{++} ions. Minority carriers (electrons) generated by illumination are trapped at these ionized atoms with the formation of Te^+. These do not attract positive holes and therefore act as traps. In n-type samples at low temperatures, there are presumably Te atoms which for a short time trap minority carriers (positive holes):

$$\mathrm{Te} + \mathrm{e}^+ \rightleftharpoons \mathrm{Te}^+ \; . \tag{12.2.7}$$

The Te^+ formed in this process, however, quickly attracts negatively charged electrons

$$\mathrm{Te}^+ + \mathrm{e}^- \rightleftharpoons \mathrm{Te} \tag{12.2.8}$$

and consequently acts as a recombination center. Fig. 12.9 shows the photo-conductive decay time vs temperature for p-type germanium doped with Te.

The spectral sensitivity at 77 K of n and p-type germanium doped with various transition metals is shown in Figs. 12.10a,b. The largest difference between n- and p-type samples is found in Mn-doped samples; this agrees with the position of the levels indicated in Fig.3.7 [12.2,7].

A rather exotic impurity in Ge is gold. Fig. 3.7 shows that it may be either a donor 0.05 eV above the *valence* band or an acceptor with three different levels.

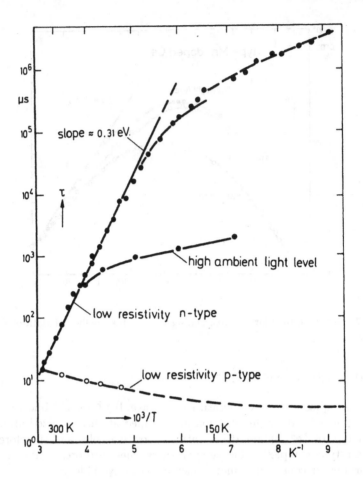

Fig. 12.8: Photoconductive response time as a function of the inverse temperature for Mn-doped germanium (after [12.3])

As with the double-acceptors discussed above, the nature of the impurity level depends on the kind of shallow-level impurities which are present in addition to the deep-level impurity. From a statistical point of view one may say that it depends on the position of the Fermi level which is close to the conduction band in an n-type semiconductor and close to the valence band in a p-type semiconductor (Chap.3). Fig. 12.11 shows again on the lhs the four gold levels in germanium and on the rhs the charge on the gold center depending on the position of the Fermi level. Since the lowest level is a donor, 0.05 eV above the valence band, and the next-highest level is an acceptor, 0.15eV above the valence band, a transition between these two levels is possible if the Fermi level is located between these levels and the photon energy equals the energy difference. Table 12.1 shows data for various Ge photodetectors.

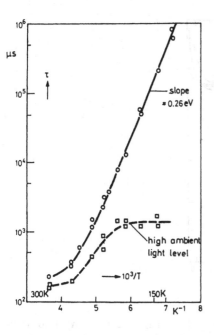

Fig. 12.9: Photoconductive response time as a function of the inverse temperature for tellurium-doped p-type germanium (after [12.3])

Table 12.1.

Ge photodetectors [12.3] (vb: valence band; cb: conduction band)

Impurity	Level [eV]	Operating temperature[K]	Long-wavelength limit[μm]
Zn	0.03 above vb	<15	>40
Zn	0.09 above vb	60	18
Te	0.10 below cb	60	15
Au	0.05 below cb	20	12
Au	0.05 above vb	20	25
Au	0.15 above vb	77	12
Cu	0.04 above vb	20	27
Mn	0.16 above vb	65	8.5

12.3 Trapping Cross Section of an Acceptor

For a determination of the capture cross section σ_p of a negatively-charged deep acceptor, both the concentration of the deep level impurity and the concentration of shallow donors must be known. Instead of pulling a crystal from the melt doped with the double acceptor, the doping is done by diffusion of, e.g.,

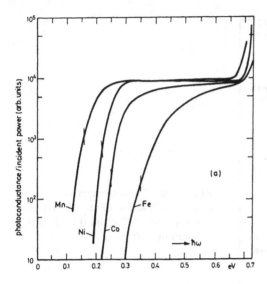

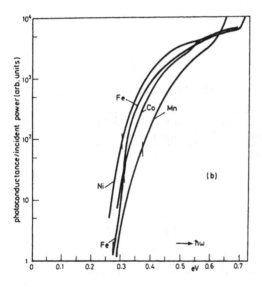

Fig. 12.10: (a) Photoconductivity spectra at 77 K for p-type germanium with deep-level impurities. The vertical bars indicate the thermal ionization energies (after [12.3]). (b) Photoconductivity spectra for n-type germanium with deep-level impurities at 77 K except for Mn which is at 196 K. The vertical bars indicate the thermal ionization energies (after [12.3])

Ni at 1123 K for times long enough to saturate the concentration of Ni into germanium crystals, having a predetermined concentration of, e.g., As. The occupation fraction f of the Ni levels is calculated from the known concentrations of As and Ni: $f = N_{As}/N_{Ni}$. The first acceptor level of Ni is 0.22 eV above the valence band (Fig.3.7). For $f > 1$ the samples are p-type. The product $f N_{Ni}$ is the concentration of negatively-charged Ni acceptors. Denoting the thermal

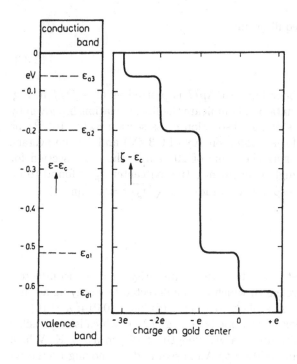

Fig. 12.11: Gold levels in germanium and the dependence of the charge per gold atom on the Fermi level at low temperatures (after [12.8, 12.9])

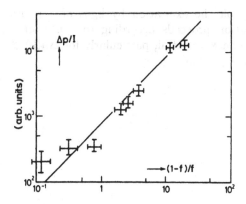

Fig. 12.12: Photo-yield at a photon energy of 0.3 eV for Ni-doped p-type germanium at 77 K, as a function of $(1 - f)/f$ (after [12.3, p. 91])

velocity of holes by v, the life-time is given by

$$\tau_p = (\sigma_p \, v \, f \, N_{\mathrm{Ni}})^{-1} \; . \tag{12.3.1}$$

Introducing the photo-ionization cross section σ_A of a neutral acceptor by

$$G = \sigma_A (1 - f) \, N_{\mathrm{Ni}} \, I/\hbar\omega \; , \tag{12.3.2}$$

we obtain from (12.1.3) for equilibrium

$$\frac{\Delta p}{I} = \frac{\sigma_A}{\sigma_p} \frac{1}{\hbar \omega v} \frac{(1-f)}{f} . \tag{12.3.3}$$

The photo-yield which is proportional to $\Delta p/I$ is plotted vs $(1-f)/f$ in Fig. 12.12 on a log-log scale. The data points indicate the linear relationship given by (12.3.3). From measurements of the absorption cross section $\sigma_A (\approx 10^{-16} \, \text{cm}^2$ for Ni in p-Ge at 77 K and for a photon energy of 0.3 eV) and for the average thermal velocity of a hole, using the value of 10^7 cm/s, the cross section for the capture of a hole by a singly-charged negative Ni center σ_p is found to be $10^{-13} \, \text{cm}^2$ at 77 K. This corresponds to a radius of $\sqrt{\sigma_p/\pi} = 1.8$ nm.

Problems

12.1. From the ground state of a hydrogenic impurity in a semiconductor, determine the maximum wavelength which can be detected by applying a semiconductor with this impurity as a photoconductor $(m/m_0 = 0.014; \kappa = 16)$.

12.2. Determine the transition in Au doped Ge (Fig. 12.11) which is responsible for its photoconductive sensitivity at a wavelength of 12 μm at 77 K. Which kind and amount of doping in addition to Au is required for moving the Fermi level to an appropriate position?

12.3. Assume that after switching off the excitation by light of a photoconductor at $t = 0$, the recombination proceeds according to $-d\Delta n/dt = C(n\,p - n_i^2)$. Show that $n(t)$ is very nonexponential, particularly for $\Delta n(0) \gg n_0$, resembling the behavior shown in Fig. 12.3.

13. Light Generation by Semiconductors

Electroluminescent devices emit incoherent visible or infrared light with typical line widths of about 10 nm while the coherent radiation emitted by the semiconductor laser may have a line width as low as 10^{-2} nm. These devices, together with photovoltaic diodes and solar cells (Sects. 5.8, 9, Chap. 12), are called *optoelectronic devices*. While the former convert electrical energy into optical radiation, the latter do the inverse process. In this chapter we will consider light-emitting diodes (LED), a light-emitting transistor (LET), and diode lasers [13.1].

There is a variety of luminescence effects depending on the kind of excitation, such as photoluminescence (excitation by optical radiation), cathodoluminescence (by cathode rays, i.e., electron beams), radioluminescence (by other fast particles including X-rays and γ-rays) and electroluminescence (by dc or ac electric fields) [13.2]. If there are traps involved in the emission process, the response times may be quite large, sometimes many seconds or even hours; the term *phosphorescence* is then used instead of *luminescence*. By gradually raising the temperature, the variation of the light output with temperature is recorded; this is called a *glow curve*. Typical phosphors are the zinc and cadmium sulfides, selenides, and tellurides [Ref. 13.2, Chap. 4]; the phenomena in these substances are, however, so complex that we have to exclude these materials from the present considerations. In passing we will also mention the *Destriau effect* [13.3] which is the excitation of light in microcrystalline, nominally insulating, solids embedded in a dielectric medium, as the result of an applied ac electric field of a frequency of typically about 100 Hz. An unexpected discovery was the photoluminescence of *porous silicon* in the visible spectrum ($\lambda \approx 750$nm) at room temperature during ultraviolet excitation [13.4], while single-crystal silicon because of its band gap of ≈ 1.1 eV shows luminescence only in the infrared range at 300 K. Porous silicon is produced from the single crystal by electrochemical etching. Whether the effect is due to a quantum confinement of electrons in the extremely small pore walls or to the formation of siloxene $Si_6O_3H_6$ covering the surface of the etched silicon, is still under debate. The luminescence spectra of porous silicon and siloxene are nearly equal.

13.1 Light Emitting Diode and Transistor

Radiative electronic transitions following an excitation are possible (a) by inter-band transitions where $\hbar\omega \gg \varepsilon_G$ (b) either by transitions from the conduction band to an acceptor or from a donor to the valence band or from a donor to an acceptor, and (c) by intraband transitions involving hot carriers (*deceleration emission*). A typical nonradiative transition is the *Auger effect* where the transition energy is transferred to another electron which is excited in this process either in a band or at an impurity and then loses its energy in a cascade of small photons or phonons [13.5].

For a calculation of the luminescence efficiency, let us consider the following rate equations. The rate of change of electron density in the conduction band is given by

$$dn/dt = G - \sigma_t v_n n N_t (1 - f_t) + A N_t f_t - \sigma_\ell v_n n p_\ell , \qquad (13.1.1)$$

where the first three terms on the rhs are essentially the same as in (12.1.3) with Δn given by (12.1.12) except that now they refer to traps (subscript t), while the last term describes the *radiative* recombination through the capture of an electron by a luminescent center, p_ℓ being the density of holes in the luminescent centers. σ_t and σ_ℓ are the cross sections of traps and luminescence centers, respectively. A is a constant. The probability of an electron staying in a trap f_t is given as the solution of the differential equation

$$d f_t/dt = \sigma_t v_n n (1 - f_t) - \sigma_r v_p p f_t - A f_t , \qquad (13.1.2)$$

where the second term on the rhs describes *nonradiative* recombination through the capture of a hole from the valence band, v_n and v_p are the electron and hole velocities, respectively. Under steady-state conditions, the time derivatives in (13.1.1, 2) vanish. Eliminating the trap cross section σ_t from both equations and solving for G yields for equilibrium ($dn/dt = d f_t/dt = 0$):

$$G = \sigma_\ell v_n n p_\ell + \sigma_r v_p N_t f_t p . \qquad (13.1.3)$$

The luminescence efficiency η is the ratio of the radiative recombination to the total recombination; it can be written as

$$\eta = \frac{\sigma_\ell v_n n p_\ell}{G} = \left(1 + \frac{\sigma_r v_p N_t f_t p}{\sigma_\ell v_n n p_\ell}\right)^{-1} . \qquad (13.1.4)$$

If the Fermi level is located between the electron trap level at energy ε_t and the luminescent level at energy ε_ℓ, we find for the concentration of trapped electrons relative to the concentration of holes in luminescent centers of concentration N_ℓ:

$$N_t f_t/p_\ell = (N_t/N_\ell) \exp[-(\varepsilon_t - \varepsilon_\ell)/k_B T] . \qquad (13.1.5)$$

This finally yields for the luminescence efficiency

$$\eta = \left[1 + \frac{N_t}{N_\ell} \frac{\sigma_r}{\sigma_\ell} \frac{v_p}{v_n} \frac{p}{n} \exp\left(-\frac{\varepsilon_t - \varepsilon_l}{k_B T}\right)\right]^{-1} . \qquad (13.1.6)$$

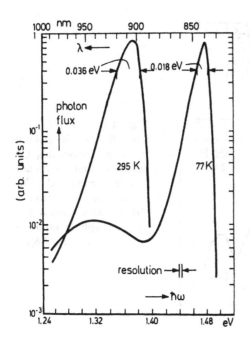

Fig. 13.1: Emission spectra observed on a forward biased gallium arsenide diode at 77 K and 295 K (after [13.6])

It is high if $N_\ell \gg N_t$ and $\varepsilon_t - \varepsilon_\ell \gg k_B T$, i.e., at low temperatures. The highest efficiency reported is $\eta = 40\%$; it has been obtained in a GaAs diode at 20 K [13.6] (GaAs is a *direct* semiconductor, Sect. 11.2). At room temperature η is not more than 7%. The response time is typically 1 ns or less. The emission spectra for 295 and 77 K are shown in Fig. 13.1. For a *direct semiconductor*, the probability of a radiative transition from the conduction band to the valence band is high. Even at low temperatures the emission is in the infrared part of the spectrum, however.[1]

The light emitting diode (LED) is manufactured by the diffusion of Zn which is an acceptor; hence, it is a p^+n junction. Because of the band tail in the heavily doped p^+-side which increases absorption, more light is emitted out of the n-side. Another production method is *amphoteric doping*: Silicon acts as an acceptor when it substitutes an As atom, and as a donor when it replaces a Ga atom. By a careful thermal treatment, closely compensated structures are obtained which are particularly effective for emission of radiation.

Large band-gap semiconductors such as GaN, GaP, and SiC emit visible light. GaN is a direct semiconductor having an energy gap ε_G of 3.7 eV at room temperature. p-n junctions are being produced since 1996 (Sect. 13.2). Schottky barrier diodes emit blue and violet light. The emission spectra of GaP and SiC are shown in Figs. 13.2, 13.3, respectively. Both substances are indirect

---

[1] Device application as a decoupling circuit in combination with a silicon detector. Nick Holonyak Jr. and S.F. Bevaqua gave the first practical demonstration of LEDs [13.7]

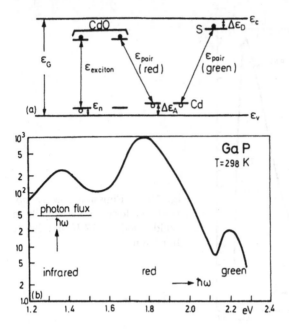

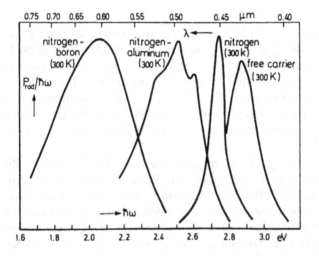

Fig. 13.2: (a) Energy band diagram for gallium phosphide containing Cd acceptor and S donor type impurities and a Cd-O complex (after [13.9]). (b) Emission from a gallium phosphide luminescent diode (after [13.8])

Fig. 13.3: Observed emission spectra from silicon carbide luminescent diodes (crystal structure: 6 H polytype) with n-regions containing the impurities given next to the curves (after [13.11])

semiconductors which explains their low efficiencies. The GaP p-n junction is doped with Cd and oxygen and contains a Cd-O complex. The emitted red light is due to a transition between the bound-exciton level at this complex and a Cd acceptor level, while the green light emission is due to a transition between an S donor level and the Cd acceptor level [13.8]. Excitons are also bound to the isoelectric impurity nitrogen (at phosphorus sites in GaP) and yield a green emission which appears as bright as the red emission since the human eye is more sensitive to green light [13.10].

The SiC emission spectrum also depends on the type of impurity (N, Al or B, Fig. 13.3) [13.11]. The light output is proportional to the current which is of course determined by the forward bias. The response time is 0.5 μs. The efficiency at room temperature is usually only about 10^{-3} % but efficiencies up to 1% have also been reported. The efficiency is, however, remarkably stable: diodes having an area of 1 mm^2 have been tested for 2 years at temperatures ranging from room temperature to 400°C and at forward currents of 50–200 mA with no measurable deterioration (other LEDs degrade more quickly because of the high photon-induced stress). The brightness level is about 100 foot-lambert [13.12]. The main application of silicon carbide diodes are to numeric and alphanumeric display devices. In this respect it may be of interest that crystals of more than 1 cm diameter have been grown. Depending on the crystallographic form (cubic or one of the polytypes), the gap varies between 2.2 and 3.1 eV. p-n junctions are produced by epitaxial growth from carbon saturated liquid silicon at 1500–1850 °C on a p-type SiC substrate at a growth rate of 0.5 μm/min. A nitrogen atmosphere serves for doping with a donor. Contacts are made by a gold-tantalum alloying process. SiO$_2$ masks can be grown on SiC, and the photo-etch resist technique is applicable just as in silicon (Sect. 5.7).

The manufacture of signal lamps and displays was based on GaAs$_x$P$_{1-x}$ which for $x \leq 0.55$ is *direct* with $\hbar\omega = 1.98$ eV (red light). For production, an epitaxial process on a GaAs substrate was preferred.

Before coming to the light-*emitting* transistor, let us briefly discuss the light-*detecting* transistor since both have a common structure. The latter is known since the late 1990s. It consists of a heterojunction of typically GaN(n)/GaN(p)-/GaN(i)/Al$_{0.20}$ Ga$_{0.80}$/N(n) where (n), (p), (i) stand for n-type, p-type, and intrinsic, respectively. Electrical contacts are deposited on collector and emitter, while the base is left floating. The substrate is sapphire, through which the incident UV light penetrates into the device. After crossing through the Al$_{0.20}$Ga$_{0.80}$N(n) layer, it is absorbed in the GaN(i)layer. The electric field in the i-region separates the electron - hole pairs. The electrons drift towards the base, the holes towards the collector. A current gain of more than 10^5 is obtained by the accumulation of holes in the floating base which increases the injection of electrons coming from the emitter. The recombination rate of injected electrons with holes is, however, reduced due to the presence of trapping centers. If a bias voltage "reset" pulse (7 to 10 V) is applied to the detector, so that holes are driven into the emitter, recombination is enhanced [13.12]. The bias voltage is 3 to 4 V. As a function of optical power, the responsibility is decreasing. The gain depends strongly on the frequency, and the gain-bandwidth product is thus constant. The UV vs. visible contrast is more than eight orders of magnitude and thus really excellent at a 10 Hz modulation frequency of the input optical signal.

After the invention of the light-detecting transistor came the light-emitting transistor [13.13]. The basic idea is again that at the "radiative" recombination of an electron with a hole, the resulting energy is emitted as a photon, as discussed before. The base ("B") structure is a graded direct-gap InGaP/GaAs heterojunction ("H") called HBT ("T" for transistor). The dimensions of the emitter are 1 μm $\times$ 16 μm. Hafez et al. [13.14, 13.15] have shown that the InP HBT operates at a speed above 450 GHz. In the present problem, however, the base current is modulated in transistor operation at only 1 MHz. While in prior work the base current recombination has been known as a source of heat, it is now identified as a source of light. The emission is detected as white spots on the screen of a silicon CCD microscope applied for looking at the device. For a base current of e.g. 4 mA and a collector current of 57 mA, the strongest light emission occurs at a collector-to-emitter voltage of about 0.8 V. The optical emission at a wavelength of about 885 nm is due to the composition-graded InGaAs base (1.4 percent Indium). The HTB has been operated also at a combined dc and ac modulation at frequencies of 100 kHz and 1 MHz. The output signal, detected with a Si APD and displayed with a sampling oscilloscope, has a peak-to-peak amplitude of 400 μV at 100 kHz and of 375 μV at 1 MHz. These results indicate the perfect function of the device (Feng, Holonyak Jr., Hafez)[13.13].

13.2 The Semiconductor Laser

The acronym *laser* has been coined from *Light Amplification by Stimulated Emission of Radiation*. We shall consider here only semiconducting solids, not gases or liquids. The difference is that in solids we have broad energy bands, in contrast to e.g. gases where the energy of atoms or molecules are represented by sharp levels. John von Neumann [13.16] seems to have been the first to suggest the generation and amplification of infrared radiation by injecting electrons into a semiconductor p-n junction by connecting it to an external current source in the forward direction. Nick Holonyak Jr. and S.F. Bevacqua [13.7] were among the first to notice a slight narrowing of the spontaneous radiation spectrum which could be attributed to stimulated transitions characteristic for "lasing". The first lasers for operation required liquid-gas cooling. For the basic treatment of stimulated emission, we go along with Casey and Panish [13.17].

Let us consider a simplified solution of the Schrödinger equation (2.1.1) as given by (2.1.6) but introduce the space coordinates x, y, z and replace α by k_x, k_y, k_z and u by E_x, E_y, E_z,

$$\partial^2 E_x / \partial z^2 + k_x^2 E_x = 0 \qquad (13.2.1)$$

where $k_x = 2\pi/\lambda$ and λ is the DeBroglie wavelength (see p.12, footnote 2). A multiple m_x of λ is the sample length L in x-direction with $m_x = (0; \pm1; \pm2; \ldots)$ and $k_x = 2\pi m_x/L$. For a cubic enclosure, the 3-dimensional wave equation with periodic boundary conditions extends the one-dimensional case to

$$k_x = 2\pi m_x/L; \quad k_y = 2\pi m_y/L; \quad k_z = 2\pi m_z/L$$

where m_x, m_y, m_z are integers $(0; \pm 1; \pm 2; \ldots)$. These discrete values of k_x, k_y, k_z give discrete values of E which are called "modes".

The particle-like nature of electromagnetic radiation is represented by photons with energy

$$\epsilon = h\nu = |p|c \tag{13.2.2}$$

with c = velocity of light, and momentum p with components $p_x = \hbar k_x$; $p_y = \hbar k_y$; $p_z = \hbar k_z$, where $\hbar$ is Planck's constant divided by 2π. The discrete values of the components of k give discrete photon energies. The allowed solutions are called "states" and the number of states per volume is the density of states. A cube with equal axis $2\pi/L$ in x-, y-, and z-direction has a volume $(2\pi/L)^3$. The number of allowed values of k in any volume V_k in k-space is the number of cubes of side $2\pi/L$ in that volume. The reciprocal of the unit volume $(2\pi L)^3$ is the unit density in k-space. $4\pi k^2 dk$ is the unit volume of a thin spherical shell of thickness dk. We thus obtain for the differential value of the density $dN(k)$

$$dN(k) = 2(L/2\pi)^3 4\pi k^2 dk \tag{13.2.3}$$

where the factor of 2 accounts for the two different states of polarization for a photon, transverse electric and transverse magnetic. Now we have to consider the relation between the wave vector k and the frequency ν in a dielectric with refractive index n (Sect. 11.1): $k = n\omega/c$ where $\omega = 2\pi\nu$. Hence, we obtain for dk:

$$dk = 2\pi(n/c)d\nu \left[1 + (\nu/n)dn/d\nu\right] \tag{13.2.4}$$

where the term in brackets is the refractive index dispersion which for free space is unity. Inserting this expression into the previous equation yields

$$dN = (8\pi n^3 \nu^2/c^3)[1 + (\nu/n)dn/d\nu]d\nu \tag{13.2.5}$$

For a system of photons, i.e. identical particles with integral spin, the Bose - Einstein distribution

$$f(\nu) = [\exp(h\nu/k_B T) - 1]^{-1}$$

has to be applied. Therefore, the photon density distribution $dD = f(\nu)dN$ with energy ϵ given by $d\nu = d\epsilon/h$:

$$dD(\epsilon) = \frac{8\pi n^3 \epsilon^2}{h^3 c^3} \frac{1 + (\epsilon/n)dn/d\epsilon}{\exp(\epsilon/k_B T) - 1} d\epsilon = P(\epsilon)d\epsilon \tag{13.2.6}$$

where $P(\epsilon) = dD(\epsilon)/d\epsilon$ and $\epsilon = h\nu$ with ν being a photon frequency.

Let us consider the absorption of a photon with the effect of an electron transition from state ϵ_1 within the valence band, to a state ϵ_2 within the conduction band such that the difference $\epsilon_2 - \epsilon_1 = \epsilon_{21}$ is equal to $h\nu$. The upward (in energy) transition rate r_{12} is the probability B_{12} that the transition can occur, times the probability f_1 that the state ϵ_1 contains an electron, times the probability $(1 - f_2)$ that the state ϵ_2 is empty, times the density $P(\epsilon_{21})$ of photons of energy ϵ_{21}:

$$r_{12} = B_{12}f_1[1 - f_2]P(\epsilon_{21}) \tag{13.2.7}$$

The return rate

$$r_{21} = B_{21}f_2[1 - f_1]P(\epsilon_{21}) \tag{13.2.8}$$

is obtained by considering a photon which, instead of being absorbed, stimulates the emission of a similar photon by a transition of an electron from energy level ϵ_2 to ϵ_1, which, of course, now is downward. B_{21} is the transition probability, f_2 is the probability that ϵ_2 is occupied, and $[1 - f_1]$ the probability that ϵ_1 is empty.

Finally, with a probability A_{21}, electrons at ϵ_2 can spontaneously return to ϵ_1 without interaction with the radiation field $P(\epsilon_{21})$. This spontaneous emission rate $r_{21}(spon)$ is given by

$$r_{21}(spon) = A_{21}f_2[1 - f_1] \tag{13.2.9}$$

Let us finally introduce a quasi-Fermi level F_1 for the valence band at non-equilibrium, and a corresponding F_2 for the conduction band so we can write for f_1

$$f_1 = \frac{1}{\exp[(\epsilon_1 - F_1)/k_BT] + 1} \tag{13.2.10}$$

and for f_2

$$f_2 = \frac{1}{\exp[(\epsilon_2 - F_2)/k_BT] + 1} \tag{13.2.11}$$

At thermal equilibrium, the upward transition rate must equal the total downward transition rate:

$$r_{12} = r_{21} + r_{21}(spon) \tag{13.2.12}$$

and $F_1 = F_2$. With these equations, one can easily prove

$$P(\epsilon_{21}) = \frac{A_{21}f_2(1 - f_1)}{B_{12}f_1(1 - f_2) - B_{21}f_2(1 - f_1)} \tag{13.2.13}$$

which, after division by $f_2(1 - f_1)$ of both numerator and denominator can be written

$$P(\epsilon_{21}) = \frac{A_{21}}{B_{12}\exp(\epsilon_{21}/k_BT) - B_{21}}. \tag{13.2.14}$$

This is easily transformed by means of the previous equations

$$\frac{8\pi n^3 \epsilon_{21}^2}{h^3 c^3} \frac{1 + [(\epsilon_{21}/n)dn/d\epsilon_{21}]}{\exp(\epsilon_{21}/k_B T) - 1} = \frac{A_{21}}{B_{12}\exp(\epsilon_{21}/k_B T) - B_{21}} \qquad (13.2.15)$$

where, if we neglect the term in brackets because of its smallness, we obtain

$$\frac{8\pi n^3 \epsilon_{21}^2}{h^3 c^3}[B_{12}\exp(\epsilon_{21}/k_B T) - B_{21}] = A_{21}\exp(\epsilon_{21}/k_B T) - A_{21} \qquad (13.2.16)$$

This equation may be separated in a temperature dependent term and a temperature independent term:

$$\frac{8\pi n^3 \epsilon_{21}^2}{h^3 c^3}B_{21} = A_{21} \qquad (13.2.17)$$

and a similar equation with B_{21} replaced by B_{12}. A comparison of the two equations reveals that

$$B_{12} = B_{21} \qquad (13.2.18)$$

The last two equations are known as the Einstein relations. They indicate that the spontaneous emission probability is related to the absorption and the stimulated emission probability. M.G.A. Bernard and G. Duraffourg [13.18] proved that *stimulated* emission in excess of absorption occurs when a photon is more likely to cause a downward transition of an electron from the conduction band with the emission of a photon than the upward transition from the valence band to the conduction band with absorption of the photon, i.e. $r_{21} > r_{12}$:

$$B_{21}f_2(1 - f_1)P(\epsilon_{21}) > B_{12}f_1(1 - f_2)P(\epsilon_{21}) \qquad (13.2.19)$$

which, with $B_{12} = B_{21}$ is simply $f_2 > f_1$. Comparing this with (13.2.10) and (13.2.11) we obtain

$$F_2 - F_1 > \epsilon_2 - \epsilon_1 \qquad (13.2.20)$$

We see that the separation of the quasi-Fermi levels must exceed the photon emission energy for the downward stimulated emission rate to exceed the upward absorption rate.

We shall now deal with artificial semiconductor crystal structures, known as Heterostructures and Double-Heterostructures including Quantum Cascade (QC) lasers. The equivalent case of Quantum Wire and Quantum Dot lasers will be presented in Sect. 14.5.

In a single crystal a junction between two dissimilar semiconductors is called a heterojunction. The components have different energy gaps in their band structures. R.L. Anderson [13.19] prepared Ge/GaAs tunnel diodes which probably were the first high-quality heterojunctions with different energy gaps and different dielectric constants on each side of the junction. With the invention of the alloy $Al_xGa_{1-x}As$, the production of the N-p-N double heterojunction GaAs-$Al_xGa_{1-x}As$ with x=0.3 was the breakthrough in the field. Fig. 13.4

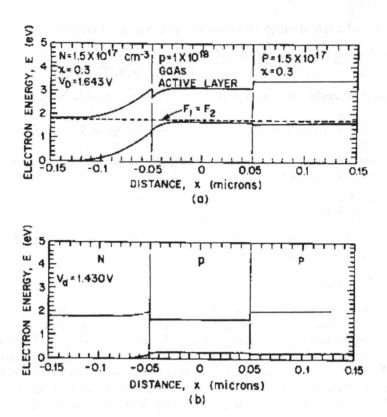

Fig. 13.4: Energy band diagram for (a) GaAs-GaAlAs N-p-P doubleheterostructure at a zero applied bias (b) at a forward bias of 1.43 V. (after H.C. Casey, Jr., and M.B. Panish [13.17]).

shows the energy band diagram for zero applied bias and for a forward bias of 1.43 V for a N-p-P DH laser. The conduction band discontinuity ΔE_c provides a barrier to electrons at the p-P heterojunction and thus confines the injected electrons to the GaAs active layer where the radiative recombination takes place. The valence band discontinuity ΔE_v plus the remaining built-in potential after forward biasing provides a potential barrier to holes at the n-N heterojunction and thus prevents hole injection into the N-layer. These barriers result in both minority and majority carrier confinement by the N-p-P double heterostructure.

Fig. 13.5 shows the corresponding diagrams for the N-n-P DH laser. Most of the built-in potential is at the n-P heterojunction, but some band-bending also occurs at the N-n heterojunction. Current continuity determines the V_a distribution between the N-n and n-P heterojunctions. The bias applied to the n-P heterojunction was 1.28 V and to the N-n heterojunction 0.15 V. This n-

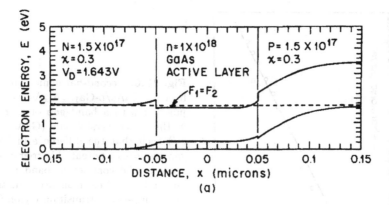

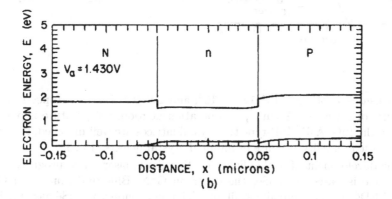

Fig. 13.5: Energy band diagram for (a) GaAs-GaAlAs N-n-P doublehetero-structure at a zero applied bias (b) at a forward bias of 1.43 V. (after H.C. Casey, Jr., M.B. Panish [13.17]).

N heterojunction is reverse biased but must break down because there is no evidence for an excess voltage. Comparison of Figs. 13.4, 13.7, and 13.5 shows that the energy band diagrams for N-p-P and N-n-P are essentially the same at the large forward bias encountered for the two types of GaAs-Al$_x$Ga$_{1-x}$As lasers.

Injection lasers are being applied more and more for optical communication by pulse-code modulation techniques via silica fibers. The attenuation of the fibers is low in the spectral range between 1 and 1.5 μm, and laser emission in this range is therefore of particular interest [13.1]. Room temperature lasers have been produced in this range from In$_{0.88}$Ga$_{0.12}$As$_{0.23}$P$_{0.77}$ − InP heterojunctions (wavelength 1.1 μm) [13.20], In$_{0.68}$Ga$_{0.32}$P − In$_{0.16}$Ga$_{0.84}$As (1.075 μm) [13.21], and GaAs$_{0.88}$Sb$_{0.12}$ − Ga$_{0.6}$Al$_{0.4}$As$_{0.88}$Sb$_{0.12}$ ($\approx$ 1 μm) [13.22] at threshold cur-

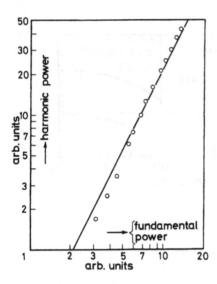

Fig. **13.6**: Second-harmonic emission from a GaAs/p-Ga$_{1-x}$Al$_x$As hetero-junction with a band gap of $\approx$ 2 eV at the mixed crystal side (GaAs: 1.4 eV) such that the difference in band gap produces a potential barrier essentially in the conduction band resulting in injected-electron accumulation in a thin p-GaAs transition region (after J.A. Armstrong et al. [13.29])

rent densities j_t of $2.8 \times 10^3, 14.9 \times 10^3$, and 2×10^3 A cm^{-2}, respectively. At j_t values of a few 10^3 A cm^{-2}, cw operation at room temperature is possible. Since in the InGaAsP−InP-type the crystal lattices are well matched, this offers the highest reliability because of minimal crystal defects [13.23].

The development of a blue-light emitting diode laser made progress in two ways: one is based on ZnSe, the other on GaN. Blue and blue-green GaN-based LEDs are commercially available. The semiconductor ZnSe has only been available with n-type conductivity. It was not until a few years ago that it became possible to produce a p-type side on the n-type material in order to finally have a p-n junction. Nitrogen acceptors in the form of free radicals were generated by a high-frequency plasma discharge during MBE growth of the crystal. The material components in a diode laser structure are actually not exactly ZnSe but ZnCdSe and ZnMgSSe [13.24]. The problem in operating the laser is the large voltage drop across the Schottky barrier between the p-type side and the metallic current lead during laser operation. This causes unwanted heat generation and a fast degradation [13.25].Therefore, no ZnSe laser has yet been commercially produced. With the GaN laser the situation is much better.

Gallium nitride has a gap of 3.4 eV and - important for heat dissipation - a high thermal conductivity of 1.3 W/cmK. The atomic bond is covalent which is stronger than the ionic bond in ZnSe. The high atomic mass difference between its components Ga and N hinders energy loss to acoustic phonons and hence leads to higher optical efficiencies and an insensibility to lattice defects. GaN is normally n-type, and the p-doping required for the formation of a p-n junction can at best be achieved with magnesium by irradiating the sample with an electron beam immediately after deposition [13.26]. The growth of GaN

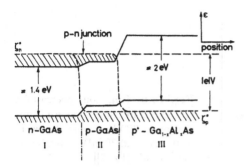

Fig. 13.7: Energy band diagram of a GaAs-Ga$_{1-x}$Al$_x$As heterojunction with a limited active volume (II) (after I. Hayashi et al. [13.29])

thin films from trimethyl-gallium and nitrogen gas in the form of a plasma produced by microwave electron-cyclotron resonance [13.27, 28] allows metalorganic vapor phase epitaxy (MOVPE) at low temperatures where strain and defects produced by thermal mismatch with the substrate during postgrowth cooling are nearly eliminated [13.30]. Epitaxially laterally overgrown GaN (ELOG) on sapphire further reduces the number of threading dislocations in the GaN epitaxial layer [13.31]. One deposits a series of doped semiconductor layers on top of one another, starting with the substrate and buffer layers. The active region at the center of the stack is a multiple-quantum-well structure where electrons and holes recombine to emit photons. On either side are light-guiding layers and cladding layers to confine the light and the carriers near the active region. Electrical contacts are made at the top and bottom of the stack. The recent development of a violet-light emitting diode laser (wavelength 408nm) with a maximum output power of 0.42 W continuous wave operation at room temperature (spectrum: Fig.13.14) at an estimated lifetime of 104 h contains two 4-nm-thick undoped In$_{0.15}$Ga$_{0.85}$N well layers separated by 10-nm-thick Si-doped In$_{0.02}$Ga$_{0.98}$N barrier layers as the active medium [13.32]. The threshold for lasing is 40 mA which corresponds to a threshold current density of 3 kA/cm^2. The 0.42 W maximum output was obtained at a current of 0.49A and a voltage of 8V. An amber LED (wavelength 594 nm) [13.33], a blue LED (470nm) [13.31], and a superbright green LED (520nm) [13.34] have also been produced by the same group. Emission of white light due to luminescence in organic and inorganic materials excited with blue LEDs has been investigated [13.35]. There are interesting applications for such LEDs in e.g. traffic lights because their lifetime is much longer than that of incandescent lamps.

Narrow-gap semiconductor diode lasers Pb$_{1-x}$Sn$_x$Te and Pb$_{1-x}$Sn$_x$Se have been produced [13.36], the former also by proton bombardment [13.37]. Operation is possible only at low temperatures. Because of the narrow gap, the radiation wavelength is very long: from the telluride system at $x = 0.315$, a wavelength of 31.8 μm has been obtained at 40 K, the threshold being at 640 A cm^{-2}. The single-mode, cw diode laser can be compositionally tailored to emit any desired wavelength in the range 6.5 to 31.8 μm [13.36]. In Fig.2.21, the variation of the energy band gap with composition in Cd$_x$Hg$_{1-x}$Te is shown. It is

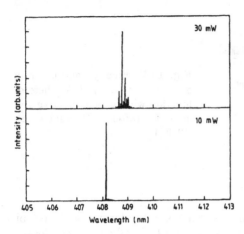

Fig. 13.8: Emission spectra of a GaN laser in *continuous wave* operation at 300 K at output powers of 10 mW and 30 mW [after Nakamura et al. [13.32]]

similar to the selenide system where the zero-gap composition is at $x = 0.15$. At $\lambda \approx 80$ μm ($h\nu = 16$ meV), optical phonon-photon interaction should become quite interesting. A serious problem in very-far infrared semiconductor lasers is the self-absorption of the radiation in the crystal by free carriers (Sect.11.10).

A type of laser for infrared and far-infrared radiation is the Quantum Cascade ("QC") laser. The first experimental demonstration was made in 1994 by Faist et al.[13.38]. In principle, it is based on intersubband transitions in a multiple quantum well heterostructure of e.g. GaAs or any other convenient semiconductor and may be tailored to emit even in the THz range at 67 μm wavelength [13.39]. The emission wavelength is primarily dependent on the quantum well thickness and practically independent of the material band gap. The "cascading" process in which tens of photons per electron are generated, provides high power for lasing. Ultrafast carrier dynamics are allowed by intersubband transitions with their LO phonon emission and optical transparency at energies below and above the optical transparency (Claire Gmachl et al. [13.40]). This laser is well suited for trace gas sensing applications [13.41].

Figure 13.9 shows one of the simplest designs of a QC laser to which an external voltage is applied (not shown) with the effect that the many quantum wells rise from the left to right. Two active regions are shown with (schematically) intermediate between them an injector region. The quantum wells consist of InGaAs and the barriers of AlInAs, lattice matched to a InP substrate. The drawings of the energy levels are based on solutions of the Schrödinger equation for this potential. Only the first three confined states of each active region are drawn. For producing the injector region, the semiconductor in this range has been doped with silicon of a level of about 4×10^{16}/cm^3. The quantum well thicknesses in the active region have been chosen as 6.0 nm and 4.7 nm. The intermediate barrier was 1.6 nm thick. In this way the energy levels no.1 and no.2 were 37 meV apart, the levels no.2 and no.3 were 207 meV apart, which corresponds to a wavelength of 6 μm. The value of 37 meV was chosen to be

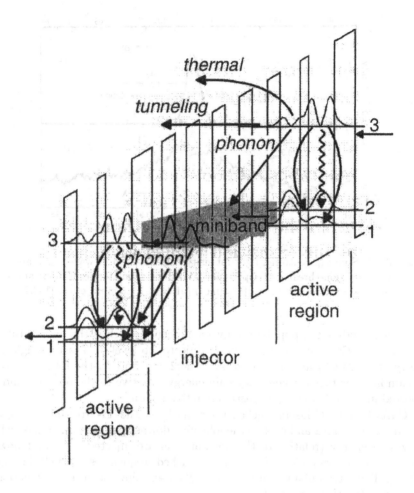

Fig. 13.9: Schematic conduction band diagram of two QC laser active regions with the intermediate injector region and the moduli squared of the wave functions involved in the laser transition (labeled 1, 2, and 3). The laser transition is indicated by the wavy arrows, and non-radiative processes by smooth arrows (after C. Gmachl et al. [13.41]).

near the LO phonon energy of the InGa/AlInAs/InP components. The bias applied corresponds to an electric field of 62 kV/cm or 0.29 V per stage of active region and injector. From the injector region the electrons tunnel to level no.3 and continue their motion by emitting LO phonons to the level no.2 and to level no.1 in rapid succession. The scattering times are $\tau_{32} = 2.2$ ps and $\tau_{31} = 2.1$ ps which yields a total upper state lifetime of 1.1 ps. The scattering time is $\tau_{21} = 0.3$ ps between levels no.2 and no.1. From $\tau_{32} \gg \tau_{21}$ it is obvious that between

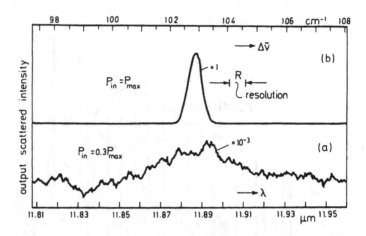

Fig. 13.10: Spin-flip scattered light intensity as a function of wavelength above the stimulated emission threshold (upper curve, R = spectrometer resolution) and below threshold (lower curve, magnified 10^3 times) (after S. Nakamura et al. [13.32])

levels no.3 and no.2 a population inversion occurs which is the prerequisite for laser action. Electrons tunnel from the injector to level no.3 fast enough (P.L. Derry et al.[13.42]) and exit from levels no.2 and no.1 to the following downstream injector region where they gain energy (relative to the bottom) and are injected into the following downstream active region.

GaAs based QC lasers reached wavelengths in the THz regime [13.43] and exit from levels no.2 and no.1 to the following downstream injector region where they gain energy (relative to the bottom) and are injected into the following downstream active region. These lasers reached frequencies in the THz regime [13.44]. The intersubband nature of the QC laser allows for the construction of multi-wavelength lasers [13.45].

An interesting modification of the semiconductor laser is the tunable *spin-flip Raman laser* (Fig. 13.10). In contrast to the injection laser, it requires another laser for a pump such as a CO_2 gas laser ($\lambda \approx 10.6\,\mu m$; $P \approx 3$ kW). The pump radiation is incident on a sample of n-type InSb ($n = 3 \times 10^{16}$ cm^{-3}) at 25–30 K in a magnetic field of 3–10 T. The Landau levels in the conduction band of InSb are spin-split, according to (9.2.21), the spacing between the spin sublevels $g\mu_B B$ being proportional to the magnetic field strength. The angular frequency of the scattered light ω_s is different from that of the incident light ω_0 by just this amount:

$$\omega_s = \omega_0 - g\mu_B B/\hbar \ . \tag{13.2.21}$$

Since the inelastic scattering process is similar to Raman scattering, it is also called a *Raman* process although it involves no phonons. In contrast to the usual Raman process, the frequency of the scattered light can be tuned by a variation

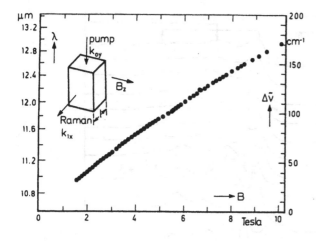

Fig. 13.11: Stimulated spin-flip scattering wavelength vs magnetic field. Laser geometry shown by the inset (after C.N.K. Patel et al. [13.46])

of the magnetic field strength. Lasing action is obtained if the pump power exceeds a certain threshold. Figure 13.15 shows the scattered light above and below threshold (upper and lower curve, respectively) [13.46]. Under stimulated-emission conditions the line is narrow (width estimated $\leq$ 0.03 cm^{-1}). In Fig.13.11 the stimulated Raman scattering wavelength is plotted vs the magnetic field strength [13.46]. Tuning has been achieved in the range from 10.9 to 13.0 μm. For a peak input power of 1.5 kW, the peak output was 30–100 W, the pulse length $\approx$ 100 ns (about one third of the pump pulse) and the repetition rate 120/s. The effect was predicted by Yafet [13.47] in an extension of earlier work by Wolff [13.48] on a Raman laser based on Landau transitions.

13.3 Optical Properties of Superlattices

Most of the experiments with superlattices (Sect. 9.1) have been performed as observations of photocurrent because of better signal-to-noise ratio than reflection, luminescence, or excitation spectra [13.53]. Moreover, being proportional to the optical absorption, the photocurrent is easier to interpret than luminescence. In one experiment the superlattice contained 60 periods of alternating GaAs and Ga$_{0.65}$Al$_{0.35}$As layers, terminated by 60 nm of Ga$_{0.65}$Al$_{0.35}$As on each side. These undoped structures constitute the intrinsic regions of $p^+ - i - n^+$ diodes grown by molecular-beam epitaxy. The well thicknesses were, depending on the sample, about 4 nm, the barrier thicknesses about 3 nm. The uniform field in the superlattice was produced by applying an external voltage on the p^+ region (relative to the n^+ side). Excitation of electrons was achieved by low-power radiation from a krypton laser-pumped dye laser near the fundamental gap of GaAs. The experimental results of photocurrent vs photon energy $\hbar\omega$

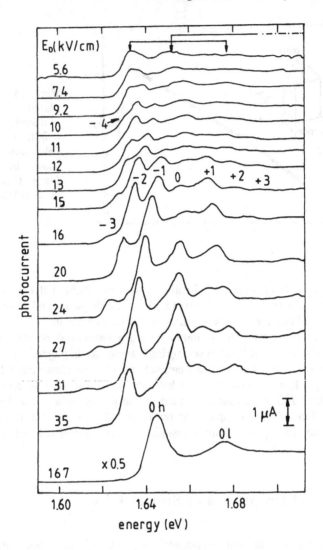

Fig. 13.12: Photocurrent spectra (vertically offset for clarity) for selected values of the dc electric field applied in the growth direction of the superlattice. Labels at the peaks give the Wannier-Stark ladder index of electrons in the heavy-hole (h) or light-hole (l) to conduction band transitions (after [13.53])

at various dc field strengths E_0 at 5 K are displayed in Fig. 13.17 [13.53]. On the 5.6 kV/cm curve, peaks indicated by arrows correspond to transitions from the heavy-hole and light-hole minibands to the conduction miniband (bottom and top). At fields from about 10 to 35 kV/cm, the system is in the Wannier-Stark ladder regime. The labels ranging from -3 to +3 correspond to the ladder

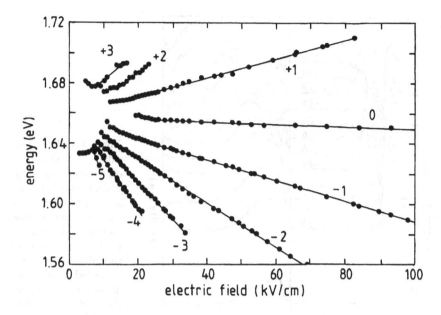

Fig. 13.13: Transition energies as a function of the electric field (after [13.54])

indices in the conduction band. The transitions with positive labels are always weaker than those with negative labels and those with higher labels which represent transitions between very distant wells are also weaker. A fan chart of the transition energies from the heavy-hole miniband as a function of the electric field for a 4 nm/2 nm superlattice is plotted in Fig. 13.18. The linearity between photon energy $\hbar\omega$ and field intensity E_0 is as expected from (9.1.26). The slope increasing with absolute value of the label number, $|n|$, is as it should be according to (9.1.28). Although Fig. 9.11 predicts steps at photon energies resonant with transitions between Wannier-Stark states, peaks are observed, a phenomenon which is also present in bulk semiconductors [13.54] and still lacks an explanation. Similar results have been obtained from transitions involving light holes. It is worth noting that in the Franz-Keldysh effect, (11.5.9), the field dependence is as the 3/2-power of E_0.

The detection of radiation emitted from Bloch-oscillating carriers in a superlattice subject to a dc electric field has been reported by Waschke et al. [13.55]. The emission could be tuned by varying the bias applied to the superlattice structure in a range from 0.5 THz to 2 THz, corresponding to a wavelength from 600 μm to 150 μm. The sample consists of 35 periods of GaAs (9.7 nm)/$Ga_{0.7}Al_{0.3}As$ (1.7 nm) with thick $Ga_{0.7}Al_{0.3}As$ buffer layers at both ends. A reverse bias field could be applied between the doped substrate and a semitransparent Schottky contact on the sample surface. The sample which is kept at 15 K, is excited by 0.1 ps Ti:sapphire-laser pulses at a repetition rate of 76 MHz which are incident at an angle of 45° to the surface. The optically

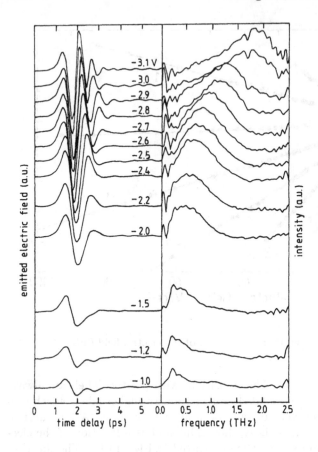

Fig. 13.14: Left side: Observed electromagnetic radiation emitted from a superlattice at various reverse-bias voltages. Right side: Fourier transform of time-domain data, corrected for frequency sensitivity of the detector (after [13.55])

excited electron concentration was less than $5 \times 10^9/\text{cm}^2$. The THz radiation, emitted collinearly with the reflected optical excitation beam through a high-resistivity silicon window, is projected via a pair of off-axis paraboloidal mirrors onto a fast dipole antenna which is gated by a second time-delayed part of the laser pulse.

Figure 13.19 shows on the lhs the detected radiation intensity as a function of delay time for various bias voltages. On the rhs of Fig. 13.19, the Fourier spectra of these curves are displayed. At low biases, there is only one peak in the spectrum whose frequency is constant. With negative bias increasing beyond 2 V, the peak shifts to higher frequencies. This result is compared with photoconduction and electroreflectance spectra of the same sample under identical excitation conditions. Again, the onset of the Wannier-Stark ladder is shifted to higher bias voltages. In all these experiments the results are comparable. They can be interpreted as follows: For reverse biases below -2.5 V, the external bias is quite effectively screened by accumulated charge carriers. Above this bias, the emission frequency depends linearly on voltage in a range, where the

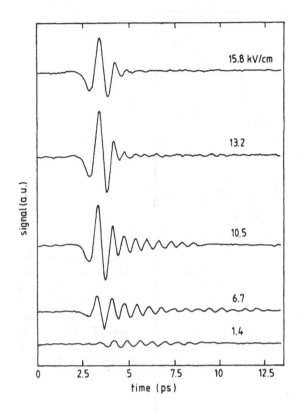

Fig. 13.15: Observed coherent electromagnetic radiation emitted from a coupled quantum well at various bias fields for $\hbar\omega = 1.54$ eV (after [13.56])

reflectance spectra also reveal the Wannier-Stark regime. The relation (9.1.29) where $E_0 = U/L$ and U is the voltage, shifted by an offset to account for field screening, and L is the thickness of the biased region, is closely obeyed. In the emission experiment, at least three transitions are excited: 0, -1, and -2, all from the heavy-hole level (Fig. 9.10a). In an earlier experiment by Roskos et al. [13.55] with the same arrangement, but with a double-well structure instead of a superlattice, the results, Fig.13.20, are rather similar to those obtained with the superlattice. The Fourier spectra, however, reveal a difference: Although the low-bias single peak is also there, the broad-band signal which appears at large biases, moves to smaller frequencies with increasing bias and is believed to be due to the creation of polarized electron-hole pairs in an electric field. In two variations a and b of a superlattice structure $Al_{0.48}In_{0.52}As/Ga_{0.47}In_{0.53}As$ (1.7nm/3.7nm, 14 periods in a; 2.3 nm/3.6 nm, 9 periods in b), either one embedded in a collector region of a p-n-p transistor as reported above, the collector current was observed at 15 K as a function of the collector bias at a constant emitter current (Fig.13.21). Particularly in sample a, the current decreases as a function of bias, i.e., a *negative differential conductivity* (n.d.c.) occurs over a wide bias range. This has been considered as a transport manifestation of

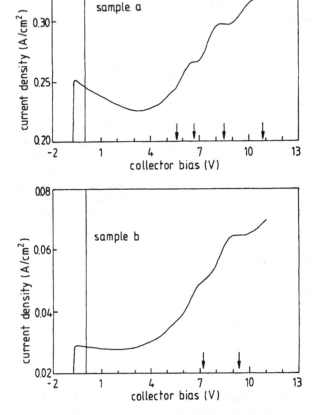

Fig. 13.16: Super-lattice current density as a function of collector bias at a constant emitter current of 0.39 A/cm at 15 K for two samples (after [13.57])

localization at a field of about 3 kV/cm in both samples, consistent with a relaxation (coherence) time of 4×10^{-13} s. The peaks on the increasing region of the characteristic are caused by carrier tunneling coherently for a certain number of superlattice periods depending on the electron mean free path (about 30 nm).

As already reported in (9.1.29), for Bloch oscillations to occur the condition is $\omega_B \tau > 1$, otherwise the eigenstates are broadened due to the electron interaction with scatterers such as interface defects, impurities, or LO phonons. Bastard et al. [13.58] present results of a calculation for these scattering events as they influence τ, plotted in Fig. 13.22 as $\omega_B \tau$ vs $2/f$ where f is the reduced field strength $e\, E_0\, d/(\Delta/2)$ as defined above, for a $GaAs/Ga_{0.7}Al_{0.3}As$ (3nm/3nm) superlattice of infinite extension. Down to the lowest field of about 5 kV/cm, $\omega_B \tau > 1$ is valid. In the phonon curve, resonances occur whenever the phonon energy $\hbar\omega_0$ equals an integer number of level distances on the Wannier-

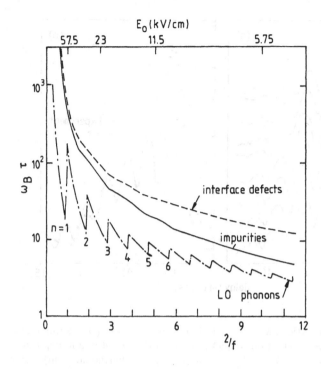

Fig. 13.17: Product of Bloch frequency and relaxation time vs field strength (upper scale) and $2/f$ (lower scale) for a compositional superlattice (after [13.58])

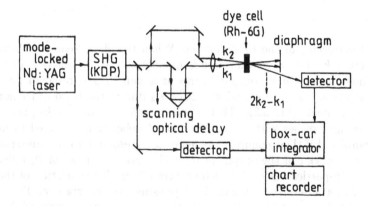

Fig. 13.18: Experimental arrangement for four-wave mixing (after [13.59])

Stark ladder, an effect comparable to the magnetophonon equivalent (9.4.1). An experiment first reported by Yajima et al. [13.59], and later dubbed *four-wave mixing* (FWM), used a dye cell rather than a superlattice for a sample. Let us consider the optical principle together with the experiment. A nonlinear resonant four-photon interaction of light pulses is applied for obtaining information on the ultrashort relaxation (dephasing) time constants associated with excited

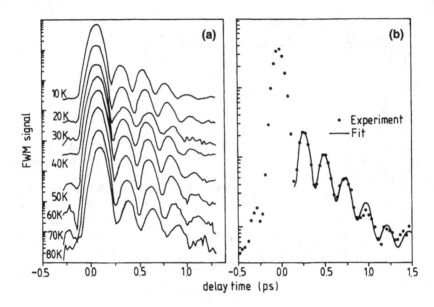

Fig. 13.19: (a) Bloch oscillations at various lattice temperatures. (b) Example of a data fit (curve) with a Bloch oscillation frequency of 4.2 THz, an interband dephasing time T_2 of 1.1 ps, and $T_{BO} = 4.2$ ps (decay of superimposed harmonic term) (after [13.60])

states of materials, in the *time domain*. When two light beams with different wave vectors k_1 and k_2 at the same frequency ω, (i.e., in different directions) are incident on a sample, output beams at ω can be generated in new directions, namely $k_1=2k_2 - k_1$, and $k_4=2k_1 - k_2$ due to the third-order nonlinear polarization effect, (11.15.3). This is called optical mixing in k-space. If two ultrashort light pulses with a variable mutual delay time are used as incident light beams and if their common frequency ω is resonant with a material transition, the behavior of the output beam with wave vector k_3 or k_4 reflects the relaxation properties associated with this transition. The variation of the delay time serves for a determination of the dephasing time constant τ. The authors give a detailed account of the theory [13.53]. A schematic diagram of the experimental arrangement is shown in Fig.13.23. Here a pulsed laser in connection with a KDP second harmonic generator has been used as a light source. The output was divided into two beams and after giving a variable delay to one beam, they were combined at the sample at an angle at 3° through a focussing lens, Fig.13.23. Absorbing the transmitted light beams with wave vector k_1 and k_2 by a diaphragm, the output beam k_3 was detected with a photomultiplier and recorded as a function of delay time applying a box-car integrator. Usually the sample is located on a cold finger of a cryostat.

Figure 13.24 shows results for a GaAs/Ga$_{0.7}$Al$_{0.3}$As (6.7nm/1.7nm) super-

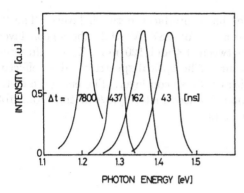

Fig. 13.20: Photoluminescence spectra (not drawn to
scale) at various time intervals after the excitation ($T =$
1.6 K; 20 n- and p-type layers:
$N_D = N_A = 2\times10^{18}$ cm^{-3}; d_n
$= d_p = 25$ nm) (after [13.61])

lattice at temperatures between 10 K and 80 K for a heavy-hole to (n=-1)
conduction band Wannier-Stark level. The light source is a titanium-sapphire
laser with a pulse duration of 0.13 ps and a repetition rate of 82 MHz. The k_3
beam is detected with a CCD camera. The excitation was kept low (10^8/cm^2)
so that the dephasing time is independent of the excitation level. On the lhs
of Fig.13.24 the FWM signal is plotted vs the delay time (curves are vertically
shifted for clarity). The rhs of Fig.13.24 shows the data of one of the set of data
points and a curve fitted according to a law

$$\gamma = \gamma_0 + \gamma_1 T + \gamma_2 \exp(-36\,meV/k_B T) \tag{13.3.1}$$

where T is the temperature; $\gamma_0 = 0.526/ps$; $\gamma_1 = 0.0026/\text{K} \cdot \text{ps}$; $\gamma_2 = 90$/ps.
The constant term dominates the other two terms below 60 K. The term linear
in T is assumed to represent acoustic phonon scattering, the last term optical
phonon scattering. The interpretation of these contributions to the damping of
the Bloch oscillations is still under discussion.

Besides the compositional type of superlattice there is also the doping type
(n-i-p-i) which was mentioned in Sect. 9.1. It requires a doping level of about
10^{12}/cm^2. Because of the high impurity level and therefore impurity scattering
of carriers there is no chance to observe Bloch oscillations. The semiconductor
in nearly all experiments is gallium arsenide.

Time-resolved photoluminescence has been investigated by taking the spectra at different time intervals Δt after the excitation. Figure 13.25 shows the
spectra as they shift to lower photon energies with increasing Δt. All spectra
are normalized for equal peak intensity. The shift is roughly proportional to the
logarithm of Δt [13.61].

Amorphous semiconductor superlattices although only of the compositional
type have also been produced [13.62]. The alternating layers consist of hydrogen-
saturated amorphous silicon a-Si : H [13.63] made from pure SiH$_4$, and silicon
nitride a-SiN$_x$: H made from 20 vol. % SiH$_4$ and 80 vol. % NH$_3$. A gas
discharge results in chemical vapor deposition ("plasma-assisted CVD") if the
quartz substrate is kept at elevated temperatures (493 K). The gases were ex-

changed rapidly enough to achieve sharp interfaces between layers. The 2.7 nm wide silicon nitride layers have a gap of 3.87 eV while the silicon layers ($\varepsilon_G = 1.77$ eV) have a width of between 0.8 and 120 nm. X-ray diffraction clearly shows the superlattice structure. The optical spectra yield an effective gap decreasing with increasing width of the silicon layer. The energetic distribution of localized states in amorphous semiconductors plays an important role also in the interpretation of these results.

Problems

13.1. *Recombination radiation in heavily doped semiconductors.* For a heavily doped degenerate p-type semiconductor at temperatures below 1 K, what is the frequency of maximum recombination radiation? Neglect impurity levels and band tails.

13.2. *Donor acceptor transitions.* Considering Coulomb interaction between donor and acceptor in a pair, determine the transition frequency as a function of the donor acceptor distance. Plot the transition in an energy band diagram.

13.3. Calculate the minimum doping concentration for GaAs at room temperature to achieve population inversion in a laser diode with band-to-band transitions. Assume $N_A = N_D$, a gap energy of 1.43 eV and an intrinsic carrier concentration of $n_i \approx 10^7$ cm^{-3}.

14. Surface and Interface Properties and the Quantum Hall Effect

So far we have dealt with the bulk properties of semiconductors and tacitly assumed that the crystal is extended infinitely. We will now briefly discuss the influence of the crystal surface and interfaces between semiconductors on the transport properties of semiconductors.

14.1 Surface States

The quantum mechanical treatment of the linear array of atoms extended to infinity by de Kronig and Penney (Sect. 2.1) was first adapted to an array of finite length by Tamm [14.1]. As a result, additional states appear which are located at the surface and therefore called *surface states*. Inside the crystal, the electron wave function shows damped oscillations with increasing distance from the surface while outside the crystal it decreases exponentially. Shockley [14.2] treated a linear chain of 8 atoms. Figure 14.1 shows the energy levels as a function of the atomic distance. As in Fig. 2.8c for 3-dimensional diamond, for example, there are energy bands and gaps. But the two ends of the linear chain introduce, in addition, two states in the middle of each gap which are the surface states.

In a 3-dimensional crystal one would expect one surface state per surface atom. Surface states can also be recognized as being due to the dangling bonds of the surface atoms. The density of surface atoms has an order of magnitude of 10^{15} cm^{-2}. There are steps at the surface, and between 1% and 20% of the surface atoms are located at these steps. Careful annealing reduces the number of steps. A *real* surface is covered by impurity atoms except when special precautions are taken. A *clean* surface may be obtained by cleavage of a crystal in an ultrahigh vacuum such as, e.g., 3×10^{-10} torr, although pressure bursts of 10^{-7} torr upon cleavage have been observed. During several hours, a monatomic layer forms on a clean surface in an ultrahigh vacuum which consists mainly of oxygen atoms. What is considered a *clean surface* depends very much on the type of measurement. For a given observation, a surface is often considered clean if the experimental results are not changed by further purification of the

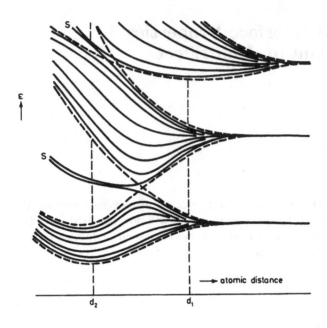

Fig. 14.1: Calculated energy bands and surface states (label "S") of a one-dimensional crystal containing eight atoms, as a function of the atomic distance (after [14.2])

surface.

If the surface under consideration is formed by the tip of a needle of radius 10^{-4} mm, a field intensity of about 10^8 V/cm can be produced. Such a field is sufficient to tear off atoms, including impurity atoms, by the electrostatic force when the needle is positively charged. This method is used in the field electron microscope [14.3]. Other methods of purification include argon bombardment and subsequent annealing in an ultrahigh vacuum. A sensitive method for the detection of surface impurities and dislocations is low-energy electron diffraction (LEED) (Ref. [14.4, Chap. 4]).

Impurity atoms on a surface of a semiconductor may be ionized. A monatomic layer would yield about 10^{15} elementary charges per cm^2 assuming that each impurity atom is ionized. However, it is not possible to have such a tremendous charge at the surface. Even with only 5×10^{13} carriers/cm^2, the electrostatic field energy is already equal to the surface energy of the crystal lattice. Observed surface charges usually vary between 10^{11} and 10^{13} carriers/cm^2.

Experimental evidence of surface states first came from the rectification properties of silicon–metal contacts. Meyerhof [14.5] observed that these were practically independent of the difference in work function between the metal and the silicon. Bardeen [14.6] explained this observation by assuming surface states which are due to impurities on the interface between the metal and the semiconductor. As a rule of thumb, such interface states are more pronounced in smaller-gap materials. The interface is often formed by an oxide film, especially if the semiconductor has been etched with an oxidizing agent (e.g., $HF+HNO_3$) before making the metallic contact. An energy band diagram for this case

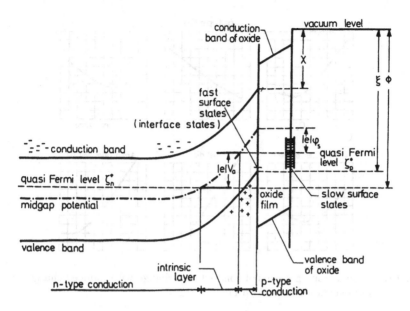

Fig. 14.2: Energy band and surface structure at semiconductor surfaces; χ = electron affinity; φ = work function; ξ = photoelectric threshold (after [14.7], p. 139)

is shown in Fig. 14.2. There are interface states called *fast states* and outer surface states called *slow states*, according to their different response times to the application of a strong electric field perpendicular to the surface. The surface in Fig. 14.2 is negatively charged while the bulk is n-type. The conduction electrons are repelled from the negative surface charge with the formation of either a depletion layer or, as shown here, of a p-type inversion layer if the valence band is bent above the bulk quasi-Fermi level. For positively charged surface states, an accumulation layer is formed.

The following effects have been applied for surface investigations: surface conductance, field effect, photoelectric methods, surface recombination, field emission, photo-surface conductance, optical absorption and reflection.

14.2 Surface Transport and Photoemission

Since the thickness d of the charge accumulation layer or inversion layer at the surface is not known, the conductance $d/\varrho = \sigma_h$ is obtained in units of "mho/square" (which is sometimes written ohm^{-1}), rather than the conductivity σ.

How can the surface conductivity be distinguished from the bulk value? This

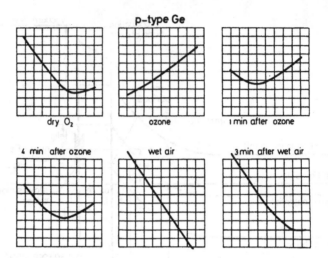

Fig. 14.3: Oscilloscope patterns of conductance change with induced charge for various gaseous ambients (after [14.7] p.111)

is possible by field effect measurements. The sample surface and the surface of a metal plate (*gate*) are carefully polished flat, a thin Mylar foil (e.g., 6 μm thick) is put in between, and the assembly is pressed tightly together. For silicon the foil may be replaced by a thermally grown oxide layer (see Sect. 5.7 on MOSFETs). A voltage V of variable magnitude and polarity is applied between the plate and one of the current probes at the sample. If the voltage is such that the charge on the plate in magnitude equals the surface charge and has, of course, the opposite sign, there is no mobile space charge, the bands are flat and the conductance is at a minimum. As shown in Fig. 14.3, exposure of the sample to gases such as oxygen, ozone or water vapor affects the surface state occupancy as well since ions are absorbed. The oscilloscope traces were obtained by applying an ac voltage. Oxygen produces an n-type surface while water vapor makes it p-type. Under certain conditions (e.g., freshly exposed to ozone), it is not possible to find the minimum even with high field strengths since dielectric breakdown in mylar at about 10^7 V/cm limits the voltage that can be applied. For a length L and a width W of the sample, the surface conductance is

$$\Delta\sigma_\square = \frac{L}{W}\left(\frac{1}{R(V)} - \frac{1}{R(0)}\right) , \tag{14.2.1}$$

where $R(0)$ is the sample resistance at the minimum. If the capacitance C per unit surface area can be determined, a *field effect mobility* μ_{FE} can be calculated from the conductance change with applied voltage:

$$\mu_{FE} = \frac{1}{C}\frac{d\sigma}{dV} . \tag{14.2.2}$$

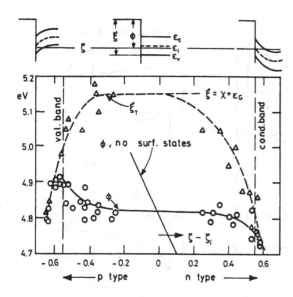

Fig. 14.4: Work function φ (full curve) and photoelectric threshold ξ (dashed) vs the difference between the Fermi energy and its intrinsic value for cleaved silicon surfaces. Indicated on top are the three energy band diagrams for extreme p- and n-type doping and the flat-band condition (after [14.10])

This is not, in general, the actual mobility of all types of carriers , however, but it approaches μ_n or μ_p when $\Delta\sigma_\square$ is much larger than the minimum value in the respective direction. Of course, due to surface scattering, these mobilities are lower than the bulk values. In Ge, field effect mobilities of the order of 10^2 cm^2/Vs have been observed [14.8]. On cleaved silicon, both contact potential and photoemission have been measured [14.9]. Fig. 14.4 shows the results evaluated in terms of the work function φ and photo-electric threshold ξ as functions of the bulk potential $\zeta - \zeta_i$, where ζ is the actual Fermi energy and ζ_i the intrinsic Fermi energy.

The work function changes only from 4.76 to 4.92 eV while the bulk Fermi level moves by 1.2 eV. This demonstrates the strong band bending effect by the surface states. For quite a range around the flat-band state, the photoelectric threshold is nearly constant and about 0.3 eV above the work function. Obviously the Fermi level is pinned at 0.3 eV above the valence band edge at the surface; i.e., here is where the surface states are located on a cleaved silicon surface.

In contrast to silicon and germanium, for sufficiently perfect surfaces of III–V compounds, no surface pinning occurs [14.10]. On a (110) surface after cleaving, atoms are rearranged within the surface unit cell moving as much as 0.05 nm from their bulk positions (*surface reconstruction*). Photoemission studies reveal that about 20% of a monolayer adsorption of either oxygen or a metal such as, e.g., Cs or Au, pin the Fermi level at the surface at about the same position. Figure 14.5 shows the surface state levels for three compounds. Circles represent n-type material and triangles p-type. From the fact that atoms with such different atomic orbitals induce states with equal energy, the conclusion is

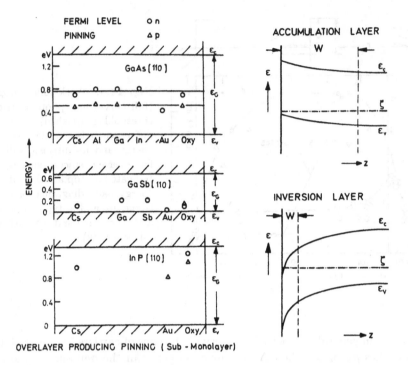

Fig. 14.5: Saturation values of the Fermi level as more and more atoms of the kind indicated are adsorbed at a III-V compound surface. Circles: n-type bulk; triangles: p-type bulk (after [14.11])

Fig. 14.6: Bent bands at the surface of a semiconductor subject to a strong electric field perpendicular to the surface ($\approx 10^5$ V/cm)

drawn that the adsorption induces lattice defects which represent the surface state rather than the adsorbed atom itself [14.12, 13].

14.3 Surface Quantization

In the course of the development of the transistor and subsequent work on surface states, J.R. Schrieffer in 1957 [Ref. 14.7, p. 55] remarked that the motion of carriers in a narrow inversion channel on the surface of a semiconductor should be limited by a quantum condition. Figure 14.6 shows the carrier energy ε as a function of the distance z from the surface. Due to an electric field perpendicular to the surface, the bands are bent and there is a n-type layer on

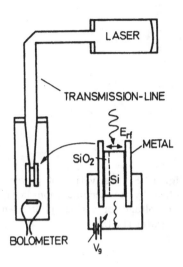

Fig. 14.7: Experimental arrangement for the observation of transitions between electric subbands in surface space charge layers [after [14.15]]

the p-type semiconductor [1] if the field is strong enough to move the conduction band edge close to the Fermi level ζ. If the width w of the inversion layer is of the order of the de Broglie wavelength λ of the carriers, which is typically 1 to 10 nm, a relation $w = \nu\lambda/2$ with ν being 1, 2, 3,..., is expected, leading to discrete values for the carrier energy perpendicular to the surface. Assuming a surface field strength E_s independent of z for simplicity, one has to solve the same type of equation (11.5.1) as was obtained for the Franz-Keldysh effect. For the solution (11.5.5), the argument of the sine function is an integer multiple ν of π for the carrier ψ function to vanish at the boundary. This yields for the carrier energy

$$\varepsilon_\nu = \left[\left(\nu - \frac{1}{4}\right)\frac{3\pi}{2}\frac{\hbar e E_s}{\sqrt{2m_z}}\right]^{2/3} \qquad \nu = 1, 2, 3, \dots \ , \tag{14.3.1}$$

where m_z is the effective mass in the z direction. (A rigorous solution takes the Poisson equation, as well as the Schrödinger equation into account [14.14].)

The energy levels have been determined by far-infrared absorption techniques [14.15]. Figure 14.7 shows the experimental arrangement. On one side of a 10 Ωcm n-Si slab, a 0.21 μm SiO$_2$ layer has been thermally grown on top of which, as well as on the back of the sample, aluminum layers have been evaporated. The sample, placed between the two copper plates of a transmission line, was irradiated at 4.2 K with the polarized 220-, 171-, or 118-μm radiation of a D$_2$O or H$_2$O laser with the polarization E_{rf} being perpendicular to the plates. Band bending on the Si-SiO$_2$ interface was achieved by means of a dc voltage V_g applied between the plates. In Fig. 14.8, the change of the transmitted laser power with V_g is plotted versus V_g for the three photon en-

---

[1] Of course, a p-type layer may also exist on an n-type semiconductor.

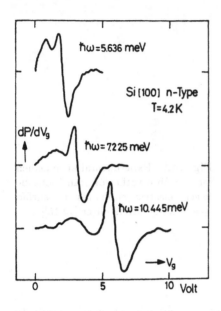

Fig. 14.8: Derivative of power absorbed by an accumulation layer vs gate voltage for various photon energies (after [14.15])

ergies. The resonances are transitions between the occupied first sub-band and the higher-lying, unoccupied sub-band.

The carrier ensemble in the inversion layer of an interface thin enough that the carrier energy component perpendicular to this layer is quantized, is called a two-dimensional electron gas (2-DEG). Such a *quantum-size effect* has also been observed in heterojunctions such as n-type GaAs/AlGaAs, where AlGaAs with its larger band gap loses electrons to GaAs, and band bending occurs at the interface. The effective density of states, given by (3.1.7) for the usual 3-dimensional electron gas (3-DEG), may be calculated similarly for the 2-DEG:

$$g_{2D}(\varepsilon)d\varepsilon = 2\frac{V d^2(\hbar k)}{(2\pi\hbar)^2} = 2\frac{V}{(2\pi)^2}2\pi k\,dk = 2\frac{Vm}{2\pi\hbar^2}d\varepsilon \tag{14.3.2}$$

where in contrast to (3.1.7) V is now the unit area, and $d^2k = 2\pi k\,dk$ involves an integration over the azimuth angle. It is remarkable that here the effective density of states is independent of energy, depending only on the effective density-of-states mass m of the carriers. Assuming a constant value for m, i.e. a parabolic band structure $\varepsilon(k) \propto k^2$, size quantization converts the band into a series of equidistant subbands with number $i = 1, 2, 3 \ldots$, as mentioned above, each of which has the same effective density of states. After summation over all subbands, the density of states of the conduction band is given by

$$g_{2D}(\varepsilon) = \sum_{i,s}\frac{Vm}{2\pi\hbar^2}\int_{-\infty}^{\varepsilon}\delta(\varepsilon' - \varepsilon_{i,s})d\varepsilon' = \frac{Vm}{2\pi\hbar^2}\sum_{i,s}\theta(\varepsilon - \varepsilon_{i,s}) \tag{14.3.3}$$

where the Heaviside step-function $\theta(\varepsilon) = 1$ for $\varepsilon > 0$ and zero otherwise. The

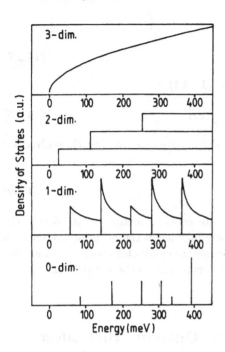

Fig. 14.9: Density of states as a function of energy (schematic) for a three-, two-, one-, and zero-dimensional electron gas with subbands resulting from size quantization (after [14.16])

band occupation is determined by the Fermi-Dirac distribution function $f(\varepsilon)$ given by (3.1.4).

In Fig. 14.9 the energy dependence of the density of states is sketched schematically not only for the three-dimensional electron gas (3-DEG) and the 2-DEG, but also for a 1-DEG (*quantum wire*) and a 0-DEG (*quantum dot*) to be discussed below [14.16].

For low temperatures T where $\exp[(\varepsilon-\zeta_n)/k_BT] \ll 1$ is approximately given by $f_0 \approx \Theta(\zeta - \varepsilon)$, ("degenerate limit"), the equilibrium number of electrons per unit area is

$$n = \int \frac{m}{\pi\hbar^2}\Theta(\varepsilon - \varepsilon_s)d\varepsilon = \frac{m}{\pi\hbar^2}(\zeta_n - \varepsilon_s) \qquad (14.3.4)$$

At low temperatures the conductance is determined by electrons close to the Fermi energy ζ corresponding to a Fermi wavenumber k_ζ which is

$$k_\zeta = \sqrt{(2\pi n)} \qquad (14.3.5)$$

and the wavelength is

$$\lambda_\zeta = \frac{2\pi}{k_\zeta} = \sqrt{(2\pi/n)} \qquad (14.3.6)$$

In (4.10.12) we calculated the *Einstein relation* for a nondegenerate electron gas. Let us now calculate the corresponding relation for a degenerate electron gas. The diffusion current

$$j = -D_n \nabla_r n \tag{14.3.7}$$

with the carrier concentration n given by (14.3.4) is

$$j = -eD_n \frac{m_n}{\pi \hbar^2} \nabla_r (\zeta_n - \varepsilon_s) = eD_n \frac{m_n}{\pi \hbar^2} eE \tag{14.3.8}$$

A relation between the conductivity σ_n and the diffusion constant D_n is obtained by comparing with $j = \sigma_n E$:

$$\sigma_n = e^2 \frac{m_n}{\pi \hbar^2} D_n \tag{14.3.9}$$

This is the *Einstein relation* for degenerate semiconductors. Considering the constance of m_n, the conductivity can only increase if the diffusion increases in this case. Remember (4.10.12) for nondegenerate conductors where the mobility is proportional to the ratio of the diffusion constant to the temperature.

14.4 Ballistic Transport and the Quantum Hall Effect

We consider the effect of a strong magnetic field on a 2-DEG. In Sect. 9.2 magnetic quantization of a 3-DEG and its effect on carrier transport have been treated. The first Hall and magnetoresistance experiments in very strong magnetic fields (13.9 T to 18 T) perpendicular to the applied electric field at low temperatures (≈ 1.5 K) were reported in 1980 by K. von Klitzing, G. Dorda and M. Pepper with the title *New Method for High-Accuracy Determination of the Fine-Structure Constant Based on Quantized Hall Resistance* [14.17]. The samples were silicon MOSFETs (Fig. 5.15), but with Hall probes for Hall effect measurements in strong magnetic fields. The two-dimensional electron gas (2-DEG) is located under the gate electrode forming the channel between source and drain. The voltage drop V_p in the source-drain current direction and the Hall voltage V_y in the transverse direction were both observed as a function of the gate voltage which determines the channel current I_x.

To his surprise, von Klitzing observed plateaus in the Hall voltage curve and, in the range of these plateaus, there was a zero voltage drop $V_p = 0$ in the middle of the sample where the Hall voltage was also observed (Fig. 14.10 (a)). The plateau voltages V_y could be expressed by the relation

$$V_y = \frac{1}{i} \frac{h}{e^2} I_x ; \qquad i = 1, 2, 3 \ldots \tag{14.4.1}$$

where h is Planck's constant ($2\pi \hbar$) and e is the elementary charge. The same phenomenon was found two years later in a GaAs/AlGaAs heterostructure by *D. C. Tsui, H. L. Störmer, and A. C. Gossard* [14.18] where, because of the higher

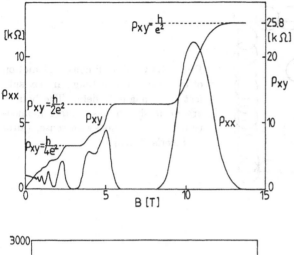

Fig. 14.10: (a) Longitudinal resistance ϱ_{xx} and Hall resistance ϱ_{xy} observed on a 2–DEG in a GaAs/Al$_{0.3}$Ga$_{0.7}$As heterostructure as a function of the magnetic field at a temperature of ≤ 1.5 K (after [14.20])

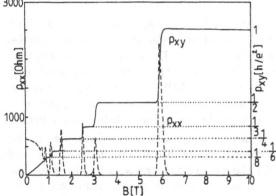

Fig. 14.10: (b) Same as Fig.14.10 (a) except that at a temperature of 0.06 K a different sample was investigated (after [14.21])

mobility of the electrons, the plateaus are even more pronounced, and in other heterojunctions such as, e.g., In$_x$Ga$_{1-x}$As/InP [14.19]. Figure 14.10 (b) shows the corresponding resistances $\varrho_{xx} = V_p/I_x$ and $\varrho_{xy} = V_y/I_x$ as a function of the magnetic field strength B at a temperature of 0.06 K, (Fig. 14.10 (a): 1.5 K). As indicated by (14.4.1), the plateau resistances V_y/I_x are independent of the type of semiconductor [14.22, 23], of the sample geometry, of the impurity concentration, of the carrier mobility, and of the type of carrier; they involve only fundamental constants. Because of the high precision of $\approx 10^{-7}$ (at the higher magnetic field) with which (14.4.1) is obeyed along the longest plateau, a resistance standard $h/e^2 = 25\,812.807\,\Omega$ could be obtained from these and later measurements and denominated the *von Klitzing constant* [14.24]. The high accuracy of this constant led to an increased accuracy of the fine-structure constant. The *integer quantum Hall effect*, as it is now called because of the integer values of i, is used by standards laboratories around the world to maintain the resistivity unit *Ohm*(Ω). The discovery earned von Klitzing the Physics Nobel

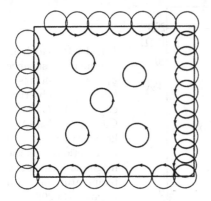

Fig. 14.11: Two-dimensional motion of electrons in a rectangular box subject to a magnetic field. The boundary orbit shown has a large momentum in an anti-clockwise sense; (after R. Peierls [14.25])

Prize of 1985.

There are two quantum effects involved: the sample width in the direction, e.g. z, of the applied magnetic field is of the order of the deBroglie wavelength, so that the electron gas is two-dimensional (2-DEG) in the x, y plain, the temperature is low enough and the magnetic field strong enough to involve magnetic quantization as has been considered for a 3-DEG in Sect. 9.2, and because of the high sample purity in connection with the extremely low temperature, there is practically no carrier scattering inside the sample ("ballistic conduction").

For the usual transport of carriers along a bar-shaped sample an essential requirement are collisions of the carriers with e.g. impurities, lattice defects, and the sample surface. Under the conditions mentioned for the quantum Hall effect, there are no such collisions with one exception: collisions at the two sample edges which are parallel to the applied electric field. Rudolf Peierls [14.25] has shown the two-dimensional motion of electrons in a rectangular box in a magnetic field (Fig. 14.11). Let us disregard the closed circles within the square shaped sample which represent carriers which do not contribute to charge transport. We also disregard any electron motion outside the sample with the result of "skipping orbits" [14.26]. Let us assume that only one pair of sample edges which are located opposite to each other are connected to metallic contact pads, and there is a voltage applied between the two pads. The arrows at the remaining circle pieces show a carrier motion along one sample edge in one direction and a second motion at the opposite edge in the opposite direction. There is no interaction of carriers in the middle of the sample. This situation is occasionally compared with the motion of cars on a divided highway: they should not collide. Of course, the circle radius must be much smaller than the dimension of the sample perpendicular to the motion which is very effectively provided here by the extremely strong magnetic field. This regime of collision-free transport inside the conductor is called "ballistic transport". It is only the scattering of carriers at the sample boundaries which limits the current, not impurity scattering. Of course, highly purified samples are a prerequisite for

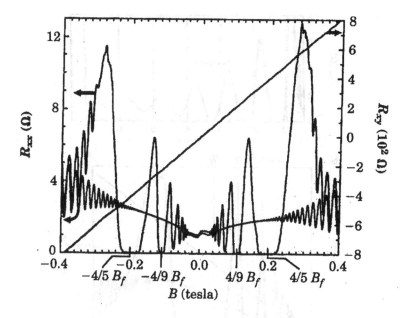

Fig. 14.12: (a) Resistance of a two-dimensional electron gas with very high mobility under microwave irradiation. Schubnikov - deHaas oscillations are seen in the longitudinal resistance R_{xx} for fields above 0.2 T. Below that, R_{xx} oscillates dramatically (curve with arrow towards left) although the transverse Hall minima are proportional to $B_f = \omega\, m_n/e$, where ω is the microwave frequency and m_n is the effective electron mass. (after R.G. Mani et al. [14.32]).

a successful quantum Hall experiment, and these were provided by G. Dorda and M. Pepper who coauthored von Klitzing's paper. Furthermore, a sample temperature in the milli - kelvin range helped to provide the required reduction of other scattering events.

Let us denote for a calculation the Fermi energy at one side, now called the left side, by ζ_L, and at the right side by ζ_R. The current I is given by

$$I = 2\frac{e}{2\pi\hbar}\, M\,(\zeta_L - \zeta_R) \tag{14.4.2}$$

where M is the number of edge states at the Fermi energy which is equal to the number of bulk Landau levels below the Fermi energy. The quantity M takes on integer values that decrease with increasing magnetic field strength, with the formation of plateaus in the resistivity-vs-magnetic field strength diagram (Figs. 14.10 (a) and (b)). The Hall voltage (subscript H) is in this case

$$V_H = \frac{2\pi\hbar}{2e^2}\,\frac{1}{M}; \qquad M = 1, 2, 3, \ldots \tag{14.4.3}$$

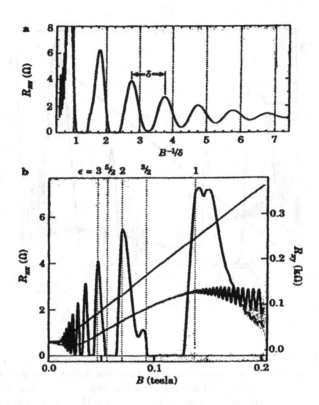

Fig. 14.12: **(b)** Upper diagram **a**:   Microwave - induced oscillations in the resistance are periodic in $1/B$, where B is the applied magnetic field. The period δ agrees with B_f^{-1} (as defined in the previous Fig. 14.12(a)) to within experimental error. The maxima are found at $B_f/B = \varepsilon = j + 1/4$ for j an integer, the minima at j - $1/4$. Lower diagram **b**: Maxima in the microwave response occur at integer ratios of ε minima below ratios of $j + 1/2$. The lowest curve is the resistance without microwaves, the remaining curve the transverse Hall resistance (after M.A. Zudov et al. [14.33].)

and the Hall resistance $R_H = V_H/I = (\zeta_L - \zeta_R)/eI$ may be written

$$2MR_H = 2\pi\hbar/e^2 \tag{14.4.4}$$

The longitudinal voltage measured along one side of the sample, V_L (subscript L for longitudinal) is

$$V_L = 0 \tag{14.4.5}$$

In contrast to a ballistic conductor which is not quantized by a strong magnetic field, we have here a spatial separation of the forward and the backward propagating states, with the result of an extremely precise quantization especially along the longest plateau.

High-precision measurements have been performed by B. Jeckelmann et al. [14.27]. They find that the quantum Hall resistance is independent of the material, device, and step number to within 3.5 parts in 10^{10} if each of the Ohmic contacts to the 2DEG used in the measurement has a resistance of less than one Ω. Their results do not prove the relation $R_H(i) = h/i\,e^2$. The correctness of this equation can only be shown in a comparison of R_H with an independent realization of h/e^2, they claim. But their demonstration of the universality of $R(i)$ at high precision adds considerable weight to the supposition that the equality exists. The pure edge-state formalism of contacts developed by M. Büttiker [14.28] and by H. Hirai et al. [14.29] is no longer valid in this domain, and the experiments do not support this theory. The quantum Hall resistance deviations caused by non-ideal contacts decrease with increasing temperature and current. They vary as $\exp(-\alpha T)$ for temperatures T in the range 0.3 - 1.2 K and as $1/I$ for a range of current I smaller than the device critical current. Anomalous behavior of quantum Hall plateaus reported by other groups can probably be ascribed to bad contacts.

The question of the high-frequency limit for inducing a quantum Hall effect by electromagnetic waves has first been investigated by *Kuchar* et al. [14.30] and *Volkov* et al. [14.31]. In the Faraday rotation of microwaves at a magnetic field of 6 T the plateaus still exist at a frequency of 62 GHz but not at 110 GHz. More recent investigations [14.32, 14.33] with GaAs/AlGaAs heterostructures of extremely high mobility (1.5×10^7 cm^2/Vs and 2.4×10^7 cm^2/Vs) revealed that with microwaves at a frequency between 3 GHz and 150 GHz dramatic resistance oscillations appear, i.e. regions of resistance that are zero to within experimental accuracy (Figs. (14.12(a); 14.12(b)). The oscillations grow in amplitude as the microwave power is increased. The temperature dependence is represented by $\exp(-\epsilon_A/k_B T)$ with an activation energy ϵ_A of the order of $\epsilon_A/k_B \simeq 10$ K, almost an order of magnitude larger than the Landau level spacing or the microwave photon energy. The relationship between ϵ_A and the magnetic field B where the minima occur, is roughly linear. (For a discussion of theoretical aspects see e.g. R. Fitzgerald, Physics Today April 2003 pp. 24 - 27)

For an application of the quantum Hall effect it is interesting that the perfect plateaus $i = 1, 2, \ldots$ are in the resistance range of about $10^3 \, \Omega$. National Bureaus of Standards have to keep a resistance standard with the highest possible precision. The quantum Hall effect provides such a standard which does not change with time as does e.g. a piece of metallic wire. There are quite a few Bureaus of Standards around the world, and for comparisons usual practice is to exchange standards which for transport should have low weight. Fortunately n-type HgTe quantum wells with an electron effective mass of only 0.02 m_0 require a low-weight magnet with a maximum field of only $B = 4$ T for the development of the $i = 1$ quantum Hall plateau as shown in Fig. 14.13 [14.36].

As mentioned before, the effect discussed so far became known as the *Integer Quantum Hall Effect* (IQHE) after a corresponding phenomenon with fractional values of the filling factor was observed by Daniel Tsui, Horst Störmer,

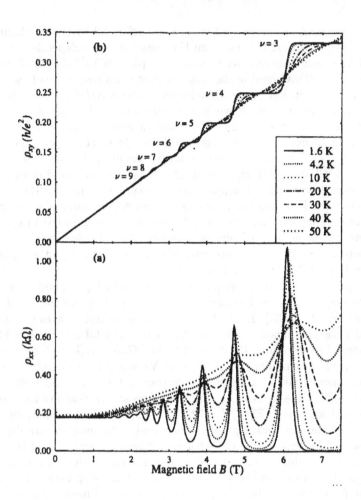

Fig. 14.13: (a) Shubnikov-deHaas oscillations, (b) quantum Hall effect for a n-type HgTe quantum well dependent on the magnetic field strength for various values of the temperature, after [14.36]

and Arthur Gossard [14.18] in 1982. The observation of this *Fractional Quantum Hall Effect* (FQHE) required even lower temperatures and samples of the heterojunction with smaller electron concentration, i.e. of higher purity. The cyclotron diameter is then smaller than the electron-electron separation. For a while it remained a mystery until Jain [14.35] in 1989 applied the concept of *composite fermions* for an explanation. The theoretician Robert Laughlin together with Horst Störmer and Daniel Tsui already mentioned above shared the 1998 Nobel Prize in physics for their discovery and interpretation of this effect. The complete theory unfortunately is beyond the scope of this book.

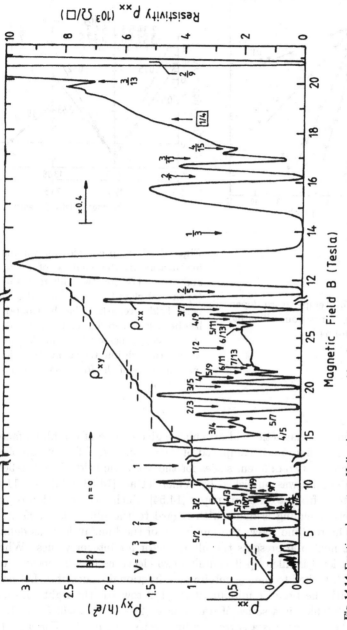

Fig. 14.14. Fractional quantum Hall resistance ρ_{xy} and longitudinal resistance ρ_{xx} at lower temperatures and higher sample purity than in Fig. 14.10. The numbers marked n indicate the Landau levels, the vertical arrows marked with integer or rational ν values represent the filling factor values. Lhs after [14.53], rhs after [14.54]. For fields higher than 14 Tesla (rhs) data are divided by 2.5

Figure 14.14 gives an overview of Hall resistance ϱ_{xy} and longitudinal resistance ϱ_{xx} with emphasis on the FQHE range. Depending on the carrier

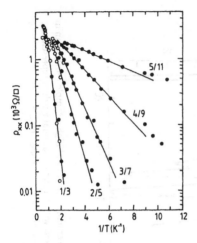

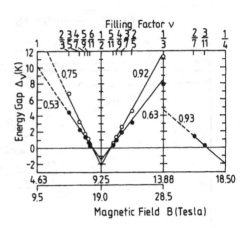

Fig. 14.15: Temperature dependence of ϱ_{xx} for various filling factors (after[14.54])

Fig. 14.16: Gap energies for various filling factors in the vicinity of $\nu = 1/2$ and $\nu = 1/4$ for two different samples (open and closed circles) plotted vs magnetic field strength. The horizontal axis has been scaled so that equivalent fractions (top scale) coincide. The number associated with each line represents the effective mass in units of the free-electron mass (after [14.54])

concentration, the magnetic field at which a given value of the filling factor occurs varies. The diagram is composed of data from two different samples, and therefore there are two different scales for the magnetic field. The 0-to-30-Tesla part of the diagram was published by Willett et al. [14.53] while the 11-to-21-Tesla part stems from a Du et al. paper [14.54]. Such a combination is possible when the actual scaling is made with regard to the filling factor as it is the case here. The top line shows the number n of the Landau level as defined by (9.2.4). The next line presents the filling factor for integer values. When the $n=0$ Landau level is filled up it contains two electrons ($\nu =2$) because of spin degeneracy. Attached to the ϱ_{xx} minima are fractional values of ν. Let us count beginning with the broad minimum at $\nu =1/2$ towards the right: 6/13; 5/11; etc. until 1/3 which is a series of fractions $p/(2p+1)$ for $p=6$; 5; ... 1. From $\nu =1/2$ to the left there is a corresponding series $p/(2p - 1)$. For $p \gg 1$ both series converge to $\nu =1/2$. Similar series are indicated which converge towards 1/4; 3/2; 5/2, the latter even within the $n=1$ Landau level. The section of data stretching from $\nu =1/2$ to 1/3 look like regular Shubnikov-de Haas oscillations (Fig. 9.13) starting from $\nu =1/2$ and exhibiting minima periodic in the inverse of the horizontal scale, i.e. $B\text{-}B_{1/2}$ where $B_{1/2}$ is the magnetic field corresponding to $\nu = 1/2$ for the particular sample. The oscillations to the left of $\nu = 1/2$ behave like a mirror image.

The energy gap Δ_ν between Landau levels at a filling factor ν (of the highest partially filled level) has been deduced from the exponential temperature dependence of the magnetoresistance $\varrho_{xx} \propto \exp(\Delta_\nu/2k_BT)$. Figure 14.15 shows log ϱ_{xx} plotted vs $1/T$ at five values of ν [14.54]. The values of $\Delta\nu$ determined from the slopes of the straight lines drawn through the data points are plotted in Fig. 14.16 vs the magnetic field $B_{p/(2p\pm1)}$ at which the minima occur. Straight lines can be drawn through these points. B scales are different for the two samples but have been adjusted for equal values of ν. There are linear relations between Δ_ν and B. Extrapolation of the lines to $\nu = 1/2$ and $1/4$, respectively, determines $B_{1/2}$ for $\nu = 1/2$ and $B_{1/4}$ for $\nu = 1/4$. Hence the ϱ_{xx} minima occur for the $\nu = 1/2$ series at

$$B_p^* = |B_{p/(2p\pm1)} - B_{1/2}| \qquad (14.4.6)$$

Since the Δ_ν are linear in B_p^*, they can be characterized by a *cyclotron effective mass*

$$M_p^* = \frac{\hbar e B_p^*}{\hbar \omega_c} = \frac{\hbar e B_p^*}{\Delta_\nu} \qquad (14.4.7)$$

Values of M_p^*/m_0 where m_0 is the free electron mass, have been attached to each line in Fig. 14.16. The effective masses M_p^* are an order of magnitude larger than the band effective mass 0.07 m_0 of the conduction electrons in GaAs.

Let us now consider these surprising results in the light of the composite-fermion model [14.35]. The basic idea is to attach to each electron of a two-dimensional electron gas a *magnetic flux tube*. The quantum unit of magnetic flux in a one-electron system is $\Phi_0 = h/e$ (see Aharonov-Bohm effect discussed in the text following (14.5.17); for Cooper pairs, e has to be replaced by $2e$). For a Landau level filling factor $\nu = 1/2$ there are exactly two flux quanta per electron. This unit, called *composite fermion* (CF), acts as if there was no magnetic field present. If the external magnetic field is somewhat smaller or larger than equivalent to the 1/2 filling, then the effective field given by (14.4.6) is active on the CF which reacts with its effective mass M_p^* given by (14.4.7). Theory predicts that CF masses are determined by electron-electron interaction and are therefore expected to scale as $l_B^{-1} \propto \sqrt{B}$ where l_B is the magnetic length (see footnote 2 of Sect. 9.2). Comparing the two samples of Fig. 14.16 for which carrier densities $n_\square$ and hence magnetic field scales differ by a factor of 2.05, the squares of CF mass ratios which are $(0.92/0.63)^2 = 2.1$ and $(0.75/0.53) = 2.0$, agree with theory [14.33, 14.34].

In contrast to the FQHE theory for $\nu = 1/2$, an explanation of experimental data for $\nu = 3/2$ requires CFs with spin. Here the lower spin state of the $n=0$ Landau level is totally occupied while the upper spin state is half filled. Rather than considering the electrons in this state, Du et al. [14.34] calculate with $\nu' = 2 - \nu$ states made up of holes. Away from $\nu = 3/2$, the CFs experience an

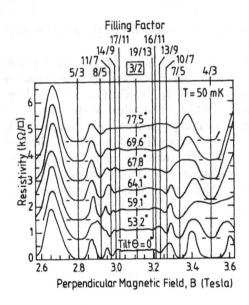

Fig. 14.17: Longitudinal resistivity as a function of the perpendicular component $B_{\text{tot}} \cos \Theta$ of a tilted magnetic field (tilt angle Θ relative to the sample normal), for various values of Θ and filling factors in the vicinity of $\nu = 3/2$, at a temperature of 0.05 K (after [14.37])

effective field

$$B^* = 3|B - B_{3/2}| \ , \tag{14.4.8}$$

where, as before, the magnetic field is perpendicular to a two-dimensional carrier gas (2-DEG) or, if the magnetic field is tilted at an angle θ relative to the normal of the 2-DEG, it is

$$B = B_{\text{tot}} \cos \theta \ , \tag{14.4.9}$$

where B_{tot} is the total field strength. The application of magnetic fields at various angles allows a distinction of orbital states of the 2-DEG which depend on the perpendicular component B and from spin states which depend on the total field B_{tot} . Observations of ϱ_{xx} vs the perpendicular component B at various tilt angles θ in a B range varying for the corresponding filling factor $\nu = 2$ to $\nu = 1$ are shown in Fig. 14.17.

The pattern is reminiscent of the angular dependence of Shubnikov-de Haas oscillations of regular 2D electrons in tilted magnetic fields [14.38, 39]. There, every second minimum disappears when, for a given angle, the Zeeman energy $g\mu_B \, B_{\text{tot}}$ equals the cyclotron energy $e\hbar B/m$ and the gap at the Fermi energy vanishes. At around $\nu = 3/2$, the situation is more complex. For one set of levels, the energy scale $e\hbar B^*/M_p^*$ is determined by B^*. The energy scale of the other set of intercepting levels depends on the total field B_{tot} which predominantly affects the Zeeman energy. A maximum of ϱ_{xx} appears at an angle where a spin level of one CF Landau level coincides with a spin level of another CF Landau level. The result for a GaAs/GaAlAs heterojunction is that around $\nu = 3/2$ the effective CF mass is $M_p^*/m_0 = 0.43$ and the effective g-factor $g^* \approx 0.61$ is the same for all spin polarizations of a given FQHE state. The latter is essentially the g-factor of the electron component of the CF particle,

$$g^* = 0.42[1 + 1.5(n_\downarrow - n_\uparrow)/(n_\downarrow + n_\uparrow)] \ . \tag{14.4.10}$$

The g-factor of pure GaAs is 0.32 [9.55].

Other FQHE studies have been performed by observing the reduction in the velocity of the surface acoustic waves on GaAs/GaAlAs heterojunctions subject to a strong magnetic field [14.40, 41]. The frequency of the acoustic waves was in the microwave regime and the temperature and magnetic field typical for FQHE observations. Other experiments include Hall and magnetoresistance measurements in these heterojunctions which contained *antidots* [14.42, 43]. These are mechanical holes in a periodic array with a lattice constant of 300 nm etched into the crystal.

14.5 Quantum Wires and Dots

In *quantum wires* free motion of carriers is possible only along the wire axis in the z-direction, while the motion in the x- and y-directions is quantized by confinement corresponding to the 2-DEG size quantization introduced in Sect. 14.3. The Schrödinger equation in this case is given by

$$\left[\frac{\hbar^2}{2m}(k_x^2 + k_y^2 + k_z^2) + V(x,y) \right] \psi(x,y,z) = \varepsilon \, \psi(x,y,z) \tag{14.5.1}$$

The potential V does not depend on z. Therefore the wave function is a plane wave in the z - direction:

$$\psi(x,y,z) = \frac{1}{\sqrt{L_z}} f(x,y) \exp(i \, k_z \, z) \tag{14.5.2}$$

The function $f(x, y)$ is a solution of

$$\left[\frac{\hbar^2}{2m}(k_x^2 + k_y^2) + V(x,y) \right] f(x,y) = \varepsilon \, f(x,y) \tag{14.5.3}$$

Now $f(x, y)$ represents a two - dimensional state with energy

$$\varepsilon = \varepsilon_i + \hbar^2 k_z^2/2m \tag{14.5.4}$$

We solved this problem already in Sect. 9.2 for the case of magnetic quantization and obtained for the density of states (9.2.5):

$$g(\varepsilon) = g_s \frac{L_z \sqrt{2m}}{2\pi\hbar\sqrt{\varepsilon - \varepsilon_i}} \tag{14.5.5}$$

where we have introduced a factor $g_s = 2$ for the spin degeneracy, which was omitted in Sect. 9.2 for simplicity's sake. We remember that L_z represents the sample width in the z-direction. We assume zero temperature T which simplifies the Fermi distribution function

$$[1 + \exp{(\varepsilon - \zeta)}/k_B T]^{-1} = \begin{cases} 1 & \text{for} & \varepsilon < \zeta \\ 1/2 & \text{for} & \varepsilon = \zeta \\ 0 & \text{for} & \varepsilon > \zeta \end{cases} \tag{14.5.6}$$

i.e. it becomes a step function at $\varepsilon = \zeta$. The product of this function with $g(\varepsilon)$ integrated over ε from zero to infinity yields the number of electrons in subband No. i per unit length L_z as

$$N_i = \frac{g_s \sqrt{2m}}{2\pi\hbar} \int_0^\zeta \frac{d\varepsilon}{\sqrt{\varepsilon - \varepsilon_i}} = \frac{g_s}{\pi\hbar} \sqrt{2m(\zeta - \varepsilon_i)} \tag{14.5.7}$$

Remember from (4.1.28) that the electron velocity $v = \hbar^{-1} d\varepsilon/dk$ with k being here k_z and $d\varepsilon/dk_z = \hbar^2 k_z/m$. The current density turns out to be

$$j_{i,z} = \frac{e\hbar k_z}{mL_z} \tag{14.5.8}$$

Of course, the current is due to a voltage applied between the ends of the sample in z-direction (simplified as the *channel*). The energetic difference between states $k_z > 0$ and $k_z < 0$ is thus eV and the current I_i is given by

$$I_i = \frac{e\hbar k_\zeta}{m} \frac{g(\zeta)}{2} eV \tag{14.5.9}$$

where $g(\zeta)$ from (14.5.5) with $\varepsilon = \zeta$ is obtained. The conductance G_i of channel No. i is thus

$$G_i = \frac{I_i}{V} = \frac{e^2}{2\pi\hbar} g_s \tag{14.5.10}$$

Each channel i contributes the same amount to the conductance. For very long wires with a cross section $L_x \times L_y$ the function $f(x, y)$ in (14.5.2) is

$$f(x, y) = \frac{1}{2\sqrt{L_x L_y}} \sin\left(\frac{n_x \pi x}{L_x}\right) \sin\left(\frac{n_y \pi y}{L_y}\right) \tag{14.5.11}$$

with $n_x = 0, 1, 2, \ldots$ and $n_y = 0, 1, 2, \ldots$ but $n_x^2 + n_y^2 \neq 0$. The energy values

$$\varepsilon_i = \frac{\hbar^2}{2m} \left(\frac{n_x^2 \pi^2}{L_x} + \frac{n_y^2 \pi^2}{L_y}\right) \tag{14.5.12}$$

for e.g. $L_x \gg L_y$ there is a ladder of x-type levels between each two adjacent y-type levels.

The actual quantum wires have, in contrast to the above calculation, a finite length. Because of the finite temperature, electron scattering causes noise. Therefore observations of quantization parameters have an experimental accuracy of only about 10^{-2}. The case of a multi-terminal ballistic transport has been treated by Landauer [14.75], Büttiker [14.76], and Picciotto et al.[14.44].

Fig. 14.18: Scanning electron micrograph of a 100–nm–wire with voltage probes defined by the etching only through the GaAs cap layer to a depth of 17 nm (after Scherer et al. [14.45])

Quantum boxes called *dots* with all three dimensions of the order of the deBroglie wavelength of the electrons, do not allow any motion of the carriers at all. If we assume that dots are isolated from each other, the carriers can, similar to electrons in an atom, take on energies $\varepsilon_{i,j,k}$, and the density of states is therefore

$$g_{0D}d\varepsilon = 2 \sum_{i,j,k} \delta(\varepsilon - \varepsilon_{i,j,k})d\varepsilon \ , \qquad (14.5.13)$$

assuming spin degeneracy for simplification. The number of carriers per dot volume V is given by

$$n_{0D} = \frac{2}{V} \sum_{i,j,k} \frac{1}{1 + \exp[(\varepsilon_{i,j,k} - \zeta)/k_B T]} \ . \qquad (14.5.14)$$

What are the methods for producing wires and dots and what is the interest in these devices? The technology was developed by *Petroff* et al. [14.46] applying a technique of photolithography and chemical etching of GaAs/GaAlAs heterostructures. On top of the "wire" with a triangular cross section ($\approx 4 \times 10^2$ to 10^4nm^2) a GaAlAs cap layer of 300 nm thickness was deposited. The length of these wires was several cm. Improvements were made by ion beam assisted etching with chlorine which reduced the damage thickness [14.45]. Figure 14.18 in a schematic way depicts the scanning electron micrograph of a 100 nm wide wire with voltage probes. The wire length is about 10 μm. Similar results have been obtained by *Thoms* et al. [14.47].

The development of dots went along similar lines. Figure 14.19 exhibits the schematic cross-sectional view of a typical sample [14.48]. On top of a semi-insulating GaAs substrate followed by buffer layers of GaAs and GaAlAs, a superlattice of twenty 2 nm GaAs wells with 10 nm GaAlAs barriers has been grown. A cap layer similar to the combined buffer layers finishes the structure. A conventional chip production technique including electron beam writing, metal plating, and reactive ion etching yielded an array of dots, as shown in Fig. 14.20 [14.48]. The figure was made from a scanning electron micrograph.

(a)

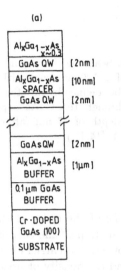

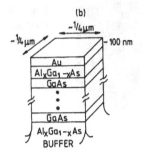

Fig. 14.19: (a) Schematic cross section view of the GaAs quantum well MBE sample. (b) Schematic cross sectional view of a quantum dot structure (after M.A. Reed et al. [14.48])

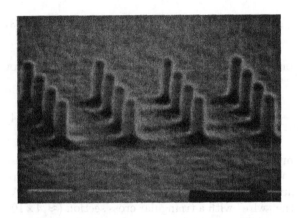

Fig. 14.20: Scanning electron micrograph of anisotropically etched columns containing quantum dots (after M.A. Reed et al.[14.49])

Present-day production of quantum dots is based on self-organization effects. Stranski and Krastanow [14.50] in 1937 were the first who considered the possibility of island formation on a flat heteroepitaxial surface for the growth of *lattice matched* ionic crystals that had different charges. Using the initials of these authors, *S K growth* now means the growth of islands as *strained* heterostructures [14.51]. Goldstein et al. [14.52] were the first who observed a regular pattern formation in an InAs/GaAs superlattice. About half of the strain energy which is due to the lattice mismatch between InAs and GaAs is supposed to be relaxed in this process. In the years 1993 and 1994 it was discovered that the islands were very uniform in size [14.55]. Now devices with more than $10^{11}/cm^2$ dots of 10 nm size arranged in arrays and stacks are applied as low-threshold lasers operated at room temperature.

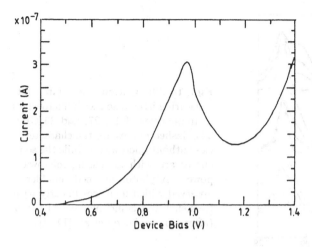

Fig. 14.21: I–V characteristic at 300K of a single quantum dot having lateral dimensions 0.25 μm $\times$ 0.15 μm (after [14.48])

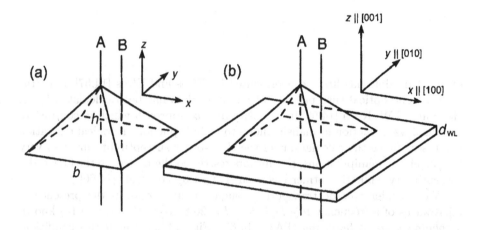

Fig. 14.22: (a) Notations of a pyramid (b) Pyramid on a wetting layer quantum well; h is the height, b is the base width, d_{WL} is the thickness of the wetting layer (after D. Bimberg et al.[14.56])

Figure 14.21 presents the current voltage characteristic at 4.2 K for a dot of lateral dimensions 0.25 μm $\times$ 0.15 μm. The diagram shows a range of negative differential resistance (NDR), as discussed in Sect. 9.1 for resonant tunneling along the superlattice inside the dot.

Figure 14.22 shows an InAs pyramidal dot on a GaAs substrate (surface crystallographic direction [001]) with an intermediate *wetting layer* [14.76]. This InAs layer has a thickness of only one or two monolayers. Its surface energy is very probably different from that of the bulk material. The theory

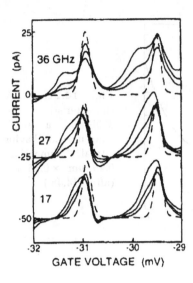

Fig. 14.23: Current vs. gate voltage during irradiation with microwaves at frequencies of 17, 27, and 36 GHz. The dashed curves are the characteristics without microwaves while the solid curves are with increasing microwave power. A photon-induced shoulder is observed which increases linearly with frequency but is independent of power (after Kouwenhoven et al. [14.61])

of quantum dots has first been presented by Efros and Efros [14.57] and Brus [14.58]. The optical susceptibility of quantum dots has been calculated by H. Haug and S.W. Koch [14.59] in a density matrix approach which in contrast to the usual wavefunction formalism allows to include phenomenological relaxation processes. Since the calculations of the third-order susceptibility involves very long algebraic manipulations, the authors restricted themselves to the first-order susceptibility while the former has been treated by Hu et al. [14.60].

We consider now dc transport through quantum dots in the presence of microwaves of a frequency $\nu \geq k_B T/\hbar \approx T \times 20.8$ GHz /K (Sect. 4.14) known as *photon assisted tunneling* (PAT) [14.61], Fig. 14.23. Under this condition, single photon processes can be observed. Another quantity of interest is the time Γ required for tunneling of an electron through the barrier ($\approx$ 2ps) depending e.g. on the source-drain voltage.

In Fig. 14.23 the current vs gate voltage is plotted for a quantum dot which is irradiated with microwaves of a frequency between 17 and 36 GHz at varies power levels. The dashed curves represent data obtained without microwaves. There is a photon induced shoulder in the curves at a position which is independent of the power but linearly dependent on the frequency. This is taken as a proof for a single-photon effect.

In recent experiments a quantum dot is required to be highly isolated from its surrounding. The system can then be charged only by optical excitation of electrons from deep levels. Such a system is investigated either by applying FIR spectroscopy or by voltage-capacity measurements. The structure may e.g. be a GaAs/GaAlAs heterostructure where the electrons are confined in the GaAs at the interface, forming a 2-DEG. The effective diameter of the 2-DEG system is less than the geometrical size because some of the electrons contributed by the donors are trapped by surface states on the side walls and reject the mobile electrons. The top metal plating which is semi-transparent for FIR radiation, serves as a gate electrode, with an insulating photoresist underneath. The gate voltage is applied between the gate and a semi-transparent contact layer in the semi-insulating GaAs substrate. For observing a detectable signal, dot arrays with 10^8 nearly identical dots covering an area of about $10\,mm^2$ are required. For noise reduction, experiments are performed at a temperature of 4 K.

Recently quantum dot arrays have been applied as the active medium of lasers. This arrangement may be thought of a (multi-) quantum well laser diode. Following Marius Grundmann [14.62] we will discuss briefly the principles of multi-level laser arrangements.

"Laser diodes" are bipolar devices which rely on interband-transitions between conduction-band and valence-band states (electrons and holes), typically used for infra-red, visible, and UV applications. They consist of a p-n diode with the active medium in the depletion layer. (Quantum) "Cascade" lasers consist of a unipolar tunneling structure, rely on intra-subband transitions (in case of quantum dots: on inter-sublevel transitions) which typically involve only conduction band states. There two major types: (i) edge emitting lasers, and (ii) vertical cavity surface emitting lasers (VC-SELs). In edge emitters a Fabry-Perot cavity is formed by cleaved facets on opposite sides of the chip. The facets can be as-cleaved (with a reflectance R of typically 0.3) or additionally coated for high or low reflectivity. In the vertical (growth) direction a wave-guide is formed by a heterostructure (or graded layers) and the corresponding profile of the refractive index. The laser emission originates from the side of the chip, parallel to the growth surface. In a VC-SEL the light is emitted vertically to the growth plane. The active medium is placed within a microcavity, formed by two highly reflective mirrors, typically realized using dielectric Bragg mirrors (λ/4 stacks).

By the application of a magnetic field B up to 5 T, additional information is obtained. Because of the small effective mass m of electrons in GaAs, the cyclotron energy $e\hbar B/m$ is comparable to the confining energy in the dot. Figure 14.24 shows the variation of the resonant IR wave number with the external magnetic field strength. For $B = 0$ there is only one resonance line at about 30 cm^{-1} corresponding to 330 μm wavelength or 3.7 meV photon energy. This came as a surprise because at the gate voltage chosen about 50 electrons formed the charge in the dot, and should not the resonant frequency depend on the charge? However, *Kumar* et al. [14.63] demonstrated that the electrostatic fields acting on the 2-DEG define the equilibrium position of these electrons in the dot center and that any electron displacement in the plane perpendicular to the dot axis obeys Hooke's law with a characteristic frequency Ω_0. The corresponding potential is parabolic. The application of a magnetic field characterized by the cyclotron frequency ω_c produces dipole-allowed transitions with Zeeman split

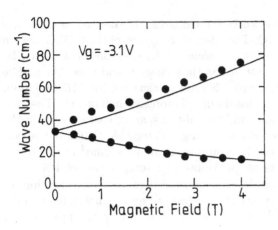

Fig. 14.24: Spectral resonance vs magnetic field strength of electrostaticly defined quantum dots on GaAs. Curves are calculated from (14.56). At the applied gate voltage of −3.1 V there are still about 50 electrons in each quantum dot (after Lorke et. al. [14.64])

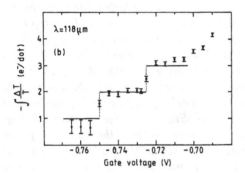

Fig. 14.25: Integrated absorption strength at $\lambda = 118$ μm wavelength at various gate voltages V_g for a quantum dot array with period 200 nm, for a series of spectra. The steps indicate the incremental occupation of the dots with 1, 2, and 3 electrons per dot (after Meurer et. al.[14.65])

energies:

$$\Delta\varepsilon^{\pm} = \hbar^2\Omega_0^2 + (\hbar\omega_c/2)^2 \pm \hbar\omega_c/2 \ . \tag{14.5.15}$$

The two signs refer to the two circular polarizations of light relative to **B**. The calculation leading to (14.5.15) is due to *Fock* [14.66]. Spin splitting is neglected.

It is interesting to compare this result with the classical equation (11.12.13) where, however, the density of carriers influences the resonance via the plasma frequency. The independence of the quantum result from the number of electrons is known as the generalized Kohn theorem [14.67] which says that the cyclotron resonance frequency in a translationally invariant system is not affected by electron–electron interactions. This makes clear why spectroscopy on parabolically confined quantum dots is much less specific than it is in ordinary atoms: one cannot determine the number of electrons in a dot from its resonant frequency. The number can, however, be determined from the absorption behavior for **B** = 0. The experimental result is shown in Fig. 14.25 [14.65]. The absorption increases stepwise upon increasing the gate voltage. This is interpreted as an increase in charge of each dot by one electron at the same gate voltage. The solution of this puzzle lies in the large charging energy required to put another electron into the dot. A rough estimate may be obtained classically. The square of the extent of the wavefunction in a harmonic oscillator representing an electron in the dot is about $(50 \text{ nm})^2$. From this and the distance between the gate and the back contact one estimates an extremely small capacitance of 5×10^{-18} F and a charging energy $e^2/2C \approx 15$ meV. We take from Fig. 14.25 a change in gate voltage ΔV_g for one step of about 30 mV. The capacitance C assuming one step for one electron equals $e/\Delta V_g$ and the energy increase $e^2/2C = e\Delta V_g/2 = 15$ meV which is the same as obtained from the capacitor geometry. Hence, the assumption seems to be correct. The injection or extraction of an electron means that the dot cannot be totally confined but that the confining barriers are made a little "leaky". That is, one must allow a little – but not too much – electron tunneling through the dot.

Interesting experiments to do this without having to produce the quantum dot by etching, etc. were published by *van Wees* et al. [14.68] after the pioneering work done by *Thornton* et al. [14.69] and *Zheng* et al. [14.70]. Two point contacts are simply defined by electrostatic depletion of the 2-DEG in the GaAs/GaAlAs heterojunction underneath a gate. Ballistic transport is achieved if the constrictions have a width w and length l much smaller than the mean free path l_e of the electrons. This requires cooling of the device below 1 K. The electron density was 3.56×10^{11} /cm^2 and the electron mobility at 0.6 K was 8.5×10^5 cm^2/Vs. The split metal gate on top of the heterostructure is illustrated by the inset of Fig. 14.26 where $w = 250$ nm and $l= 1$ μm are indicated. At a gate voltage V_g of –0.6 V the 2-DEG underneath the gate is depleted and the point contacts have their maximum width. By further decreasing V_g, the width of the point contacts can be gradually driven into pinch-off at $V_g = -2.2$ V. Figure 14.26 shows a plot of the conductance vs gate voltage, calculated from the resistance measured at 0.6 K, after subtraction of a lead resistance of 400 Ω . There are plateaus at integer multiples of $2\,e^2/h$. Notice that no magnetic field was applied. At 4.2 K the plateaus have nearly disappeared. Similar results have been found by *Wharam* et al. [14.71]. Concerning the theory of this effect we follow *van Wees* et al. [14.68], *Kouwenhoven* et al. [14.72], and *Beenakker* [14.73] in a simplified version. The current $I = dQ/dt$ changes the charge Q of a capacitor in a continuous way since Q may be an arbitrarily small fraction of an elementary charge e caused by a small shift of the electrons relative to the positive ionic background. Tunneling through a junction, however, results in a sudden charge or discharge by e. The change in Coulomb energy $Q^2/2C$, where C is the junction capacitance, by a tunneling event is

$$\Delta\varepsilon = Q^2/2C - (Q - e)^2/2C = e(Q - e/2)/C \ . \tag{14.5.16}$$

At $T = 0$ tunneling can occur only if $\Delta\varepsilon > 0$ which implies a *Coulomb blockade* for tunneling for $Q \leq e/2$. The current–voltage characteristic shows zero current for $-e/2C < V < e/2C$. The current source charges the capacitor until the threshold charge $e/2$ is reached. Then a tunnel transition occurs, the charge is now $Q = -e/2$, and a new charging cycle begins. This

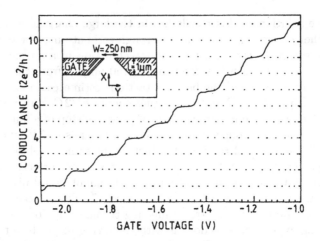

Fig. 14.26: Quantized conductance of a quantum point contact at 0.6 K. The conductance was obtained from the measured resistance after subtraction of an additional series resistance of 400 Ω. Point contact layout (after van Wees et al.[14.68])

repeated sudden charging is called *single electron tunneling* (SET) oscillation of the voltage with the fundamental frequency I/e. Deviations from this relation arise from finite-temperature effects, electron heating, co-tunneling, and moving background charges. In co-tunneling events an electron tunnels to the dot while a second electron simultaneously leaves the dot across the tunnel junction on the opposite side. Also the above analysis assumes that the tunnel junction is independent of its electromagnetic environment which, however, has an impedance of the order of the impedance of free space which is about 377 Ω. For an electron energy required from the voltage source to join the electron with the dot, the impedance should be much larger than the resistance quantum $R_K \approx 25.8 \, k\Omega$. Only then a Coulomb blockade of tunneling may be observed [14.74]. The reference to R_K takes care of the fact that in ballistic transport Ohm's law is no longer valid and electron drift between two reservoirs obeys the laws of electromagnetic wave transport in a waveguide with the electromagnetic wave replaced by the deBroglie wave of the electron, roughly speaking. Ballistic transport means that there are no phase breaking collisions during the electron motion between the reservoirs and therefore the usual relaxation time approximation in the Boltzmann equation is not applicable. *Landauer* [14.75] and later *Büttiker* [14.76] developed such a theory of conduction based on transmission and reflection of the deBroglie waves at entrance and exit ports of the reservoirs. We take the shortest derivation for the simplest case here only [14.77]. The product of the group velocity v_n and the 1-DEG density of states $g_n = \pi^{-1}$ $(d \, \varepsilon_n/dk)^{-1}$ is the current per unit energy interval. From the text after (2.1.20) we take $v_n = \hbar^{-1} d\varepsilon_n/dk$ and obtain $(\pi\hbar)^{-1}$ for the product which is the same for each subband $n = 1, 2, 3 \ldots N$. The sum over all subbands yields a factor of N therefore. The electrical current is the electron charge e times the particle current and therefore $N e(\pi\hbar)^{-1}$ times the difference in quasi-Fermi energies between both sides which is equal to eV, where V is the applied voltage.

Hence, the conductance $G = I/V$ is given by

$$G = 2e^2 N/h \quad N = 1, 2, 3, \ldots \tag{14.5.17}$$

For simplicity, the transmission was tacitly assumed to be 100% and the reflection as zero. The geometry of the waveguide does not enter. This quantum type of conduction of electric charge is called *mesoscopic* because the size of the device elements is smaller than macroscopic but larger than atomic (microscopic).

Let us now consider the effect of a magnetic field B on mesoscopic conduction in a ring. B is perpendicular to the plane of e.g. a semiconducting ring (diameter 1.9 μm in the 2-DEG of a GaAs/GaAlAs heterojunction, mobility 10^6 cm^2/Vs, scattering lengh 5 μm, temperature 300 mK) [14.78]. The current enters the ring at one side, divides equally for the two ring halves, and leaves at the opposite side. At a variation of the magnetic field a Shubnikov-de Haas oscillation of the resistance is observed with a period - and that makes it different from Shubnikov-de Haas - of h/e, if we take the Fourier transform of the ring resistance oscillation with the magnetic field strength . From this periodicity it is even possible to deduce the ring diameter which agrees within 2% with the actual lithographic diameter. The effect is called *Aharonov-Bohm effect*, after Y. Aharonov and David Bohm who predicted it [14.79] before it was first observed in metallic rings [14.80]. From (11.2.13) we notice that the magnetic field enters the Schrödinger equation in the form of the vector potential A. Neglecting an A^2 term for a non-relativistic theory and solving for the wave function $\psi(r)$, we notice that the A-dependence in ψ goes by a phase factor

$$\exp\left[-ie\hbar^{-1}\int (Ads)\right] = \exp(-i\,2\pi\,\Phi/\Phi_0) \tag{14.5.18}$$

where $\Phi = \int(A\,ds)$ is the magnetic flux and $\Phi_0 = h/e = 4.136 \times 10^{-15}$ Vs is the flux quantum. The phase of the wave traveling halfway around the ring inside which the flux is Φ, changes by $\pi\Phi/\Phi_0$ while the equivalent for the wave traveling around the other half of the ring changes by the negative of the above. After interference the sum of the two partial wave amplitudes is proportional to $\exp(-i\pi\Phi/\Phi_0) + \exp(i\pi\Phi/\Phi_0) = 2\cos(\pi\Phi/\Phi_0)$ and the intensity proportional to the square which has a component $\cos(2\pi i\Phi/\Phi_0)$. Hence, the frequency is the inverse of Φ_0 which is e/h. Oscillations as large as 50% of the total resistance were observed, with the period independent of the average resistance. More than 10^3 periods with no apparent attenuation in the amplitude were registered. In one of the first experiments polycrystalline gold in a ring of diameter 825 nm and width of 41 nm was investigated at 60 mK temperature [14.81]. This means that the Aharonov-Bohm effect is not particular to a specific material or its dimension (as long as these are comparabel to the electron DeBroglie wavelength), comparable to conditions of the quantum Hall effect, just general to any conductor under mesoscopic conditions with either a 3-DEG or a 2-DEG.

Besides the quantized absorption of electromagnetic radiation discussed so far there is also emission in the form of continuous-wave laser radiation from quantum dots even at room temperature which suggests the possibility of technical applications. The first quantum dot semiconductor laser has been reported in 1994 by Kirstaedter et al. [14.82]. In contrast to this and similar lasers, which were grown by molecular beam epitaxy (MBE), a recent development is based on metalorganic chemical vapor deposition (MOCVD) [14.83]. By this technique, InAs dots as well as binary/ternary dots (BTQDs) have been produced either as single sheets or as vertically stacked sheets. Without going into details of the fabrication, it has been shown that the BTQDs demonstrate much superior laser performance. A threshold current density of 1.3 kA/cm^2 under pulsed mode current injection has been achieved with a fivefold InAs QD stack. The photon energy was 1.188 eV for a threefold–stack. The threshold current was 24 mA for a stripe geometry of 8 μm $\times$ 500 μm. By combining InAs and In$_{0.3}$Ga$_{0.7}$As deposition, this result was achieved. It represents an increase of the luminescence efficiency by a factor of three.

Problems

14.1. Calculate the potential distribution in the direction perpendicular to the surface of a semiconductor with surface states for the case of strong inversion. Assume n-type Ge at room temperature with 10^{12}cm^{-2} positively charged surface states and 10^{15}cm^{-3} donors in the bulk (dielectric constant $\kappa = 16$; $n_i = 3 \times 10^{13}$cm^{-3}). For no applied field, the field strength at the surface $E_s = 10^5$ V/cm. How much are the bands bent (*diffusion voltage*) and what is the width of the inversion layer? Neglect surface quantization for simplicity. *Hints*: introduce the normalized potential $u = e\varphi/k_BT$. For the accumulation layer $0 \leq z'$, consider only holes in the space charge on the rhs of the Poisson equation. Since $p = n_i \exp(-u)$, introduce a Debye length $L_D = (\kappa\kappa_0 k_B T/n_i e^2)^{1/2}$. For the depletion range $z' \leq z \leq L$, consider only the ionized donors, i.e., introduce

$$L_D = (\kappa\kappa_0 k_B T/N_D e^2)^{1/2}. \tag{14.0.1}$$

The boundary conditions are $(du/dz)_L = 0$; $(du/dz)_0 = -eE_s/k_BT$; $(du^{(I)}/dz)_{z'} = (du^{(II)}/dz)_{z'}$ and $u^{(I)}(z') = u^{(II)}(z')$, where $u^{(I)}$ and $u^{(II)}$ are the solutions for the ranges considered.

14.2. Show that the application of an external voltage V perpendicular to the surface of the semiconductor of problem 14.1 is described by replacing n_i by $n_i \exp(-eV/2k_BT)$ in the equations of this problem (*field effect*).

Further Reading:

Supriyo Datta: Electronic Transport in Mesoscopic Systems, Cambridge University Press, (UK), 1997

Thomas Heinzel: Mesoscopic Electronics in Solid State Nanostructures, Wiley VCH Weinheim, (Germany), 2003

Marius Grundmann (Ed.): Nano-Optoelectronics, Concepts, Physics and Devices, Springer Verlag Heidelberg, (Germany) 2002

B. Deveaud, A. Quattropani, P. Schwedimann: Electron and Photon Confinement in Semiconductor Nanostructures, Proc. Int'l School of Physics of the Italian Physical Society, IOS Press Amsterdam, (Netherlands) 2003

15. Miscellaneous Semiconductors

In semiconductor physics, two fields have made a come-back in recent years: amorphous silicon through the discovery of hydrogen saturation of dangling bonds, and organic semiconductors through the manufacture of a p-n junction from a polymer, polyacetylene. For production purposes, amorphous silicon may be of interest for solar cells for terrestrial applications if the efficiency of these cells can be doubled from the present $\approx 5\%$ [Ref. [15.1], Chap. 10]. The doping of some polymers produces both n- and p-type behavior up to metallic conductivities. This led to the discussion of new conduction mechanisms (e.g., solitons [15.2]) and stimulated practical applications, such as a new type of a rechargeable battery.

15.1 Amorphous Semiconductors

Liquids and amorphous solids are similar in that the arrangement of their molecules, at least on a macroscopic scale, is not regular. Some semiconductors, e.g., selenium, remain semiconducting upon melting and are obtained as amorphous solids, i.e., in a vitreous (glassy) state, by quenching (supercooling) the melt. During this process the conductivity does not change appreciably. Amorphous materials are also obtained by condensation from the vapor phase on a cooled substrate, by sputtering, or by chemical vapor deposition. Ion implantation can produce an amorphous surface layer on a crystal.

The electronic properties of amorphous materials depend strongly on the presence of localized states in the energy gap. Figure 15.1 shows the density-of-states functions $g(\varepsilon)$ for amorphous silicon (a-Si) prepared in two different ways: the upper curve was obtained from a thin film on an insulating substrate which was achieved by the condensation of silicon vapor. The lower curve represents material produced by a glow discharge. The vertical short lines indicate the typical Fermi level position for an undoped material [15.4]. For a-Ge, the conductivity is plotted vs. $T^{-1/4}$ in Fig. 15.2.

Since the density of gap states is high as shown in Fig. 15.1, linear behavior in a $\log \sigma$ vs $T^{-1/4}$ plot as evidence for variable-range hopping is to be expected [15.6]. Further evidence for hopping is also indicated by the frequency depen-

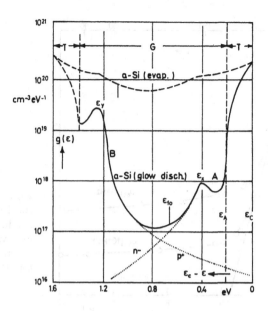

Fig. 15.1: Density-of-states function for two types of amorphous Si. Upper curve: Sample obtained by condensation from the vapor phase. Lower curve: Sample made by a rf-type glow discharge decomposition of silane on a substrate heated to 520 K. Measurement by field effect. T = tail states; G = gas states; A: acceptor-like; B: donor-like (after [15.3])

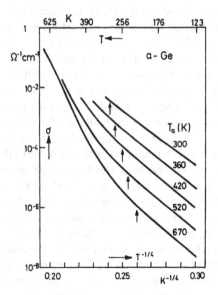

Fig. 15.2: Temperature dependence of conductivity for a-Ge (preparation: evaporation on a quartz substrate held at 300 K; annealing for 15 min at temperatures T_a indicated) (after [15.5])

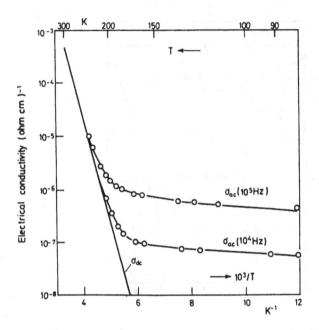

Fig. 15.3: For a-As$_2$Te$_3$, the ac and dc conductivity are plotted vs the reciprocal temperature (after [15.7])

dence of the ac conductivity. Figure 15.3 shows the temperature dependence of the ac conductivity for an a-As$_2$Te$_3$ film for two frequencies, together with the dc conductivity. At low temperatures the ac conductivity is much higher than the dc conductivity and proportional to T. The low-temperature ac conductivity increases almost linearly with the frequency, at least up to 10^{10} Hz. One of the main methods of investigating a band structure is the observation of fundamental absorption. Figure 15.4a shows a comparison of the reflectivities for amorphous and crystalline germanium while the imaginary parts of the dielectric constant are compared in Fig. 15.4b. The sharp reflectivity maximum at 4.5 eV is completely washed out in the amorphous state while the 2 eV peak is considerably broadened. Since the 4.5 eV peak is due to a transition on the ⟨100⟩ axis in the Brillouin zone while the nearest neighbor in the diamond lattice is found in the ⟨111⟩ direction, it is concluded that the long-range order is strongly disturbed (⟨100⟩ and neighboring directions) while the short-range order is only weakly disturbed (⟨111⟩ direction). As indicated by Fig. 15.4b where κ_i is proportional to the absorption coefficient in the amorphous state, the optical absorption starts at somewhat lower frequencies than in the crystalline state. This is not due to transitions between localized states in the band gap. In a model suggested by Fritzsche [15.9], the localization of carriers occurs in potential troughs of the strongly irregular bands. Since the irregularities of the valence and conduction bands are such that the gap energy is the same at all positions in the crystal, the optical absorption is independent of these irregularities. Of course it is highly questionable if the band model developed for

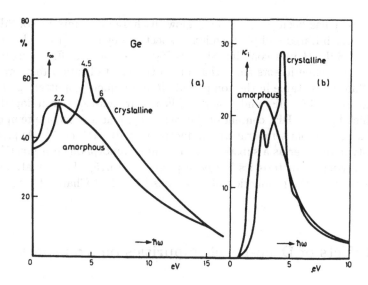

Fig. 15.4: (a) Reflectivity and (b) imaginary part of the dielectric constant, $\kappa_i = 2nk$, of crystalline and amorphous Ge (after [15.8])

the periodic arrangement of molecules in the crystalline state is applicable to the highly irregular arrangement in the amorphous state.

In the chalcogenide glasses, structural changes occur during irradiation with high-energy photons at a high flux level [15.10]. This has been observed as changes in the optical band gap, refractive index, X-ray diffraction, chemical reactivity, and film thickness. The effect can be reversed by annealing.

Hydrogenated amorphous silicon (a-Si:H) was first investigated by Chittik et al. in 1969 [15.11, 12]. It is produced by a glow discharge in silane at substrate temperatures below 100°C. It contains SiH, SiH$_2$ and SiH$_3$ groups and possibly a polymer (SiH$_2$)$_x$, as concluded from infrared studies. Annealing at temperatures below 600°C produces nearly hydrogen-free a-Si. Reabsorption of atomic hydrogen occurs at 500°C, e.g., by exposure to a hydrogen plasma [Ref. 15.1, p. 199]. As seen from Fig. 15.1, the density of gap states is much smaller in glow discharge a-Si than in vapor deposited material. The former material may contain mostly point defects while the properties of evaporated or sputtered material may be dominated by microvoids containing many dangling bonds. Doping of a-Si:H is done by mixing either diborane (for p-type) or phosphine to silane before the glow discharge. For 10% mixtures, room temperature conductivities of $10^{-2}\Omega^{-1}\text{cm}^{-1}$ are obtained, while for undoped a-Si:H, values of about $10^{-9}\Omega^{-1}\text{cm}^{-1}$ have been found, the latter being defect controlled. A minimum conductivity of $10^{-12}\Omega^{-1}$ cm^{-1} is measured for a ratio of 10^{-4} for B$_2$H$_6$/SiH$_4$. By adding fluorine to a-Si:H, a room temperature conductivity of $\approx \Omega^{-1}\text{cm}^{-1}$ has been observed with phosphorus doped material ([15.13] and [Ref. 15.1,

Chap. 9]). It has been prepared by a glow discharge in a mixture of SiF_4, H_2, and PH_3. By infrared and photoemission spectra, Ley et al. [15.14] have shown that the a-Si:F, H film contains Si-F, Si-F_2, Si-F_3, and Si-F_4 configurations. Gas evolution experiments show that up to the crystallization temperature of 600°C, fluorine is stable in the compound. Radiofrequency sputtering of silicon in a mixture of SiF_4, BF_3 and Argon yields a hydrogen-free boron doped a-Si:F film, which for a 5% BF_3 content of the gas mixture at a substrate temperature of 550 K, has a room temperature conductivity of $10^{-3}\Omega^{-1}cm^{-1}$. Applying a 2 Ω-cm n-type Si wafer as a substrate, a crystalline-amorphous p-n junction with good rectifying characteristics has been prepared [15.15]. These materials may be of interest for the production of solar cells [Ref. 15.1,Chap.10].

15.2 Effects of Deep-Level Impurities on Transport

We have seen deep impurity levels to arise from transition metals. Typical cases are, e.g., gold in germanium (Fig. 3.7) or chromium in gallium arsenide. Their effects on photoconductivity have been dealt with in Sect. 12.2. They do, however, influence transport phenomena quite often in such a way that the usual assumption of a homogeneous distribution of the electric field along a filamentary sample is incorrect. Trapped charges are introduced, particularly so at low temperatures, which prevent steady-state conditions to be restored within a reasonable time interval after injection. If the resistivity is very large the dielectric relaxation time τ_d can be much larger than the carrier lifetime. This state is referred to as the "relaxation case", in contrast to the usual "lifetime case" of low-resistivity semiconductors.

    Queisser et al. [15.16] observed the current-voltage characteristics of an oxygen doped GaAs p^+ -n junction with the p^+ contact being produced by diffusion of zinc. Forward ("f") and reverse ("r") currents are plotted in Fig. 15.5. Oxygen in GaAs acts as a deep-level impurity. The carrier concentration at room temperature is only 3×10^7 cm^{-3}, the mobility 4500 cm^2/Vs. The dielectric relaxation time is 10^{-4} s, which is large compared with a lifetime of less than 10^{-8} s. The forward and reverse currents are equal at low voltages. Here the main resistance contribution comes from the space charge region next to the p-n junction. At a higher reverse bias the space charge region widens because of majority carrier depletion, and the slope of the curve decreases. When the entire sample is depleted, a linear and finally a superlinear increase (3 r and 4 r in Fig. 15.5) are observed as a result of a space-charge limitation of the current. In the forward direction injected holes deplete the *n*- region by recombination with electrons. As a consequence, there is a sublinear current rise with increasing voltage (2 f). At even higher current values (4 f) double injection occurs with the formation of a negative differential resistance region.

    An interesting phenomenon studied in thin epitaxial layers of GaAs on Cr-

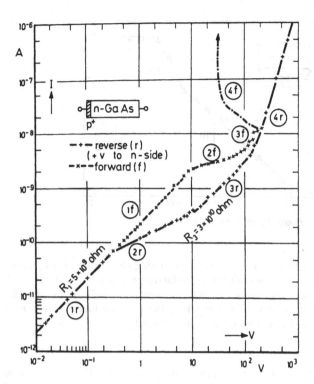

Fig. 15.5: Current-voltage charac-teristics (forward and reverse) of a relaxation case semi-conductor: oxygen doped GaAs with a p^+n junction at room temperature (after [15.16])

doped semi-insulating GaAs substrates is "persistent photocurrent": a photoin-duced conductivity increment persists *after* the illumination, often with immea-surably long time constants [15.17, 18]. Hall-effect measurements show a linear increase of carrier concentration $n_\square$ per cm^2 (layer thickness: 0.3 μm) with the photon dose (from 10^{12} up to 10^{16} photon/cm^2) with light of 815 nm wave-length, which just can create pairs across the band gap. The phenomenon is explained by assuming that a fraction of the pairs is separated in the potential barrier between the substrate and the layer, the holes are immobilized at deep Cr traps in the substrate, and the recombination after the illumination is pre-cluded. Above a dose of 10^{16} photons/cm^2 the carrier concentration becomes constant. The mobility which is determined by ionized impurity scattering, is increased up to 9 % with increasing dose, falling to a 3 % increase at saturation of $\Delta n_\square$. From the Brooks-Herring theory (Sect. 6.3) the 3-dimensional carrier concentration has been estimated and found to increase by far not as much as the area density $n_\square$. The conclusion is that most of the persistent photocurrent is caused by a widening of the conductive layer at the expense of the adjacent nonconductive space-charge region [15.17].

In another experiment two epitaxial layers, E and S_1, were on top of one another, with S_2 being the substrate as before (Fig. 15.6, inset). The layers were 0.5 and 0.2 μm thick, respectively. S_1 was Cr doped just as the substrate

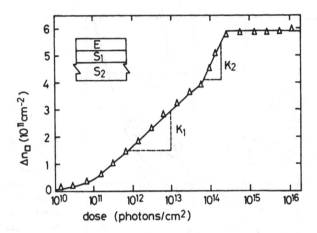

Fig. 15.6: Increase of the density per unit area of persisting electrons in the epitaxial n-GaAs layer E measured in the dark after a cumulative photon dose ($\triangle$). The full line represents a theoretical curve with two straight sections of slope K_1 and K_2, respectively. *Inset*: Sample cross section indicating the semiconducting layer E, a semi-insulating layer S_1, and the semi-insulating substrate S_2 (after [15.19])

but with a smaller concentration. The curve relating $\Delta n_\square$ and the photon dose shows two straight regions with different slopes K_1 and K_2. At the beginning of illumination holes are trapping at Cr traps in the layer S_1. When the layer S_1 has been filled up, holes have to find Cr traps in the substrate with its higher Cr concentration (hence the steep slope of the curve). Obviously the method is valuable for finding depth profiles of deep traps in high-resistivity substrates, particularly since it is non-destructive [15.19].

Corresponding observations have been made with silicon. A thin layer at the surface was sulfur diffused. Sulfur acts as a double donor with a level 0.32 eV below the conduction band edge for the singly ionized atom and a level of 0.59 eV below the conduction-band edge for the doubly ionized atom. The photo-generated electrons are trapped in the surface layer while the holes remain mobile in the p-type bulk. At the low temperature of the experiment (45 K) silicon is very photosensitive because of carrier freeze-out. After a total dose of 10^{18} photons/cm^2 at a wavelength of 638 nm, the concentration $p_\square$ is more than doubled and the mobility almost triplicate. The slopes of the curves are increasing with the dose since the sulfur concentration increases with distance from the junction [15.20].

Space charges due to deep traps particularly in thin layers, on surfaces or at interfaces with a high resistance and/or a large dielectric constant where the dielectric relaxation time is large, can easily lead to a misinterpretation of experimental results. While in some materials it yields a persistent photo-conduction, it may in others cause a reduced Hall voltage or dose-dependent photovoltaic effect (e.g. Staebler-Wronski effect in amorphous-silicon photo-voltaic cells [15.21]). Useful applications are charge-coupled devices (Sect. 5.7) and memories.

15.3 Organic Semiconductors

For many years one of the best known organic semiconductors had been an-thracene $C_{14}H_{10}$ consisting of three benzene rings in series [15.22]. It had been purified well enough that even cyclotron resonance could be observed at a tem-perature of 2 K resulting in a hole effective mass of 11 m_0 and a scattering time of about 10^{-10} s, depending somewhat on the particular sample [15.23]. Almost all organic semiconductors may actually be classified as insulators.

Somewhat later, polyacetylene $(CH)_x$, a polymer known for a long time, was doped with, e.g., AsF_5 resulting in a room temperature conductivity of the order of $10^3 \, \Omega^{-1}cm^{-1}$ [15.24]. This is the highest known conductivity of any organic material at room temperature. By doping, a range of 12 orders of magnitude has been covered [15.25]. The thermopower of highly doped material is like that of a metal, with a value of 10 $\mu V/K$ at room temperature and strictly proportional to T between 4 K and 300 K. In the optical reflectivity spectrum, there is a plasma edge at about 1 eV which is consistent with the assumption of about 10^{21} carriers per cm^3. The Hall effect is positive for AsF_5 doping which would agree with a reaction $3 \, AsF_5 + 2e^- \rightarrow 2AsF_6^- + AsF_3$ where the electrons are removed from the $(CH)_x$ with resulting p-type conductivity. By doping $(CH)_x$ with sodium naphthalide, n-type conductivity is obtained. An n-type sample in contact with a p-type sample has an I-V characteristic typical of a p-n junction except for a hysteresis in the forward direction [15.26]. Depositing a thin layer of iodine doped $(CH)_x$ on n-type silicon yields a solar cell with an open-circuit voltage of 0.395 V, a fill factor of 23% and an efficiency of 1.5% for 57.3 mW/cm^2 incident light at a cell area of 0.1 cm^2. The short circuit current is 1 mA [15.27]. An interesting application is the use of polymers as electrodes in organic batteries [15.28]. A very stimulating subject which we cannot discuss here in detail is the structure of the polymer both on a microscopic (cis and trans) and a macroscopic scale (fibers) [15.25]. It certainly influences galvanomagnetic phenomena even on the order of magnitude [15.29]. Of particular interest is the discovery of superconductivity in organic compounds which are salts of TMTSF [short for di $(\Delta^{2,2}$-bi-4,5-dimethyl-1,3-diselenole)] such as $(TMTSF)_2PF_6$ [15.30]. These and similar materials (e.g., TTF-TCNQ) belong to the group of *one-dimensional* conductors. New conduction mechanisms such as charge density waves and solitons have been considered in this group of materials. Duke and Schein raised the question as to whether energy band theory is enough for organic solids [15.31].

A two-layer photocell has been produced with I-V characteristics shown

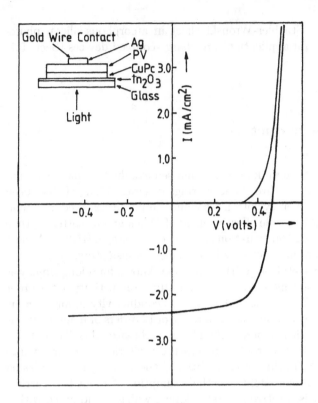

Fig. 15.7: *I − V* characteristics and configuration of ITO/CuPc/PV/Ag diode in the dark (upper curve) and under *Air − Mass* 2 illumination (after [15.32])

in Fig. 15.7 both in the dark and under simulated AM2-illumination which corresponds to a sun-light intensity of 75 mW cm^2 [15.32]. Indium tin oxide (ITO)– coated glass was used as the transparent conducting substrate on which 30 nm of copper phthalocyanine (CuPc) were evaporated. A second organic layer is a perylene tetracarbolic derivative (PV), about 50 nm thick, evaporated on top of the previous layer, and finally an opaque Ag layer evaporated on top of the PV layer. The forward bias direction corresponds to the positive voltage on the ITO electrode. The *I-V* characteristics obey (5.3.28) with an ideality factor (see footnote 3 on the same page as (5.3.28)) of 1.6 ± 0.1. The photovoltaic fill factor (5.9.9) under AM2 conditions is only 65 % which is 10 % less than the one of a silicon solar cell, and it decreases by about 1/3 in the course of a five-day test. Conjugated organic polymers may be applied for the production of large-area light-emitting diodes (LED) in the green-yellow part of the spectrum [15.33]. *M. Schwoerer* [15.34] reported on LED's on the basis of poly-para-phenylene- vinylene (PPV) with an active area of up to 50 cm^2. For emitting in the visible part of the spectrum, a voltage of less than 3 V is required. The structures can also be applied as photodiodes with a quantum efficiency of up to 1 %. The ideality factor is here 2 ± 0.2.

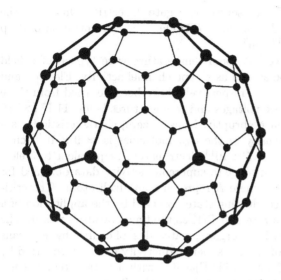

Fig. 15.8: Arrangement of sixty carbon atoms in a cage–like structure forming a fullerene

15.4 Fullerenes

In 1985 the chemists H.W. Kroto, J.R. Heath, S.C. O'Brien, R.F. Curl, and R.E. Smalley discovered another modification of carbon (besides the well known diamond and graphite) called C_{60} or Fullerene (or colloquial Buckyball) [15.36]. Sixty carbon atoms are arranged forming a ball shaped cage structure as shown in Fig. 15.8. Kroto, Curl, and Smalley shared the 1996 Chemistry Nobel Prize for the discovery.

The material is produced by the evaporation of carbon when a strong laser pulse hits the carbon surface in a helium atmosphere and the gas is expanded in a vacuum chamber and cooled to a temperature of a few degrees K. Besides C_{60} with its 60 carbon atoms, there are also C_{70} molecules although with a smaller abundance. Single crystals of several mm^3 were grown. It was also possible to place metal atoms inside these structures. A modification of the ball-type structure are tubes. The name Fullerene was coined by comparing the half-cage structure with a dome-shaped roof designed by the architect Buckminster Fuller at Montreal.

Undoped C_{60} has a band gap of ≈ 2 eV and is therefore insulating. After doping with three alkali metal atoms per C_{60} molecule, it exhibits metallic conductivity and, at low temperatures, is superconducting. The inter-molecular bond is of the van der Waals type and thus dangling bonds are absent. The Fermi level can be easily shifted across the band gap leading to n-type as well as p-type conductivity. Source and drain contacts were produced on the single crystals by gold evaporation through a shadow mask [15.36]. Sputtered Al_2O_3

with a capacitance of 185 nF/cm^2 served as a gate dielectric. On top of the
dielectric, gold was evaporated as a gate electrode. The concentration of deep
levels was estimated as $3 \times 10^{12}/cm^3$.

The characteristics observed at room temperature were typical of a field
effect transistor but for n-type as well as p-type channel activity. Electron and
hole mobilities of 2.1 and 1.8 cm^2, respectively, have been deduced from these
data. At very high positive gate voltages and temperatures below 11 K (> 180
V at 5 K) the channel resistance abruptly drops to zero, characteristic for superconductivity. The area density of the C_{60} molecules is $\approx 9 \times 10^{13}/cm^2$.
The gate charge corresponds to 2.7×10^{13} electrons/cm^2, equivalent to 3 electrons per molecule. This is equivalent to superconductivity data obtained for
chemical doping in bulk A_3C_{60} where A represents an alkali atom. The critical temperature for the superconducting state observed by the application of a
magnetic field B decreases at a rate of 5 T/K as the temperature is raised. The
appearance of superconductivy in an organic material is, of course, by no means
trivial. The first observation of organic superconductivity has been reported by
Jerome et al. [15.30] in 1980 for $(TMTSF)_2ClO_4$ with a critical temperature
of $T_c = 1.25$ K where TMTSF stands for Tetramethyl-Tetra-Thiofulvalinium.
Its room temperature conductivity is about $10^3(\Omega \; cm)^{-1}$ in the chain direction
with lower values in the other directions.

15.5 Carbon Nanotubes

After the invention of the ball - like fullerenes, research in this field was stimulated by the detection of carbon tubules with diameters of the order of one
nanometer (10^{-9} m) which are now called nanotubes. The length is of the
order of up to 700 mm. The first observations made with transmission electron microscopy (TEM) showed coaxial tubes with a hollow core [15.37]. The
typical interlayer distance is 0.34 nm. Scanning tunneling microscopy (STM)
and atomic force microscopy (AFM) [15.38] have been applied for investigations of the electronic density of states and the elastic properties of these tubes,
respectively. Russian scientists [15.39] produced carbon tubules and bundles,
also called ropes with length-to-diameter ratios of 10 or less which they called
barrelenes.

The production methods are similar to those for fullerenes, i.e. by an arc
between two graphite rods in helium gas at a pressure of about 200 to 500 torr,
employing a dc current of 50 to 100 amperes and a voltage of 20 to 25 V at a
rod distance of 1 mm. The 6 to 7 mm diameter positive electrode (anode) is
consumed while on the 9 to 20 mm diameter movable negative electrode (cathode) the deposit forms mostly on its inner region at an estimated temperature
of 2500 to 3000°C. The deposit consists of bundles of at least 10 to 100 aligned
nanotubes of about the same length (Fig. 15.9). The growth chamber must, of

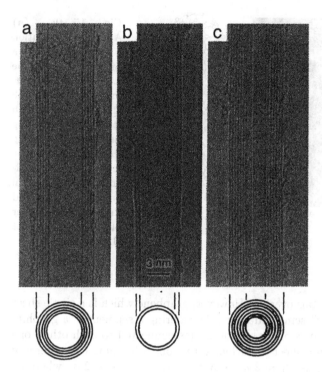

Fig. 15.9: The observation of N coaxial carbon tubules with various inner diameters d_i and outer diameters d_0 reported by Iijima using TEM: (a) $N = 5$, $d_0 = 6.7$ nm, (b) $N = 2$, $d_0 = 5.5$ nm, (c) $N = 7$, $d_i = 2.3$ nm, $d_0 = 6.5$ nm (after Iijima [15.37])

course, be cooled adequately for an optimal number of tubules. For more details of this process as well as elastic and electric properties including superconductivity see [15.40]. Carbon nanotubes (CNs) combine nanometer size with particular electrical properties and can be both semiconducting and metallic. The nanotube mantle is built out of a hexagonal arrangement of carbon atoms known as "honeycomb lattice". A tube consisting of a single shell is called "single wall carbon nanotube" or simply SWNT. The multi wall nanotube, MWNT, consists of multiple concentric shells. The electronic properties have been reviewed by Charlier et al. [15.41] and their growth by [15.42].

Bundles of nanotubes have diameters of several nanometers. The most efficient process to produce SWNTs is laser ablation developed at Rice University [15.43]. The chirality[1] of the tubes cannot be controlled by means of temperature during the growth process.

CNs are discussed as being extremely sensible detectors because their conductivity can be influenced significantly by their operation in various gas environments. Graphite has a metallic character because between different planes of

---

[1] E.g. the fullerene C_{76} has a chiral structure [15.44]. In general see Ernest L. Eliel. Samuel H. Wilen: Stereochemistry of Organic Compounds, Wiley Interscience, New York, 1994

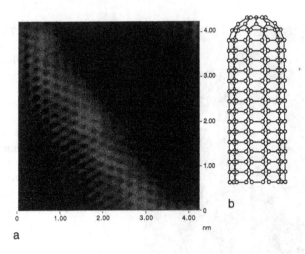

Fig. 15.10: (a) Atomic resolution STM image of a carbon nanotube 3.5 nm in diameter. In addition to the atomic honeycomb structure, a zigzag superpattern along the tube axis can be seen. **(b)** A ball-and-stick structural model of a C_{60}-based carbon tubule. (after M.S. Dresselhaus et al. [15.40])

carbon atoms, the π-orbitals overlap. However, graphene which is a single sheet of graphite, is a "gapless" semiconductor. Considering a segment of a graphite sheet which consists of rings of six carbon atoms connected to each other, one could think of rolling up a sheet in various directions, obtaining circumference vectors of the tube such as $C_1 = 3\,a_1 + 3\,a_2$ or $C_2 = 4\,a_1 + 2\,a_2$ where a_1 and a_2 are unit vectors of equal length but differently oriented. Any kind of (n, m) tubes with n and m being natural numbers such as $(3, 3)$ or $(4, 2)$ in the example just mentioned define a tube structure [15.42]. Three electrons per carbon atom form the covalent bonds in the carbon plane, one electron per atom may fill the π-band. In the form of a tube, rather than a sheet, there is a selection of allowed k-values according to $(k \cdot C) = 2\pi \cdot i$ with i being an integer parameter. For $n - m = 3i$ carbon nanotubes are metallic, for $n - m \neq 3i$ they are semiconducting with gaps of several tenths of an eV. The metallic character is true actually only for $n = m$ because of the curvature of the sheet in the cylindrical shape of a tubule. Otherwise an energy gap of about 0.6 eV exists at a tubule diameter of typically 1.4 nanometer which for room temperature T is to be compared with $k_B T \approx 25$ meV. For scattering events, phase space is too small and electron transport is probably ballistic over a length of several micrometers at room temperature. Metallic nanotubes allow current transport up to several μA without breakdown.

A problem are electrical contacts to nanotubes. The presence of metal contacts to a tubule, even if it is metallic, may produce barriers inside [15.45]. Applying both top- and bottom-contacts to a bundle (≈ 100) of metallic tubules might perhaps solve the problem [15.46] and the sample resistivity may be ten times smaller than that of copper wires which are used in chips. In 1998 at Delft University [15.48] and at IBM [15.49] scientists applied a single-wall nanotube

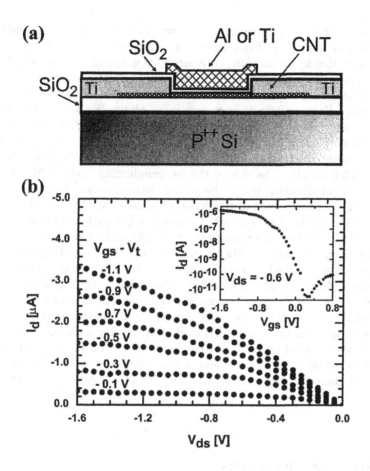

Fig. 15.11: (a) Schematic cross section of the top-gated CNFET. (b) Output characteristic of a p-type device with a gate oxide thickness of 15 nm. The inset shows the corresponding transfer characteristics. (after S.J. Wind et al. [15.47])

for the channel of a field effect transistor as it is schematically displayed in Fig. (15.11). On a heavily doped silicon substrate covered with an insulating silicon dioxide layer of 140 to 300 nm thickness, carbon nanotubes were dispersed and gold or platinum electrodes with a separation of between 300 nm and 500 nm were applied as source and drain. For simplicity the tubule is sketched as being located underneath the contacts. By modulating the gate field, the conductance of the nanotube which serves as a channel, can be varied by modulating the gate field by more than five orders of magnitude. Currents of μA can be driven through the device for sufficiently negative gate voltages to the p - type carbon field effect transistor (CNFET). Obviously Schottky barriers exist and play a dominant role for carrier transport through a metal/ nanotube/metal system [15.50]. For a sufficiently negative gate voltage, a p - type CNFET is in the on - state. However, although the characteristics of CNFETs resemble those of conventional MOSFETs, the physics behind are significantly different. Today's nanotube transistors behave as ultra - thin body Schottky barrier CNFETs [15.50; 15.51]. The size of the semiconducting body does not only influence the field situation inside the actual channel region but is also very important for the exact shape of the Schottky barriers at the contact/channel interface [15.52]. In the case of a carbon nanotube, the extremely small diameter reduces the Schottky barrier thickness drastically and allows for a significant current flow from the source to the drain in the CNFET on - state. Since the characteristic is not related to a scattering inside the channel as in the case of a conventional MOSFET, it is not easily possible to extract properties such as e.g. mobility from CNFET characteristics. The performance of today's CNFETs is determined by the response of the Schottky barriers at the contacts. In a geometry where a zirconium ZrO_2 oxide layer having a dielectric constant of about 25, is deposited on top of the contacted nanotube and a metal gate is aligned on top of the device, which allows gate biases of the order of one volt, the electrical characteristics as shown in Fig. 15.11 are really excellent (S.J. Wind et al. [15.47]).

15.6 Molecular Electronics

The development of new experimental tools in nanotechnology raised the question of producing electrical contacts on the few-nanometer scale and, as a consequence, of applying molecules as basic components of electronic devices. Long molecules in the form of thin films are available in organic chemistry. The best investigated covalent link between an organic molecule and a metal as a contact is the bond between a thiol (sulfur) group on the molecule and a gold substrate. Gold is favorable due to its proper and non-oxidizing surface. The thiol endgroup is one of the rather rare possibilities to form a covalent bond with the nobel metal gold. The bond at room temperature must be loose enough to allow rearranging of the molecules for ordering as a monolayer. Van der Waals force acts, in particular, between Langmuir - Blodgett films of organic lipophiles and planar substrates. The contact at the opposite side of the molecules is made of gold also.

The very first molecular electronic device considered, the rectifier by A. Aviram and M.R. Ratner [15.53], was based on a donor and an acceptor π-

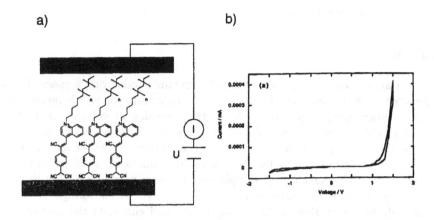

Fig. 15.12: a) Schematic presentation of a rectifying device based on a Langmuir-Blodgett film of a donor-acceptor molecule. b) The I - V curve of the sandwiched Langmuir-Blodgett monolayer displaying rectifying character, after Fischer et al.[15.54]

system linked together by a spacer. A transfer of electrons from the cathode to the acceptor and subsequently from the donor to the anode should be feasible with less applied voltage than the transport in the opposite direction. However, although this first attempt for such an electronic device with rectifying ability was successful, it lacked the required stability because of the weak van der Waals interaction between the electrode and the molecules. In the 1990s the rectifying principle was in fact realized by C.M. Fischer, M. Burghard, S. Roth, and K. von Klitzing [15.54] and by R.M. Metzger et al [15.55]. A Langmuir-Blodgett film of molecules consisting of an electron - rich π-system as donor linked to an electron-poor π-system as acceptor which is decorated with a long alkyl chain, has been sandwiched between two electrodes (Fig. 15.12a). A torsion angle between the A and D π-systems reduces the overlap of both π-systems and allows a zwitterionic ground state, T - D^+ - π - A^- where T is the hexadecyl 'tail' serving to support the Langmuir - Blodgett film formation. Fig. 15.12b shows the current-vs.-voltage curve displaying the rectifying performance of the device [15.56]. It was studied theoretically by O. Kwon et al. [15.57]. An approach towards real integrated devices on the deposition of catenanes and rotaxanes on a substrate, with carbon nanotubes ['Sect. 15.5] as bottom electrodes, may serve as a switch between two configurations by voltage pulses. The state of the switch may subsequently be read out with the tunnel current at low voltage [15.58].

This chapter requires, in contrast to the previous chapters, a certain knowledge of organic chemistry. Actually its main purpose is to indicate the kind of problems one encounters in trying to reduce the size of the transistor to that of molecules. Whether this will ever be successful for mass production cannot be predicted at the present stage.

Problems

15.1. In amorphous and in high-resistivity organic conductors, a space charge quite often limits the current. Assume a carrier concentration which varies along the filamentary sample, $p = p(x)$. Neglecting diffusion at voltages $\gg k_BT/e$, the local field strength $E = -d\varphi/dx = j/(e\mu_p p)$. The derivative $dE/dx = -(j\,dp/dx)/(e\mu_p p^2)$ is given by the Poisson equation. Solve for dx and integrate. Take L for the sample length and assume that because of the injection of carriers at the contact at one sample end, the carrier concentration there is much larger than at the other contact. Calculate the total voltage V across the sample by integration of $d\varphi = (d\varphi/dx)dx$ and eliminate the carrier concentration from this equation and the one for L. The resulting current voltage characteristics is known as Child's law in gaseous electronics [15.35].

15.2. Because of band tails in amorphous solids, carriers in the gap contribute somewhat to the conductivity since they are not really localized, and it is reasonable to introduce a differential conductivity $\sigma(\varepsilon)\,d\varepsilon$ valid throughout the gap and the adjacent bands. The thermopower is thus written as

$$\frac{d\Theta}{dT} = \frac{k_B}{e} \int_0^\infty \frac{\varepsilon - \zeta}{k_BT} \frac{\sigma(\varepsilon)}{\sigma} \left(\frac{-\partial f_0}{\partial \varepsilon}\right) d\varepsilon \tag{15.0.1}$$

where σ is the integral conductivity. Show that this definition makes sense and that it includes the usual definition for a crystalline well-behaved semiconductor (4.9.14).

Further reading:

M.S Dresselhouse, G. Dresselhouse, P.C. Eklund, Science of Fullerenes and Carbon Nanotubes, Academic Press, San Diego (U.S.A.) (1995)

V. Derycke, J. Appenzeller in "Fundamentals of Nanoelectronics" Forschungszentrum Jülich (Germany)(2002)

Appendix A

Table A : Physical Constants

Quantity	Symbol	Value
Avogadro constant	N_{Av}	$6.02204 \times 10^{23}\,\mathrm{mol}^{-1}$
Bohr magneton	$\mu_B = e\,\hbar/2\,m_0$	$5.78832 \times 10^{-5}\,\mathrm{eVT}^{-1}$
Bohr radius	$a_B = 4\,\pi\,\kappa_0\,\hbar^2/m_0 e^2$	0.052917 nm
Boltzmann constant	k_B	$1.3806513 \times 10^{-23}\,\mathrm{WsK}^{-1}$
		$8.61573 \times 10^{-5}\,\mathrm{eVK}^{-1}$
Elementary charge	$|e|$	$1.60219 \times 10^{-19}\,\mathrm{As}$
Fine structure constant (reciprocal)	$\alpha^{-1} = 4\,\pi\,\kappa_0\,\hbar/e^2$	137.036
Free electron rest mass	m_0	$0.910956 \times 10^{-30}\,\mathrm{kg}$
		$0.910956 \times 10^{-34}\,\mathrm{W\,s^3 cm}^{-2}$
Free electron charge per mass	$|e|/m_0$	$1.75880 \times 10^{15}\,\mathrm{cm^2\,V^{-1}s^{-2}}$
		$1.75880 \times 10^{11}\,\mathrm{s^{-1}T^{-1}}$
Impedance of free space	$1/\kappa_0 c = \mu_0 c$	$376.732\ \Omega$
Permeability in vacuum	μ_0	$1.25663 \times 10^{-8}\,\mathrm{VsA^{-1}cm^{-1}}$
		$(4\pi \times 10^{-9})$
Permittivity in vacuum	κ_0	$8.85418 \times 10^{-14}\,\mathrm{AsV^{-1}cm^{-1}}$
Planck constant	h	$6.62619 \times 10^{-34}\,\mathrm{Ws^2}$
		$4.13567 \times 10^{-15}\,\mathrm{eVs}$
Proton rest mass	M_P	$1.67264 \times 10^{-27}\,\mathrm{kg}$
Reduced Planck constant	$\hbar = h/2\pi$	$1.0545727 \times 10^{-34}\,\mathrm{Ws^2}$
		$= 6.582 \times 10^{-16}\,\mathrm{eV\,s}$
Rydberg energy	$Ry = m_0 e^4/2(4\pi\,\kappa_0\,\hbar)^2$	13.6060 eV
Speed of light in vacuum	c	$2.99792 \times 10^{10}\,\mathrm{cm\,s^{-1}}$
Von Klitzing constant	$R_K = h/e^2$	$2.5812807 \times 10^4\ \Omega$

Units of length: 1 nm $= 10^{-9}$ m; 1 Å$= 10^{-10}$ m $= 0.1$ nm

B. Envelope Wave Function for Quantum Wells

The following treatise has been adapted from [A1]. Assume a layer of a semiconductor A between two thick layers of another semiconductor B such as in the pairs of materials GaAs/GaAlAs, GaInAsP/InP, GaInAs/AlInAs or GaSb/AlSb etc. The energy levels in the conduction band in Fig. B.1 will be calculated in the *envelope function approximation*. The electron wave function is approximately

$$\psi = \sum_{A,B} \exp(i\,\mathbf{k}_\perp \mathbf{r}) u_{c\mathbf{k}}^{A,B}(\mathbf{r}) \chi_n(z) \tag{B.1}$$

where $\mathbf{k}_\perp$ is the transverse wave vector, $u_{c\mathbf{k}}^{A,B}$ is the Bloch wave function in material A and B, respectively. The envelope wave function $\chi_n(z)$ in the case of a quantum well of finite depth V_0 for two bound energy levels as shown in Fig. B.1 can be solved for the lower level by

$$\chi_n(z) = A\cos(kz) \qquad \text{for } |z| < L/2$$

$$\chi_n(z) = B\exp[-\kappa(z - L/2)] \qquad \text{for } z > L/2 \tag{B.2}$$

$$\chi_n(z) = B\exp[+\kappa(z + L/2)] \qquad \text{for } z < L/2$$

for the equation

$$\left(-\frac{\hbar^2}{2m(z)}\frac{\partial^2}{\partial z^2} + V_0(z)\right)\chi_n(z) = \epsilon_n\,\chi_n(z) \tag{B.3}$$

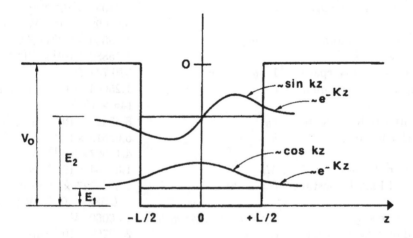

Fig. B .1: First two bound energy levels and wave functions in a finite quantum well

Continuity for $\chi_n(z)$ at $z = \pm L/2$ is achieved by assuming

$$A \cos(k\, L/2) = B \tag{B.4}$$

and for $d\chi_n(z)/dz$ by

$$(k/m_A)\, A \sin(k\, L/2) = \kappa\, B\, /m_B\ . \tag{B.5}$$

Eliminating B and A from (B.4) and (B.5) yields for the lower level

$$(k/m_A)\tan(k\, L/2) = \kappa/m_B\ . \tag{B.6}$$

For the upper level we have

$$\chi_n(z) = A \sin(k\, z) \qquad\qquad \text{for } |z| < L/2 \tag{B.7}$$

and otherwise the same as in (B.2), and in correspondence to (B.6):

$$(k/m_A) \cot(k\, L/2) = -(\kappa/m_B) \tag{B.8}$$

for the upper level. In an approximation $m_A = m_B = m$ the equations are easier to solve by introducing $k_0 = 2\, m\, V_0/\hbar^2$:

$$\cos(kL/2) = k/k_0 \qquad\qquad \text{for } \tan(k\, L/2) > 0 \tag{B.9}$$

$$\sin(kL/2) = k/k_0 \qquad\qquad \text{for } \tan(k\, L/2) < 0 \tag{B.10}$$

The number of bound states is $1 + \text{Int}[(L/\pi)\sqrt{k_0}]$ where $\text{Int}[x]$ is the *integer part of x*.

Appendix C

Table C : Semiconductor and Semimetal Data

semicon- ductor	crystal structure	lattice const. at 300 K [Å]	ε_G(300K) (in)direct [eV]	mobility electrons μ_n	$[\frac{cm^2}{Vs}]$ holes μ_p	diel. $\kappa(0)$	const. $\kappa(\infty)$
diamond	diamond	3.56683	(i) 5.49	2000	2100	5.70	
Si	diamond	5.43102	(i) 1.1242	1450	435	11.9	
Ge	diamond	5.65791	(i) 0.661	3850	1845	16.2	
Sn(grey)	diamond	6.4897	(d) 0.0	2940	2990	24.0	
SiC (3C)	zincblende	4.3596	(i) 2.2	900		9.72	6.52
BN	zincblende	3.6157	(i) 6-8			7.1	4.5
BP	zincblende	4.5383	(i) 2-2.4	120	500	11	
AlN	zincblende	3.11	(d) 6.25		14	9.1	4.84
AlP	zincblende	5.467	(i) 2.45	80		9.8	7.5
AlAs	zincblende	5.660	(i) 2.239	294		10.06	8.16
AlSb	zincblende	6.1355	(i) 1.62	200	400	12.04	10.24
GaN	wurtzite	3.175	(d) 3.44	440		11.3	5.5
GaP	zincblende	5.4512	(i) 2.272	190	150	11.1	9.11
GaAs	zincblende	5.6532	(d) 1.424	9750	450	12.8	10.75
GaSb	zincblende	6.0959	(d) 0.74	7700	1000	15.69	14.44
InN	zincblende	3.5446	(d) 1.99	250		9.3	
InP	zincblende	5.8687	(d) 1.344	5400	150	12.6	9.61
InAs	zincblende	6.0583	(d) 0.356	33000	450	15.15	12.31
InSb	zincblende	6.4794	(d) 0.18	77000	850	17.4	15.68
ZnO	wurtzite	3.2501	(d) 3.44	167, 150		8.75	3.73
ZnS	zincblende	5.4077	(d) 3.71	230	40	8.5	5.35
ZnS	wurtzite	3.8227	(d) 3.93	100	15	9.6	5.4
ZnSe	zincblende	5.664	(d) 2.72	560	110	8.6	5.7
ZnSe	wurtzite	4.00					
ZnTe	zincblende	6.1004	(d) 2.2	1500	960	9.1	6.75
ZnTe	wurtzite	4.273	(d) 2.81				
CdS	zincblende	5.811	(d) 2.53				
CdS	wurtzite	4.136	(d) 2.514	330	48	8.53	5.26
CdSe	zincblende	6.078	(d) 1.7			8.41	
CdSe	wurtzite	4.300	(d) 1.747	660	40	9.73	6.35
CdTe	zincblende	6.481	(d) 1.482	1000	60	10.2	7.25
HgTe	zincblende	6.462	(d) -0.160	$\leq 10^5$		21	10
PbS	rocksalt	5.936	(i) 0.41	600	700	17.0	
PbTe	rocksalt	6.468	(i) 0.31	6000	4000	30.0	
CuCl	zincblende	5.416				6.3	3.7
CuBr	zincblende	5.691				7.0	4.4
CuI		6.09	2.26		110	7.1	5.5
AgI	zincblende	6.486					

References

Chapter 1

1.1   C.K. Chiang, C.R. Fincher, Jr., Y.W. Park, A.J. Heeger, H. Shirakawa, E.J. Louis, S.C. Gau, A.G. MacDiarmid, Phys. Rev. Lett. **39**, 1098 (1977)
1.2   W.C. Dash, J. Appl. Phys. **29**, 736 (1958), **30**, 459 (1959)
1.3   J. Czochralski, Z. Phys. Chem. **92**, 219 (1918); J.C. Brice, Crystal Growth Processes, Blackie (London) 1973
1.4   P.H. Keck, M.J.E. Golay, Phys. Rev. **89**, 1297 (1953)
1.5   M.A. Hermann, H. Sitter, Molecular Beam Epitaxy Springer - Verlag Heidelberg, 1989
1.6   C.T. Foxon, B.A. Joyce, in R.A. Stradling, P.C. Klipstein, (eds.), Growth and Characterization of Semiconductors, (Hilger Bristol) (1990), p.35
1.7   J.M. Meese, Neutron Transmutation Doping in Semiconductors, (Plenum, New York 1979)
1.8   G. Bertolini, A Coche, Semiconductor Detectors, (North-Holland, Amsterdam 1968)
1.9   G. Dearnally, D.C. Northrop, Semiconductor Counters for Nuclear Radiation, (Spon, London 1966)
1.10   G. Mandel, Phys. Rev. **134**, A1073 (1964)

Chapter 2

2.1   R. de Laer Kronig, W.J. Penney, Proc. Roy. Soc. London, **130**, 499 (1930)
2.2   A. Nussbaum, Semiconductor Device Physics,(Prentice-Hall, Englewood Cliffs, N.J. 1962) Sects. 1.9 and 1.13
2.3   R.A. Smith, Wave Mechanics of Crystalline Solids, 2nd edn. (Chapman and Hall, London 1969) Sect. 6.2
2.4   J.P. McKelvey, Solid-State and Semiconductor Physics, (Harper and Row, New York 1966) Sect. 8.6
2.5   E.O. Kane, J. Phys. Chem. Solids 1, 249 (1957)
2.6   W.A. Harrison, Electronic Structure and the Properties of Solids (Freeman, San Francisco 1980). M.L. Cohen, J.R. Chelikowsky, Electronic Structure and Optical Properties of Semiconductors, 2nd edn., Springer Ser. Solid-State Sci., vol. 75 (Springer, Berlin, Heidelberg 1989)
2.7   D. Long, Energy Bands in Semiconductors (Wiley, New York 1968) p.49
2.8   D. Matz, J. Phys. Chem. Solids **28**, 373 (1967)
2.9   S. Flügge, Rechenmethoden der Quantentheorie, 3rd edn. Heidelberger Taschenbücher, Vol. 6 (Springer, Berlin, Heidelberg 1966) Problem No.91
2.10   F. Hund, B. Mrowka, Ber. Sächs. Akad. Wiss. **87**, 185, 325 (1935); A.H. Wilson, The Theory of Metals, 2nd edn. (Cambridge Univ. Press, London 1965) pp.66-70, 83
2.11   G.E. Kimball, J. Chem. Phys. **3**, 560 (1935)
2.12   F. Hund, Theorie des Aufbaues der Materie, (Teubner, Stuttgart 1961) Chap. VII

2.13   E. Wigner, F.Seitz, Phys. Rev. **43**, 804 (1933)
2.14   J.C. Slater, Phys. Rev. **51**, 846 (1937)
2.15   C. Herring, Phys. Rev. **57**, 1169 (1940)
2.16   E. Antoncik, J. Phys. Chem. Solids **10**, 314 (1959)
2.17   J.C. Phillips, L. Kleinman, Phys. Rev. **116**, 287, 880 (1959)
2.18   M.L. Cohen, V. Heine, Solid State Physics **24**, 37 (Academic, New York 1970) (review)
2.19   J. Koringa, Physica **13**, 392 (1947)
2.20   W. Kohn, N. Rostoker, Phys. Rev. **94**, 1111 (1954)
2.21   N.W. Ashcroft, N.D. Mermin, Solid State Physics (Holt, Rinehart, and Winston, New York 1976) p.202
2.22   W. Shockley, Electrons and Holes in Semiconductors (Van Nostrand, New York 1950)
2.23   C. Kittel, Introduction to Solid State Physics, 7th edn, (Wiley, New York 1995), O. Madelung, Introduction to Solid-State Theory, Springer Ser. Solid-State Sci., Vol. 2 (Springer, Berlin, Heidelberg 1978)
2.24   R.A. Smith, Wave Mechanics of Crystalline Solids (Chapman and Hall, London 1961)
2.25   H. Jones, The Theory of Brillouin Zones and Electronic States in Crystals (North-Holland, Amsterdam 1960)
2.26   F. Seitz, Modern Theory of Solids (McGraw-Hill, New York 1940)
2.27   L.P. Bouckaert, R. Smoluchowski, E.P. Wigner, Phys. Rev. **50**, 58 (1936)
2.28   J. Callaway, Energy Band Theory (Academic, New York 1964)
2.29   H.W. Streitwolf, Gruppentheorie in der Festkörperphysik (Geest und Portig, Leipzig 1967)
2.30   D. Long, Energy Bands in Semiconductors (Wiley, New York 1968)
2.31   C. Rigaux, G. Drilhorn, J. Phys. Soc. Jpn. **21**, Suppl., 193 (1966) (Proc. Int'l. Conf. Phys. Semicond. Kyoto)
2.32   J.R. Drabble, In Progress in Semiconductors, Vol. 7, ed. by A.F. Gibson, R.E. Burgess (Temple, London 1963) p.45
2.33   D.L. Greenaway, G. Harbeke, J. Phys. Chem. Solids **26**, 1585 (1965)
2.34   I. Melngailis, J. Physique, Suppl. No.11–12, **29** , C4-84 (1968)
2.35   J.O. Dimmock, I. Melngailis, A.J. Strauss, Phys. Rev. Lett. **16**, 1193 (1966)
2.36   C. Vèriè, Festkörper-Probleme/Advances in Solid State Physics, X, 1–19 (Vieweg, Braunschweig 1970)
2.37   S. Groves, W. Paul, Phys. Rev. Lett. **11**, 194 (1963)
2.38   R. Dornhaus, G. Nimtz, Springer Tracts Mod. Phys. **78**, 1–112 (Springer, Berlin, Heidelberg 1976)
2.39   M. Cardona, F.H. Pollak, Phys. Rev. **142**, 530 (1966)
2.40   J.C. Phillips, D. Brust, F. Bassani, Proc. Int'l. Conf. Phys. Semicond., Exeter (The Institute of Physics, London 1962) p.564
2.41   A.B. Pippard, Philos. Trans. A**250**, 325 (1957)
2.42   W. Brauer, Einführung in die Elektronentheorie der Metalle (Geest und Portig, Leipzig 1966)

Chapter 3

3.1   E. Spenke, Electronic Semiconductors (McGraw-Hill,New York 1965)Chap. 8
3.2   E. Schrödinger, Statistical Thermodynamics (Cambridge Univ. Press, Cambridge 1948)
3.3   J.S. Blakemore, Semiconductor Statistics (Pergamon, Oxford 1962) Appendix B; J. McDougall, E.C. Stoner, Philos. Trans. A **237**, 67 (1938); P. Rhodes, Proc. Roy. Soc. London A **204**, 396 (1950)
3.4   H.J. Goldsmid, Problems in Solid State Physics (Academic, New York 1969) p.354
3.5   H. van Cong, S. Brunet, C. Charar, J.L. Birman, M. Averons, Solid State Commun. **45**, 611 (1983)
3.6   W. Jantsch, Untersuchungen zur Ladungsträgerumbesetzung in höhere Leitungsbandminima von Gallium-Antimonid durch hohe elektrische Felder, Dissertation, Univ. Vienna, Austria (1971)
3.7   J.A. van Vechten, Handbook on Semiconductors, Vol. 3 ed. by S.P. Keller (North-Holland, Amsterdam 1980) p.l (review)
3.8   H. Fritzsche, Phys. Rev. **120**, 1120 (1960)
3.9   F. Bassani, G. Iadonisi, B. Preziosi, Rep. Prog. Phys. **37**, No.9, 1099 (1974)
3.10   S.H. Koenig, R.D. Brown III, W. Schillinger, Phys. Rev. **128**, 1668 (1962)
3.11   D. Long, C.D. Motchenbacher, J. Myers, J. Appl. Phys. **30**, 353 (1959)
3.12   J. Blakemore, Semiconductor Statistics (Pergamon , Oxford 1962) Sect. 3.2.4
3.13   J.S. Blakemore, Philos. Mag. **4**, 560 (1959)
3.14   N.B. Hannay (ed.), Semiconductors (Reinhold, New York 1959) p.31
3.15   E.M. Conwell, Proc. IRE **46**, 1281 (1958)
3.16   T.H. Geballe, In [Ref.3.13, ps. 341, 342]
3.17   R. Newman, W.W. Tyler, Solid State Physics Vol. **8**, (F. Seitz, D. Turnbull, eds)(Academic, New York 1959) p.62

Chapter 4

4.1   L. Onsager, Phys. Rev. **37**, 405 (1931); ibid. **38**, 2265 (1931); A.C. Smith, J.F. Janak, R.B. Adler, Electronic Conduction in Solids (McGraw-Hill, New York 1967) Sect. 3.5
4.2   S.R. de Groot, Thermodynamics of Irreversible Processes, (North-Holland Amsterdam 1953)
4.3   A.C. Beer, In The Hall Effect and its Applications, ed. by C.L. Chien, C.R. Westgate (Plenum, New York 1980) p.299 (review)
4.4   E.H. Putley, The Hall Effect and Related Phenomena ; (Butterworth, London 1960)

4.5   A.H. Wilson, The Theory of Metals, 2nd edn. (Cambridge Univ. Press, London 1965);Sect. 8.51

4.6   O. Madelung, H. Weiss, Z. Naturforsch. **9a**, 527 (1954)

4.7   N.I. Volokobinskaya, V.V. Galavanov, D.N. Nasledov, Fiz. Tverd. Tela **1**, 756 (1959) [Engl., Sov. Phys.-Solid State **1**, 687 (1959)]

4.8   N.Z. Lupu, N.M. Tallan, D.S. Tannhauser, Rev. Sci. Instrum. **38**, 1658 (1967)

4.9   H.H. Wieder, Laboratory Notes on Electrical and Galvanomagnetic Measurements (Elsevier, Amsterdam 1979)

4.10   L.J. van der Pauw, Philips Techn. Rundsch. **20**, 230 (1958/59); Philips Res. Rep. **13**, 1 (1958),**16**, 187 (1961)

4.11   J. Hornstra, L.J. van der Pauw, J. Electron. Control **7**, 169 (1959), describing a method for measuring electrical resistivities of anisotropic materials

4.12   H.C. Montgomery, J. Appl. Phys. **42**, 2971 (1971)

4.13   B. Schwartz (ed.), Ohmic Contacts to Semiconductors (The Electrochemical Society, New York 1968)

4.14   R.K. Willardson, T. Harman, A.C. Beer, Phys. Rev. **96**, 1512 (1954)

4.15   A. Kawabata, J. Phys. Soc. Jpn. **49**, 628 (1980)

4.16   O.M. Corbino, Phys. Z. **12**, 561 (1911)

4.17   H. Welker, H. Weiss, Z. Physik **138**, 322 (1954)

4.18   W. Röss, A. Philipp, K. Seeger, K. Ehinger, K. Menke, S. Roth, Solid State Commun. **45**, 933 (1983)

4.19   R.F. Wick, J. Appl. Phys. **25**, 741 (1954)

4.20   A.H. Thompson, G.S. Kino, J. Appl. Phys. **41**, 3064 (1976)

4.21   R.W. Rendell, S.M. Girvin, Phys. Rev. B **23**, 6610 (1981)

4.22   P.M. Morse, H. Feshbach, Methods of Theoretical Physics, Pt. I (McGraw-Hill, New York 1953) p.707

4.23   B. Neudecker, K.H. Hoffmann, Solid State Commun. **62**, 135 (1987)

4.24   J.A.M. Dikhoff, Solid-State Electronics **1**, 202 (1960)

4.25   P.R. Camp, J. Appl. Phys. **25**, 459 (1954)

4.26   F. Kuhn-Kuhnenfeld, J. Electrochem. Soc. Solid State Science and Technology **119**, 1063 (1972)

4.27   D.C. Miller, G.A. Rozgonyi, In Handbook on Semiconductors, Vol. 3, ed. by S.P. Keller (North-Holland, Amsterdam 1980) p.217 (review)

4.28   H. Rupprecht, R. Weber, H. Weiss, Z. Naturforsch. **15a**, 783 (1960)

4.29   C. Herring, J. Appl. Phys. **31**, 1939 (1960)

4.30   H. Weiss, J. Appl. Phys. Suppl. **32**, 2064 (1961)

4.31   D.J. Ryden, J. Phys. C **7**, 2655 (1974)

4.32   J.Y.W. Seto, J. Appl. Phys. **46**, 5247 (1975)

4.33   J.W. Orton, M.J. Powell, Rep. Prog. Phys. **43**, 1263 (1980) (review)

4.34   C. Yamanouchi, K. Mizuguchi, W. Sasaki, J. Phys. Soc. Jpn. **22**, 859 (1967)

4.35   C. Goldberg, R.E. Davis, Phys. Rev. **94**, 1121 (1954)

4.36   L. Grabner, Phys. Rev. **117**, 689 (1960)

4.37   A.C. Smith, J.F. Janak, R.B. Adler, Electronic Conduction in Solids (McGraw-Hill, New York 1967) Appendix I

4.38   M. Sze, Physics of Semiconductor Devices, 2nd edn. (Wiley, New York 1981)

4.39 J. Appel, Progress in Semiconductors, Vol. 5, ed. by A.F. Gibson, R.E. Burgess (Temple, London 1960) p.142 (review)

4.40 T.H. Geballe, G.W. Hull, Phys. Rev. **110**, 773 (1958)

4.41 H. Weiss, in Halbleiter und Phosphore, ed. by M. Schön, H. Welker (Vieweg, Braunschweig 1957) p.497

4.42 J. Schröder, Rev. Sci. Instrum. **34**, 615 (1963)

4.43 W. Fulkerson, J.P. Moore, R.K. Williams, R.S. Graves, D.L. McElroy, Phys. Rev. **167**, 765 (1968)

4.44 O. Madelung, Introduction to Solid-State Theory , Springer Ser. Solid-State Sci., Vol. 2 (Springer, Berlin, Heidelberg 1978)

4.45 H.M. Rosenberg, Low Temperature Solid State Physics (Oxford Univ. Press, London 1965) p.8; F. Pobell, Matter and Methods at Low Temperatures, 2nd. edn. (Springer, Berlin, Heidelberg 1996)

4.46 M.W. Zemansky, Heat and Thermodynamics, 4th edn. (McGraw-Hill, New York 1957) pp.192, 301

4.47 A.H. Wilson, The Theory of Metals, 2nd edn. (Cambridge Univ. Press, Cambridge 1965) p.205

4.48 A.V. Gold, D.K.C. McDonald, W.B. Pearson, I.M. Templeton, Philos. Mag. **5**, 765 (1960)

4.49 T.H. Geballe, G.W. Hull, Phys. Rev. **98**, 940 (1955)

4.50 T.H. Geballe, G.W. Hull, Phys. Rev. **94**, 1134 (1954)

4.51 L.J. Kroko, A.G. Milnes, Solid-State Electron. **8**, 829 (1965)

4.52 C. Herring, In Halbleiter und Phosphore, ed. by M. Schön, H. Welker (Vieweg, Braunschweig 1958) p.184

4.53 H.P.R. Frederikse, Phys. Rev. **91**, 491 (1953); ibid. **92**, 248 (1953)

4.54 L. Sosnowski, Semiconductors, Proc. Int'l School of Physics XXII (Academic, London 1963) p.436 (review)

4.55 J. Bardeen, C. Herring, in Imperfections in Nearly Perfect Crystals, ed. by F. Seitz (Wiley, New York 1952)

4.56 R. Becker, Theory of Heat, 2nd edn (Springer, Berlin, Heidelberg 1967) Chap. 95; W. Brenig, Statistical Theory of Heat (Springer, Berlin, Heidelberg 1989)

4.57 R. Bowers, R.W. Ure, Jr., J.E. Bauerle, A.J. Cornish, J. Appl. Phys. **30**, 930 (1959)

4.58 A.F. Joffe, Semiconductor Thermoelements and Thermoelectric Cooling (Infosearch, London 1958)

4.59 H.P.R. Frederikse, V.A. Johnson, W.W. Scanlon, Solid State Physics, Vol. 6B, ed. by K. Lark-Horowitz, V.A. Johnson (Academic, New York 1959) p.114

4.60 R. Wolfe, Semiconductor Products **6**, 23 (1963)

4.61 H. Wagini, Z. Naturforsch. **19a**, 1541 (1964)

4.62 A.C. Beer, J.A. Armstrong, I.N. Greenberg, Phys. Rev. **107**,1506(1957)

4.63 L. Sosnowski, Proc. Int'l Conf. Phys. Semicond. Paris 1964 (Dunod, Paris 1964) p.341

4.64 D.N. Nasledov, J. Appl. Phys. **32**, Suppl., 2140 (1961)

4.65 S. Minomura, H.G. Drickamer, J. Phys. Chem. Solids **23**, 451 (1962)

4.66 C. Kittel, Introduction to Solid State Physics, 4th edn. (Wiley, New York 1965) Chap. 4

4.67   B.K. Vainshtein, Fundamentals of Crystals, 2nd edn., Modern Crystallogra-
        phy I, Springer Ser. Solid-State Sci., Vol. 15 (Springer, Berlin, Heidelberg
        1994)
4.68   E.G.S. Paige, in Progress in Semiconductors, Vol. 8, ed. by A.F. Gibson,
        R.E. Burgess (Temple, London 1964) p.159
4.69   C.S. Smith, Solid State Physics 6, 175 (Academic, New York 1958)
4.70   W. Voigt, Lehrbuch der Kristallphysik (Teubner, Leipzig 1928)
4.71   C.S. Smith, Phys. Rev. 94, 42 (1954)
4.72   F.J. Morin, T.H. Geballe, C. Herring, Phys. Rev. 105, 525 (1957)
4.73   R.W. Keyes, Solid State Physics 11, 172 (Table V) and 176 (Table VI)
        (Academic, New York 1960), ed. by F. Seitz, D. Turnbull
4.74   E.J. Ryder, W. Shockley, Phys. Rev. 81, 139 (1951)
4.75   E.J. Ryder, Phys. Rev. 90, 766 (1953)
4.76   H. Heinrich, K. Seeger, Verh. Dtsch. Phys. Ges. (VI) 2, 26 (1967);K.
        Seeger, Acta Phys. Austriaca 27, 1 (1968)
4.77   H. Heinrich, G. Bauer, D. Kasperkovitz, Phys. Status Solidi 28, K51 (1968)
4.78   H. Heinrich, W. Jantsch, Solid State Commun. 7, 377 (1969)
4.79   A. Alberigi-Quaranta, C. Jacoboni, G. Ottaviani, Riv. Nuovo Cimento 1,
        445 (1971)
4.80   A.G.R. Evans, Electron. Lett. 8, 195 (1972)
4.81   K. Seeger, J. Appl. Phys. 63, 5439 (1988); Appl. Phys. Lett. 54, 1268
        (1989) Lithuanian Journal of Physics 36, 466 - 469 (1996)
4.82   K. Seeger, Synth. Metals 15, 361 (1986)
4.83   K. Seeger, K.F. Hess, Z. Physik 237, 252 (1970)
4.84   S. Winnerl, E. Schomburg, J. Grenzer, H.-J. Regl, A.A. Ignatov, A.D. Se-
        menov, K.F. Renk, D.G. Pavel'ev, Yu. Koschurinov, B. Melzer, V. Usti-
        nov, S. Ivanov, S. Schaposchnikov, P.S. Kop'ev, Phys. Rev. B 56, 13
        268 (1997), Yu. A. Romanov, Opt. Spectrosc. 33, 917 (1972)/Opt. Soc.
        America/Acad.Sci.SSR
4.85   J.R. Tucker, IEEE J. Quantum Electronics QE-15,1234 (1979)
4.86   A. van der Ziel, Noise, Measurement and Characterization (Prentice-Hall,
        Englewood Cliffs, NJ 1970)
4.87   D. Wolf (ed.), Noise in Physical Systems , Springer Ser. Electrophys., Vol.
        2 (Springer, Berlin, Heidelberg 1978)
4.88   H. Nyquist, Phys. Rev. 32, 110 (1928)
4.89   A.L. McWhorter, in Semiconductor Surface Physics, ed. by R.H. Kingston
        (Univ. of Pennsylvania Press, Philadelphia 1957) p.207, R.H. Kingston,
        A.L. McWhorter, Phys. Rev. 103, 534 (1956)
4.90   T.G. Maple, L. Bess, H.A. Gebbie, J. Appl. Phys. 26, 490 (1955)
4.91   D. Sautter, K. Seiler, Z. Naturforsch. 12a, 490 (1957)
4.92   W. Shockley, Electrons and Holes in Semiconductors; (Van Nostrand,
        Princeton, NJ 1950) p.345
4.93   G. Lautz, M. Pilkuhn, Naturwissenschaften 47, 198, 394 (1960)
4.94   E. Erlbach, J.B. Gunn, Proc. Int. Conf. Phys. Semicond ., Exeter (Insti-
        tute of Physics, London 1962) p.128
4.95   C.A. Bryant, Bull. Am. Phys. Soc. (Ser. II) 9, 62 (1964)
4.96   P.J. Price, in Fluctuation Phenomena in Solids, ed. by R.E. Burgess (Aca-
        demic, New York 1964) Chap. 8

Chapter 5

5.1 F. Stöckmann, Halbleiterprobleme **VI**, 279 (Vieweg, Braunschweig 1961)

5.2 R.B. Adler, A.C. Smith, R.L. Longini, Introduction to Semiconductor Physics (Wiley, New York 1964)

5.3 J.P. McKelvey, in Problems in Solid State Physics, ed. by H. Goldsmid (Academic, New York 1968) p.82

5.4 J.R. Haynes, W. Shockley, Phys. Rev. **81**, 835 (1951)

5.5 W. Schottky, Z. Physik **118**, 539 (1942)

5.6 W. Shockley, Bell Syst. Tech. J. **28**, 435 (1949)

5.7 C.T. Sah, R.N. Noyce, W. Shockley, Proc. IRE **45**, 1228 (1957)

5.8 S.M. Sze, Physics of Semiconductor Devices, 2nd edn (Wiley, New York 1981) p.92

5.9 W. Pietenpol, Phys. Rev. **82**, 120 (1951)

5.10 D.V. Lang, J. Appl. Phys. **45**, 3023 (1974); D.V. Lang, in Thermally Stimulated Relaxation in Solids, ed. by P. Bräunlich, Topics Appl. Phys., Vol. 37 (Springer , Berlin, Heidelberg 1979) Chap. 3

5.11 N.M. Johnson, D.J. Barteling, J.P. McVittie, J. Vac. Sci. Technol. **16**, 1407 (1979)

5.12 J. Frenkel, Tech. Phys. USSR **5**, 685 (1938); Phys. Rev. **54**, 647 (1938)

5.13 A.G. Milnes, Deep Impurities in Semiconductors (Wiley-Interscience, New York 1973) p.99; Impurities and Defects in Group IV Elements and III–V Compounds. Landolt–Börnstein, Vol.146;22 Pt. b (Springer, Berlin, Heidelberg 1989)

5.14 J.L. Hartke, J. Appl. Phys. **39**, 4871 (1968)

5.15 W. Keller, K. Wünstel, Appl. Phys. A**31**, 9 (1983)

5.16 J.B. Gunn, J. Electron. Control **4**, 17 (1958)

5.17 J. Bardeen, Semiconductor Research Leading to the Point Contact Transistor; H. Brattain, Surface Properties of Semiconductors; W. Shockley, Transistor Technology Evokes New Physics; in Nobel Lectures Physics 1942-1962 (Elsevier, Amsterdam 1964) pp. 318–341; 377-384; 344–374

5.18 E. Braun, S. MacDonald, Revolution in Miniature (Cambridge Univ. Press, Cambridge 1978)

5.19 M.J. Morant, Introduction to Semiconductor Devices (Addison-Wesley, Reading MA 1964)

5.20 E.H. Rhoderick, Metal-Semiconductor Contacts (Clarendon, Oxford 1978)

5.21 K. Graff, H. Fischer, In Solar Energy Conversion, ed. by B.O. Seraphin, Topics Appl. Phys., Vol.146;31 (Springer, Berlin, Heidelberg, New York 1979)

5.22 J. Bardeen, W.H. Brattain, Phys. Rev. **74**, 230 (1948)

5.23 H.A. Gebbie, P.C. Banbury, C.A. Hogarth, Proc. Phys. Soc. London **63**B, 371 (1950)

5.24 F. Rosenberger, Fundamentals of Crystal Growth I, Springer Ser. Solid-State Sci., Vol. 5 (Springer, Berlin, Heidelberg, 1978)

5.25 M. Tanenbaum, D.G. Thomas, Bell Syst. Tech. J. **35**, 1 (1956)

5.26 J.A. Hoerni, IRE Electron Devices Meeting, Washington DC (1960)

5.27 C.J. Frosch, L. Derrick, J. Electrochem. Soc. **104** 547 (1957)

5.28   L.F. Thompson, R.E. Kerwin, Annu. Rev. Mater. Sci. **6**, 267 (1976)

5.29   R.A. Bartolini, in Holographic Recording Materials, ed. by H.M. Smith, Topics Appl. Phys., Vol. 20 (Springer, Berlin, Heidelberg 1977) Chap. 7

5.30   D.F. Barbe (ed.), Very Large Scale Integration VLSI, 2nd edn., Springer Ser. Electrophys., Vol. 5 (Springer, Berlin, Heidelberg 1981) Y. Tarui (ed.), VSLI Technology, Springer Ser. Electrophys., Vol. 12 (Springer, Berlin, Heidelberg 1986)

5.31   E. Spiller, R. Feder, in X-Ray Optics, ed. H.J. Queisser, Topics Appl. Phys., Vol. 22 (Springer, Berlin, Heidelberg 1977) Chap. 3

5.32   G. Stengl, R. Kaitna, H. Löschner, P. Wolf, R. Sacher, J, Vac. Sci. Technol. **16**, 1883 (1979)

5.33   M. Schulz, G. Pensl (eds.), Insulating Films on Semiconductors, Springer Ser. Electrophys., Vol. 7 (Springer, Berlin, Heidelberg 1981)

5.34   D.F. Barbe (ed.), Charge-Coupled Devices, Topics Appl. Phys., Vol. 38 (Springer, Berlin, Heidelberg 1980)

5.35   S.R. Hofstein, F.P. Heiman, Proc. IEEE **51**, 1190 (1963) 5.36; M.M. Atalla, E. Tannenbaum, E.J. Scheibner, Bell Syst. Tech. J. **38**, 749 (1959)

5.37   O.G. Folbert, IEEE J. SC-**16**, 51 (1981)

5.38   J.R. Barker, in Physics of Nonlinear Transport in Semiconductors, ed. by D.K. Ferry, J.R. Barker, C. Jacoboni (Plenum, New York 1980)

5.39   R.C. Eden, B.M. Welch, R. Zucca, S.I. Long, IEEE J. SC-**14**, 221 (1979)

5.40   R.C. Eden, Proc. IEEE **70**, 5 (1982)

5.41   H. Dember, Phys. Z. **32**, 554, 856 (1931); ibid. **33**, 207 (1932)

5.42   J. Frenkel, Nature **132**, 312 (1933)

5.43   W. van Roosbroeck, J. Appl. Phys. **26**, 380 (1955)

5.44   W. van Roosbroeck, Phys. Rev. **101**, 1713 (1956)

5.45   M.B. Prince, J. Appl. Phys. **26**, 534 (1955)

5.46   D.M. Chapin, C.S. Fuller, G.L. Pearson, J. Appl. Phys. **25**, 676 (1954)

5.47   B.O. Seraphin, (ed.),Solar Energy Conversion, Topics Appl. Phys., Vol. 31 (Springer, Berlin, Heidelberg 1979)

5.48   T.L. Chu, S.S. Chu, K.Y. Duh, H.I. Yoo, Proc. 12th IEEE Photovoltaic Specialists Conf. (IEEE, New York 1976) p.74

5.49   J.K. Hirvonen (ed.), Ion Implantation, Treatise Mater. Sci. Techn., Vol. 18 (Academic, New York 1980)

5.50   C.W. White, P.S. Peercy (eds.), Laser and Electron Beam Processing of Materials (Academic, New York, 1980)

5.51   D.E. Carlson, C.R. Wronski, in Amorphous Semiconductors, 2nd edn., ed. by M.H. Brodsky, Topics Appl. Phys., Vol. 36 (Springer , Berlin, Heidelberg, New York 1985) Chap. 10, J. Electron. Mater. **6**, 95 (1977)

5.52   K.W. Böer, in 2nd Photovoltaic Solar Energy Conf., ed. by R. Van Overstraeton, W. Palz (Reidel, Dordrecht 1979) p.671

5.53   G.W. Cullen, C.C. Wang (eds.), Heteroepitaxial Semiconductors for Electron Devices (Springer, New York 1978) (review)

5.54   J. Javetski, Electronics (July 19, 1979) p.105 (review)

5.55   F.A. Shirland, P. Rai-Choudhury, Rep. Prog. Phys. **41**, 1839 (1978)

5.56   F.R. Kalhammer, Sci. Am. **241**, 42 (1979)

Chapter 6

6.1   L.I. Schiff, Quantum Mechanics, 3rd edn. (McGraw-Hill, New York 1968)

6.2   H.S.W. Massey, E.H.S. Burhop, Electronic and Ionic Impact Phenomena, Vol. 1 (Clarendon, Oxford 1969) Sect. 6.3

6.3   C. Ramsauer, Ann. Phys. **64**, 513 (1921); ibid. **66**, 545 (1921)

6.4   C. Erginsoy, Phys. Rev. **79**, 1013 (1950)

6.5   S.H. Koenig, R.D. Brown III, W. Schillinger, Phys. Rev. **128**, 1668 (1962)

6.6   E. Otsuka, K. Murase, J. Iseki, J. Phys. Soc. Jpn. **21**, 1104 (1966)

6.7   C. Schwartz, Phys. Rev. **124**, 1468 (1961)

6.8   M. Rotenberg, Ann. Phys. **19**, 262 (1962)

6.9   M. Pomerantz, Proc. IEEE **53**, 1438 (1965)

6.10   E. Fermi, Nuclear Physics (Univ. Chicago Press, Chicago 1950) p.142

6.11   R.B. Adler, A.C. Smith, R.L. Longini, Introduction to Semiconductor Physics (Wiley, New York 1964) Sects. 1.5.3 and 3.3

6.12   H. Brooks, In Advances in Electronics and Electron Physics, Vol. 7, ed. by L. Marton (Academic, New York 1955) p.85; H. Brooks, Phys. Rev. **83**, 879 (1951)

6.13   E. Conwell, V.F. Weisskopf, Phys. Rev. **77**, 388 (1950)

6.14   F.J. Blatt, J. Phys. Chem. Solids **1**, 262 (1957); F.J. Blatt, Solid State Physics **4**, 199 (Academic, New York 1957) ed. by F. Seitz, D. Turnbull

6.15   D. Long, C.D. Motchenbacher, J. Myers, J. Appl. Phys. **30**, 353 (1959)

6.16   L.M. Falicov, M. Cuevas, Phys. Rev. **164**, 1025 (1967)

6.17   D. Kranzer, E. Gornik, Solid State Commun. **9**, 1541 (1971)

6.18   J.B. Krieger, T. Meeks, Phys. Rev. B **8**, 2780 (1973)

6.19   D. Chattopadhyay, H.J. Queisser, Rev. Mod. Phys. **53**, 745 (1981)

6.20   J. Bardeen, W. Shockley, Phys. Rev. **80**, 72 (1950)

6.21   R. Peierls, Ann. Phys. (Leipzig) **12**, 154 (1932)

6.22   A.H. Wilson, The Theory of Metals (Cambridge Univ. Press, London 1965) pp. 255, 298

6.23   R.F. Greene, J. Electron. Control **3**, 387 (1957)

6.24   W. Szymanska, J.P. Maneval, Solid State Commun. **8**, 879 (1970)

6.25   W. Shockley, Bell Syst. Tech. J. **30**, 990 (1951)

6.26   K.J. Schmidt-Tiedemann, Festkörperprobleme **1**, 122 (Vieweg, Braunschweig 1962)

6.27   M. Abramowitz, I.A. Stegun, Handbook of Mathematical Functions (Dover, New York 1968)

6.28   P.P. Debye, E.M. Conwell, Phys. Rev. **93**, 693 1954)

6.29   I. Adawi, Phys. Rev. **120**, 118 (1960)

6.30   K. Seeger, Z. Physik **156**, 582 (1959)

6.31   M.S. Sodha, Phys. Rev. **107**, 1266 (1957)

6.32   K Seeger, Z. Physik **244**, 439 (1971)

6.33   G. Tschulena, Acta Phys. Austriaca **33**, 42 (1971)

6.34   G. Tschulena, R. Keil, Phys. Status Solidi (b) **49**, 191 (1972)

6.35   M. Lannoo, J. Bourgoin, Point Defects in Semiconductors I, Springer Ser. Solid State Sci., Vol. 22 (Springer, Berlin, Heidelberg 1981)

6.36   W.W. Tyler, H.H. Woodbury, Phys. Rev. **102**, 647 (1956)

6.37   D.M. Brown, R. Bray, Phys. Rev. **127**, 1593 (1962)

6.38   P.M. Eagles, D.M. Edwards; Phys. Rev. **138**, A1706 (1965)

6.39   C. Herring, E. Vogt, Phys. Rev. **101**, 944 (1956)

6.40   A.R. Hutson, J. Appl. Phys. Suppl. **32**, 2287 (1961)

6.41   H.J.G. Meyer, D. Polder, Physica **19**, 255 (1953)

6.42   Sh.M. Kogan, Fiz. Tverd. Tela **4**, 2474 (1962) [Engl. transl., Sov. Phys. Solid State **4**, 1813 (1963)]; J.D. Zook, Phys. Rev. **136**, A869 (1964); R.S. Crandall, Phys. Rev. **169**, 577, 585 (1968)

6.43   P. Brüesch, Phonons, Theory and Experiments I–III, Springer Ser. Solid-State Sci., Vols. 34, 65, 66 (Springer, Berlin, Heidelberg 1982, 1986, 1987)

6.44   H. Bilz, W. Kress, Phonon Dispersion Relations in Insulators, Springer Ser. Solid-State Sci., Vol. 10 (Springer, Berlin, Heidelberg 1979)

6.45   M. Cardona (ed.), Light Scattering in Solids, 2nd edn., Topics Appl. Phys., Vol. 8 (Springer, Berlin, Heidelberg 1982)

6.46   M. Cardona, G. Güntherodt (eds.), Light Scattering in Solids, Topics Appl. Phys., Vol. 68 (Springer, Berlin, Heidelberg 1991)

6.47   S.W. Lovesey, T. Springer (ed.), Dynamics of Solids and Liquids by Neutron Scattering, Topics Current Phys., Vol.146; 3 (Springer, Berlin, Heidelberg 1977)

6.48   B. Dörner, Coherent Inelastic Neutron Scattering in Lattice Dynamics, Springer Tracts Mod. Phys., Vol. 93 (Springer, Berlin, Heidelberg 1982)

6.49   I. Pelah, C.M. Eisenhauer, D.J. Hughes, H. Palevsky, Phys. Rev. **108**, 1091 (1957)

6.50   B.N. Brockhouse, H. Palevsky, D.J. Hughes, W. Kley, E. Tunkelo, Phys. Rev. Lett. **2**, 258 (1959)

6.51   B.N. Brockhouse, P.K. Iyengar, Phys. Rev. **111**, 747 (1958)

6.52   G. Dolling, J.L.T. Waugh, in Lattice Dynamics, ed. by R.F. Wallis (Pergamon, London 1965) p.19

6.53   R.H. Lyddane, R.G. Sachs, E. Teller, Phys. Rev. **59**, 673 (1941)

6.54   M. Born, K. Huang, Dynamical Theory of Crystal Lattices (Clarendon, Oxford 1954) p.82

6.55   H.B. Callen, Phys. Rev. **76**, 1394 (1949)

6.56   B. Szigeti, Trans. Faraday Soc. **45**, 155 (1949)

6.57   A.C. Beer, Solid State Physics **4**, 286 (Academic, New York 1963) ed. by F. Seitz, D. Turnbull

6.58   R. Stratton, Proc. Roy. Soc. London A **246**, 406 (1958); J. Phys. Soc. Jpn. **17**, 590 (1962)

6.59   E.M. Conwell, Solid State Physics **9**, Sects. V,VII ed. by F. Seitz, D. Turnbull (Academic, New York 1967)

6.60   H.J.G. Meyer, Phys. Rev. **112**, 298 (1958)

6.61   W. Pötz, P. Vogl, Phys. Rev. B **24**, 1500 (1981)

6.62   W.A. Harrison, Phys. Rev. **104**, 1281 (1956)

6.63   D.J. Howarth, E.H. Sondheimer, Proc. Roy. Soc. (London) A **219**, 53 (1953)

6.64   H. Ehrenreich, J. Appl. Phys. Suppl. **32**, 2155 (1961)

6.65   E.M. Conwell, Phys. Rev. **143**, 657 (1966)

6.66   K. Hess, H. Kahlert, J. Phys. Chem. Solids **32**, 2262 (1971)

6.67   F. Kuchar, A. Phillip, K. Seeger, Solid State Commun. **11**, 965 (1972)

6.68   G. Bauer, H. Kahlert, Phys. Rev. B **5**, 566 (1972)

6.69   H. Fröhlich, B.V. Paranjape, Proc. Phys. Soc. London B **69**, 21 (1956); H. Fröhlich, Proc. Roy. Soc. London A **160**, 230 (1937); Adv. Phys. **3**, 325 (1961)

6.70   K. Binder (ed.), The Monte Carlo Method in Condensed Matter Physics, 2nd edn. Topics Appl. Phys. Vol. 71 (Springer, Berlin, Heidelberg 1995)

6.71   T. Kurosawa, Proc. Int'l Conf. Phys. Semicond, Kyoto 1966. J. Phys. Soc. Jpn. Suppl. **21**, 424 (1966)

6.72   W. Fawcett, A.D. Boardmann, S. Swain, J. Phys. Chem. Solids **31**, 1963 (1970)

6.73   H. Budd, Proc. Int'l Conf. Phys. Semicond. Kyoto 1966. J. Phys. Soc. Jpn. Suppl. **21**, 420 (1966)

6.74   H.D. Rees, J. Phys. Chem. Solids **30**, 643 (1969); IBM J. Res. Dev. **13**, 537 (1969)

6.75   K. Seeger, H. Pötzl, Acta Phys. Austriaca, Suppl. X, 341 (1973); O. Zimmerl, Iterative solution of the Boltzmann equation for hot electrons in InSb, Thesis (in german), T.U. Vienna, Austria (1972)

6.76   O. Madelung, Introduction to Solid-State Physics, Springer Ser. Solid-State Sci., Vol. 2 (Springer, Berlin, Heidelberg 1978)

6.77   D.M. Larsen, Phys. Rev. **135**, A419 (1964); ibid, **144**, 697 (1966)

6.78   D.M. Larsen, E.J. Johnson, Proc. Int'l Conf. Phys. Semicond. Kyoto, 1966; J. Phys. Soc. Jpn. Suppl. **21**, 443 (1966)

6.79   M. Mikkor, F.C. Brown, Phys. Rev. **162**, 848 (1967)

6.80   J.W. Hodby, J.A. Borders, F.C. Brown, J. Phys. C **3**, 335 (1970)

6.81   H. Ehrenreich, J. Phys. Chem. Solids **8**, 130 (1959)

6.82   T. Holstein, Ann. Phys. NY **8**, 325, 343 (1959)

6.83   D. Emin, in Linear and Nonlinear Transport in Solids, ed. by J.T. Devrees, V.E. van Doren (Plenum, New York 1976) p.409 (review)

6.84   P. Nagels, in Amorphous Semiconductors, 2nd edn., ed. by M.H. Brodsky, Topics Appl. Phys., Vol. 36 (Springer, Berlin, Heidelberg 1985)

6.85   A.F.J. Levi, J.R. Hayes, P.M. Platzmann, W. Wiegmann, Phys. Rev. Lett. **55**, 2071 (1985)

6.86   M. Heiblum, M.I. Nathan, D.C. Thomas, C.M. Knoedler, Phys. Rev. Lett. **55**, 2200 (1985)

6.87   J.G. Ruch, IEEE Trans. ED-**19**, 652 (1972)

6.88   T.E. Bell, IEEE Spectrum **23**, 36 (August 1986); M. Heiblum, L.F. Eastman, Sci. Am. **256**, 65 (February 1987)

6.89   J. Appel, Phys. Rev. **122**, 1760 (1961)

6.90   R.T. Bate, R.D. Baxter, F.J. Reid, A.C. Beer, J. Phys. Chem. Solids **26**, 1205 (1965)

6.91   M. Kohler, Z. Physik **124**, 772 (1948); ibid, **125**, 679 (1949)

6.92   E.H. Sondheimer, Proc. Roy. Soc. London A **203**, 75 (1950)

6.93   T.P. McLean, E.G.S. Paige, J. Phys. Chem. Solids **16**, 220 (1960)

6.94   V.L. Bonch-Bruyevich, The Electronic Theory of Heavily Doped Semiconductors (American Elsevier, New York 1966)

6.95   C.M. Wolfe, G.E. Stillman, Semiconductors and Semimetals **10**, 135 (Academic New York 1974)

6.96   R.A. Abram, G.J. Rees, B.L.H. Wilson, Adv. Phys. **27**, 799 (1978)
6.97   C. Yamanouchi, K. Mizuguchi, W. Sasaki, J. Phys. Soc. Jpn. **22**, 859 (1967)
6.98   H. Fritzsche, J. Phys. Chem. Solids **6**, 69 (1958)
6.99   W. Klein, D. Geist, Z. Physik **201**, 411 (1967)
6.100   A. Miller, E. Abrahams, Phys. Rev. **120**, 745 (1960); B.I. Shklovskii, A.L. Efros, Electronic Properties of Doped Semiconductors, Springer Ser. Solid-State Sci., Vol. 45 (Springer, Berlin, Heidelberg 1984)
6.101   M. Pollak, T.H. Geballe, Phys. Rev. **122**, 1742 (1961)
6.102   J.P. McKelvey, Solid-State and Semiconductor Physics (Harper and Row, New York 1966)
6.103   J. Friedel, Dislocations (Pergamon, Oxford 1964)
6.104   S. Amelincks, in Solid State Physics, Suppl. 6, ed. by F. Seitz, D. Turnbull (Academic, New York 1964)
6.105   G.L. Pearson, W.T. Read, Jr., F.J. Morin, Phys. Rev. **93**, 93 (1954)
6.106   W. Schröter, Phys. Status Solidi **21**, 211 (1967)
6.107   W. Schröter, R. Labusch, Phys. Status Solidi **36**, 539 (1969)
6.108   W.T. Read, Jr., Philos. Mag. **46**, 111 (1955)
6.109   V.L. Bonch-Bruevich, Sh.M. Kogan, Fiz. Tverd. Tela **1**, 1221 (1959) [Engl. transl., Sov. Phys. - Solid-State **1**, 1118 (1959)]
6.110   B. Pödör, Acta Phys. Acad. Sci. Hung. **23**, 393 (1967); Phys. Status Solidi **16**, K167 (1966)
6.111   F. Düster, R. Labusch, Phys. Status Solidi (b) **60**, 161 (1973)
6.112   J.H.P. van Weeren, R. Struikmans, J. Blok, Phys. Status Solidi **19**, K107 (1967)
6.113   A.F. Gibson, J. Phys. Chem. Solids **8**, 147 (1959)

Chapter 7

7.1   O. Madelung, Introduction to Solid-State Theory, Springer Ser. Solid-State Sci., Vol. 2 (Springer, Berlin, Heidelberg 1978) Chap. 4
7.2   C. Herring, Bell. Syst. Tech. J. **34**, 237 (1955)
7.3   C. Herring, E. Vogt, Phys. Rev. **101**, 944 (1956); Erratum, Phys. Rev. **105**, 1933 (1957)
7.4   S.H. Koenig, Proc. Int'l School Phys. Enrico Fermi XXII, 515 (1961)
7.5   J.E. Smith, Jr., Appl. Phys. Lett. **12**, 233 (1968)
7.6   E.G.S. Paige, In Progress in Semiconductors, Vol. 8, ed. by A.F. Gibson, R.E. Burgess (Temple, London 1960) p.158, Fig. 36
7.7   H. Heinrich, M. Kriechbaum, J. Phys. Chem. Solids **31**, 927 (1970)
7.8   K. Bulthuis, Philips Res. Rep. **23**, 25 (1968)
7.9   M. Shibuya, Phys. Rev. **95**, 1385 (1954); Physica **20**, 971 (1954)
7.10   A.C. Beer, Solid State Physics Vol. **4**, p.228 (Academic, New York 1963) ed. by F. Seitz, D. Turnbull
7.11   G.L. Pearson, H. Suhl, Phys. Rev. **83**, 768 (1951)
7.12   G.L. Pearson, C. Herring, Physica **20**, 975 (1954)

7.13   F. Seitz, Phys. Rev. **79**, 372 (1950)
7.14   H. Miyazawa, Proc. Int'l Conf. Phys. Semicond. Exeter, 1962 (Institute of Physics, London 1962) p.636
7.15   R.J. Stirn, W.M. Becker, Phys. Rev. **141**, 621 (1966)
7.16   C. Herring, T.H. Geballe, J.E. Kunzler, Bell Syst. Tech. J. **38**, 657 (1959)
7.17   P.M. Eagles, D.M. Edward, Phys. Rev. **138**, A1706 (1965)
7.18   D. Long, Phys. Rev. **120**, 2024 (1960)
7.19   R.A. Laff, H.Y. Fan, Phys. Rev. **112**, 317 (1958)
7.20   H.O. Haller, 33-GHz-Mikrowellen-Faraday-Effekt in isotropen und anisotropen Halbleitern, Thesis, Univ. Vienna, Austria (1972); L.J. Neuringer, Proc. Int'l Conf. Phys. Semicond., Paris 1964 (Dunod, Paris 1964) p.379
7.21   H. Bruns, Z. Naturforsch. **19a**, 533 (1964)
7.22   R.A. Laff, H.Y. Fan, Phys. Rev. **112**, 317 (1958)
7.23   C. Goldberg, W.E. Howard, Phys. Rev. **110**, 1035 (1953)
7.24   H.W. Streitwolf, Phys. Status Solidi **37**, K47 (1970)
7.25   D. Long, Phys. Rev. **120**, 2024 (1960)
7.26   M. Asche, B.L. Boichenko, V.M. Bondar, O.G. Sarbej, Proc. Int'l Conf. Phys. Semicond., Moscow 1968 (Nauka, Leningrad 1968) p.793
7.27   W.A. Harrison, Phys. Rev. **104**, 1281 (1956). Showing that optical intervalley scattering in n-Si is negligible
7.28   E.M. Conwell, In Solid State Physics Suppl. 9, ed. by F. Seitz, D. Turnbull, H. Ehrenreich (Academic, New York 1967)
7.29   W. Sasaki, M. Shibuya, K. Mizuguchi, J. Phys. Soc. Jpn. **13**, 456 (1958); W. Sasaki, M. Shibuya, K. Mizuguchi, G. Hatoyama, J. Phys. Chem. Solids **8**, 250 (1959)
7.30   M. Shibuya, Phys. Rev. **99**, 1189 (1955)
7.31   H. Heinrich, K. Lischka, M. Kriechbaum, Phys. Rev. B. **2**, 2009 (1970)
7.32   D.E. Aspnes, Phys. Rev. B. **14**, 5331 (1976)
7.33   J.E. Smith, M.I. Nathan, J.C. McGroddy, S.A. Porowski, W. Paul, Appl. Phys. Lett. **15**, 242 (1969)
7.34   B.K. Ridley, T.B. Watkins, Proc. Phys. Soc. London **78**, 293 (1961); B.K. Ridley, Proc. Phys. Soc. London **82**, 954 (1963)
7.35   C. Hilsum, Proc. IRE **50**, 185 (1962)
7.36   J.B. Gunn, Solid State Commun. **1**, 88 (1963); Plasma Effects in Solids, ed. by J. Bok (Dunod, Paris 1964) p.199
7.37   J.B. Gunn, Int'l J. Sci. Technol. **46**, 43 (1965) (historical review)
7.38   J.E. Carroll, Hot Electron Microwave Generators (Arnold, London 1970) p.105ff
7.39   D.E. McCumber, A.G. Chynoweth, IEEE Trans. ED-13 , 4 (1966)
7.40   P.N. Butcher, W. Fawcett, C. Hilsum, Brit. J. Appl. Phys. **17**, 841 (1966)
7.41   P.N. Butcher, W. Fawcett, Brit. J. Appl. Phys. **17**, 1425 (1966)
7.42   J.B. Gunn, J. Phys. Soc. Jpn. Suppl. **21**, 505 (1966)
7.43   B.K. Ridley, Phys. Lett. **16**, 105 (1965)
7.44   A.R. Hutson, A. Jayaraman, A.G. Chynoweth, A.S. Corriel, A.L. Feldman, Phys. Rev. Lett. **14**, 639 (1965)

7.45   J.W. Allen, M. Shyam, Y.S. Chen, G.L. Pearson, Appl. Phys. Lett. **7**, 78 (1965)

7.46   P.J. Vinson, C. Pickering, A.R. Adams, W. Fawcett, G.D. Pitt, Proc. 13th Int'l Conf. Phys. Semicond, Rome 1976, ed. by F.G. Fumi (North-Holland, Amsterdam 1976) p.1243

7.47   R.W. Keyes, Solid State Physics Vol. **11**, ed. by F. Seitz, D. Turnbull (Academic, New York 1960) p.149

7.48   H.W. Thim, M.R. Barber, B.W. Hakki, S. Knight, M. Uenohara, Appl. Phys. Lett. **7**, 167 (1965)

7.49   J.A. Copeland, J. Appl. Phys. **38**, 3096 (1967)

7.50   G.A. Acket, Festkörper-Probleme **IX**, 282 (Vieweg, Braunschweig 1969)

7.51   B. Bott, C. Hilsum, IEEE Trans. ED-**14**, 492 (1967)

7.52   S. Kataoka, H. Tateno, M. Kawashima, Y. Komamiya,7th Int'l Conf. Microwave Optical Generation and Amplification, Hamburg 1968 (VDE Berlin 1968)

7.53   G.W. Ludwig, IEEE Trans. ED-**14**, 547 (1967)

7.54   J.C. McGroddy, M.R. Lorenz, T.S. Plaskett, Solid State Commun. **7**, 901 (1969)

7.55   J.W. Allen, M. Shyam, G.L. Pearson, Appl. Phys. Lett. **11**, 253 (1967)

7.56   J.C. McGroddy, M.I. Nathan, J.E. Smith, Jr., IBM J. Res. Dev. **13**, 543 (1969)

7.57   E.G.S. Paige, IBM J. Res. Dev. **13**, 562 (1969)

7.58   J.E. Smith, Jr., Appl. Phys. Lett. **12**, 233 (1968)

7.59   J.E. Smith, Jr., J.C. McGroddy, M.I. Nathan, Proc. Int'l Conf. Phys. Semicond. Moscow 1968 (Nauka, Leningrad 1968) p.950

7.60   T.K. Gaylord, P.L. Shah, T.A. Rabson, IEEE Trans. ED-**15**, 777 (1968); ibid. ED-**16**, 490 (1969)

7.61   A.M. Barnett, IBM J. Res. Dev. **13**, 522 (1969)

7.62   J.D. Maines, E.G.S. Paige, J. Physique C **2**, 175 (1969)

7.63   G. Weinreich, T.M. Sanders, Jr., H.G. White, Phys. Rev. **114**, 33 (1959)

7.64   N.G. Einspruch, Solid State Physics Vol. **17**, ed. by F. Seitz, D. Turnbull (Academic, New York 1965) p.243

7.65   M. Pomerantz, Proc. IEEE **53**, 1438 (1965)

7.66   D.L. White, J. Appl. Phys. **33**, 2547 (1962); A.R. Hutson, D.L. White, J. Appl. Phys. **33**, 40 (1962)

7.67   N.I. Meyer, M.H. Jörgensen, Festkörper-Probleme **X**, 21 (Vieweg, Braunschweig 1970)

7.68   D.L. White, E.T. Handelman, J.T. Hanlon, Proc. IEEE **53**, 2157 (1965)

7.69   J.H. McFee, J. Appl. Phys. **34**, 1548 (1963)

7.70   M. Cardona (ed.), Light Scattering in Solids, 2nd edn., Topics Appl. Phys., Vol. 8 (Springer, Berlin, Heidelberg 1982)

7.71   M. Cardona, G. Güntherodt (eds.), Light Scattering in Solids II and III, Topics Appl. Phys., Vols. 50 and 51 (Springer, Berlin, Heidelberg 1982)

7.72   W. Wettling, in II-VI Semiconducting Compounds, ed. by D.G. Thomas (Benjamin, New York 1967) p.928

7.73   K. Hess, H. Kuzmany, Proc. Int'l Conf. Phys. Semicond, Warsaw 1972, ed. by M. Miasek (PWN-Polish Scientific Publ., Warsaw 1972) p.1233

7.74   P.K. Tien, Phys. Rev. **171**, 970 (1968)

Chapter 8

8.1   L. Pincherle, in Proc. Int'l School of Physics XXII, ed. by R.A. Smith (Academic, New York 1963) p.43

8.2   J.G. Mavroides, in Optical Properties of Solids, ed. by F. Abeles (North-Holland, Amsterdam 1972) p.394

8.3   G. Dresselhaus, A.F. Kip, C. Kittel, Phys. Rev. **98**, 368 (1955)

8.4   B.W. Levinger, D.R. Frankl, J. Phys. Chem. Solids **20**, 281 (1961)

8.5   G.E. Pikus, G.L. Bir, Fiz. Tverd. Tela **1**, 1642 (1959) [Engl. transl., Sov. Phys. - Solid State **1**, 1502 (1959)]

8.6   K. Bulthuis, Philips Res. Rep. **23**, 25 (1968)

8.7   S.H. Koenig, in Proc. Int'l School of Physics XXII, ed. by R.A. Smith (Academic, New York 1963) p.515

8.8   B. Lax, J.G. Mavroides, Phys. Rev. **100**, 1650 (1955)

8.9   J.W. McClure, Phys. Rev. **101**, 1642 (1956)

8.10   J.G. Mavroides, B. Lax, Phys. Rev. **107**, 1530 (1957); ibid. **108**, 1648 (1957)

8.11   P. Lawaetz, Phys. Rev. **174**, 867 (1968) footnote p.875

8.12   P. Lawaetz, Some Transport Properties of Holes in Germanium, Thesis, Tech. Univ. Copenhagen (1967) p.138

8.13   A.C. Beer, Solid State Physics Vol. **4**, Chap. 20b (Academic, New York 1963) ed. by F. Seitz, D. Turnbull

8.14   G.L. Pearson, H. Suhl, Phys. Rev.**83**, 768 (1951)

8.15   G.L. Pearson, C. Herring, Physica **20**, 975 (1954)

8.16   D.M. Brown, R. Bray, Phys. Rev. **127**, 1593 (1962)

8.17   D.M.S. Bagguley, R.A. Stradling, Proc. Phys. Soc. London **78**, 1078 (1961)

8.18   A.C. Beer, R.K. Willardson, Phys. Rev. **110**, 1286 (1958)

8.19   E.O. Kane, J. Phys. Chem. Solids **1**, 82 (1956)

8.20   D. Matz, J. Phys. Chem. Solids **28**, 373 (1967)

8.21   M. Asche, J. von Borzeszkowski, Phys. Status Solidi **37**, 433 (1970)

8.22   P. Hauge, The Microwave Magneto-Kerr Effect in Germanium and Silicon and its Applications to Studies of Carrier Relaxation Time and Effective Mass. Ph.D. Thesis, Univ. of Minnesota, USA (1967)

8.23   M. Costato, L. Reggiani, Lett. Nuovo Cimento Ser.I., **3**, 239 (1970)

8.24   G. Persky, D.J. Bartelink, IBM J. Res. Dev. **13**, 607 (1969)

8.25   W. Fawcett, J.G. Ruch, Appl. Phys. Lett. **15**, 369 (1969)

8.26   H. Miyazawa, K. Suzuki, H. Maeda, Phys. Rev. **131** , 2442 (1963)

8.27   D. Long, Phys. Rev. **107**, 672 (1957)

8.28   R. Bray, D.M. Brown, Proc. Int'l Conf. Phys. Semicond, Prague 1960 (Czech. Acad. Sciences, Prague 1960) p.82

8.29   A.C. Prior, Proc. Phys. Soc. London **76**, 465 (1960)

8.30   W.E.K. Gibbs, J. Appl.Phys. **33**, 3369 (1962)

Chapter 9

9.1   L. Esaki, Phys. Rev. **109**, 603 (1958)

9.2   W.K. Chow, Principles of Tunnel Diode Circuits (Wiley, New York 1964) (review)
9.3   S.M. Sze, Physics of Semiconductor Devices (Wiley, New York 1969)
9.4   D. Bohm, Quantum Theory (Prentice Hall, Englewood Cliffs, NJ 1951)
9.5   E.O. Kane, J. Appl. Phys. **32**, 83 (1961); J. Phys. Chem. Solids **2**, 181 (1960)
9.6   J. Karlovsky, Phys. Rev. **127**, 419 (1962)
9.7   E.O. Kane, Phys. Rev. **131**, 79 (1963)
9.8   P. Thomas, H.J. Queisser, Phys. Rev. **175**, 983 (1968)
9.9   P. Thomas, Fachberichte DPG (Teubner, Stuttgart 1968) p.114 (review); H. Zetsche, Fachberichte DPG (Teubner, Stuttgart 1970) p.172
9.10   C. Zener, Proc. Roy. Soc. London **145**, 523 (1934)
9.11   K.B. McAfee, E.J. Ryder, W. Shockley, M. Sparks, Phys. Rev. **83**, 650 (1951)
9.12   S.L. Miller. Phys. Rev. **99**, 1234 (1955)
9.13   L. Esaki, R. Tsu, IBM J. Res. Dev. **14**, 61 (1970)
9.14   A.Y. Cho, Thin Solid Films **100**, 291 (1983)
9.15   P.D. Dapkus, J. Crystal Growth **68**, 345 (1984)
9.16   W. Richter, Festkörperprobleme/Advances in Solid State Physics **26**, 335, ed. by P. Grosse (Vieweg, Braunschweig 1986)
9.17   R. Tsu, L. Esaki, Appl. Phys. Lett. **22**, 562 (1973)
9.18   H. Morkoc, J. Chen, U.K. Reddy, T. Henderson, S. Luryi, Appl. Phys. Lett. **49**, 70 (1986)
9.19   S. Luryi, in High-Speed Semiconductor Devices, ed. by S.M. Sze, (Wiley, New York 1990)
9.20   S. Luryi, in Heterojunction Band - Discontinuities, Physics and Device Applications, ed. by F. Capasso, G. Margaritondo, (Elsevier, Amsterdam 1987) p.489
9.21   T.C.L.G. Sollner, W.D. Goodhue, P.E. Tannenwald, C.D. Parker, D.D. Peck, Appl. Phys. Lett. **43**, 588 (1983)
9.22   T.C.L.G. Sollner, P.E. Tannenwald, D.D. Peck, W.D. Goodhue, Appl. Phys. Lett. **45**, 1319 (1984)
9.23   F. Capasso, R.A. Kiehl, J. Appl. Phys. **58**, 1366 (1985)
9.24   F. Capasso, S. Sen, A.C. Gossard, A.L. Hutchinson, J.E. English, IEEE EDL-**7**, 573 (1986)
9.25   M. Jonson, in Quantum Transport in Semiconductors ed. by D.K. Ferry, C. Jakoboni (Plenum, New York 1991) p.193
9.26   B.D. McCombe, A. Petrou, in Handbook on Semiconductors ed. by T.S. Moss, M. Balkanski (Elsevier, Amsterdam 1994) Vol. 2, p.285 ff
9.27   T.P. Pearsall, J.M. Vandenberg, R. Hall, J. Bonar, Phys. Rev. Lett. **63**, 2104 (1989)
9.28   F.H. Pollak, in Handbook on Semiconductors ed. by T.S. Moss (Elsevier, Amsterdam 1994) Vol. 2, p.527
9.29   W. Wegscheider, J. Olajos, U. Menczigar, W. Dondl, G. Abstreiter, J. Crystal Growth **123**, 75 (1992)
9.30   G. Abstreiter, J. Olajos, R. Schorer, P. Vogl, W. Wegscheider, Semicond. Sci. Technol. **8**, S6 (1993)

9.31   J. Allegre, J. Calayud, B. Gil, H. Mathieu, H. Tuffigo, G. Lentz, N. Magnea,
        H. Mariette, Phys. Rev. B **41**, 8195 (1990)
9.32   G.H. Döhler, Phys. Status. Solidi (b) **53**, 73, 533 (1972)
9.33   K.K. Choi, B.F. Levine, R.J. Malik, J. Walker, C.G. Bethea, Phys. Rev. B
        35, 4172 (1987)
9.34   J. Kastrup, R. Hey, K.H. Ploog, H.T. Grahn, L.L. Bonilla, M. Kindelau,
        M. Moscoso, A. Wacker, J. Galan, Phys. Rev. B **55**, 2476 (1997)
9.35   G.H. Wannier, Elements of Solid State Theory (Cambridge Univ. Press,
        Cambridge (England) 1959); Rev. Mod. Phys. **34**, 645 (1962); Phys. Rev.
        181, 1364 (1969)
9.36   G. Bastard, J.A. Brum, R. Ferreira, Solid State Physics Vol. **44**, ed. by H.
        Ehrenreich, D.Turnbull (Academic, New York 1991) p.316
9.37   E.E. Mendez, F. Agullò-Rueda, J.M. Hong, Phys. Rev. Lett. **60**, 2426
        (1988)
9.38   G. von Plessen, P. Thomas, Phys. Rev. B **45**, 9185 (1992)
9.39   J. J. Bleuse, G. Bastard, P. Voisin, Phys. Rev. Lett. **60**, 220 (1988)
9.40   P. Voisin, G. Bastard, M. Voos, Phys. Rev. B **29**, 935 (1984)
9.41   E.E. Mendez, G. Bastard, Phys. Today **46**, 34 (June 1993)
9.42   G. Bastard, Wave Mechanics Applied to Semiconductor Heterostructures
        (Les Editions de Physique, Les Ulis Cedex, France, 1987) p.26
9.43   K. Unterrainer, B.J. Keay, M.C.Wanke, S.J. Allen, D. Leonhard, G.
        Medeiros-Ribeiro, U. Bhattacharya, M.J.W. Rodwell, Phys. Rev. Lett.
        76, 2973 (1996)
9.44   A.A. Ignatov, A.P. Jauho, J. Appl. Phys. **85**, 3643 (1999)
9.45   S. Winnerl, E. Schomburg, J. Grenzer, H.J. Regl, A.A. Ignatov, A.D. Se-
        menov, K.F. Renk, D.G. Pavel'ev, Yu. Koscharinov, B. Melzer, V. Ustinov,
        S. Ivanov, S. Schaposchnikov, P.S. Kop'ev, Phys. Rev. B **56**, 10303 (1997)
9.46   Yu.A. Romanov, Opt.Spectrosc. **33**, 917 (1972)
9.47   L.D. Landau, Z. Physik **64**, 629 (1930)
9.48   O. Madelung, Festkörper-Probleme **V**, 87 (Vieweg, Braunschweig 1960)
9.49   R.G. Chambers, Can. J. Phys. **34**, 1395 (1956)
9.50   G. Bauer, H. Kahlert, Phys. Rev. B **5**, 566 (1972); Proc. Int'l Conf. Phys.
        Semicond., Cambridge MA, 1970 (USAEC, Oak Ridge, TN 1970) p.65
9.51   L. Shubnikov, W.J. de Haas, Leiden Commun. 207a, 207c, 207d, 210a
        (1930)
9.52   E.N. Adams, T.D. Holstein, J. Phys. Chem. Solids **10**, 254 (1959)
9.53   L.I. Schiff, Quantum Mechanics (McGraw-Hill, New York 1968) p.441
9.54   M. Cardona, J. Phys. Chem. Solids **24**, 1543 (1963)
9.55   R.B. Dingle, Proc. Roy. Soc. London A **211**, 517 (1952)
9.56   R.A. Isaacson, F. Bridges, Solid State Commun. **4**, 635 (1966)
9.57   G. Bauer, H. Kahlert, Phys. Lett. **41**, A351 (1972)
9.58   L.M. Roth, P.N. Argyres, Semiconductors and Semimetals **1**, ed. by R.K.
        Willardson, A.C. Beer (Academic, New York 1966) p.159; S.M. Puri, T.H.
        Geballe, Semiconductors and Semimetals **1**, ed. by R.K. Willardson, A.C.
        Beer (Academic, New York 1966) p.203 (thermomagnetic effects)
9.59   W.J. de Haas, P.M. van Alphen, Leiden Commun. 208d, 212a (1930); 220d
        (1933)

9.60   R.A. Smith, The Physical Principles of Thermodynamics (Chapman and
          Hall, London 1952) p.154; R.A. Smith, Wave Mechanics of Crystalline Solids
          (Chapman and Hall, London 1961) p.384

9.61   J.M. Ziman, Principles of the Theory of Solids (Cambridge Univ. Press,
          Cambridge 1964) Sect.9.7

9.62   D. Shoenberg, in The Physics of Metals 1. Electrons ed. by J.M. Ziman
          (Cambridge Univ. Press, Cambridge 1969)

9.63   Y. Yafet, R.W. Keyes, E.N. Adams, J. Phys. Chem. Solids **1**, 137 (1956)

9.64   R.W. Keyes, R.J. Sladek, J. Phys. Chem. Solids **1**, 143 (1956)

9.65   E.H. Putley, Proc. Phys. Soc. London **76**, 802 (1960); J. Phys. Chem.
          Solids **22**, 241 (1961)

9.66   L.J. Neuringer, Proc. Int'l Conf. Phys. Semicond., Moscow 1968 (Nauka,
          Leningrad 1968) p.715

9.67   E.H. Putley, Semiconductors and Semimetals 1, ed. by R.K. Willardson,
          A.C. Beer (Academic, New York 1966) p.289; E.H. Putley, Phys. Status
          Solidi **6**, 571 (1964)

9.68   E.R. Brown, M.J. Wengler, T.G. Phillips, J. Appl. Phys. **58**, 2051 (1985)

9.69   V.L. Gurevich, Yu.A. Firsov, Zh. Eksp. Teor. Fiz. **40**, 199 (1961) [Engl.
          transl., Sov. Phys. - JETP **13**, 137 (1961)]

9.70   M.I. Klinger, Fiz. Tverd. Tela **3**, 1342 (1961) [Engl. transl., Sov. Phys. -
          Solid State **3**, 974 (1961)]

9.71   Yu.A. Firsov, V.L. Gurevich, R.V. Parfeniev, S.S. Shalyt, Phys. Rev. Lett.
          12, 660 (1964) for a degenerate electron gas

9.72   A.L. Efros, Fiz. Tverd. Tela **3**, 2848 (1961) [Engl. transl. Sov. Phys. -
          Solid State **3**, 2079 (1962)]

9.73   L. Eaves, R.A. Stradling, R.A. Wood, Proc. Int'l Conf. Phys. Semicond.,
          Cambridge MA, 1970, ed. by S.P. Keller, J.C. Hensel, F. Stern (USAEC,
          Oak Ridge, TN 1970) p.816

9.74   V.L. Gurevich, Yu.A. Firsov, Zh. Eksp. Teor. Fiz. **47**, 734 (1964) [Engl.
          transl., Sov. Phys. - JETP **20**, 489 (1964)]

9.75   R.A. Stradling, R.A. Wood, J. Physique C **1**, 1711 (1968)

9.76   A.L. Mears, R.A. Stradling, E.K. Inall, J. Physique C **1**, 821 (1968)

9.77   R.A. Reynolds, Solid State Electron. **11**, 385 (1968)

9.78   R.A. Wood, R.A. Stradling, I.P. Molodyan, J. Phys. Paris C **3**, L154 (1970)

9.79   S.M. Puri, T.H. Geballe, Semiconductors and Semimetals 1, 203 (Academic,
          New York 1966)

9.80   J. Bardeen, L.N. Cooper, J.R. Schrieffer, Phys. Rev. **108**, 1157 (1957)

9.81   J.R. Tucker, IEEE J. QE-15, 1234 (1979)

Chapter 10

10.1   G. Lautz, Halbleiterprobleme **VI**, 21 (Vieweg, Braunschweig 1961)

10.2   S.H. Koenig, G.R. Gunther-Mohr, J. Phys. Chem. Solids **2**, 268 (1957)

10.3   N. Sclar, E. Burstein, J. Phys. Chem. Solids **2**, 1 (1967)

| 10.4 | K. Baumann, M. Kriechbaum, H. Kahlert, J. Phys. Chem. Solids **31**, 1163 (1970) |

10.4   K. Baumann, M. Kriechbaum, H. Kahlert, J. Phys. Chem. Solids **31**, 1163 (1970)

10.5   S.H. Koenig, in Proc. Int'l Conf. Solid State Physics, Brussels 1958, ed. by M. Desirant (Academic, London 1960) p.422

10.6   M. Lax, J. Phys. Chem. Solids **8**, 66 (1959)

10.7   G. Bauer, F. Kuchar, Phys. Status Solidi (a) **13**, 169 (1972)

10.8   W.P. Dumke, Phys. Rev. **167**, 783 (1968)

10.9   R.C. Curby, D.K. Ferry, Phys. Status Solidi (a) **15**, 319 (1973)

10.10   G. Nimtz, in Proc. Int'l. Conf. Phys. Semicond. Cambridge MA 1970, ed. by S.P. Keller, J.C. Hensel, F.Stern (USAEC, Oak Ridge, TN 1970) p.396

10.11   A.L. McWhorter, R.H. Rediker, Proc. Int'l Conf. Phys. Semicond., Prague 1960 (Czech. Acad. Sciences, Prague 1960) p.134

10.12   B.K. Ridley, Proc. Phys. Soc., London **81**, 996 (1963)

10.13   A.M. Barnett, IBM J. Res. Dev. **13**, 522 (1969)

10.14   R.F. Kazarinov, V.G. Skobov, Zh. Eksp. Teor. Fiz. **42**, 1047 (1962) [Engl. transl., Sov. Phys. - JETP **15**, 726 (1962)]

10.15   B. Ancker-Johnson, in Semiconductors and Semimetals, Vol. 1, ed. by R.K. Willardson, A.C. Beer (Academic, New York 1966)

10.16   A.G. Chynoweth, K.G. McKay, Phys. Rev. **108**, 29 (1957)

10.17   A.G. Chynoweth, Semiconductors and Semimetals Vol. 4, ed. by R.K. Willardson, A.C. Beer (Academic, New York 1968) p.263

10.18   C.A. Lee, R.A. Logan, R.L. Batdorf, J.J. Kleimack, W. Wiegman, Phys. Rev. **134**, A761 (1964)

10.19   S.M. Sze, Physics of Semiconductor Devices, (Wiley, New York 1969) p.60

10.20   G.A. Baraff, Phys. Rev. **128**, 2507 (1962); ibid, **133**, A26 (1964)

10.21   J.E. Carroll, Hot Electron Microwave Generators (Arnold, London 1970)

10.22   A.F. Gibson, J.W. Granville, E.G.S. Paige, J. Phys. Chem. Solids **19**, 198 (1961)

10.23   A.E. Michel, M.I. Nathan, J.C. Marinace, J. Appl. Phys. **35**, 3543 (1964)

10.24   W. Shockley, Bell Syst. Tech. J. **33**, 799 (1954)

10.25   W.T. Read, Bell. Syst. Tech. J. **37**, 401 (1958)

10.26   H. Hartnagel, Semiconductor Plasma Instabilities (Heinemann, London 1969)

Chapter 11

11.1   Landolt-Börnstein, Group 3, Vol. 17, Pt.a, Physics of Group IV Elements and III–V Compounds (Springer, Berlin, Heidelberg 1982) Sect. 3.2.5; M. Balkanski (ed.), Handbook on Semiconductors, Vol. 2, Optical Properties of Solids (North-Holland, Amsterdam 1981); J.I. Pankove, Optical Processes in Semiconductors (Prentice Hall, Englewood Cliffs, NJ 1971) (reviews)

11.2   R.E. Lavilla, H. Mendlowitz, J. Appl. Phys. **40**, 3297 (1969)

11.3   H.R. Philipp, E.A. Taft, Phys. Rev. **113**, 1002 (1959)

11.4   L. Marton, J. Toots, Phys. Rev. **160**, 602 (1967)

11.5   W.R. Hunter, in Optical Properties and Electronic Structure of Metals and Alloys, ed. by F. Abeles (North-Holland, Amsterdam 1966) p.136

11.6   R. de Laer Kronig, J. Opt. Soc. Am. **12**, 547 (1926); H.A. Kramers, Atti Congr. Fisici Como (1927) p.545

11.7   D.L. Greenaway, G. Harbeke, Optical Properties of Semiconductors (Pergamon, New York 1968)

11.8   D. Brust, J.C. Phillips, G.F. Bassani, Phys. Rev. Lett. **9**, 94 (1962)

11.9   M. Cardona, F.H. Pollak, Phys. Rev. **142**, 530 (1966)

11.10   O. Madelung, Introduction to Solid-State Theory, Springer Ser. Solid-Sci., Vol. 2 (Springer, Berlin, Heidelberg 1978) J. Callaway, Energy Band Theory (Academic, New York 1964)

11.11   J. Tauc, A. Abraham, J. Phys. Chem. Solids **20**, 190 (1961)

11.12   T.S. Moss, T.D. Hawkins, Infrared Phys. **1**, 111 (1962)

11.13   R. Newman, W.W. Tyler, Solid State Physics Vol. **8**, ed. by F. Seitz, D. Turnbull (Academic, New York 1959) p.49

11.14   G.G. Macfarlane, T.P. McLean, J.E. Quarrington, V. Roberts, Phys. Rev. **108**, 1377 (1957)

11.15   P.J. Dean, D.G. Thomas, Phys. Rev. **150,** 690 (1966)

11.16   J. Kudman, T. Seidel, J. Appl. Phys. **33**, 771 (1962)

11.17   J.A. Stratton, Electromagnetic Theory (McGraw-Hill, New York 1941) p.23

11.18   D.F. Edwards, T.E. Slykhouse, H.G. Drickamer, J. Phys. Chem. Solids **11**, 140 (1959)

11.19   E.J. Johnson, Semiconductors and Semimetals, Vol. **3**, ed. by R.K. Willardson, A.C. Beer (Academic, New York 1967) Table II, p.200

11.20   D.E. Aspnes, Phys. Rev. B **14**, 5331 (1976)

11.21   R. Braunstein, A.R. Moore, F. Herman, Phys. Rev. **109**, 695 (1958)

11.22   H.J. Hrostowski, G.H. Wheatley, W.F. Flood, Phys. Rev. **95**, 1683 (1954)

11.23   R.G. Breckenridge, R.F. Blunt, W.R. Hosler, H.P.R. Frederikse, J.H. Becker, W. Oshinsky, Phys. Rev. **96**, 571 (1954)

11.24   W. Kaiser, H.Y. Fan, Phys. Rev. **98**, 966 (1955)

11.25   E. Burstein, Phys. Rev. **93**, 632 (1954)

11.26   K. Cho (ed.), Excitons, Topics Curr. Phys., Vol. 14 (Springer, Berlin, Heidelberg 1979)

11.27   S. Nikitine, J. Bielmann, J.L. Deiss, M. Grossmann, J.B. Grun, R. Ringeissen, G. Schwab, M. Siesskind, L. Wursteisen, Proc. Int'l Conf. Phys. Semicond., Exeter 1962 (Institute of Physics, London 1962)

11.28   M.D. Sturge, Phys. Rev. **127,** 768 (1962)

11.29   G.H. Wannier, Phys. Rev. **52**, 191 (1937)

11.30   J. Frenkel, Phys. Rev. **37**, 17, 1276 (1931)

11.31   D.G. Thomas, J.J. Hopfield, Phys. Rev. **124**, 657 (1961)

11.32   D.L. Dexter, R.S. Knox, Excitons (Wiley, New York 1965)

11.33   R.S. Knox, in Solid State Physics, Suppl.5, ed. by F. Seitz, D. Turnbull (Academic, New York 1963)

11.34   Ya.E. Pokrovskii, K.I. Svistunova, Sov. Phys. Semicond. **4**, 409 (1970); Sov. Phys. - JETP Lett. **13**, 212 (1971)

11.35   J.R. Haynes, Phys. Rev. Lett. **17**, 860 (1966) for Si

11.36   L.V. Keldysh, Proc. 9th Int'l Conf. Phys. Semicond., Moscow 1968 (Nauka, Leningrad 1968) p.1303, particularly p.1306

11.37   C. Benoit a la Guillaume, M. Voos, F. Salvan, J.M. Laurant, A. Bonnot, C.R. Acad. Sci. Ser. B **272**, 236 (1971)

11.38   G.A. Thomas, T.M. Rice, J.C. Hensel, Phys. Rev. Lett. **33**, 219 (1974)

11.39   E. Burstein, G.S. Picus, H.A. Gebbie, F. Blatt, Phys. Rev. **103**, 826 (1956)

11.40   S. Zwerdling, R.J. Keyes, S. Foner, H.H. Kolm, B. Lax, Phys. Rev. **104**, 1805 (1956)

11.41   S. Zwerdling, B. Lax, L.M. Roth, Phys. Rev. **108** , 1402 (1957)

11.42   S. Zwerdling, B. Lax, Phys. Rev. **106**, 51 (1957)

11.43   L.M. Roth, B. Lax, S. Zwerdling, Phys. Rev. **114** , 90 (1959)

11.44   W. Franz, Z. Naturforsch. A **13**, 484 (1958)

11.45   L.V. Keldysh, Zh. Eksp. Teor. Fiz. **34**, 1138 (1958) [Engl. transl. Sov. Phys. - JETP 7, 788 (1958)]

11.46   M. Abramowitz, I.A. Stegun, Handbook of Mathematical Functions (Dover, New York 1965) pp.475 - 477 (Airy function); K. Tharmalingam, Phys. Rev. **130**, 2204 (1963)

11.47   T.S. Moss, J. Appl. Phys. Suppl. **32**, 2136 (1961); C. Coriasso, D. Campi, C. Cacciatore, C. Alibert, S. Gaillard, B. Lambert, A. Regreny, Europhys. Lett. **16**, 591 (1991)

11.48   Y. Hamakawa, F. Germano, P. Handler, Proc. Int'l Conf. Phys. Semicond., Kyoto 1966 (The Physical Society of Japan, Tokyo 1966) p.111

11.49   B.O. Seraphin, R.B. Hess, Phys. Rev. Lett. **14**, 138 (1965)

11.50   E.J. Johnson, Semiconductors and Semimetals **3**, 153 (R.K. Willardson, A.C. Beer, eds.) (Academic, New York 1967)

11.51   E.J. Johnson, H.Y. Fan, Phys. Rev. **139**, A1991 (1965)

11.52   R. Kaiser, H.Y. Fan, Phys. Rev. A **138**, 156 (1956)

11.53   E.B. Owens, A.J. Strauss, Proc. Int'l Conf. Ultrapurif. Semicond. Materials (Macmillan, New York 1961) p.340

11.54   D. Effer, P.J. Etter, J. Phys. Chem. Solids **25**, 451 (1964)

11.55   P. Vogl, Festkörper-Probleme/Advances in Solid State Physics **21**, 191 (Vieweg, Braunschweig; Pergamon, London 1981) p.191

11.56   H.J. Hrostowski, in Semiconductors, ed. by N.B. Hannay (Reinhold, New York 1959)

11.57   H.J. Hrostowski, R.H. Kaiser, Phys. Rev. **107**, 966 (1957)

11.58   W.S. Boyle, J. Phys. Chem. Solids **8**, 321 (1959)

11.59   J.M. Luttinger, W. Kohn, Phys. Rev. **97**, 869 (1955)

11.60   R.R. Haering, Can. J. Phys. **36**, 1161 (1958)

11.61   J.H. van Vleck, The Theory of Electric and Magnetic Susceptibilities(Oxford Univ. Press, London 1932) p.178

11.62   T.H. Geballe, F.J. Morin, Phys. Rev. **95**, 1805 (1954)

11.63   R.E. Howard, H. Hasegawa, Bull. Am. Phys. Soc. **5**, 178 (1960)

11.64   W.S. Boyle, R.E. Howard, J. Phys. Chem. Solids **19**, 181 (1961)

11.65   B. Lax, S. Zwerdling, Progress in Semiconductors **5**, 269 (Temple, London 1960)

11.66   W.J. Turner, W.E. Reese, Phys. Rev. **127**, 126 (1962)

11.67   M. Hass, Semiconductors and Semimetals Vol. **3**, ed. by R.K. Willardson, A.C. Beer (Academic, New York 1967) p.7

11.68   A. Mooradian, G.B. Wright, Solid State Commun. **4**, 431 (1966)

11.69   A. Mooradian, Festkörper-Probleme **IX**, 74 (Vieweg, Braunschweig 1969)
         (review where surface scattering and Raman scattering by thermal and hot
         solid state plasmas are included)

11.70   M. Cardona, G. Güntherodt (eds.), Light Scattering in Solids II, Topics
         Appl. Phys., Vol. 50 (Springer, Berlin, Heidelberg 1982)

11.71   K. Huang, Proc. Roy. Soc. London A **208**, 352 (1951)

11.72   C.H. Henry, J.J. Hopfield, Phys. Rev. Lett. **15**, 964 (1965)

11.73   J.J. Hopfield, Phys. Rev. **112**, 1555 (1958)

11.74   U. Fano, Phys. Rev. **103**, 1202 (1956)

11.75   W.G. Spitzer, Semiconductors and Semimetals Vol. **3**, ed. by R.K. Willard-
         son, A.C. Beer (Academic, New York 1967) p.17

11.76   F.A. Johnson, Proc. Phys. Soc. London B**73**, 265 (1959)

11.77   B.N. Brockhouse, Phys. Rev. Lett. **2**, 256 (1959)

11.78   R. Braunstein, Phys. Rev. **130**, 879 (1963)

11.79   G.F. Bassani, Proc. Int'l School of Physics, Vol. 34, ed. by J. Tauc
         (Academic, New York 1966) p.33

11.79a   W. Heitler, Quantum Theory of Radiation (Clarendon, Oxford 1954)

11.80   L.I. Schiff, Quantum Mechanics, 3rd edn. (McGraw-Hill New York 1968)

11.81   A.H. Wilson, The Theory of Metals (Cambridge Univ. Press, London 1953)
         p.46

11.82   M. Cardona, in Problems in Solid State Physics, ed. by H.J. Goldsmid
         (Pion, London; Academic, New York 1968) p.425

11.83   L. van Hove, Phys. Rev. **89**, 1189 (1953)

11.84   J.C. Phillips, Phys. Rev. **104**, 1263 (1956)

11.85   D.J. Chadi, M.L. Cohen, Phys. Lett. **49A**, 381 (1974)

11.86   P. Drude, Wied. Ann. **46**, 353 (1882); ibid. **48**, 122 (1883); and ibid. **49**,
         690 (1883)

11.87   P. Grosse, Freie Elektronen in Festkörpern (Springer, Berlin, Heidelberg
         1979)

11.88   J.R. Dixon, Proc. Int'l Conf. Phys. Semicond., Prague 1960 (Czech. Acad.
         Sci., Prague 1960) p.366

11.89   A. Yariv, Quantum Electronics (Wiley, New York 1968) Sect. 21.7

11.90   H.Y. Fan, W. Spitzer, R.J. Collins, Phys. Rev. **101**, 566 (1956); H.Y. Fan,
         Rep. Prog. Phys. **14**, 119 (1956)

11.91   H.J.G. Meyer, Phys. Rev. **112**, 298 (1958); J. Phys. Chem. Solids **8**, 264
         (1959); R. Rosenberg, M. Lax, Phys. Rev. **112**, 843 (1958)

11.92   V.L. Gurevich, I.G. Lang, Yu.A. Firsov, Fiz. Tverd. Tela **4**, 1252 (1962)
         [Engl. transl., Sov. Phys. - Solid State **4**, 918 (1962)]

11.93   W.M. König, Acta Phys. Austriaca **33**, 275 (1971)

11.94   D. Kranzer, W.M. König, Phys. Status Solidi (b) **48**, K133 (1971)

11.95   R. Wolfe, Proc. Phys. Soc. London A **67**, 74 (1954)

11.96   S. Visvanathan, Phys. Rev. **120**, 379 (1960)

11.97   A. Sommerfeld, Ann. Phys. (Leipzig) **11**, 257 (1931)

11.98   W.G. Spitzer, M. Gershenzon, C.J. Frosch, D.F. Gibbs, J. Phys. Chem.
         Solids **11**, 339 (1959)

11.99	W.G. Spitzer, J.M. Whelan, Phys. Rev. **114**, 59 (1959)
11.100	I. Balslev, Phys. Rev. **173**, 762 (1968) (non-equivalent valley transitions)
11.101	W.J. Turner, W.E. Reese, Phys. Rev. **117**, 1003 (1960)
11.102	R. Braunstein, E.O. Kane, J. Phys. Chem. Solids **23**, 1423 (1962)
11.103	I. Balslev, Phys. Rev. **177**, 1173 (1969)
11.104	W.G. Spitzer, H.Y. Fan, Phys. Rev. **106**, 882 (1957)
11.105	B. Lax, Proc. Int'l School of Physics, Vol.XXII, ed. by R.A. Smith (Academic, New York 1963) p.240 (review)
11.106	B. Lax, H.J. Zeiger, R.N. Dexter, Physica **20**, 818 (1954)
11.107	C.J. Rauch, J.J. Stickler, H. Zeiger, G. Heller, Phys. Rev. Lett. **4**, 64 (1960)
11.108	G. Dresselhaus, A.F. Kip, C. Kittel, Phys. Rev. **98**, 368 (1955)
11.109	W. Shockley, Phys. Rev. **79**, 191 (1950)
11.110	E.D. Palik, G.S. Picus, S. Teitler, R.F. Wallis, Phys. Rev. **122**, 475 (1961); E.D. Palik, G.B. Wright, Semiconductors and Semimetals Vol. **3**, ed. by R.K. Willardson, A.C. Beer (Academic, New York 1967) p.421
11.111	G. Bemski, Phys. Rev. Lett. **4**, 62 (1960)
11.112	E.D. Palik, S. Teitler, B.W. Henvis, R.F. Wallis, Proc. Int'l Conf. Phys. Semicond.,Exeter 1962, ed. by A.G. Stickland (Institute of Physics, London 1962) p.288
11.113	H. Piller, J. Appl. Phys. **37**, 763 (1966)
11.114	H.Y. Fan, in Semiconductors and Semimetals, Vol. 3, ed. by R.K. Willardson, A.C. Beer (Academic, New York 1967); the "optical effective mass" is the mass which appears in the expression for the free-carrier contribution to the dielectric constant
11.115	M.J. Stephen, A.B. Lidiard, J. Phys. Chem. Solids **9**, 43 (1959)
11.116	B. Donovan, J. Webster, Proc. Phys. Soc. London **79**, 46, 1081 (1962); ibid. **81**, 90 (1963)
11.117	R.R. Rau, M.E. Caspari, Phys. Rev. **100**, 632 (1955)
11.118	A. Bouwknegt, Microwave Galvanomagnetic Effects in n-type Germanium, Dissertation, Univ. Utrecht (1965) (review of experimental methods)
11.119	C.R. Pidgeon, Infrared Magneto-optical Phenomena in Semiconductors, Ph.D. Thesis, Univ. Reading (1962)
11.120	E.D. Palik, Appl. Opt. **2**, 527 (1963)
11.121	G.B. Wright, B. Lax, J. Appl. Phys. Suppl. **32**, 2113 (1961)
11.122	M. Born, E. Wolf, Principles of Optics, (Pergamon, London 1959) p.26
11.123	S. Teitler, E.D. Palik, R.F. Wallis, Phys. Rev. **123**, 1631 (1961) S. Teitler, E.D. Palik, Phys. Rev. Lett. **5**, 546 (1960)
11.124	Y. Nishina, J. Kolodziejczak, B. Lax, Proc. Int'l Conf. Phys. Semicond., Paris 1964, ed. by J. Bok (Dunod, Paris 1964) p.867
11.125	J. Halpern, B. Lax, Y. Nishina, Phys. Rev. **134** , A140 (1964)
11.126	J. Halpern, J. Phys. Chem. Solids **27**, 1505 (1966)
11.127	M. Glicksman, Solid State Physics **26**, 275 (Academic, New York 1971) ed. by F. Seitz, D. Turnbull
11.128	P.M. Platzman, P.A. Wolff, in Solid State Physics, Suppl. 13, ed. by H. Ehrenreich, F. Seitz, D. Turnbull (Academic, New York 1972)

11.129   A. Libchaber, R. Veilex, Phys. Rev. **127**, 774 (1962)
11.130   N. Perrin, B. Perrin, W. Mercouroff, in Plasma Effects in Solids, ed. by J. Bok (Dunod, Paris 1964) p.37
11.131   I. Rosenman, Solid State Commun. **3**, 405 (1965)
11.132   N. Bloembergen, Nonlinear Optics (Benjamin, New York 1965); Y.R. Shen (ed.); Nonlinear Infrared Generation, Topics Appl. Phys., Vol. 16 (Springer, Berlin, Heidelberg 1977); A.F. Gibson, Sci. Prog. **56**, 479 (1968)
11.133   J.H. McFee, Appl. Phys. Lett. **11**, 228 (1967)
11.134   C.K.N. Patel, R.E. Slusher, P.A. Fleury, Phys. Rev. Lett. **17**, 1011 (1966)
11.135   L. Stenflo, Phys. Rev. B **1**, 2821 (1970)
11.136   P.A. Wolff, G.A. Pearson, Phys. Rev. Lett. **17**, 1015 (1966); J. Kolodziejczak, Proc. Int'l Conf. Phys. Semicond., Moscow 1968 (Nauka, Leningrad 1968) p.233
11.137   B. Lax, W. Zawadzki, M.H. Weiler, Phys. Rev. Lett. **18**, 462 (1967)
11.138   R. Claus, L. Merten, J. Brandmüller, Light Scattering by Phonon-Polaritons, Springer Tracts Mod. Phys., Vol. 75 (Springer, Berlin, Heidelberg 1975); R. Claus, Festkörper-Probleme **XII**, 381 (Vieweg, Braunschweig 1972)

Chapter 12

12.1   S.M. Sze, Physics of Semiconductor Devices, 2nd edn. (Wiley, New York 1981)
12.2   H.J. Queisser, in Festkörper-Probleme, Vol. XI, ed. by O. Madelung (Vieweg, Braunschweig 1971)
12.3   R. Newman, W.W. Tyler, in Solid State Physics, Vol. 8, ed. by F. Seitz, D. Turnbull (Academic, New York 1959)
12.4   C.D. Thurmond, in Semiconductors, ed. by N.B. Hannay (Reinhold, New York 1960) p.145
12.5   W.W. Tyler, H.H. Woodbury, Phys. Rev. **96**, 874 (1954)
12.6   E.M. Pell, J. Appl. Phys. **26**, 658 (1955)
12.7   A.G. Milnes, Deep Impurities in Semiconductors (Wiley, New York 1973)
12.8   W.C. Dunlap, in Progress in Semiconductors, Vol. 2, ed. by A.F. Gibson (Temple, London 1957)
12.9   J.S. Blakemore, Semiconductor Statistics (Pergamon, London 1962)

Chapter 13

13.1   T.S. Moss, G.J. Burrell, B. Ellis, Semiconductor Optoelectronics (Butterworth, London 1973)
13.2   J.I. Pankove (ed.), Electroluminescence, Topics Appl. Phys., Vol. 17 (Springer, Berlin, Heidelberg 1977)
13.3   G. Destriau, J. Chim. Phys. **33**, 587 (1936)

13.4   L. Canham, Appl. Phys. Lett. **57**, 1046 (1990)

13.5   P.J. Dean, D.C. Herbert, in Excitons, ed. by K. Cho, Topics Curr. Phys., Vol. 14 (Springer, Berlin, Heidelberg 1979) Sect. 3.3.3

13.6   W.N. Carr, IEEE Trans. ED-12, 531 (1965)

13.7   N. Holonyak Jr., F. Bevaqua, Appl. Phys. Lett. **1**, 82 (1962)

13.8   M. Gershenzon, Bell. Syst. Tech. J. **45** , 1599 (1966)

13.9   C.H. Henry, P.J. Dean, J.D. Cuthbert, Phys. Rev. **166**, 754 (1968)

13.10   D.L. MacAdam, Color Measurement, 2nd edn., Springer Ser. Opt. Sci., Vol. 27 (Springer, Berlin, Heidelberg (1985)

13.11   R.W. Brander, Mater. Res. Bull. **4**, 187 (1969)

13.12   A.A. Bergh, P.J. Dean, Proc. IEEE **60** , 156 (1972)

13.13   M. Feng, N. Holonyak Jr., W. Hafez, Appl. Phys. Lett. **84**, 151 (2004)

13.14   W. Hafez, J.W. Lai, M. Feng, Electron Lett. **39** , 1475 (2003)

13.15   W. Hafez, J.W. Lai, M. Feng, IEEE Electron Device Lett. **24** , 436 (2003)

13.16   J. Bardeen in "Collected Works of John Von Neumann" Vol.5, p. 420, Pergamon, New York 1963

13.17   H.C. Casey Jr., M.B. Panish, Heterostructure Lasers,Part A, Academic Press, New York, 1978

13.18   M.G.A. Bernard, G. Duraffourg, Phys. Stat Sol. **1**, 699 (1961)

13.19   R.L. Anderson, IBM J. Res. Dev. **4**, 283 (1960)

13.20   J.J. Hsieh, Appl. Phys. Lett. **28**, 283 (1976)

13.21   C.J. Nuese, G. Olsen, Appl. Phys. Lett. **26**, 528 (1975)

13.22   R.E. Nahory, M.A. Pollak, E.D. Beebe, J.C. De Winter, R.W. Dixon, Appl. Phys. Lett. **28**, 19 (1976)

13.23   H. Kressel, in Handbook on Semiconductors, Vol. 4, ed. by C. Hilsum (North-Holland, Amsterdam 1981) p.617

13.24   A.V. Nurmikko, R.L. Gunshor, In Proc. 22nd Int'l Conf. Phys. Semicond., Vancouver (1994) p.27, ed. by D.J. Lockwood (World Scientific, Singapore 1995)

13.25   D.J. Olego, Europhys. News **26**, 112 (1995)

13.26   H. Amano, M. Kito, K. Hiramatsu, I. Akasaki, Jpn. J. Appl. Phys. **28**, L2112 (1998)

13.27   S. Nakamura, G. Fasol, The Blue Laser Diode, GaN based Light Emitters and Lasers (Springer-Verlag, Berlin, Heidelberg 1997)

13.28   H. Morkoç, S. Srite, G.B. Gao, M.E. Lin, B. Sverdlov, M. Burus, J. Appl. Phys. **76**, 1363 (1994)

13.29   S. Nakamura, Y. Harada, M. Seno, Appl. Phys. Lett. **58**, 2021 (1991)

13.30   I. Hayashi, M.B. Panish, J. Appl. Phys. **41**, 150 (1950)

13.31   T. Mukai, K. Takegawa, S. Nakamura, Jpn. J. Appl. Phys. **37**, L 839 (1998)

13.32   S. Nakamura, M. Senoh, S. Nagahama, N. Iwasa, T. Yamada, T. Matsushita, H. Kiyoku, Y. Sugimoto, T. Kozaki, H. Umemoto, M. Sano, K. Chocho, Jpn. J. Appl. Phys. **37**, L 627 (1998)

13.33   T. Mukai, H. Narimatsu, S. Nakamura, Jpn. J. Appl. Phys. **37**, L 479 (1998)

13.34   S. Nakamura, M. Senoh, N. Iwasa, S. Nagahama, T. Yamada, T. Mukai, Jpn. J. Appl. Phys. **34**, L 1332 (1995)

13.35   J. Baur, P. Schlotter, J. Schneider,Festkörperprobleme/ Advances in Solid State Physics, Vol. **37**, ed. by R. Helbig (Vieweg, Wiesbaden 1997) p.67

13.36   T.C. Harman, in The Physics of Semimetals and Narrow-Gap Semiconductors, ed. by D.L. Carter, R.T. Bate (Pergamon, Oxford 1971) p.363

13.37   J.F. Faist, F. Capasso, D.L. Sivco, C. Sirtori, A.L. Hutchinson, A.Y. Cho, Science **264**, 553 (1994)

13.38   R. Köhler, A. Tredicucci, F. Beltram, H.E. Beere, E.H. Linfield, A.G. Davies, D.A. Ritchi, R.C. Iotti, F. Rossi, Nature **417**, 157 (2002)

13.39   C. Gmachl et al., Rep.Prog.Phys. **64**, 1533 (2001)

13.40   C. Gmachl et al., IEEE J. Quantum Electron. **38**, 569 (2002)

13.41   C. Gmachl, R. Colombelli, F. Capasso, D.L. Sivco, A.Y. Cho, Proc. Int'l School of Physics, Course CL, IOS Press Amsterdam (2003)

13.42   D.L. Derry, A. Yariv, K.Y. Lau, N. Bar-Chaim, K. Lee, J. Rosenberg, Appl. Phys. Lett. **50**, 1773 (1987)

13.43   C.K.N. Patel, E.D. Shaw, Phys. Rev. Lett. **24**, 451 (1970)

13.44   C. Sirtori, A. Tredicucci, F. Capasso, J. Faist, D.L. Sivco, A.L. Hutchinson, A.Y. Cho, Opt. Lett. **23** 463 (1998)

13.45   C.K.N. Patel, E.D. Shaw, R.J. Kerl, Phys. Rev. Lett. **25** 8 (1970)

13.46   Y. Yafet, Phys. Rev. **152** 858 (1966)

13.47   P.A. Wolff, Phys. Rev. Lett. **16** 225 (1966); IEEE J. QE. **2** 659 (1966)

13.48   F. Agulló - Rueda, E.E. Mendez, J.M. Hong, Phys. Rev. B **40**, 1357 (1989)

13.49   E.E. Mendez, G. Bastard, Physics Today, June 1993, p.34

13.50   C. Waschke, H.G. Roskos, R. Schwedler, K. Leo, H. Kurz, Phys. Rev. Lett. **70**, 3319 (1993)

13.51   H.G. Roskos, C. Nuss, J. Shah, K. Leo, D.A.B. Miller, A.M. Fox, S. Schmitt-Rink, K. Köhler, Phys. Rev. Lett. **68**, 2216 (1992)

13.52   F. Beltram, F. Capasso, D.L. Sivco, A.L. Hutchinson, S.N.G. Chu, A.Y. Cho, Phys. Rev. Lett. **64**, 3167 (1990)

13.53   G. Bastard, A. Brum, R. Ferreira, Solid State Physics, Vol. 44, ed. by H. Ehrenreich, D. Turnbull (Academic, New York 1991) p.229

13.54   T. Yajima, Y. Taira, J. Phys. Soc. Jpn. **47**, 1620 (1979)

13.55   G. Valusis, V.G. Lyssenko, D. Klatt, F. Löser, K.H. Pantke, K. Leo, K. Köhler, 23rd Int'l Conf. Phys. Semicond. Vol. 3, ed. by M. Scheffler, R. Zimmermann (World Scientific, Singapore 1996) p.1783

13.56   W. Rehm, H. Künzel, G.H. Döhler, K. Ploog, P. Ruden, Physica **117B**,and **118B**, 732 (1983)

13.57   B. Abeles, T. Tiedje, Phys. Rev. Lett. **51**, 2003 (1983)

13.58   J.D. Joannopoulos, G. Lucovsky (eds.), The Physics of Hydrogenated Amorphous Silicon I and II, Topics Appl. Phys., Vols. 55 and 56 (Springer, Berlin, Heidelberg 1984)

Chapter 14

14.1   I. Tamm, Z. Phys. **76**, 849 (1932); Phys. Z. Sowjetunion **1**, 733 (1932); Zh. Eksp. Teor. Fiz. **3**, 34 (1933)

14.2   W. Shockley, Phys. Rev. **56**, 317 (1939)

14.3   E.W. Müller, Ergebnisse der Exakten Naturwissenschaften **XXVII**, 290 (Springer, Berlin, Göttingen 1953)

14.4   H. Ibach (ed.), Electron Spectroscopy for Surface Analysis, Topics Curr. Phys., **4** (Springer, Berlin, Heidelberg 1977)

14.5   W.E. Meyerhof, Phys. Rev. **71**, 727 (1947)

14.6   J. Bardeen, Phys. Rev. **71**, 717 (1947)

14.7   R.H. Kingston (ed.), Semiconductor Surface Physics; (Univ. of Pennsylvania Press, Philadelphia 1957)

14.8   M. Henzler, Festkörper-Probleme **XI**, 187 (Vieweg, Braunschweig 1971)

14.9   L. Ley, M. Cardona, R. Pollak, in Photoemission in Solids II, ed. by L. Ley, M. Cardona, Topics Appl. Phys., Vol. 27 (Springer, Berlin, Heidelberg 1979)

14.10   F.G. Allen, G.W. Gobeli, Phys. Rev. **127**, 150 (1962)

14.11   W.E. Spicer, P. Skeath, C.Y. Su, I. Lindau, J. Phys. Soc. Jpn. **49**, Suppl. A, 1079 (1980)

14.12   B. Feuerbacher, B. Fitton, In [Ref. 14.4, Chap. 5]

14.13   F. Garcia-Moliner, F. Flores, Introduction to the Theory of Solid Surfaces (Cambridge Univ. Press, London 1979); and all papers in J. Vac. Sci. Technol. **17**, (1980) No. 5

14.14   F. Stern, Phys. Rev. B **5**, 4891 (1972); F. Stern, W.E. Howard, Phys. Rev. **163**, 816 (1967)

14.15   A. Kamgar, P. Kneschaurek, G. Dorda, J.F. Koch, Phys. Rev. Lett. **32**, 1251 (1974)

14.16   A. Forchel, H. Leier, B.E. Maile, R. Germann, Festkörperprobleme / Advances in Solid State Physics **28**, ed. by U. Rössler (Vieweg, Braunschweig 1988) p.99

14.17   K.v. Klitzing, G. Dorda, M. Pepper, Phys. Rev. Lett. **45**, 494 (1980)

14.18   D.C. Tsui, H.L. Störmer, A.C. Gossard, Phys. Rev. Lett. **48**, 1559 (1982)

14.19   M. Razeghi, J.P. Duchemin, J.C. Portal, L. Dmowski, G. Remeni, R.J. Nicholas, A. Briggs, Appl. Phys. Lett. **48**, 712 (1986)

14.20   K.v. Klitzing, B. Tausendfreund, H. Obloh, T. Herzog, Lecture Notes Phys. **177**, p.1, Fig. 6. (Springer, Berlin, Heidelberg 1987) ed. by G. Landwehr

14.21   D.C. Tsui, in M. Janssen, O. Viehweger, U. Fastenrath, J. Hajdu, Introduction to the Theory of the Integer Quantum Hall Effect, VCH, Weinheim, 1994 p.16

14.22   I. Yye, E.E. Mendez, W.I. Wang, L. Esaki, Phys. Rev. B **33**, 5854 (1986)

14.23   G. Landwehr, in High magnetic Fields in Semiconductors ed. by G. Landwehr, Springer Ser. Solid-State Sci., Vol. 71 (Springer , Berlin, Heidelberg 1987) p.295

14.24   M.E. Cage, R.F. Dziuba, B.F. Field, IEEE Trans. IM-**34**, 301 (1985)

14.25   R. Peierls, Surprises in Theoretical Physics, Princeton University Press 1979, Section 4.3

14.26   C.W. Beenakker, H. van Houten, Superlattices and Microstructures **5**, 127 (1989)

14.27   B. Jeckelmann, B. Jeanneret, D. Inglis, Phys.Rev. B **55**, 13124 (1997)

14.28   M. Büttiker, Phys. Rev. B **38**, 9375 (1988)

14.29   H. Hirai, S. Komiyama, J. Appl. Phys. **68**, 655 (1990); S. Koiyama, H. Hirai, Phys. Rev. B (40), 7767 (1989)

14.30   F. Kuchar, R. Meisels, G. Weimann, W. Schlapp, Phys. Rev. B. **33**, 2965 (1986) RC

14.31   V.A. Volkov, D.V. Galchenkov, L.A. Galchenkov, I.M. Grodnenski, O.R. Matov, S.A. Mikhailov, A.P. Senichkin, K.V. Strostin, Pis'ma Zh. Eksp. Teor. Fiz. **43**, 255(1986)[Engl. transl. JETP Lett. **43**, 326 (1986)]

14.32   R.G. Mani, J.H. Smet, K. von Klitzing, V. Narayanamurti, W.B. Johnson, V. Umansky, Nature **420**, 646 (2002)

14.33   M.A. Zudov, R.R. Du, J.A. Simmons, J.L. Reno, Phys.Rev. B **64**, 201311 (2001)

14.34   R.R. Du, A.S. Yeh, H.L. Störmer, D.C. Tsui, L.N. Pfeiffer, K.W. West, Phys. Rev. Lett. **75**, 3926 (1995)

14.35   J.K. Jain, Phys. Rev. Lett. **63**, 19 (1989); O. Heinonen (ed.), Composite Fermions (World Scientific, Singapore 1998)

14.36   G. Landwehr, J. Gerschütz, S. Oehling, A. Pfeuffer - Jeschke, V. Latussek, C.R. Becker, Physica E **6**, 713 (2000)

14.37   D.R. Leadley, M van der Burgt, R.J. Nicolas, C.T. Foxon, J.J. Harris, Phys. Rev. B **53**, 2057 (1996)

14.38   T. Englert, D.C. Tsui, A.C. Gossard, Ch. Uihlein, Surf. Sci. **113**, 295 (1982)

14.39   R.J. Nicholas, R.J. Haug, K. von Klitzing, G. Weimann, Phys. Rev. B **37**, 1294 (1988)

14.40   R.L. Willett, M.A. Paalanen, R.R. Ruel, K.W. West, L.N. Pfeiffer, D.J. Bishop, Phys. Rev. Lett. **65**, 112(1990)

14.41   R.L. Willett, R.R. Ruel, M.A. Paalanen, K.W. West, L.N. Pfeiffer, Phys. Rev. B **47**, 7344 (1993)

14.42   D. Weiss, M.L. Roukes, A. Menschig, P. Grambow, K. v. Klitzing, G. Weimann, Phys. Rev. Lett. **66**, 2790 (1991)

14.43   S. Uryu, T. Ando, Physica B **227**, 138 (1996)

14.44   R. de Picciotto, H.L. Störmer, A. Yacoby, K.W. Baldwin, L.N. Pfeiffer, K.W. West, Physica E**6**, 514 (2000)

14.45 A. Scherer, H.G. Craighead, M.L. Roukes, J.P. Harbison, J.Vac. Sci. Technol. B **6**, 277 (1988)

14.46 P.M. Petroff, A.C. Gossard, R.A. Logan, W. Wiegmann, Appl. Phys. Lett. **41**, 635 (1982)

14.47 S. Thoms, I. McIntyre, S.P. Beaumont, M. Al-Mudares, R. Cheung, C.D.W. Wilkinson, J. Vac. Sci. Technol. B **6**, 127 (1988)

14.48 M.A. Reed, R.A. Bate, K. Bradshaw, W.M. Duncan, W.R. Frensley, J.W.Lee, H.D. Shih, J. Vac. Sci. Technol. B **4**, 358 (1986)

14.49 J.N. Randall, M.A. Reed, T.M. Moore, R.J. Matyi, J.W. Lee, J. Vac. Sci. Technol. B **6**, 302 (1988)

14.50 I.N. Stranski, L. Krastanow, Sitzungsber. d. Akad. d. Wiss. Wien, Abt.IIb, Bd.**146**, 797 (1937)

14.51 E. Bauer, Z. Kristallogr. **110**, 372 (1958)

14.52 L. Goldstein, F. Glas, J.Y. Marzin, M.N. Charasse, G. LeRom, Appl. Phys. Lett. **47**, 1099 (1985)

14.53 R. Willett, J.P. Eisenstein, H.L. Störmer, D.C. Tsui, A.C. Gossard, J.H. English, Phys. Rev. Lett. **59**, 1776 (1987)

14.54 R.R. Du, H.L. Störmer, D.C. Tsui, L.N. Pfeiffer, K.W. West, Phys. Rev. Lett. **70**, 2944 (1993)

14.55 D. Leonard, M Krishnamirthy, C.M. Reaves, S.P. DenBaars, P.M. Petroff, Appl. Phys. Lett. **63**, 3203 (1993)

14.56 D. Bimberg, M. Grundmann, N.N. Ledentsov, Quantum Dot Heterostructures, Wiley, New York (1999)

14.57 Al.L. Efros, A.L. Efros, Sov.Phys.Semicond. **16**, 772 (1982)

14.58 L.E. Brus, J.Chem.Phys. **80**, 4403 (1984)

14.59 H. Haug, S.W. Koch, Quantum Theory of the Optical and Electronic Properties of Semiconductors, World Sci., Singapore (1993)

14.60 Y.Z. Hu, M. Lindberg, S.W. Koch, Phys.Rev. B **42**, 1713 (1990)

14.61 L.P. Kouwenhoven, N.C. van der Vaart, Yu.V. Nazarov, S. Nauhar, D. Dixon, K. McCormick, J. Orenstein, P.L. McEuen, Y. Nagamune, J. Motohisa, H. Sakaki, Surf.Sci. **361/362**, 591 (1996)

14.62 M. Grundmann (ed.), Nano-Optoelectronics, Concepts, Physics, and Devices, Springer Heidelberg 2002, p. 299, M. Grundmann: Theory of Quantum Dot Lasers.

14.63 A. Kumar, S.E. Laux, F. Stern, Phys. Rev. B **42**, 5166 (1990)

14.64 A. Lorke, J.P. Kotthaus, K. Ploog, Phys. Rev. Lett. **64**, 2559 (1990)

14.65 B. Meurer, D. Heitmann, K. Ploog, Phys. Rev. Lett. **68**, 1371 (1992)

14.66 V. Fock, Z. Physik **47**, 446 (1928)

14.67 W. Kohn, Phys. Rev. **123**, 1242 (1961)

14.68 B.J. van Wees, H. van Houten, C.W. J. Beenakker, J.G. Williamson, L.P. Kouwenhoven, D. van der Marel, C.T. Foxon, Phys. Rev. Lett. **60**, 848 (1988)

14.69 T.J. Thornton, M. Pepper, H. Ahmed, D. Andrews, G.J. Davies, Phys. Rev. Lett. **56**, 1198 (1986)

14.70 H.Z. Zheng, H.P. Wei, D.C. Tsui, G. Weimann, Phys. Rev. B **34**, 5635 (1986)

14.71   D.A. Wharam, T.J. Thornton, R. Newbury, M. Pepper, H. Ahmed,
         J.E.F. Frost, D.G. Hasko, D.C. Peacock, D.A. Ritchie, G.A.C. Jones,
         J. Phys. C **21**, L 209 (1988)
14.72   L.P. Kouwenhoven, N.C. van der Vaart, A.T. Johnson, W. Kool,
         C.J.P.M. Harmans, J.G. Williamson, A.A.M. Staring, C.T. Foxon,
         Z. Physik B **85**, 367 (1991)
14.73   C.W.J. Beenakker, Phys. Rev. B **44**, 1646 (1991)
14.74   H. Grabert, Z. Physik B **85**, 319 (1991); (P. Wyrowski, H. Grabert,
         Z. Physik B **85**, 443 (1991))
14.75   R. Landauer, IBM-J. Res. Dev. **1957**; Z. Physik B **21**, 247 ()1975;
         Analogies in Optics and Micro - Electronics, ed. W. van Haeringen,
         D. Lenstra, (Kluwer, Dordrecht 1990) p.243
14.76   M. Büttiker, IBM J. Res. Develop. **32**, 317 (1988)
14.77   H. van Houten, C.W.J. Beenakker, B.J. van Wees, Semiconductors
         and Semimetals **35**, (Academic, New York, (1992)p.9)
14.78   P.M. Mankiewich, R.E. Behringer, R.E. Howard, A.M. Chang, B.
         Chelluri, J. Cunningham, G. Timp, J. Vac. Sci. Technol. B **6**, 131
         (1988)
14.79   Y. Aharonov, D. Bohm, Phys. Rev. **115**, 485 (1959)
14.80   R.G. Chambers, Phys. Rev. Lett. **5**, 3 (1960)
14.81   R.A. Webb, S. Washburn, C.P. Umbach, R.B. Laibowitz, J. Magn.
         Magnetic Materials **54–57**, 1423 (1986)
14.82   N. Kirstaedter, N.N. Ledentsov, M. Grundmann, D. Bimberg, V.M.
         Ustinov, S.S. Ruvimov, M. V. Maximov, P.S. Kop'ev, Zh. I. Alferov,
         U. Richter, P. Werner, U. Gösele, J. Heydenreich, Electron. Lett. **30**,
         1416 (1994)
14.83   F. Heinrichsdorff, M.-H. Mao, N. Kirstaedter, A. Krost, D. Bimberg,
         A.O. Kosogov, P. Werner, Appl. Phys. Lett. **71**, 22 (1997)

Chapter 15

15.1   M.H. Brodsky (ed.), Amorphous Semiconductors, 2nd edn, Topics
         Appl. Phys., Vol. 36 (Springer, Berlin, Heidelberg 1985)
15.2   R.K. Bullough, P.J. Caudrey (eds.), Solitons, Topics Curr. Phys. Vol.
         17 (Springer, Berlin, Heidelberg 1980)
15.3   W.E. Spear, P.G. Le Comber, A.J. Snell, Philos. Mag. B **38**, 303
         (1978)
15.4   G.A.N. Connell, R.A. Street, Handbook on Semiconductors, Vol. 3
         (North Holland, Amsterdam 1980) p.689
15.5   H. Mell, in Proc. 5th Int'l Conf. Amorphous and Liquid Semiconduc-
         tors, ed by J. Stuke, W. Brenig (Taylor and Francis, London 1974)
         p.203

15.6   A.H. Clark, Phys. Rev. **154**, 750 (1967); N.F. Mott, Philos. Mag.
         19, 835 (1969), (theory); N.F. Mott, E.A. Davis, Electronic Processes
         in Non-Crystalline Materials (Oxford Univ. Press, London 1971)
15.7   E.B. Iokin, B.T. Kolomiets, J. Non-Cryst. Solids **3**, 41 (1970)
15.8   J. Tauc. A. Abraham, L. Pajasova, R. Grigorovici, A. Vancu, Proc.
         Conf. Phys. Non-Crystalline Solids (Delft) 1964, ed. by J.A. Prins
         (North-Holland, Amsterdam 1965) p.606; For optical properties of
         polycrystalline Si films, see, e.g., G. Harbeke (ed.), Polycrystalline
         Semiconductors, Springer Ser. Solid-State Sci., Vol. 57 (Springer,
         Berlin, Heidelberg 1985) pp.156-169
15.9   H. Fritzsche, J. Non-Cryst. Solids **6**, 49 (1971)
15.10   J.P. de Neufville, in Optical Properties of Solids - New Developments,
         ed. by B.O. Seraphin (North-Holland, Amsterdam 1976) p.437
15.11   R.C. Chittick, J.H. Alexander, H.F. Sterling, J. Electrochem. Soc.
         116, 77 (1969)
15.12   F. Yonezawa (ed.), Fundamental Physics of Amorphous Semiconduc-
         tors, Springer Ser. Solid-State Sci., Vol. 25 (Springer, Berlin, Heidel-
         berg 1981)
15.13   A. Madan, S.R. Ovshinsky, J. Non-Cryst. Solids **35/36**, 171 (1980)
15.14   L. Ley, H.R. Shanks, C.J. Fang, K.F. Gruntz, M. Cardona, J. Phys.
         Soc. Jpn. **49**, Suppl. A, 1241 (1980)
15.15   Dai Guo-cai, Song Xue-wen, Chen You-peng, J. Phys. Soc. Jpn. **49**,
         Suppl. A, 1257 (1980); R.A. Gibson, P.G. Le Comber, W.E. Spear,
         Appl. Phys. **21**, 307 (1980); A.J. Snell, W.E. Spear, P.E. Le Comber,
         K. Mackenzie, Appl. Phys. A **26**, 83 (1981)
15.16   H.-J. Queisser, H.C. Casey, Jr., W. Van Roosbroeck, Phys. Rev. Lett.
         26, 551 (1971)
15.17   H.-J. Queisser, D.E. Theodorou, Phys. Rev. Lett. **43**, 401 (1979)
15.18   M.K. Sheinkman, A.Ya. Shik, Sov. Phys. -Semicond. **10**, 128 (1976)
15.19   D.E. Theodorou, H.-J. Queisser, E. Bauser, Appl. Phys. Lett. **41**,
         628 (1982)
15.20   H.-J. Queisser, D.E. Theodorou, Solid State Commun. **51**, 875 (1984)
15.21   D.L. Staebler, C.R. Wronski, J. Appl. Phys. **51**, 3263 (1980); For
         changes in the properties of n-i-p-i superlattices due to this effect, see
         J. Kakalios, H. Fritzsche, in Proc. 17th Int'l Conf. Phys. Semicond.,
         San Francisco 1984, ed. by J.D. Chadi, W.A. Harrison, R.Z. Bachrach
         (Springer, New York 1985)
15.22   H. Meier, Organic Semiconductors (Chemie-Verlag , Weinheim 1974)
15.23   D.M. Burland, Phys. Rev. Lett. **33**, 833 (1974)
15.24   C.K. Chiang, C.R. Fincher, Jr., Y.W. Park, A.J. Heeger, H. Shi-
         rakawa, E.J. Louis, S.C. Gau, A.G. MacDiarmid, Phys. Rev. Lett.
         39, 1098 (!977)
15.25   K. Seeger, W.D. Gill, Colloid Polym. Sci. **258**, 252 (1980) (review)
15.26   C.K. Chiang, S.C. Gau, C.R. Fincher, Jr., Y.W. Park, A.G. MacDi-
         armid, A.J. Heeger, Appl. Phys. Lett. **33**, 18 (1978)

15.27   H. Shirakawa, Y. Kobayashi, A. Nagai, S. Ikeda, I. Shijima, M. Konagai, K. Takahashi, Polym. Prepr. **28**, 467 (1979)

15.28   D. MacInnes Jr., M.A. Druy, P.J. Nigrey, D.P. Nairns, A.G. MacDiarmid, A.J. Heeger, J. Chem. Soc., Chem. Commun. **317** (1981)

15.29   K. Seeger. W.D. Gill, T.C. Clarke, G.B. Street, Solid State Commun. **28**, 873 (1978)

15.30   D. Jerome, A. Mazaud, M. Ribault, K. Bechgaard, J. Physique Lett. **41**, L-95 (1980); D. Jerome, J. Phys. Soc. Jpn. **49**, Suppl. A., 845 (1980)

15.31   C. Duke, L. Schein, Phys. Today **33**, 42 (February 1980)

15.32   C.W. Tang, Appl. Phys. Lett. **48**, 183 (1986)

15.33   J.H. Burroughes, D.D.C. Bradley, A.R. Brown, R.N. Marks, K. Mackay, R.H. Friend, P.L. Burns, A.B. Holmes, Nature **347**, 539 (1990)

15.34   M. Schwoerer, Phys. Bl. **49**, 52 (1994); J. Gmeiner, S. Karg, M. Meier, W. Rieβ, P. Strohriegl, M. Schwoerer, Acta Polymer. **44**, 201 (1993); W. Schmid, R. Dankesreiter, J. Gmeiner, Th. Vogtmann, M. Schwoerer, Acta Polymer. **44**, 208 (1993); S. Karg, W. Rieβ, V. Dyakonov, M. Schwoerer, Synth. Met, **54**, 427 (1993)

15.35   M.A. Lampert, P. Mark, Current Injection in Solids (Academic, New York 1970)

15.36   H.W. Kroto, J.R. Heath, S.C. O'Brien, R.F. Curl, R.E. Smalley, Nature **318**, 162 (1985)

15.37   S. Iijima, Nature **354**, 56 (1991); M. Endo, H. Fujiwara, E. Fukunaga, Japan. Carbon Soc. Meeting, 34-35 (1991)

15.38   K. Sattler, Carbon **33**, 915 (1995)

15.39   Z.Y. Kosakovskaya et al., JETP Lett. **56**, 26 (1992); E.G. Gal'pern et al., JETP Lett. **55**, 483 (1992)

15.40   M.S. Dresselhaus, G. Dresselhaus, P.C. Eklund, Science of Fullerenes and Carbon Nanotubes, Academic Press, San Diego (1995)

15.41   J.C. Charlier, J.P. Issi, Appl. Phys. A **67**, 79 (1998); L. Forro et al., Proc. of Nanotubes **99**, (1999)

15.42   C. Journet, P. Bernier, Appl. Phys. A **67**, 1 (1998)

15.43   A. Thess et al., Science **273**, 483 (1996)

15.44   U. Schneider et al., Chem. Phys. Lett. **210**, 165 (1993)

15.45   M. Bockrath et al. Nature **397**, 598 (1999)

15.46   J. Appenzeller, R. Martel, Ph. Avouris, H. Stahl, B. Lengler, Appl Phys. Lett. **78**, 3313 (2001)

15.47   S.J. Wind, J. Appenzeller, R. Martel, V. Deryke, Ph. Avouris, Appl. Phys. Lett. **80**, 3817 (2002)

15.48   S.J. Tans, A. Verschueren, C. Dekker, Nature **393**, 49, (1998)

15.49   R. Martel, T. Schmidt, H.R. Shea, T. Hertel, Ph. Avouris, Appl. Phys. Lett. **73**, 2447, (1998)

15.50   J. Appenzeller et al., Phys.Rev.Lett. **89**, 126801, (2002)

15.51   J. Appenzeller et al., IEDM Technical Digest 2002 **2002**, 285

15.52   J. Knoch, J. Appenzeller, Appl.Phys.Lett. **81**, 3082, (2002)

15.53   A. Aviram, M.R. Ratner, Chem. Phys. Lett. **29**, 277 (1974)

15.54   C.M. Fischer, M. Burghard, S. Roth, K. von Klitzing, Appl. Phys. Lett. **66**, 3331 (1995)

15.55   R.M. Metzger, B. Chen, U. Höpfner, M.V. Lakshmikantham, D. Vuillaume, T. Kawai, X. Wu, H. Tachibana, T.V. Hughes, H. Sakurai, J.W. Baldwin, C. Hosch, M.P. Cava, L. Brehmer, G.J. Ashwell, J. Am. Chem. Soc. **119**, 10455 (1997)

15.56   R.M. Metzger, B. Chen, D. Vuillaume, M.V. Lakshmikantam, U. Höpfner, T. Kawai, J.W. Baldwin, X. Wu, H. Tachibana, H. Sakurai, M.P. Cava, Thin Solid Films, **327-329**, 326 (1998)

15.57   O. Kwon, M.L. McKee, R.M. Metzger, Chem.Phys.Lett., **313**, 321 (1999)

15.58   C.P. Collier, G. Mattersteig, E.W. Wong, Y. Luo, K. Beverly, J. Sampaio, F.M. Raymo, J.P. Stoddart, J.R. Heath, Science, **289**, 1172 (2000)

B1   G. Bastard, Phys.Rev. **B24**, 5693 (1981); Phys.Rev. **B25**, 7594 (1982); C. Weisbuch, B. Vinter, Quantum Semiconductor Structures (Academic, New York 1991)

Subject Index

About the Author

Born (1927) and educated at Bad Nauheim (Germany). After World War II, workmanship at the local William G. Kerckhoff Balneological Institute. September 1945 to 1950: study of physics, chemistry, and mathematics at the University of Göttingen (Germany) which included experimental work at the local Max-Planck Institute (Head: Werner Heisenberg, 1932 Nobel Physics prize winner). 1951 – 1957: University of Heidelberg (Germany), Institute of Physics II (Head: Otto Haxel). 1955: PhD. 1957 – 1958: University of Illinois at Urbana (Ill., U.S.A.) Physics Dept. (Head: John Bardeen, the only winner of two Nobel Prizes in Physics): Microwave investigations of hot electrons in semiconductors. 1958 – 1966: University of Heidelberg, Dozent (Lecturer), both lecturing and experimental research on semiconductors. Since 1966: University Wien (Vienna, Austria), Applied Physics Dept. and Ludwig Boltzmann Institut for Solid State Physics (Vienna) (Head). Since 1995: Professor Emeritus. Experimental research on semiconductors and organic metals (including microwave investigations) as well as on high-temperature superconductors and charge density waves, development of semiconductor solar cells (supported by the Shell Oil Co., Vienna Section). 1977/1978: Experimental work as Senior Scientist at IBM San Jose, (California, U.S.A.).

Scientific awards: Erwin Schrödinger Prize 1992 (Austrian Academy of Sciences); 1988: Ernst Mach Medal (Ernst Mach Institut, Freiburg Brsg., Germany).

Memberships: Austrian Academy of Sciences; Austrian Physical Society; Chemical Physical Society (Austria); German Physical Society.

半导体物理学

（西德）K. Seeger 著

徐 乐 钱建业 译
叶良修 校

人民教育出版社

Karlheinz Seeger

Halbleiterphysik
Eine Einführung

Band I

Mit 180 Abbildungen

vieweg

セミコンダクターの
物理学
（第4版）　（上）

K．ジーガー著
山本恵一
林　真至
青木和徳
共訳

物理学叢書
60

К. Зеегер

ФИЗИКА
ПОЛУПРОВОДНИКОВ

Перевод с английского
Р. БРАЗИСА, А. МАТУЛЕНИСА и А. ТЕТНРВОВА
Под редакцией
Ю. К. ПОЖЕЛЫ

ИЗДАТЕЛЬСТВО «МИР»
МОСКВА 1977

Previous versions of the present book have also appeared as foreign–language editions